2018 IEEE Symposium on VLSI Technology

Honolulu, Hawaii, USA
18-22 June 2018

IEEE Catalog Number: CFP18VTS-POD
ISBN: 978-1-5386-4219-1

**Copyright © 2018 by the Institute of Electrical and Electronics Engineers, Inc.
All Rights Reserved**

Copyright and Reprint Permissions: Abstracting is permitted with credit to the source. Libraries are permitted to photocopy beyond the limit of U.S. copyright law for private use of patrons those articles in this volume that carry a code at the bottom of the first page, provided the per-copy fee indicated in the code is paid through Copyright Clearance Center, 222 Rosewood Drive, Danvers, MA 01923.

For other copying, reprint or republication permission, write to IEEE Copyrights Manager, IEEE Service Center, 445 Hoes Lane, Piscataway, NJ 08854. All rights reserved.

*** This is a print representation of what appears in the IEEE Digital Library. Some format issues inherent in the e-media version may also appear in this print version.

IEEE Catalog Number:	CFP18VTS-POD
ISBN (Print-On-Demand):	978-1-5386-4219-1
ISBN (Online):	978-1-5386-4218-4
ISSN:	0743-1562

Additional Copies of This Publication Are Available From:

Curran Associates, Inc
57 Morehouse Lane
Red Hook, NY 12571 USA
Phone: (845) 758-0400
Fax: (845) 758-2633
E-mail: curran@proceedings.com
Web: www.proceedings.com

2018 IEEE Symposium on VLSI Technology

Honolulu, Hawaii, USA
18-22 June 2018

IEEE Catalog Number: CFP18VTS-POD
ISBN: 978-1-5386-4219-1

TABLE OF CONTENTS

MEMORY TECHNOLOGY: THE CORE TO ENABLE FUTURE COMPUTING SYSTEMS.................................3
Scott Deboer

REVOLUTIONIZING CANCER GENOMIC MEDICINE BY AI AND SUPERCOMPUTER WITH BIG DATA 7
Satoru Miyano

SHAPING CIRCUIT ENVIRONMENT TO FACE THE THERMAL CHALLENGE INNOVATIVE TECHNOLOGIES FROM LOW TO HIGH POWER ELECTRONICS.................................15
P. Coudrain ; J.-P. Colonna ; L.-M. Collin ; R. Prieto ; L.G. Fréchette ; J. Barrau ; G. Savelli ; P. Vivet ; Q. Struss ; J. Widiez ; K. Vladimirova ; K. Triantopoulos ; H. Beckrich-Ros ; M. Vilarrubí ; G. Laguna ; H. Azarkish ; M. Shirazi ; J. Michailos

THERMAL MANAGEMENT RESEARCH – FROM POWER ELECTRONICS TO PORTABLES 17
Ki Wook Jung ; Chi Zhang ; Tanya Liu ; Mehdi Asheghi ; Kenneth E. Goodson

ELECTROMIGRATION EFFECTS IN POWER GRIDS CHARACTERIZED USING AN ON-CHIP TEST STRUCTURE WITH POLY HEATERS AND VOLTAGE TAPPING POINTS.................................19
Chen Zhou ; Richard Wong ; Shi-Jie Wen ; Chris H. Kim

LOW THERMAL BUDGET AMORPHOUS INDIUM TUNGSTEN OXIDE NANO-SHEET JUNCTIONLESS TRANSISTORS WITH NEAR IDEAL SUBTHRESHOLD SWING.................................21
Po-Yi Kuo ; Chien-Min Chang ; Po-Tsun Liu

CAPACITOR-BASED CROSS-POINT ARRAY FOR ANALOG NEURAL NETWORK WITH RECORD SYMMETRY AND LINEARITY 25
Y. Li ; S. Kim ; X. Sun ; P. Solomon ; T. Gokmen ; H. Tsai ; S. Koswatta ; Z. Ren ; R. Mo ; C. C. Yeh ; W. Haensch ; E. Leobandung

ANALOG SPIKE PROCESSING WITH HIGH SCALABILITY AND LOW ENERGY CONSUMPTION USING THERMAL DEGREE OF FREEDOM IN PHASE TRANSITION MATERIALS.................................27
T. Yajima ; T. Nishimura ; A. Toriumi

AN ENERGY EFFICIENT FINFET-BASED FIELD PROGRAMMABLE SYNAPSE ARRAY (FPSA) FEASIBLE FOR ONE-SHOT LEARNING ON EDGE AI 29
J. L. Kuo ; H. W. Chen ; E. R. Hsieh ; Steve S. Chung ; T. P. Chen ; S. A. Huang ; J. Chen ; Osbert Cheng

NOVEL IN-MEMORY MATRIX-MATRIX MULTIPLICATION WITH RESISTIVE CROSS-POINT ARRAYS.................................31
Yan Liao ; Huaqiang Wu ; Weier Wan ; Wenqiang Zhang ; Bin Gao ; H.-S. Philip Wong ; He Qian

SENSORS AND RELATED DEVICES FOR IOT, MEDICINE AND SMART-LIVING 35
T. Ernst ; R. Guillemaud ; P. Mailley ; J.P. Polizzi ; A. Koenig ; S. Boisseau ; E. Pauliac-Vaujour ; C. Plantier ; G. Delapierre ; E. Saoutieff ; R. Gerbelot-Barillon ; E. Calvanese Strinati ; S. Hentz ; E. Colinet ; O. Thomas ; P. Boisseau ; P. Jallon

DEVELOPMENT OF A MULTISITE, CLOSED-LOOP NEUROMODULATOR FOR THE THERANOSIS OF NEURAL DEGENERATIVE DISEASES.................................37
Hsin Chen ; Yen-Chung Chang ; Shih-Rung Yeh ; Chih-Cheng Hsieh ; Kea-Tiong Tang ; Ping-Hsuan Hsieh ; Yu-Te Liao ; Ramesh Perumel ; Ji-Feng Chuang ; Ching-Chih Chang ; Yu-Chieh Chen ; Shih-Hsin Chen ; Sung-En Hsieh ; Yen-Peng Chen ; Ye-Ting Chen ; Tzu-Hao Liu ; Yu-Ming Chang ; Wei-Chih Lai ; Chuang-Yi Wu ; Yu-Hsin Chen ; Yi-Chin Weng

HIGH PERFORMANCE HIGH DENSITY GAS-FET ARRAY IN STANDARD CMOS 39
Qian Yu ; Xiaopeng Zhong ; Farid Boussaid ; Amine Bermak ; Cy Tsui

HIGH-SENSITIVITY AND LOW-POWER INERTIAL MEMS-ON-CMOS SENSORS USING LOW-TEMPERATURE-DEPOSITED POLY-SIGE FILM FOR THE IOT ERA.................................41
Hideyuki Tomizawa ; Yoshihiko Kurui ; Ippei Akita ; Akira Fujimoto ; Tomohiro Saito ; Akihiro Kojima ; Hideki Shibata

A COMPREHENSIVE STUDY OF POLYMORPHIC PHASE DISTRIBUTION OF FERROELECTRIC-DIELECTRICS AND INTERFACIAL LAYER EFFECTS ON NEGATIVE CAPACITANCE FETS FOR SUB-5 NM NODE.................................45
Y.-T Tang ; C.-J. Su ; Y.-S. Wang ; K.-H. Kao ; T.-L. Wu ; P.-J. Sung ; F.-J. Hou ; C.-J. Wang ; M.-S. Yeh ; Y.-J. Lee ; W.-F. Wu ; G.-W. Huang ; J.-M. Shieh ; W.-K. Yeh ; Y.-H. Wang

FIRST EXPERIMENTAL DEMONSTRATION OF NEGATIVE CAPACITANCE INGAAS MOSFETS WITH $HF_{0.5}ZR_{0.5}O_2$ FERROELECTRIC GATE STACK.................................47
Q. H. Luc ; C. C. Fan-Chiang ; S. H. Huynh ; P. Huang ; H. B. Do ; M. T. H. Ha ; Y. D. Jin ; T. A. Nguyen ; K. Y. Zhang ; H. C. Wang ; Y. K. Lin ; Y. C. Lin ; C. Hu ; H. Iwai ; E. Y. Chang

RESPONSE SPEED OF NEGATIVE CAPACITANCE FINFETS 49

Daewoong Kwon ; Yu-Hung Liao ; Yen-Kai Lin ; Juan Pablo Duarte ; Korok Chatterjee ; Ava J. Tan ; Ajay K. Yadav ; Chenming Hu ; Zoran Krivokapic ; Sayeef Salahuddin

FERROELECTRIC SWITCHING DELAY AS CAUSE OF NEGATIVE CAPACITANCE AND THE IMPLICATIONS TO NCFETS 51

B. Obradovic ; T. Rakshit ; R. Hatcher ; J. A. Kittl ; M. S. Rodder

NEGATIVE CAPACITANCE, N-CHANNEL, SI FINFETS: BI-DIRECTIONAL SUB-60 MV/DEC, NEGATIVE DIBL, NEGATIVE DIFFERENTIAL RESISTANCE AND IMPROVED SHORT CHANNEL EFFECT 53

Hong Zhou ; Daewoong Kwon ; Angada B. Sachid ; Yuhung Liao ; Korok Chatterjee ; Ava J. Tan ; Ajay K. Yadav ; Chenming Hu ; Sayeef Salahuddin

TRUE 7NM PLATFORM TECHNOLOGY FEATURING SMALLEST FINFET AND SMALLEST SRAM CELL BY EUV, SPECIAL CONSTRUCTS AND 3RDGENERATION SINGLE DIFFUSION BREAK 59

WC Jeong ; S. Maeda ; HJ Lee ; KW Lee ; TJ Lee ; DW Park ; BS Kim ; JH Do ; T Fukai ; DJ Kwon ; KJ Nam ; WJ Rim ; MS Jang ; HT Kim ; YW Lee ; JS Park ; EC Lee ; DW Ha ; C.H. Park ; H.-J. Cho ; S.-M. Jung ; H.K. Kang

NANOSECOND LASER ANNEAL FOR BEOL PERFORMANCE BOOST IN ADVANCED FINFETS 61

Rinus T.P. Lee ; N. Petrov ; J. Kassim ; M. Gribelyuk ; J. Yang ; L. Cao ; K.B. Yeap ; T. Shen ; A. N. Zainuddin ; A. Chandrashekar ; S. Ray ; E. Ramanathan ; A. S. Mahalingam ; R. Chaudhuri ; J. Mody ; D. Damjanovic ; Z. Sun ; R. Sporer ; T. J. Tang ; H. Liu ; J. Liu ; B. Krishnan

FROM MEMORY TO SENSOR: ULTRA-LOW POWER AND HIGH SELECTIVITY HYDROGEN SENSOR BASED ON RERAM TECHNOLOGY 63

Zhiqiang Wei ; Kazunari Homma ; Koji Katayama ; Ken Kawai ; Satoru Fujii ; Yasuhisa Naitoh ; Hisashi Shima ; Hiroyuki Akinaga ; Satoru Ito ; Shinichi Yoneda

DEMONSTRATION OF ULTRA-LOW VOLTAGE AND ULTRA LOW POWER STT-MRAM DESIGNED FOR COMPATIBILITY WITH 0X NODE EMBEDDED LLC APPLICATIONS 65

Guenole Jan ; Luc Thomas ; Son Le ; Yuan-Jen Lee ; Huanlong Liu ; Jian Zhu ; Jodi Iwata-Harms ; Sahil Patel ; Ru-Ying Tong ; Vignesh Sundar ; Santiago Serrano-Guisan ; Dongna Shen ; Renren He ; Jesmin Haq ; Zhongjian Jeffrey Teng ; Vinh Lam ; Yi Yang ; Yu-Jen Wang ; Tom Zhong ; Hideaki Fukuzawa ; Po-Kang Wang

3D SEQUENTIAL STACKED PLANAR DEVICES ON 300 MM WAFERS FEATURING REPLACEMENT METAL GATE JUNCTION-LESS TOP DEVICES PROCESSED AT 525⁰ C WITH IMPROVED RELIABILITY 69

A. Vandooren ; J. Franco ; B. Parvais ; Z. Wu ; L. Witters ; A. Walke ; W. Li ; L. Peng ; V. Deshpande ; F.M. Bufler ; N. Rassoul ; G. Hellings ; G. Jamieson ; F. Inoue ; G. Verbinnen ; K. Devriendt ; L. Teugels ; N. Heylen ; E. Vecchio ; T. Zheng ; E. Rosseel ; W. Vanherle ; A. Hikavyy ; B. T. Chan ; R. Ritzenthaler ; G. Besnard ; W. Schwarzenbach ; G. Gaudin ; I. Radu ; B.-Y. Nguyen ; N. Waldron ; V. De Heyn ; D. Mocuta ; N. Collaert

AN OVER 120 DB WIDE-DYNAMIC-RANGE 3.0 μM PIXEL IMAGE SENSOR WITH IN-PIXEL CAPACITOR OF 41.7 FF/UM² AND HIGH RELIABILITY ENABLED BY BEOL 3D CAPACITOR PROCESS 71

M. Takase ; S. Isono ; Y. Tomekawa ; T. Koyanagi ; T. Tokuhara ; M. Harada ; Y. Inoue

SELECTIVE PORE-SEALING OF HIGHLY POROUS ULTRALOW-K DIELECTRICS FOR ULSI INTERCONNECTS BY CYCLIC INITIATED CHEMICAL VAPOR DEPOSITION PROCESS 73

Seong Jun Yoon ; Kwanyong Pak ; Hyun Jun Ahn ; Alexander Yoon ; Sung Gap Im ; Byung Jin Cho

PERFORMANCE AND RELIABILITY OF A FULLY INTEGRATED 3D SEQUENTIAL TECHNOLOGY 75

A. Tsiara ; X. Garros ; L. Brunet ; P. Batude ; C. Fenouillet-Béranger ; K. Triantopoulos ; M. Cassé ; M. Vinet ; F. Gaillard ; G. Ghibaudo

METAL/P-TYPE GESN CONTACTS WITH SPECIFIC CONTACT RESISTIVITY DOWN TO 4.4×10⁻¹⁰Ω-CM2 77

Ying Wu ; Wei Wang ; Saeid Masudy-Panah ; Yang Li ; Kaizhen Han ; Liuhuiquan He ; Zheng Zhang ; Dian Lei ; Shengqiang Xu ; Yuye Kang ; Xiao Gong ; Yee-Chia Yeo

MULTIPLE WORKFUNCTION HIGH PERFORMANCE FINFETS FOR ULTRA-LOW VOLTAGE OPERATION 81

M. Togo ; R. Asra ; P. Balasubramaniam ; X. Zhang ; H. Yu ; S. Yamaguchi ; E. Geiss ; H. S. Yang ; B. Cohen ; H-C. Lo ; O. Hu ; H. Lazar ; O. Kwon ; D. Burnett ; J. Versaggi ; E. Banghart ; M. K. Hassan ; E. Bazizi ; L. Pantisano ; J. G. Lee ; S. B. Samavedam ; D. K. Sohn

AN IN-DEPTH STUDY OF HIGH-PERFORMING STRAINED GERMANIUM NANOWIRES PFETS 83

J. Mitard ; D. Jang ; G. Eneman ; H. Arimura ; B. Parvais ; O. Richard ; P. Van Marcke ; L. Witters ; E. Capogreco ; H. Bender ; R. Ritzenthaler ; H. Mertens ; A. Hikavyy ; R. Loo ; H. Dekkers ; F. Sebaai ; A. Milenin ; N. Horiguchi ; A. Mocuta ; D. Mocuta ; N. Collaert

SI/SIGE SUPERLATTICE I/O FINFETS IN A VERTICALLY-STACKED GATE-ALL-AROUND HORIZONTAL NANOWIRE TECHNOLOGY 85

G. Hellings ; H. Mertens ; A. Subirats ; E. Simoen ; T. Schram ; L.-A. Ragnarsson ; M. Simicic ; S.-H. Chen ; B. Parvais ; D. Boudier ; B. Cretu ; J. Machillot ; V. Pena ; S. Sun ; N. Yoshida ; N. Kim ; A. Mocuta ; D. Linten ; N. Horiguchi

LEAKAGE AWARE SI/SIGE CMOS FINFET FOR LOW POWER APPLICATIONS 87

Gen Tsutsui ; Curtis Durfee ; Miaomiao Wang ; Aniruddha Konar ; Heng Wu ; Shogo Mochizuki ; Ruqiang Bao ; Stephen Bedell ; Juntao Li ; Huimei Zhou ; Daniel Schmidt ; Chun Ju Yang ; James Kelly ; Koji Watanabe ; Theodore Levin ; Walter Kleemeier ; Dechao Guo ; Devendra Sadana ; Dinesh Gupta ; Andreas Knorr ; Huiming Bu

FIRST DIRECT EXPERIMENTAL STUDIES OF $HF_{0.5}ZR_{0.5}O_2$ FERROELECTRIC POLARIZATION SWITCHING DOWN TO 100-PICOSECOND IN SUB-60MV/DEC GERMANIUM FERROELECTRIC NANOWIRE FETS 89

Wonil Chung ; Mengwei Si ; Pragya R. Shrestha ; Jason P. Campbell ; Kin P. Cheung ; Peide D. Ye

$10\mu W/CM^2$-CLASS HIGH POWER DENSITY PLANAR SI-NANOWIRE THERMOELECTRIC ENERGY HARVESTER COMPATIBLE WITH CMOS-VLSI TECHNOLOGY 93

M. Tomita ; S. Oba ; Y. Himeda ; R. Yamato ; K. Shima ; T. Kumada ; M. Xu ; H. Takezawa ; K. Mesaki ; K. Tsuda ; S. Hashimoto ; T. Zhan ; H. Zhang ; Y. Kamakura ; Y. Suzuki ; H. Inokawa ; H. Ikeda ; T. Matsukawa ; T. Matsuki ; T. Watanabe

A LOW-POWER AND HIGH-SPEED TRUE RANDOM NUMBER GENERATOR USING GENERATED RTN 95

James Brown ; Rui Gao ; Zhigang Ji ; Jiezhi Chen ; Jixuan Wu ; Jianfu Zhang ; Bo Zhou ; Qi Shi ; Jacob Crowford ; Weidong Zhang

ULTRAHIGH-SENSITIVE AND CMOS COMPATIBLE ISFET DEVELOPED IN BEOL OF INDUSTRIAL UTBB FDSOI 97

Getenet Tesega Ayele ; Stephane Monfray ; Serge Ecoffey ; Frederic Boeuf ; Romain Bon ; Jean-Pierre Cloarec ; Dominique Drouin ; Abdelkader Souifi

RX-PUF: LOW POWER, DENSE, RELIABLE, AND RESILIENT PHYSICALLY UNCLONABLE FUNCTIONS BASED ON ANALOG PASSIVE RRAM CROSSBAR ARRAYS 99

Mohammad Reza Mahmoodi ; Hussein Nili ; Dmitri. B. Strukov

A METHODOLOGY TO IMPROVE LINEARITY OF ANALOG RRAM FOR NEUROMORPHIC COMPUTING 103

Wei Wu ; Huaqiang Wu ; Bin Gao ; Peng Yao ; Xiang Zhang ; Xiaochen Peng ; Shimeng Yu ; He Qian

NON-VOLATILE TERNARY CONTENT ADDRESSABLE MEMORY (TCAM) WITH TWO $HFO_2/AL_2O_3/GEO_X/GE$ MOS DIODES 105

Yi Zhang ; Bing Chen ; Wenfeng Dong ; Wei Liu ; Shun Xu ; Ran Cheng ; Shiuh-Wuu Lee ; Yi Zhao

SELECTOR REQUIREMENTS FOR TERA-BIT ULTRA-HIGH-DENSITY 3D VERTICAL RRAM 107

Zizhen Jiang ; Shengjun Qin ; Haitong Li ; Shosuke Fujii ; Dongjin Lee ; Simon Wong ; H.-S. Philip Wong

5X RELIABILITY ENHANCED 40NM TAOX APPROXIMATE-RERAM WITH DOMAIN-SPECIFIC COMPUTING FOR REAL-TIME IMAGE RECOGNITION OF IOT EDGE DEVICES 109

Yusuke Yamaga ; Yoshiaki Deguchi ; Shouhei Fukuyama ; Ken Takeuchi

COMPREHENSIVE THERMAL SPICE MODELING OF FINFETS AND BEOL WITH LAYOUT FLEXIBILITY CONSIDERING FREQUENCY DEPENDENT THERMAL TIME CONSTANT, 3D HEAT FLOWS, BOUNDARY/ALLOY SCATTERING, AND INTERFACIAL THERMAL RESISTANCE WITH CIRCUIT LEVEL RELIABILITY EVALUATION 113

Jhih-Yang Yan ; Chia-Che Chung ; Sun-Rong Jan ; H. H. Lin ; W. K. Wan ; M.-T. Yang ; C. W. Liu

DIFFERENTIATED PERFORMANCE AND RELIABILITY ENABLED BY MULTI-WORK FUNCTION SOLUTION IN RMG SILICON AND SIGE MOSFETS 115

R. Bao ; R. G. Southwick ; H. Zhou ; C. H. Lee ; B. P. Linder ; T. Ando ; D. Guo ; H. Jagannathan ; V. Narayanan

PROCESS OPTIMIZATION OF PERPENDICULAR MAGNETIC TUNNEL JUNCTION ARRAYS FOR LAST-LEVEL CACHE BEYOND 7 NM NODE 117

Lin Xue ; Chi Ching ; Alex Kontos ; Jaesoo Ahn ; Xiaodong Wang ; Renu Whig ; Hsin-Wei Tseng ; James Howarth ; Sajjad Hassan ; Hao Chen ; Mangesh Bangar ; Shurong Liang ; Rongjun Wang ; Mahendra Pakala

DEPENDENCE OF RELIABILITY OF FERROELECTRIC $HFZRO_X$ ON EPITAXIAL SIGE FILM WITH VARIOUS GE CONTENT 119

Kuen-Yi Chen ; Yen-Hua Huang ; Ruei-Wen Kao ; Yan-Xiao Lin ; Yung-Hsien Wu

MODELING OF FINFET SELF-HEATING EFFECTS IN MULTIPLE FINFET TECHNOLOGY GENERATIONS WITH IMPLICATION FOR TRANSISTOR AND PRODUCT RELIABILITY 121

H. C. Sagong ; K. Choi ; J. Kim ; T. Jeong ; M. Choe ; H. Shim ; W. Kim ; J. Park ; S. Shin ; S. Pae

ALL-ELECTRICAL CONTROL OF A HYBRID ELECTRON SPIN/VALLEY QUANTUM BIT IN SOI CMOS TECHNOLOGY 125

L. Hutin ; L. Bourdet ; B. Bertrand ; A. Corna ; H. Bohuslavskyi ; A. Amisse ; A. Crippa ; R. Maurand ; S. Barraud ; M. Urdampilleta ; C. Bäuerle ; T. Meunier ; M. Sanquer ; X. Jehl ; S. De Franceschi ; Y.-M. Niquet ; M. Vinet

HIGH-DENSITY AND FAULT-TOLERANT CU ATOM SWITCH TECHNOLOGY TOWARD 28NM-NODE NONVOLATILE PROGRAMMABLE LOGIC 127

R. Nebashi ; N. Banno ; M. Miyamura ; Y. Tsuji ; A. Morioka ; X. Bai ; K. Okamoto ; N. Iguchi ; H. Numata ; H. Hada ; T. Sugibayashi ; T. Sakamoto ; M. Tada

A THRESHOLD SWITCH AUGMENTED HYBRID-FEFET (H-FEFET) WITH ENHANCED READ DISTINGUISHABILITY AND REDUCED PROGRAMMING VOLTAGE FOR NON-VOLATILE MEMORY APPLICATIONS 129

M. Jerry ; A. Aziz ; K. Ni ; S. Datta ; S. K. Gupta ; N. Shukla

A CIRCUIT COMPATIBLE ACCURATE COMPACT MODEL FOR FERROELECTRIC-FETS 131

Kai Ni ; Matthew Jerry ; Jeffrey A. Smith ; Suman Datta

RECORD 47 MV/DEC TOP-DOWN VERTICAL NANOWIRE INGAAS/GAASSB TUNNEL FETS 133

Alireza Alian ; Salim El Kazzi ; Anne Verhulst ; Alexey Milenin ; Nicolò Pinna ; Tsvetan Ivanov ; Dennis Lin ; Dan Mocuta ; Nadine Collaert

IMPROVING PERFORMANCE, POWER, AND AREA BY OPTIMIZING GEAR RATIO OF GATE-METAL PITCHES IN SUB-10NM NODE CMOS DESIGNS 137

Yongchan Ban ; Xuelian Zhu ; Jan Petykiewicz ; Jia Zeng

ACHIEVING HIGH-SCALABILITY NEGATIVE CAPACITANCE FETS WITH UNIFORM SUB-35 MV/DEC SWITCH USING DOPANT-FREE HAFNIUM OXIDE AND GATE STRAIN 139

Chia-Chi Fan ; Chun-Hu Cheng ; Chun-Yuan Tu ; Chien Liu ; Wan-Hsin Chen ; Tun-Jen Chang ; Chun-Yen Chang

THE COMPLEMENTARY FET (CFET) FOR CMOS SCALING BEYOND N3 141

J. Ryckaert ; P. Schuddinck ; P. Weckx ; G. Bouche ; B. Vincent ; J. Smith ; Y. Sherazi ; A. Mallik ; H. Mertens ; S. Demuynck ; T. Huynh Bao ; A. Veloso ; N. Horiguchi ; A. Mocuta ; D. Mocuta ; J. Boemmels

POWER-PERFORMANCE TRADE-OFFS FOR LATERAL NANOSHEETS ON ULTRA-SCALED STANDARD CELLS 143

M. Garcia Bardon ; Y. Sherazi ; D. Jang ; D. Yakimets ; P. Schuddinck ; R. Baert ; H. Mertens ; L. Mattii ; B. Parvais ; A. Mocuta ; D. Verkest

ENABLING CMOS SCALING TOWARDS 3NM AND BEYOND 147

A. Mocuta ; P. Weckx ; S. Demuynck ; D. Radisic ; Y. Oniki ; J. Ryckaert

SMART SCALING TECHNOLOGY FOR ADVANCED FINFET NODE 149

Jongwook Kye ; Hoonki Kim ; Jinyoung Lim ; Seungyoung Lee ; Jonghoon Jung ; Taejoong Song

SUB-550MV SRAM DESIGN IN 22NM FINFET LOW POWER (22FFL) TECHNOLOGY WITH SELF-INDUCED COLLAPSE WRITE ASSIST 151

Daeyeon Kim ; Jami Wiedemer ; Pramod Kolar ; Ayush Shrivastava ; Jinal Shah ; Satyanand Nalam ; Gwanghyeon Baek ; Xiaofei Wang ; Zheng Guo ; Eric Karl

DESIGN TECHNOLOGY CO-OPTIMIZATION IN ADVANCED FDSOI CMOS AROUND THE MINIMUM ENERGY POINT: BODY BIASING AND WITHIN-CELL V_T-MIXING 153

F. Andrieu ; L. Pirro ; R. Berthelon ; J. Morgan ; G. Cibrario ; M. Wiatr ; J. Hoentschel ; M. Vinet

SELF-ORGANIZED GATE STACK OF GE NANOSPHERE/SIO$_2$/SI$_{1-X}$GE$_X$ ENABLES GE-BASED MONOLITHICALLY-INTEGRATED ELECTRONICS AND PHOTONICS ON SI PLATFORM 157

P. H. Liao ; M. H. Kuo ; C. W. Tien ; Y. L. Chang ; P. Y. Hong ; T. George ; H. C. Lin ; P. W. Li

A NEAR- & SHORT-WAVE IR TUNABLE INGAAS NANOMEMBRANE PHOTOFET ON FLEXIBLE SUBSTRATE FOR LIGHTWEIGHT AND WIDE-ANGLE IMAGING APPLICATIONS 159

Yida Li ; Alireza Alian ; Li Huang ; Kah Wee Ang ; Dennis Lin ; Dan Mocuta ; Nadine Collaert ; Aaron V-Y Thean

INTEGRATION OF 2D BLACK PHOSPHORUS PHOTOTRANSISTOR AND SILICON PHOTONICS WAVEGUIDE SYSTEM TOWARDS MID-INFRARED ON-CHIP SENSING APPLICATIONS 161

Li Huang ; Bowei Dong ; Xin Guo ; Yuhua Chang ; Nan Chen ; Xin Huang ; Hong Wang ; Chengkuo Lee ; Kah-Wee Ang

NEXT-GENERATION FUNDUS CAMERA WITH FULL COLOR IMAGE ACQUISITION IN 0-LX VISIBLE LIGHT BY 1.12-MICRON SQUARE PIXEL, 4K, 30-FPS BSI CMOS IMAGE SENSOR WITH ADVANCED NIR MULTI-SPECTRAL IMAGING SYSTEM 163

Hirofumi Sumi ; Hironari Takehara ; Shunsuke Miyazaki ; Daiki Shirahige ; Kiyotaka Sasagawa ; Takashi Tokuda ; Yoshihiro Watanabe ; Norimasa Kishi ; Jun Ohta ; Masatoshi Ishikawa

INGAAS-ON-INSULATOR MOSFETS FEATURING SCALED LOGIC DEVICES AND RECORD RF PERFORMANCE 165

C. B. Zota ; C. Convertino ; V. Deshpande ; T. Merkle ; M. Sousa ; D. Caimi ; L. Czomomaz

NEUROMORPHIC TECHNOLOGY BASED ON CHARGE STORAGE MEMORY DEVICES 169

Sung-Tae Lee ; Suhwan Lim ; Nagyong Choi ; Jong-Ho Bae ; Chul-Heung Kim ; Soochang Lee ; Dong Hwan Lee ; Tackhwi Lee ; Sungyong Chung ; Byung-Gook Park ; Jong-Ho Lee

NONVOLATILE CIRCUITS-DEVICES INTERACTION FOR MEMORY, LOGIC AND ARTIFICIAL INTELLIGENCE 171

Chun-Meng Dou ; Wei-Hao Chen ; Cheng-Xin Xue ; Wei-Yu Lin ; Wei-En Lin ; Jun-Yi Li ; Huan-Ting Lin ; Meng-Fan Chang

XNOR-SRAM: IN-MEMORY COMPUTING SRAM MACRO FOR BINARY/TERNARY DEEP NEURAL NETWORKS 173

Zhewei Jiang ; Shihui Yin ; Mingoo Seok ; Jae-Sun Seo

A 4M SYNAPSES INTEGRATED ANALOG RERAM BASED 66.5 TOPS/W NEURAL-NETWORK PROCESSOR WITH CELL CURRENT CONTROLLED WRITING AND FLEXIBLE NETWORK ARCHITECTURE 175

Reiji Mochida ; Kazuyuki Kouno ; Yuriko Hayata ; Masayoshi Nakayama ; Takashi Ono ; Hitoshi Suwa ; Ryutaro Yasuhara ; Koji Katayama ; Takumi Mikawa ; Yasushi Gohou

A NOVEL 3D AND-TYPE NVM ARCHITECTURE CAPABLE OF HIGH-DENSITY, LOW-POWER IN-MEMORY SUM-OF-PRODUCT COMPUTATION FOR ARTIFICIAL INTELLIGENCE APPLICATION 177

Hang-Ting Lue ; Weichen Chen ; Hung-Sheng Chang ; Keh-Chung Wang ; Chih-Yuan Lu

EMBEDDED STT-MRAM IN 28-NM FDSOI LOGIC PROCESS FOR INDUSTRIAL MCU/IOT APPLICATION 181

Yong Kyu Lee ; Yoonjong Song ; Joochan Kim ; Sechung Oh ; Byoung-Jae Bae ; Sanghumn Lee ; Junghyuk Lee ; Unghwan Pi ; Boyoung Seo ; Hyunsung Jung ; Kilho Lee ; Hyunchul Shin ; Hyuntaek Jung ; Mark Pyo ; Artur Antonyan ; Daesop Lee ; Sohee Hwang ; Daehyun Jang ; Yongsung Ji ; Seungbae Lee ; Jungman Lim ; Kwan-Hyeob Koh ; Kihyun Hwang ; Hyeongsun Hong ; Kichul Park ; Gitae Jeong ; Jong Shik Yoon ; Es Jung

22-NM FD-SOI EMBEDDED MRAM WITH FULL SOLDER REFLOW COMPATIBILITY AND ENHANCED MAGNETIC IMMUNITY 183

K. Lee ; K. Yamane ; S. Noh ; V. B. Naik ; H. Yang ; S. H. Jang ; J. Kwon ; B. Behin-Aein ; R. Chao ; J. H. Lim ; K. W. Gan ; D. Zeng ; N. Thiyagarajah ; L. C. Goh ; B. Liu ; E. H. Toh ; B. Jung ; T. L. Wee ; T. Ling ; T. H. Chan ; N. L. Chung ; J. W. Ting ; S. Lakshmipathi ; J. S. Son ; J. Hwang ; L. Zhang ; R. Low ; R. Krishnan ; T. Kitamura ; Y. S. You ; C. S. Seet ; H. Cong ; D. Shum ; J. Wong ; S. T. Woo ; J. Lam ; E. Quek ; A. See ; S. Y. Siah

LOW RA MAGNETIC TUNNEL JUNCTION ARRAYS IN CONJUNCTION WITH LOW SWITCHING CURRENT AND HIGH BREAKDOWN VOLTAGE FOR STT-MRAM AT 10 NM AND BEYOND 185

C. Park ; H. Lee ; C. Ching ; J. Ahn ; R. Wang ; M. Pakala ; S. H. Kang

RARE-FAILURE ORIENTED STT-MRAM TECHNOLOGY OPTIMIZATION 187

Nuo Xu ; Fan Chen ; Dmytro Apalkov ; Weiyi Qi ; Jing Wang ; Zhengping Jiang ; Woosung Choi ; Dae Sin Kim

SIGNIFICANT PERFORMANCE ENHANCEMENT OF UTB GEOI PMOSFETS BY ADVANCED CHANNEL FORMATION TECHNOLOGIES 191

W. H. Chang ; T. Irisawa ; H. Ishii ; H. Hattori ; N. Uchida ; T. Maeda

FIRST DEMONSTRATION OF VERTICALLY-STACKED GATE-ALL-AROUND HIGHLY-STRAINED GERMANIUM NANOWIRE P-FETS 193

E. Capogreco ; L. Witters ; H. Arimura ; F. Sebaai ; C. Porret ; A. Hikavyy ; R. Loo ; A. P. Milenin ; G. Eneman ; P. Favia ; H. Bender ; K. Wostyn ; E. Dentoni Litta ; A. Schulze ; C. Vrancken ; A. Opdebeeck ; J. Mitard ; R. Langer ; F. Holsteyns ; N. Waldron ; K. Barla ; V. De Heyn ; D. Mocuta ; N. Collaert

HOLE MOBILITY ENHANCEMENT IN EXTREMELY-THIN-BODY STRAINED GOI AND SGOI PMOSFETS BY IMPROVED GE CONDENSATION METHOD 195

K.-W. Jo ; W.-K. Kim ; M. Takenaka ; S. Takagi

GESN P-FINFETS WITH SUB-10 NM FIN WIDTH REALIZED ON A 200 MM GESNOI SUBSTRATE: LOWEST SS OF 63 MV/DECADE, HIGHEST $G_{M,INT}$ OF 900 µS/µM, AND HIGH-FIELD μ_{EFF} OF 275 CM^2/V•S 197

Dian Lei ; Kaizhen Han ; Kwang Hong Lee ; Yi-Chiau Huang ; Wei Wang ; Sachin Yadav ; Annie Kumar ; Ying Wu ; Huiquan Heliu ; Shengqiang Xu ; Yuye Kang ; Yang Li ; Eugene Y.-J. Kong ; Chuan Seng Tan ; Xiao Gong

SPACE PROGRAM SCHEME FOR 3-D NAND FLASH MEMORY SPECIALIZED FOR THE TLC DESIGN 201

Ho-Jung Kang ; Nagyong Choi ; Dong Hwan Lee ; Tackhwi Lee ; Sungyong Chung ; Jong-Ho Bae ; Byung-Gook Park ; Jong-Ho Lee

FIRST DEMONSTRATION OF MONOCRYSTALLINE SILICON MACARONI CHANNEL FOR 3-D NAND MEMORY DEVICES 203

R. Delhougne ; A. Arreghini ; E. Rosseel ; A. Hikavyy ; E. Vecchio ; L. Zhang ; M. Pak ; L. Nyns ; T. Raymaekers ; N. Jossart ; L. Breuil ; S. S. V-Palayam ; C.-L. Tan ; G. Van Den Bosch ; A. Furnemont

HIGH ENDURANCE SELF-HEATING OTS-PCM PILLAR CELL FOR 3D STACKABLE MEMORY 205

C. W. Yeh ; W. C. Chien ; R. L. Bruce ; H. Y. Cheng ; I. T. Kuo ; C. H. Yang ; A. Ray ; H. Miyazoe ; W. Kim ; F. Carta ; E. K. Lai ; M. Brightsky ; H. L. Lung

TE-BASED BINARY OTS SELECTORS WITH EXCELLENT SELECTIVITY (>10^5), ENDURANCE (>10^8) AND THERMAL STABILITY (>450^oC) 207

Jongmyung Yoo ; Yunmo Koo ; Solomon Amsalu Chekol ; Jaehyuk Park ; Jeonghwan Song ; Hyunsang Hwang

HALF-THRESHOLD BIAS IOFFREDUCTION DOWN TO NA RANGE OF THERMALLY AND ELECTRICALLY STABLE HIGH-PERFORMANCE INTEGRATED OTS SELECTOR, OBTAINED BY SE ENRICHMENT AND N-DOPING OF THIN GESE LAYERS 209

Naga Sruti Avasarala ; G. L. Donadio ; T. Witters ; K. Opsomer ; B. Govoreanu ; A. Fantini ; S. Clima ; H. Oh ; S. Kundu ; W. Devulder ; M. H. Van Der Veen ; J. Van Houdt ; M. Heyns ; L. Goux ; G. S. Kar

HIGHLY MANUFACTURABLE LOW POWER AND HIGH PERFORMANCE 11LPP PLATFORM TECHNOLOGY FOR MOBILE AND GPU APPLICATIONS 213

H.-J. Kim ; B.H. Choi ; Y.H. Lee ; J.H. Ahn ; Y.S. Bang ; Y.D. Lim ; J.H. Do ; J.H. Jung ; T.J. Song ; Y. Yasuda-Masuoka ; K.C. Park ; S.D. Kwon ; J.S. Yoon

A 12NM FINFET TECHNOLOGY FEATURING 2ND GENERATION FINFET FOR LOW POWER AND HIGH PERFORMANCE APPLICATIONS 215

H.C. Lo ; D. Choi ; Y. Hu ; Y. Shen ; Y. Qi ; J. Peng ; D. Zhou ; M. Mohan ; C. Yong ; H. Zhan ; H. Wei ; X. He ; D. Kang ; A. Sirman ; Y. Wang ; H. Zang ; S.Y. Mun ; A. Vinslava ; W.H. Chen ; C. Gaire ; J. Liu ; X. Dou ; Y. Shi ; P. Zhao ; B. Zhu ; A. Jha ; X. Zhang ; X. Wan ; E. Lavigne ; C. Kyono ; M. Togo ; J. Versaggi ; H. Yu ; O. Hu ; J.G. Lee ; S. B. Samavedam ; D.K. Sohn

8LPP LOGIC PLATFORM TECHNOLOGY FOR COST-EFFECTIVE HIGH VOLUME MANUFACTURING 217

Hwasung Rhee ; Ilryong Kim ; Jaehun Jeong ; Nakjin Son ; Heebum Hong ; Sungil Cho ; Yongmin Park ; Dongwoo Kim ; Yunki Choi ; Jeonghoon Ahn ; Sung Gun Kang ; Kyunghwan Yeo ; Jungtae Kim ; Euncheol Lee ; Jong Mil Youn ; Jong Shik Yoon

HIGH PERFORMANCE MOBILE SOC PRODUCTIZATION WITH SECOND-GENERATION 10-NM FINFET TECHNOLOGY AND EXTENSION TO 8-NM SCALING 219

Jun Yuan ; Ken Rim ; Ying Chen ; Ming Cai ; Youseok Suh ; Jihong Choi ; Jie Deng ; Jerry Bao ; Zhimin Song ; Lixin Ge ; Hao Wang ; Xiao-Yong Wang ; Vicki Lin ; Chihwei Kuo ; Sam Yang ; Ashwin Rabindranath ; Shrihari Siva ; Prasad Bhadri ; Sungwon Kim ; Kwon Lee ; Soon Cho ; Sunggun Kang ; Saechoon Oh ; S. D. Kwon ; Xiangdong Chen ; Paul Penzes ; Parag Agashe ; William Miller ; P. R. Chidambaram

HYBRID 14NM FINFET - SILICON PHOTONICS TECHNOLOGY FOR LOW-POWER TB/S/MM^2 OPTICAL I/O 221

M. Rakowski ; Y. Ban ; P. De Heyn ; N. Pantano ; B. Snyder ; S. Balakrishnan ; S. Van Huylenbroeck ; L. Bogaerts ; C. Demeurisse ; F. Inoue ; K. J. Rebibis ; P. Nolmans ; X. Sun ; P. Bex ; A. Srinivasan ; J. De Coster ; S. Lardenois ; A. Miller ; P. Absil ; P. Verheyen ; D. Velenis ; M. Pantouvaki ; J. Van Campenhout

Author Index

FOREWORD

Welcome to the 2018 Symposium on VLSI Technology

On behalf of the organizing committees, it is our pleasure to welcome you to the 38th Symposium on VLSI Technology to be held June 18-22, 2018, at the Hilton Hawaiian Village in Honolulu, Hawaii. Since its founding in 1981, the Symposium, jointly sponsored by the IEEE Electron Devices Society and the Japan Society of Applied Physics, has been recognized as one of the premiere international conferences on VLSI & Semiconductor technology. As always, the latest technology advancements in semiconductors will be shared by leading technologists from industry and academia this year too.

Following a long tradition, the conference is held in conjunction with the Symposium on VLSI Circuits. The co-location of these two Symposia provides a unique opportunity for the attendees to span from process technology topics to System-on-Chip and System-in-Package design and integration. This cross-disciplinary setting is informal and allows technologists, scientists and designers to network and exchange ideas in an open forum. This year the theme for both Symposia is **"Technology, Circuits and Systems for Smart Living"**. It reflects tight interactions between R&D in all three areas required to achieve continuous improvement in performance, power reduction and cost. Special joint focus sessions will offer unique learning opportunities for attendees on "Emerging Memory", "In Memory and In Sensor Computing", "Power Devices and Circuits" and "Design and Technology Co-Optimization in Advanced CMOS Technology".

The conference also features *Technology Focus Sessions* with this year's emphasis on "Back-End Compatible Devices and Advanced Thermal Management" and "Sensors and Devices for IoT, Medicine and Smart Living", embedded in an outstanding program of 20 sessions with 87 excellent papers including 9 invited and 2 plenary papers. The research results encompass a broad spectrum of VLSI technology topics, including high performance and low power transistor scaling, breakthroughs in new materials and processes which provide the key elements for tomorrow's technologies, novel memory structures, neuromorphic devices, technologies for Artificial Intelligence (AI), heterogeneous integration, beyond CMOS functional devices, advanced device analysis, and VLSI manufacturing.

We are part of a constantly evolving industry where innovation on multiple fronts shapes the technologies around us, and in front of us. For attendees to learn about these semiconductor innovations and to give them a broader perspective on implications to devices and system applications, we will start the symposium with a one-day *Short Course* on June 18th, **"Device and Integration Technologies for Sub-5nm CMOS and Next Wave of Computing".** This day-long *Short Course* will address topics of interest to both Technology and Circuits attendees, consisting of eight distinguished lectures given by subject matter leaders. The *Short Course* presents a great opportunity to learn about recent trends and challenges in advanced microelectronics technology, its adaptation into new computing systems, cloud storage, wearable devices and healthcare products.

The *Plenary Session* on June 19th opens the Symposium with two stimulating invited technology talks given by distinguished speakers. The first presentation on **"Memory Technology: The Core to Enable Future Computing Systems,"** Scott J. DeBoer, Micron Technology, Inc., highlights memory technology roadmap and innovations for future system architecture. The second presentation, **"Revolutionizing Cancer Genomic Medicine by AI and Supercomputer with Big Data,"** Satoru Miyano, University of Tokyo describes how AI can be used to create clinical experts equipped with exoskeleton.

The Evening Panel Discussions at the Symposium on VLSI Technology are well known for their selection of timely topics and lively exchanges between the technical leaders on the panel themselves and the audience. This year, there are two exciting Panel Sessions. The Joint Technology/Circuits Panel **"Is the CPU Dying or Dead? Are Accelerators the Future of Computation?"** will be held on Monday, June 18th. Then, **"Storage Class Memories: Who cares? DRAM is Scaling Fine, NAND is Stacking Great,"** will be held on Tuesday evening, June 19th.

Another popular feature of the Symposia is the *Luncheon Talk* on Thursday, June 21st. This year we have **"The Hardware of Mind, from Turing to Today,"** by Grady Booch, IBM which will take us on an exciting journey to understand how neuroscience is informing engineering of silicon.

Friday full day Forum will be dedicated to "**Machine Learning Today and Tomorrow: Technology, Circuits and System View**". Several experts in the field will steer discussion on how technology and circuits will drive the future of Artificial Intelligence and Machine Learning. This session aims to provide background knowledge, as well as the preparation needed to engage in the future of machine-learning systems.

This year's excellent technical program is a result of an outstanding effort of the Technical Program Committees under the leadership of the Program Chair, Chorng-Ping Chang, Applied Materials, and Co-Chair, Shinya Yamakawa, Sony Semiconductor Solutions. We express our sincere appreciation to all committee members, who are themselves leaders in the field of VLSI technologies, for their highly-skilled efforts and dedicated support of the Symposium on VLSI Technology. Their commitment and hard work resulted in the selection of high quality papers, panels, invited talks, focus sessions and short course.

We hope you enjoy the presentations and lively discussions with your colleagues and find ample opportunities to network with them in and outside of the sessions. We certainly hope that you a have a productive and enjoyable experience in beautiful Hawaii. In 2019, the Symposium on VLSI Technology together with the Symposium on VLSI Circuits will return to Kyoto, Japan. We look forward to you joining us again next year.

Mukesh Khare   Meishoku Masahara
2018 Symposium Chair   2018 Symposium Co-Chair

2018 VLSI TECHNOLOGY SYMPOSIUM

Chair:	Mukesh Khare	IBM
Symposium Co-Chair:	Meishoku Masahara	AIST
Program Chair:	Chorng-Ping Chang	Applied Materials
Program Co-Chair:	Shinya Yamakawa	Sony Semiconductor Solutions Corp.
Secretary:	Tomas Palacios	Massachusetts Institute of Technology
Secretary Co-Chair:	Kazuyuki Tomida	Sony Semiconductor Solutions Corp.
Publicity Chair:	Malgorzata (Gosia) Jurczak	ASM
Publicity Co-Chair:	Takaaki Tsunomura	Tokyo Electron
Short Course Chair:	Willy Rachmady	Intel Corp.
Short Course Co-Chair:	Munehiro Tada	NEC Corp.
Evening Panel Chair:	Nirmal Ramaswamy	Micron
Evening Panel Co-Chair:	Masaharu Kobayashi	The University of Tokyo
Publication Chair:	Luca Selmi	University of Modena and Reggio Emilia
Publication Co-Chair:	Ken Uchida	Keio University
Focus Sessions Chair:	Eric Pop	Stanford
Focus Sessions Co-Chair:	Kazuhiko Endo	AIST
Demo Chair:	Kamel Benaissa	Texas Instruments
Demo Co-Chair:	Noboyuki Sugii	Hitachi, Ltd.
Best Student Paper:	Maud Vinet	CEA-LETI
Best Student Paper Co-Chair:	Hiroshi Morioka	Socionext Inc.

EXECUTIVE COMMITTEES

IEEE
Chair: H.S. Philip Wong — Stanford University
Members:
Ajith Amerasekera	Texas Instruments
Chorng-Ping Chang	Applied Materials
Ken Chang	Xilinc, Inc.
Vivek De	Intel Corp
Ichiro Fujimori	Broadcom Corporation
Jeffrey Gealow	Analog Devices, Inc.
Raj Jammy	Carl Zeiss
Mukesh Khare	IBM Research
Tsu-Jae King Liu	University of California, Berkeley
Stephen Kosonocky	AMD Ft. Collins
Gunther Lehman	Infineon Technologies AG
Un-Ku Moon	Oregon State University
Katsu Nakamura	Analog Devices, Inc.
Klaus Schruefer	Intel Mobile Communications GmbH
Hans Stork	ON Semiconductor
Jason Woo	University of California, Los Angeles

JSAP
Chair: Tadahiro Kuroda — Keio University
Members:
Eiji Fujii	Panasonic Corporation
Toshiro Hiramoto	The University of Tokyo
Dai Hisamoto	Hitachi, Ltd.
Makoto Ikeda	The University of Tokyo
Satoshi Inaba	Toshiba Memory Corp.
Toshihiko Kanayama	AIST
Chin-Yuan Lu	Macronix International Co., Ltd.
Meishoku Masahara	AIST
Yoshio Masubuchi	Toshiba Corp.
Yasunori Mochizuki	NEC Corp.
Tohru Mogami	PETRA
Masato Motomura	Hokkaido University
Masaaki Niwa	Tohoku University
Satoshi Shigematsu	NTT Corp.
Ken Takeuchi	Chuo University
Hirotaka Tamura	Fujitsu Laboratories Ltd.
Taku Umebayashi	Sony Corp.
Hitoshi Wakabayashi	Tokyo Institute of Technology
Shinya Yamakawa	Sony Semiconductor Solutions Corp.
Kazuo Yano	Hitachi, Ltd.
Hoi-Jun Yoo	KAIST
Kevin Zhang	TSMC

TECHNICAL PROGRAM COMMITTEES

NORTH AMERICA/EUROPE

Chair:	Chorng-Ping Chang	Applied Materials, Inc.
Members:	Kelly Baker	Freescale/NXP
	Kamel Benaissa	Texas Instruments
	Jian Chen	AMD
	Nadine Collaert	imec
	Suman Datta	University of Notre Dame
	Emmanuel Dubois	IEMN/CNRS
	Gertjan Hemink	SanDisk
	Gosia Jurczak	ASM
	Ted Letavic	Globalfoundries
	Yue Liang	nVidia
	Vijay Narayanan	IBM
	Tomas Palacios	Massachusetts Institute of Technology
	Yang Pan	Lam Research
	Eric Pop	Stanford
	Willy Rachmady	Intel
	Nirmal Ramaswamy	Micron Technology Inc.
	Luca Selmi	University of Modena and Reggio Emilia
	Thomas Skotnicki	STMicroelectronics
	Seung-Chul Song	Qualcomm, Inc.
	Maud Vinet	CEA-LETI, MINATEC
	Peide Ye	Purdue University
	Greg Yeric	ARM, Inc.

JAPAN/FAR EAST

Co-Chair:	Shinya Yamakawa	Sony Semiconductor Solutions Corp.
Members:	Kazuhiko Endo	AIST
	Osbert Cheng	United Microelectronics Corp
	Steve S. Chung	National Chiao Tung University
	Sung-Woong Chung	SK Hynix Inc.
	Masaharu Kobayashi	The University of Tokyo
	Byoung Hun Lee	Gwangju Inst. of Science and Tech
	Hang-Ting (Oliver) Lue	Macronix International Co., Ltd.
	Yuri Masuoka	Samsung Electronics Co., Ltd.
	Hidenharu Miyake	Micron Memory Japan, Inc.
	Katsura Miyashita	Toshiba Corp.
	Hiroshi Morioka	Socionext Inc.
	Yuta Shiratori	NTT Corp.
	Nobuyuki Sugii	Hitachi, Ltd.
	Munehiro Tada	NEC Corp.
	Shinichi Takagi	The University of Tokyo
	Tetsu Tanaka	Tohoku University
	Kenji Tateiwa	Tower Jazz Panasoinc Semi. Co. Ltd.
	Kazuyuki Tomida	Sony Semiconductor Solutions Corp.
	Takaaki Tsunomura	Tokyo Electron Ltd.
	Ken Uchida	Keio University
	Tomohiro Yamashita	Renesas Electronics Corp.
	Yee Chia Yeo	TSMC

2018 Symposium on VLSI Technology Digest of Technical Papers

Memory Technology: The Core to Enable Future Computing Systems

Scott DeBoer

Micron Technology, Inc.
8000 S. Federal Way
Boise, ID 83707
sdeboer@micron.com

Abstract

Roughly 300 billion gigabytes (GB) of semiconductor memory will be produced this year (2018) — 40GB for every person on the planet -- with projections to double every two years for the foreseeable future. As user demand for large amounts of instantly accessible data continues to increase, memory is becoming both a solution and a bottleneck, spurring the industry to redefine how memory is used in systems and to innovate for new types of memory. This paper discusses the scaling roadmap for NAND and DRAM memories, the introduction of new emerging memories to supplement NAND and DRAM, and opportunities for changes in system architectures to exploit the inherent capabilities of memory.

The Importance of Memory in Modern Systems

The demand for memory is projected to cross 300 billion GB (3×10^{20} bytes) this year and double at a rate of every two years going forward, as shown for NAND in Figure 1.

Memory has become a defining component of modern computing, from mobile devices to servers. The dollars spent on memory content in a smart phone is comparable to that of the CPU. More than any other component, memory often defines the performance, power, and cost of servers.

As memory becomes central to modern systems, its limitations are becoming relevant to system performance and cost. Memory performance has failed to keep pace with the CPU, a problem that has been aggravated by the emergence of multi-core CPUs and the increasing use of GPUs and other specialized, high-performance processors. Memory cost reduction, which has fueled the rapid growth of memory usage, is being challenged by the complexities of continued scaling.

To address the scaling issues, the memory semiconductor industry has been focused on pushing out the memory "scaling wall" while, in parallel, searching for new types of memory (emerging memory) to supplement or eventually replace current memory technologies. While these efforts are important, it is becoming apparent they do not address some of the fundamental limitations of memory performance and energy bottlenecks and new system architectures may be required, moving memory and compute closer together may provide a more optimal solution.

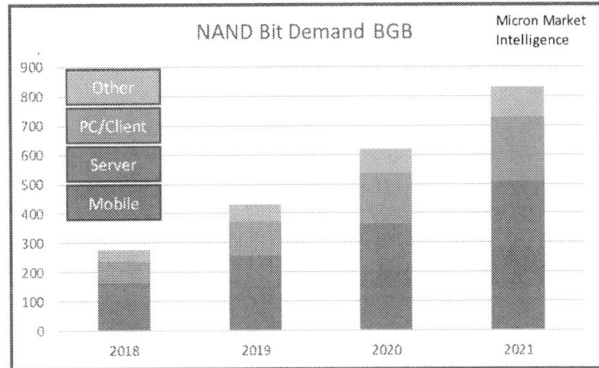

Figure 1: Industry NAND demand forecast

Pushing Out the NAND and DRAM Scaling Wall

The memories common in today's computing systems emerged in the early 1970s, at the dawn of the semiconductor industry. The electron storage physics on which these devices are based has had an amazingly long life with a phenomenal track record of scaling. It is generally agreed that today's memories are scalable for several more generations as innovation continues to circumvent the perceived limitations. It is becoming clear, however, that fundamental physical and economic limitations are quickly being approached for electron-based storage that will slow the pace of scaling and reduce memory functionality.

Planar NAND scaling ended at ~15nm because the number of electrons stored on the floating gate approached zero, resulting in numerous unsolvable technical problems [1]. The transition to 3D NAND resulted in a large cylindrical cell enabling more than an order of magnitude increase in electrons stored in the cell, which substantially mitigates scaling problems [2]. 3D NAND successfully supplanted planar NAND and greatly relaxed the lithography requirements relative to planar NAND. Currently, 3D NAND aggressively stacks many memory layers, with stacks of 64 tiers in high-volume production. 3D NAND stacks of 96 tiers are now ramping up in production and tier counts of more than 120 are under development. While the scaling path will be extremely challenging, particularly in the areas of high-aspect ratio etch and film depositions, continued 3D NAND scaling currently has no obvious roadblocks.

The DRAM scaling path is more complex than NAND's. DRAM scaling continues along a planar path following a conventional scaling strategy. In contrast to 3D NAND, DRAM requires an aggressive lithography roadmap (both resolution and alignment). As feature sizes are scaled, more advanced lithography is required. Figure 2 shows the feature sizes where the lithography transitions to more complex technology. The cost for the lithography step also increases with large error bars on the relative costs depending on both the pace of cost maturity for extreme EUV as well as the methodology used for pattern multiplication. Eventually, EUV lithography will likely play a significant role in DRAM scaling, but it is not an absolute requirement for DRAM technology roadmap enablement in the next few years. It is also possible that other technical challenges may limit DRAM scaling before EUV is finally required. As shown in Figure 2, EUV technology still requires significant progress in terms of cost relative to mature pattern multiplication technology. Clearly, a key to successful DRAM scaling will be aligning lithography technology roadmap choices carefully to ensure costs are controlled as feature sizes are reduced and overlay requirements are managed.

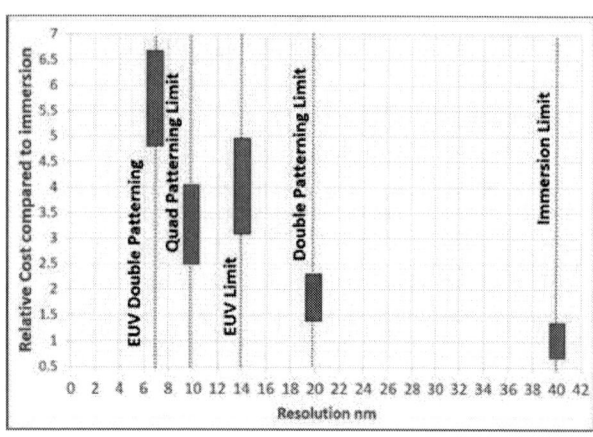

Figure 2: Lithography technology breakpoints and relative costs

Apart from lithography and cost constraints, continued DRAM scaling faces several materials and process integration challenges. Specifically, dielectric constant of the cell capacitor has steadily increased over the last several nodes to mitigate the rapidly decreasing cell capacitor area. Continued increase of dielectric constant for future nodes faces significant engineering challenges and will require tremendous innovation in the materials science area. Shrinking cell area also forces high aspect ratio capacitor etches, which become a significant challenge with further scaling, as shown in Figure 3. Resistor/capacitor (RC) constraints due to shrinking digit lines (less than 15nm) and word lines (less than 20nm) also pose a significant scaling challenge.

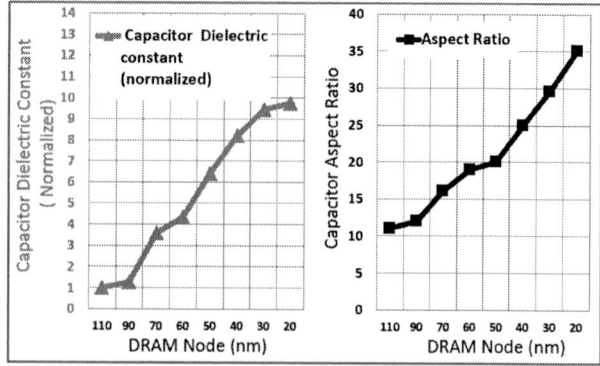

Figure 3: DRAM capacitor dielectric and aspect ratio trends

Potential Emerging Memories

To address scaling issues, industry members have been exploring new types of memories using new storage physics, as well as exploring new ways to use existing memories. New memories may offer alternative ways to use memory in systems and enable scaling beyond the limits of existing memories.

Several new memory options have been evaluated for high performance and high-density integration viability (see Figure 4) over the last decade, such as resistive memory (ReRAM), spin torque transfer memory (STTRAM), and 3D XPoint™ technology [3][6].

In resistive memories, conductance through the cell determines the cell state. Conductance can be modulated by oxygen vacancy filament (Ox-ReRAM), metal filament (M-ReRAM), or uniform ion migration. Uniform ion migrating cells typically demonstrate small read margin, poor endurance and low sense signal, undermining their performance in comparison to filamentary ReRAM. In filamentary ReRAM, M-ReRAM has been shown to significantly outperform Ox-ReRAM for various cell metrics such as read margin, endurance, retention and lower current operation. [4].

	DRAM	STTRAM	PCM/ 1T1R RRAM	Cross point RRAM	NAND
Read Latency	20ns	~ 50ns	~100ns-200ns	~100ns-200ns	~10us
Write Latency	20ns	~ 50ns	~1us	~1us	~10us
Read Endurance	>1e15	>10¹¹	>10⁷	>10⁷	>10⁷
Write Endurance	>1e15	>10¹¹	>10⁶	>10⁶	2K-100K
Write/Read Energy/bit	<10pJ/bit	~25pJ/bit	~100-200 pJ/bit	~100-200 pJ/bit	> 100pJ/bit
Alterability	~2KB	< 2KB	~10's B	~10's B	Large Blocks
Retention@RT	~milli seconds	Months	~Years	~Years	Years
Areal Density	1x				~50x

Figure 4: Comparison of various emerging memory technologies

Overall, although ReRAM technology has shown significant promise, it suffers from high variability and the lack of a viable selector. Significant work and major breakthroughs are still required to enable ReRAM as a future option for high-density memories.

STT-RAM offers significant promise for high-speed and high-endurance memory, but the narrow operating window between TDDB, low-current operation, retention,

978-1-5386-4219-1/18 $31.00 © 2018 IEEE

and array parasitics makes it a difficult choice to be competitive in terms of density and cost with leading edge DRAM.

Clearly the most successful of the emerging memories is 3D XPoint™, where the technology performance has been proven and volume production is underway. 3D XPoint™ performance and density are "midway" between DRAM and NAND which offers opportunities to greatly enhance system-level performance by augmenting existing memory technologies or even directly replacing them in some applications. Significant system-level enablement is required to exploit the full value of 3D XPoint™ memory, and this ongoing effort will take time to fully mature.

Introducing a new memory type is complex, with several challenges to overcome to achieve high-volume production [7]. New memory types will first emerge as supplements to existing memory to help overcome scaling deficiencies and address system performance and energy challenges. The new memories may also find an entry point in applications that highly value their unique set of features, such as high-performance storage devices and emerging Internet of Things (IoT) devices.

Innovating on System Architectures to Better Exploit Memory Capabilities

The way memory is used in systems -- the memory hierarchy -- has evolved over the past four decades. How hardware and software handle memory has been defined by access latency, access granularity, volatility, power, and cost of available memories. As shown in Figure 5, the memory hierarchy has evolved to a tiered system of various levels of caching in SRAM, to main memory in DRAM, to storage in SSD. This hierarchy, and the resulting hardware and software wrapped around it, is as much defined by the deficiencies of the memories as by their features, and has evolved to be a performance and energy constraint for today's systems.

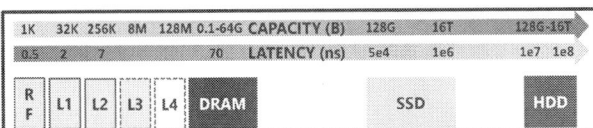

Figure 5: Memory hierarchy – CPU to HDD

Performance of a modern system is often constrained by the bandwidth of the memory. Fundamentally the memory is not able to supply data at the rate the CPU can consume data. This trend is being aggravated by the emergence of multi core processors, the use of GPUs, and specialized processors for machine learning. Memory bandwidth is constrained by the narrow bus connecting the memory to the CPU. Bandwidth internal to the memory chip itself is greater than 100X larger than the bus constrained value, as shown in Figure 6.

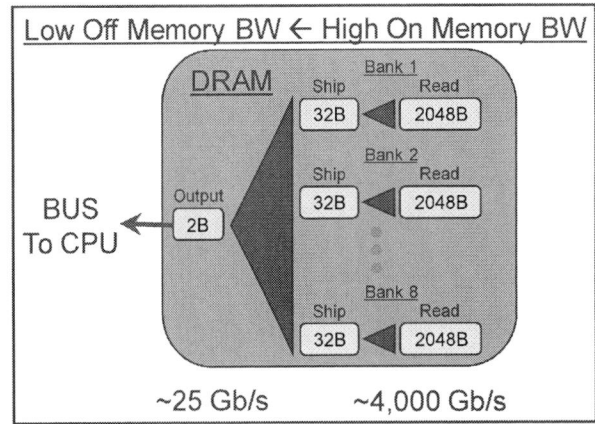

Figure 6: System performance is constrained by the narrow bus

A similar situation exists for energy, where the bus energy dominates the energy required to move data from the memory to the CPU.

Considering the significant impact of the bus on performance and power, different partitioning of where computation is done in the system relative to the memory may enable better overall system optimization. Figure 7 shows one possible implementation of this partitioning, with an abstracted interface to a memory subsystem where the subsystem takes on some compute functions in the ASIC and in the memory components, reducing power and exploiting local bandwidth [5].

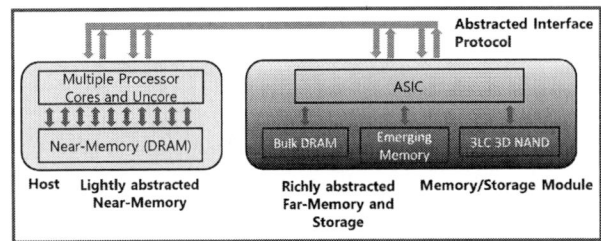

Figure 7: Abstracted memory system with embedded compute

In addition to reducing power and increasing performance, the concept of abstraction of the memory subsystem is an enabling capability for extending DRAM scaling. As the complexity of scaling increases there will be a need for more aggressive management of the DRAM bits to maintain performance and reliability. It is appropriate for much of this management to be done close to the memory bits where it can be accomplished with a high level of granularity and a short period of time. An example of this concept is the Hybrid Memory Cube (HM), which improves manageability, performance, and power of DRAM (Figure 8).

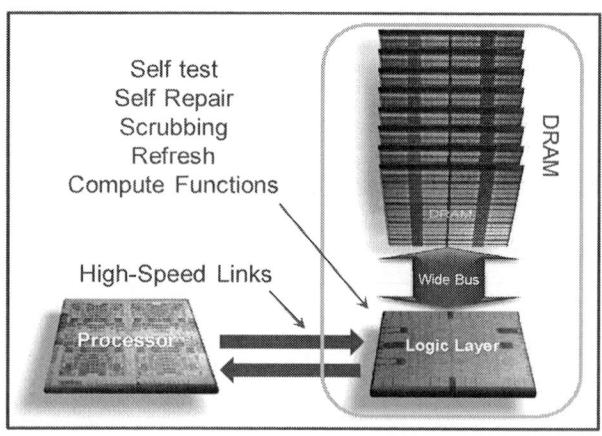

Figure 8: Hybrid Memory Cube (HMC)

Figure 7 shows an example of emerging memory in the memory subsystem box. Emerging memories have a promise to fill the large performance and density gap existing in the memory hierarchy, as shown in Figure 9.

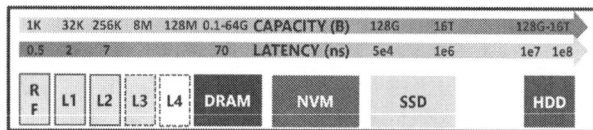

Figure 9: Adding emerging memory (NVM) to the hierarchy

Conclusion

Memory has become central to modern systems, from mobile devices to servers, often defining the system cost and performance.

Scaling of DRAM and NAND continues as innovations push out the perceived scaling wall, but fundamental economic-driven limitations are being approached. New kinds of memory -- emerging memories -- are being explored to supplement DRAM and NAND. These emerging memories begin to close the gaps in the memory hierarchy, but they do not challenge NAND cost or DRAM performance.

Modern system architectures are not exploiting the full performance and energy capabilities of the memory. Performance and energy are being constrained by the narrow link between the memory and the processor. Addressing this issue through better integration of compute and memory is a significant opportunity.

References

[1] K. Prall, "Scaling Non-volatile Memory Below 30nm," 2007 Non-volatile Semiconductor workshop, pg. 5-10

[2] A. Goda, "Opportunities and Challenges of 3-D NAND Scaling," 2013 VLSI-TSA

[3] N. Ramaswamy et.al, "3D ReRAM: Crosspoint Memory Technologies," 2017 IEDM Short Course

[4] A. Calderoni et.al," Performance Comparison of O-based and Cu-based ReRAM for High-Density Applications," 2014 IMW

[5] J. Pawlowski, "Memory as We Approach a New Horizon," 2016 MEMSYS

[6] G. Atwood, "Current and Emerging Memory Technology Landscape," 2011 Flash Memory Summit

[7] S. DeBoer et.al, "A semiconductor memory development and manufacturing perspective," 2014 ESSDERC

978-1-5386-4219-1/18 $31.00 © 2018 IEEE

Revolutionizing Cancer Genomic Medicine by AI and Supercomputer with Big Data

Satoru Miyano and IMSUT Cancer Clinical Sequence Research Team

Human Genome Center, Institute of Medical Science, University of Tokyo, Minato-ku, Tokyo, Japan
miyano@ims.u-tokyo.ac.jp

Abstract

We are running a cancer clinical sequence system based on whole genome/exome, RNA sequence and epigenome as research. When focused on hematology/oncology, it takes currently four days for a patient from signing informed consent (IC) to diagnosis. This process consists of IC, specimen collection, next-generation sequencer analysis, data analysis, interpretation/translation of mutations, determining the diagnosis combined with all pathological/clinical data and returning the result to the patient. Therapies are not only drugs but also hematopoietic stem cell transplantation. A pipeline Genomon for analyzing cancer genomes and RNA sequences by next-generation sequencers plays one of the key roles. It is running on the supercomputer system at Human Genome Center. The bottleneck of interpretation/translation was drastically resolved by employing IBM Watson for Genomics in harmony with our in-house human curation pipeline. We report how our system works as a conglomerate of oncologists, cancer biologists, bioinformatics experts augmented with Watson and Genomon.

Introduction

Cancer is a very complex disease caused by a variety of mutations in genomes that evolve spatiotemporally in our body. The knowledge about cancer is also getting deeper and broader. More than 200,000 papers with keywords tumor/cancer were published only in 2017 and rapidly increasing. More than 5,000,000 coding mutations are reported in 25,000 papers. Reported biological mechanisms in the literature are digitalized and enormous. When we sequence one patient's whole cancer genome, we face with several hundreds to millions of mutations. Pathological diagnosis is often incorrect due to the complexity of cancer. Actually, the efficacy of anti-cancer drugs specified to organs is just around 20%. Cancer genome sequencing is a savior for precision medicine. However, it is beyond human abilities to clinically interpret the patient genome sequence information with big data. The speed also matters seriously. To overcome these difficulties, we are using next-generation sequencers for genome sequencing, the supercomputer SHIROKANE for sequence data analysis, and IBM Watson for Genomics and several databases for clinical interpretation of mutations. By uploading the mutation data of a patient's specimen, driver genes and candidate drugs will be recommended together with their reasoning information. It is currently possible for the conducting doctor to return the diagnosis to the patient in four days that includes whole genome sequencing and data analysis time.

Understanding cancer is beyond human abilities

A. Cancer is very complex

It is increasingly clear that cancer is a very complex disease that occurs from accumulation of multiple genetic and epigenetic changes in individuals who carry different genetic backgrounds and have suffered from distinct carcinogen exposures (Fig. 1).

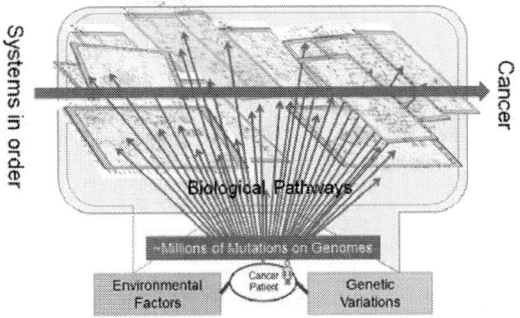

Fig. 1 Cancer is a disease of systems disorder caused by accumulated mutations.

B. Cancer genome sequence data and supercomputer

Human genome information is a collection of three billion letters of A, C, T, G ranging over chromosomes. The term "sequencing" is used as a process of determining these letters in sequence. The most commonly used next-generation sequencers (NGS) are produced by Illumina (https://www.illumina.com/). It produces shredded pieces of sequences of length about 100 from DNA (Fig. 2).

Fig. 2 "What NGS does." For whole genome sequence (WGS) analysis, we need 30 copies of germline DNA (from peripheral blood, swab) and 40 copies of cancer DNA.

For cancer genome analysis, two big jigsaw puzzles, one for germline and the other for cancer, are solved by using supercomputer (Fig. 3). Detection of mutations is the first step to investigate cancer.

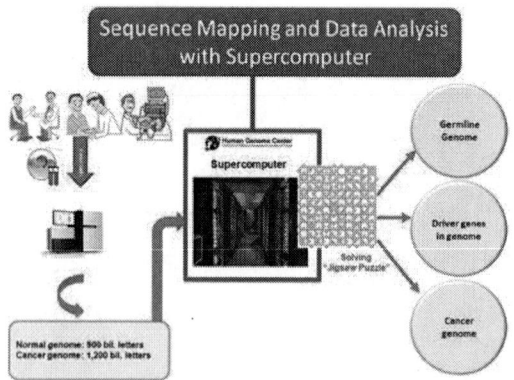

Fig. 3 Supercomputer is employed to solve big jigsaw puzzles using "reference genome sequence data."

C. Big digitalized biomedical knowledge

In order to understand cancer genomes, we use various biomedical databases.

NIH PubMed Database (https://www.ncbi.nlm.nih.gov/pubmed/) compiles more than 27 million publication abstracts in 2017. If their full papers are printed and piled, its height exceeds 4km (higher than Mt. Fuji). As estimation, it will reach 100km in 2050. The database COSMIC, the Catalogue Of Somatic Mutations In Cancer (https://cancer.sanger.ac.uk/cosmic), is the world's largest resource of somatic mutations in human cancer. In the v84 (released on the13th February 2018), 5,448,850 coding mutations are linked to 25,807 papers by human-curation. Retrieving these databases is a heavy daily burden for cancer genomic scientists. ClinVar (https://www.ncbi.nlm.nih.gov/clinvar/) collects information about genomic variation and its relationship to human health. It is updated weekly. NIH Clinical.gov provides information of more than 250,000 clinical trials. The database of molecular interactions NDEx (http://www.ndexbio.org/) includes NIH National Cancer Institute's Pathway Interaction Database (NCI-PID). It is a pathway interaction database that is a curated collection of information about known bio-molecular interactions and key cellular processes assembled into signaling pathways.

All are described with English, a natural language. Natural language processing technology should play a key role for the future of cancer genomics because no one can read them all.

D. Bottleneck

Millions of mutations can be found from the genome sequence data of one cancer patient. But their clinical/biological translation is a severe bottleneck [1]. We face with many mutations having biologically interesting features that may not have prognostic or therapeutic relevance. A major challenge is identification of clinically actionable mutations, where "actionable" means a potential target or risk factor that affects the treatment plan.

Genomon GO: Get Mutations!

Genomon (https://github.com/Genomon-Project) is a suite of bioinformatics tools for analyzing cancer genome and RNA sequencing data for Illumina NGS data (Fig. 4). It performs sensitive and accurate detection of most types of genomic variants (single nucleotide variants, short insertions/deletions, mid-size (20bp - 300bp) insertions/deletions and large scale structural variations), and transcriptomic changes (gene fusions, aberrant splicing patterns). Its good performance has been demonstrated through a large number of important cancer genome projects that led to very impactful discoveries [2-5]. Detection of mid-size insertions/deletions was considered as a "blind spot of Illumina NGS data" and Genomon solved this blind spot by sophisticated mathematical methods.

Fig. 4 Genomon for cancer genomics.

An efficient job scheduling framework realizes easy analysis of several hundreds of genome and transcriptome sequencing data simultaneously. Genomon is easy to install. After installation, users can start analysis just preparing simple sample configuration files. It is running on the supercomputer SHIROKANE at our Human Genome Center that consists of 550TFLOPS computation nodes at peak (Thin:5GB/Core; Fat: 2TB/node), 30PB Lustre File System and IBM Tape Archive System+1PB Nearline Disk (https://supcom.hgc.jp/ english/). This combination of Genomon and SHIROKANE reduced the time for data analysis to less than an hour.

Watson for Genomics

A. Cancer clinical sequence system at Institute of Medical Science University of Tokyo

By the drastic advancement of sequencing technologies, we have extended the system to whole genome sequencing, and more in 2011. The ELSI team and the genetic counseling team are involved in this system. At the research hospital, currently, clinical sequencing covers colorectal cancer and blood disorders as research. We recognized that the bottleneck of the system is "clinical interpretation/translation" and the more serious problem is a lack of people who can do this. Human Genome Center and Health Intelligence Center are responsible for sequencing and data analysis with Genomon using the supercomputer system. Decision support system and effective utilization of biomedical big data are practical key issues. This is the main motivation for introducing IBM Watson for Genomics in our system. The concept is shown in Fig. 5. Various sequencers are employed in this system for precision medicine (Fig. 6). The total system is illustrated in Fig. 7 and Fig. 8.

2018 Symposium on VLSI Technology Digest of Technical Papers

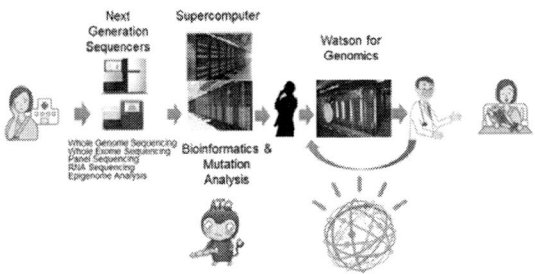

Fig. 5 The concept: from supercomputer & Watson to patients.

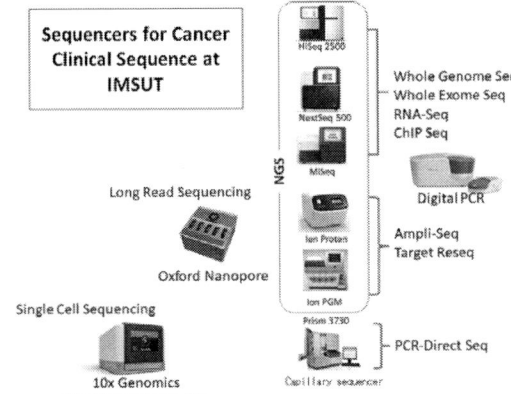

Fig. 6 Sequencers at IMSUT clinical sequencing system.

B. Before Watson

When we started whole genome clinical sequence in 2011, Prof. Yoichi Furukawa, MD, PhD (Institute of Medical Science University of Tokyo, abbreviated to IMSUT) faced with a serious problem. A mystery was on a Familial Adenomatous Polyposis (FAP) patient who has no deleterious mutations on the coding regions (exons). Whole genome sequencing was done and we analyzed the sequence data by human eyes and database searches at that time. After heavy efforts of investigations, we found a deletion of 10kb (10,000 letters) in the upstream of *APC* gene (Fig. 7) containing the promoter 1B. This deleted region causes reduced expression of APC-1B but not APC-1A by the deletion of promoter 1B that is responsible for familial adenomatous polyposis [6]. This interpretation was a result of huge labor and time. We thought

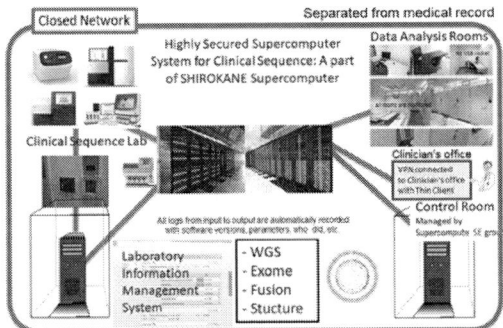

Fig. 7 Cancer clinical sequencing system of IMSUT.

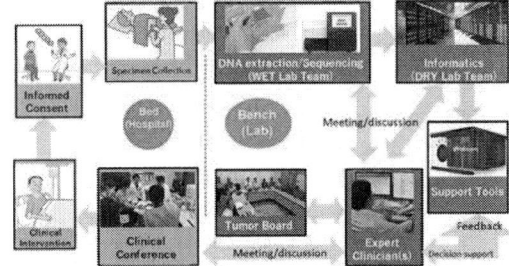

Fig. 8 Total system of oncologists, cancer biologists, bioinformatics experts augmented with Watson and Genomon.

we should have a new paradigm to go further. This was a main motivation to introduce Watson for Genomics in our clinical sequencing system.

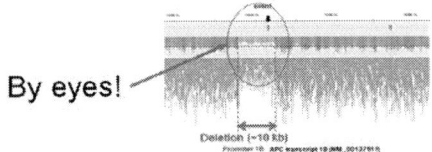

Changes the gene transcription!

Fig. 9 [Whole Genome Sequencing + Databases + Eyes] identified a large deletion in *APC* gene – This is the cause of the patient.

C. After Watson

In July of 2015, our institute introduced IBM Watson Genomic Analytics (current version is called Watson for Genomics, abbreviated to WfG) as a research on supporting cancer clinical sequencing. WfG employs technologies of natural language processing and machine learning. It was trained at New York Genome Center for solid tumors with 20 million PubMed abstracts, 15 million patent data, COSMIC (Catalogue of Somatic Mutations in Cancer, UK), ClinVar (Genomic Variation and Heath/Disease, NIH, USA) National Cancer Institute Pathways (NIH, USA), etc. From 2015, the hematology/oncology team directed by Prof. Arinobu Tojo, MD, PhD (IMSUT) joined in the project. The team started with the Illumina myeloid panel of 54 genes due to the cost and speed for diagnosis. WfG reads the list of mutations identified by Genomon as input, and suggest recommendations together with evidences in a way that clinicians can check and follow the reasons why such recommendations are generated. Initially, clinicians were skeptical about the recommendations by WfG and we found that WfG was not well-trained for blood tumors with the supervise of experts. After six months of training WfG by our hematologists/oncologists, its accuracy became the level of moderate acceptance. Simultaneously, we recognized only half of the patients could be diagnosed based on the 54 gene panel and we had to proceed to whole exome analysis. At that time, WfG covers only single nucleotide mutations. Currently, whole genome/exome, RNA sequencing as routine, and sometimes epigenetic analysis are done for diagnosis and search for drugs and therapies. Since WfG is still incomplete and under development, our in-house human

978-1-5386-4219-1/18 $31.00 © 2018 IEEE

curation pipeline is inevitable, for example, RNA sequencing data, epigenetic data, time-course data and therapies other than drugs. The turnaround time from IC to diagnosis is now about four days for whole genome analysis. This was realized by the team work of concologists, cancer biologists, bioinformatics experts and in-house curation pipeline augmented with WfG and Genomon on supercomputers.

D. An example of analysis by WfG

The following is an analysis of colorectal cancer cell line RKO [7], not a patient data. Whole exome analysis revealed 4,237 single nucleotide variants (SNVs) in this cell line. The input for WfG is the list of these SNVs. WfG analysis finished in less than 30 minutes. Fig. 10 is a molecular profile tab that suggest strong driver genes mapped on chromosomes 1-22, X and Y.

Drug tab (Fig. 11) shows target genes and their candidate drugs. Fig. 12 is pathway map that provides the reasons for recommendations from the viewpoint of biological mechanisms. All gene names and drugs are linked to literature, documents for drugs and clinical trial information so that clinician can review the recommendations from the viewpoint of experts. Before introducing WfG, it was impossible to review all genes due to the limitation of time to diagnosis and labor workload to the experts.

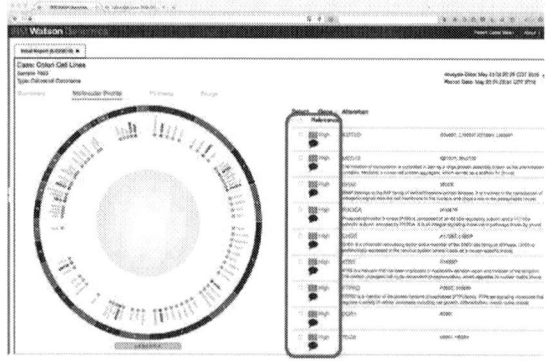

Fig. 10 Driver gene candidates were narrowed down from 4,237 mutations' list (SNVs) to 12 highly relevant genes.

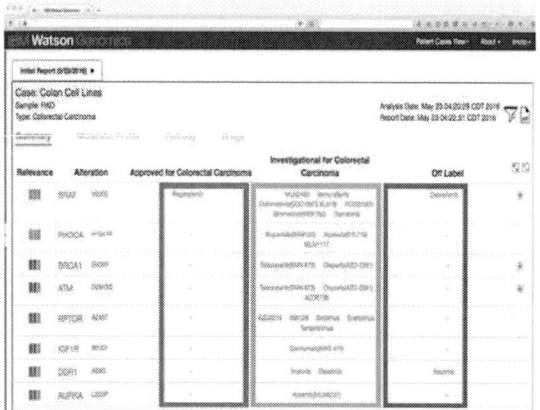

Fig. 11 Three categories of recommended drugs.: Approved for "target" cancer (red), Investigational under clinical trials (green), Off label: approved for "other" cancers (blue).

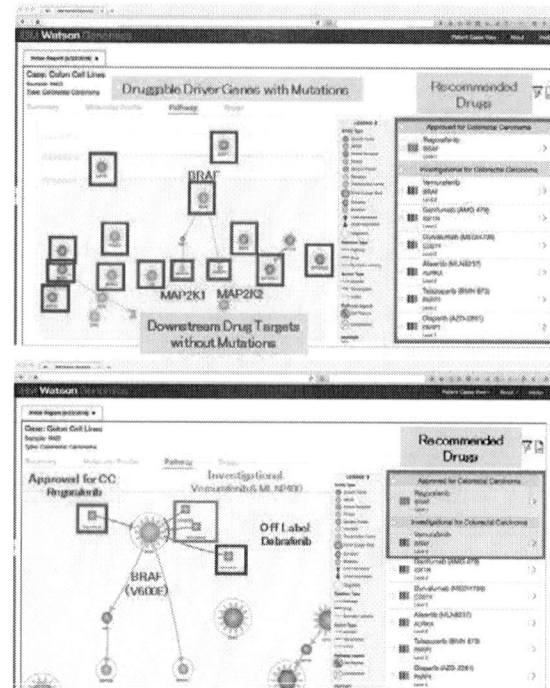

Fig. 12 Driver genes and drugs are mapped on pathways so that user (clinician) can check reason why those drugs are recommended.

Conclusion

Supercomputers reduced the time for data analysis for identifying genomic aberrations to the almost negligible level even for whole genome sequence data. However, interpretation/translation of aberrations is a sever bottleneck because of the number of mutations and the hugeness of background biomedical big data. This is getting severer year by year and already beyond human abilities. This is a reason why introduction of AI is inevitable. Diagnosis is responsible for medical doctors not AI. In our study, we learned that AI, in this study WfG, is not a technology to replace human experts but creates clinical experts equipped with AI exoskeleton.

Acknowledgments

We would like to thank all members of IMSUT cancer clinical sequence research team, especially, Prof. Yoichi Furukawa and his team for solid tumors, Prof. Arinobu Tojo and his team for hematology/oncology, Prof. Seiya Imoto and Prof. Rui Yamaguchi and their colleagues for informatics, and Prof. Yoichi Furukawa and Prof. Koichi Yuji for genetic counselling, and Prof. Kaori Muto for ELSI. This research is partly supported by MEXT grants: 15H05912, hp150265, hp160219, hp170227, and AMED grant JP18kk0205003.

References

[1] Good BM, Ainscough BJ, McMichael JF, Su AI, Griffith OL. Organizing knowledge to enable personalization of medicine in

cancer. *Genome Biol*. 2014 Aug 27;15(8):438.

[2] Kataoka K, Shiraishi Y, Takeda Y, Sakata S, Matsumoto M, Nagano S, Maeda T, Nagata Y, Kitanaka A, Mizuno S, Tanaka H, Chiba K, Ito S, Watatani Y, Kakiuchi N, Suzuki H, Yoshizato T, Yoshida K, Sanada M, Itonaga H, Imaizumi Y, Totoki Y, Munakata W, Nakamura H, Hama N, Shide K, Kubuki Y, Hidaka T, Kameda T, Masuda K, Minato N, Kashiwase K, Izutsu K, Takaori-Kondo A, Miyazaki Y, Takahashi S, Shibata T, Kawamoto H, Akatsuka Y, Shimoda K, Takeuchi K, Seya T, Miyano S, Ogawa S. Aberrant *PD-L1* expression through 3'-UTR disruption in multiple cancers. *Nature*. 2016 Jun 16;534(7607):402-406.

[3] Kataoka K, Nagata Y, Kitanaka A, Shiraishi Y, Shimamura T, Yasunaga J, Totoki Y, Chiba K, Sato-Otsubo A, Nagae G, Ishii R, Muto S, Kotani S, Watatani Y, Takeda J, Sanada M, Tanaka H, Suzuki H, Sato Y, Shiozawa Y, Yoshizato T, Yoshida K, Makishima H, Iwanaga M, Ma G, Nosaka K, Hishizawa M, Itonaga H, Imaizumi Y, Munakata W, Ogasawara H, Sato T, Sasai K, Muramoto K, Penova M, Kawaguchi T, Nakamura H, Hama N, Shide K, Kubuki Y, Hidaka T, Kameda T, Nakamaki T, Ishiyama K, Miyawaki S, Yoon SS, Tobinai K, Miyazaki Y, Takaori-Kondo A, Matsuda F, Takeuchi K, Nureki O, Aburatani H, Watanabe T, Shibata T, Matsuoka M, Miyano S, Shimoda K, Ogawa S. Integrated molecular analysis of adult T cell leukemia/lymphoma. *Nat Genet*. 2015 Nov;47(11):1304-1315.

[4] Yoshizato T, Dumitriu B, Hosokawa K, Makishima H, Yoshida K, Townsley D, Sato-Otsubo A, Sato Y, Liu D, Suzuki H, Wu CO, Shiraishi Y, Clemente MJ, Kataoka K, Shiozawa Y, Okuno Y, Chiba K, Tanaka H, Nagata Y, Katagiri T, Kon A, Sanada M, Scheinberg P, Miyano S, Maciejewski JP, Nakao S, Young NS, Ogawa S. Somatic Mutations and Clonal Hematopoiesis in Aplastic Anemia. *N Engl J Med*. 2015 Jul 2;373(1):35-47.

[5] Yoshida K, Sanada M, Shiraishi Y, Nowak D, Nagata Y, Yamamoto R, Sato Y, Sato-Otsubo A, Kon A, Nagasaki M, Chalkidis G, Suzuki Y, Shiosaka M, Kawahata R, Yamaguchi T, Otsu M, Obara N, Sakata-Yanagimoto M, Ishiyama K, Mori H, Nolte F, Hofmann WK, Miyawaki S, Sugano S, Haferlach C, Koeffler HP, Shih LY, Haferlach T, Chiba S, Nakauchi H, Miyano S, Ogawa S. *Nature*. 2011 Sep 11;478(7367):64-69.

[6] Yamaguchi K, Nagayama S, Shimizu E, Komura M, Yamaguchi R, Shibuya T, Arai M, Hatakeyama S, Ikenoue T, Ueno M, Miyano S, Imoto S, Furukawa Y. Reduced expression of APC-1B but not APC-1A by the deletion of promoter 1B is responsible for familial adenomatous polyposis. *Sci Rep*. 2016 May 24;6:26011.

[7] Mouradov D, Sloggett C, Jorissen RN, Love CG, Li S, Burgess AW, Arango D, Strausberg RL, Buchanan D, Wormald S, O'Connor L, Wilding JL, Bicknell D, Tomlinson IP, Bodmer WF, Mariadason JM, Sieber OM. Colorectal cancer cell lines are representative models of the main molecular subtypes of primary cancer. *Cancer Res*. 2014 Jun 15;74(12):3238-3247.

Gap in pagination due to formatting issues.

Pages 12-14

Shaping circuit environment to face the thermal challenge
Innovative technologies from low to high power electronics

P. Coudrain[1], J.-P. Colonna[2], L.-M. Collin[3], R. Prieto[1], L.G. Fréchette[3], J. Barrau[4], G. Savelli[2], P. Vivet[2], Q. Struss[1,2,3], J. Widiez[2],
K. Vladimirova[2], K. Triantopoulos[2], H. Beckrich-Ros[1], M. Vilarrubí[4], G. Laguna[4], H. Azarkish[3], M. Shirazi[3], J. Michailos[1]

[1]STMicroelectronics, 850 rue Jean Monnet, 38926 Crolles Cedex, France
[2]Univ. Grenoble Alpes, CEA, LETI, 38000 Grenoble, France   /   [3]Université de Sherbrooke, Sherbrooke, Canada   /   [4]Universitat de Lleida, Leida, Spain

Abstract

This paper describes evolutions of circuit environment to face an ever-increasing thermal challenge, from early design stage down to the final package. To illustrate this critical concern we give a portrayal of innovative technologies and concepts studied for efficient thermal management from low to high power electronics, with an emphasis on hot spot management.

Thermal challenges in modern circuits and systems

A. Thermal management from the design perspective

Moore's law has been accompanied by an increase in power densities. This now constitutes a critical challenge for modern ICs as the major part of dissipated power results in the generation of heat. Consecutive temperature rise modifies devices dynamical performances (Fig. 1) and increases leakage currents, exacerbating Joule effect, which results in thermal runaway [1]. Recent introduction of SOI and nanowire-based transistors with limited conduction performance even worsens the picture [2]. Lowering temperature has become a necessity to limit the risk of gate oxide breakdown, electromigration in interconnects or delaminations due to CTE mismatch. Hot spots ($Q > 300$ W/cm²) should be minimized to reduce intra-chip variability. Electrical-thermal co-design able to model the chip and its surrounding environment has become mandatory for layout optimization. Recent adaptive CPU architectures use *in-situ* temperature monitoring for thermal throttling and/or task dynamic allocation [3]. Self-adaptive robust asynchronous communication can be achieved to cope with thermal impact of hot spots in 3D Networks-on-Chip (3D-NoC) (Fig. 2) [4].

B. Multi-scale technologies for thermal management

Treating thermal challenge from a technological point of view deals with heat transport to extract the heat out of the chip and limit amplitude of hot spots (Fig. 3). In compact systems and low accessible volumes, heat is extracted by conduction through the substrate. Inversely for high cooling demand, heat is removed by convection with additional heat spreader and heatsink on backside.

3D integrations give more tricky schemes with increased power density per unit area and additional thermal interfaces. Depending on 3D integration (Fig. 4), layers strongly vary in nature and performances [5]. Despite material improvement, underfills remain responsible for large inter-tier thermal resistance in microbumping schemes [6-7]. Inversely Cu/SiO₂ hybrid bonding and 3DVLSI allow for improved thermal coupling between tiers [8].

Technologies for going beyond current limitations

Modern approaches integrate cooling solutions close to hot sources in silicon, with a technological toolbox that tends more and more to spread from packaging to 3D-integration processes.

A. Performing efficient passive heat spreading

Numerous studies have been performed to improve the thermal conductivity of package materials such as TIMs and adhesives. We extensively studied carbon-based heat spreaders in a large set of configurations of wire-bonding, flip-chip, 2D or 3D circuits from low to mid power (<20 W). Besides high conductivity up to 2000 W/m.K, pyrolytic graphite holds the advantage of an in-plane CTE close to Silicon, minimizing thermomechanical stress and allowing for thinner interface with reduced thermal resistance R_{th} [9]. Thus, graphite spreaders exhibit excellent hot spot mitigation capabilities that also benefit to low power dissipation paths through the laminate [10-11]. Integration of patterned pyrolytic graphite heat spreaders compatible with wire-bonds has been demonstrated for 3D hybrid-bonded circuits (Fig. 5 & 6) [12]. This result prefigures novel inter-tier 3D heat spreader integration approaches.

Higher thermal conductivities can be found in graphene [13] or with alternative spreading structures. Vapor chambers are independent self-adaptive systems based on the vaporization and condensation of a fluid to transport the heat (Fig. 7). Thicker than solid-state spreaders, minimal thickness demonstrated is 720 μm for Si [14], they exhibit effective thermal conductivities over 2700 W/m.K [15]. Performances mostly rely on wick structure [16].

B. Convection-based cooling towards embedded microfluidics

Air-cooling limitations in high power systems raised the need for liquid cooling with microchannel heatsinks, spray and jet cooling or hybrid jet-impingement/microchannels with stepwise-varying width sections (Fig. 8) [17]. R_{th} reduction has been achieved with the suppression of intrinsic package thermal interfaces by embedding microchannels directly in the Silicon [18]. Microchannels can be either located at chip backside or in a die attached on it (Fig. 9). This latter is interesting as the coolant is partially in contact with the dissipating chip backside without the need to further process it, reducing yield and reliability issues [19]. Microchannel geometries have been optimized to increase Nusselt number (*Nu*) while minimizing pressure drop increase. Hot spot-targeted optimizations have been proposed with narrow channels or dense pin fins in the vicinity of hot spots and wider channels on background zones (Fig. 10). Cooling of fluxes up to 1185 W/cm² has been simulated with a minimal pressure drop of 20 kPa [20]. Volumetric 3D circuit cooling with inter-tier microchannels has also been reported [21].

C. Energy-efficient active cooling

These approaches are optimized for constant flux distributions but use oversized pumping powers, as they are not able to respond to time-dependent heat loads. To overcome this limitation, design-independent and self-adaptive fluidic networks are presented in the STREAMS project (Fig. 11). Reacting to temporal variations of thermal mapping, they aim at achieving optimal cooling efficiency with minimal pumping power. A versatile microfluidic actuation is based on an array of microfluidic cells (Fig. 12) [22-23]. Coolant flow is fed in parallel to the cells to minimize pressure drop and allow local throttling of flow rate with local heat flux. Local flow rates are controlled by self-adaptive microvalves tailoring their aperture to local temperature. This approach exhibits pumping power efficiency superior to static approaches (Fig. 13) [24-27].

D. New packaging developments towards high power electronics

In the last years, many studies have been conducted towards fast power devices such as silicon Super Junction MOSFETs, SiC FETs and GaN HEMTs. To reach optimal performances converters need an efficient packaging avoiding wire bonds on topside and solder materials on backside contacts [28]. We present a new 3D packaging approach consisting in the implementation of a thick patterned wafer-level copper leadframe, acting as final interconnections for power devices of switching cells [29-30]. The patterned leadframe (Fig. 14) is bonded between two semiconductor wafers, replacing wire bonds by massive copper interconnections (Fig. 15). No other material is needed since thermo-compression bonding is employed (Fig. 16). One of the benefits of this approach is to offer a double side cooling though the metallic substrates.

Summary and Conclusions

Cooling and CMOS technologies are entering a fusion era where the thermal solution is not only taken into account from the design stage, but is also processed in the fab. Developed technologies such as self-adaptive microfluidics brings near-junction cooling schemes that are able to face the thermal challenge of future low to high power heterogeneous systems where circuit functionality and power handling will constitute time-to-market key differentiators.

2018 Symposium on VLSI Technology Digest of Technical Papers

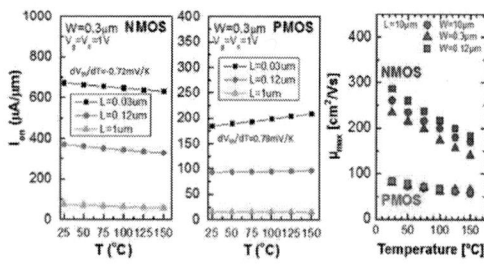

Fig. 1: Temperature dependence of FDSOI NMOS and PMOS I_{ON} as a function of gate length. Mobility vs. Temperature for long channel NMOS and PMOS transistors [1]

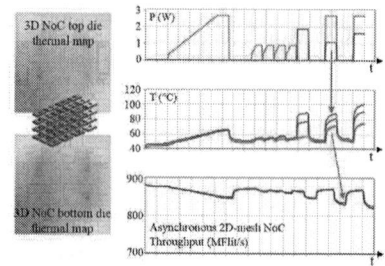
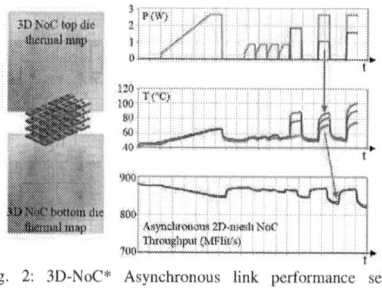

Fig. 2: 3D-NoC* Asynchronous link performance self-adaptation with temperature [4]. While injected power profile creates temperature fluctuations in top and bottom tiers, Asynchronous NoC performance is kept optimized

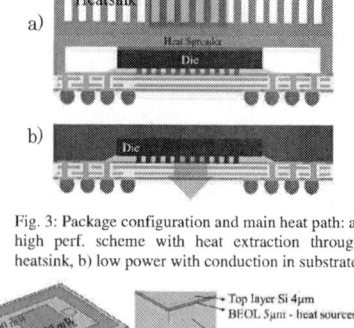

Fig. 3: Package configuration and main heat path: a) high perf. scheme with heat extraction through heatsink, b) low power with conduction in substrate

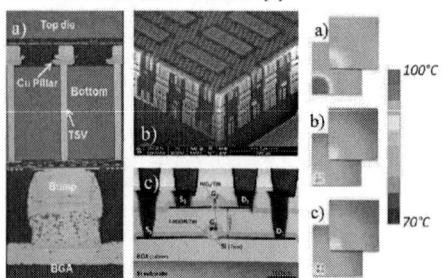

Fig. 4: Left: 3D integration approaches: a) micro-bumping, b) SiO_2/Cu hybrid bonding, c) 3DVLSI. Right: associated simulation of temperature mappings in 4th and 8th tier for a 8-tier stack with a corner hotspot on 2nd tier [8]

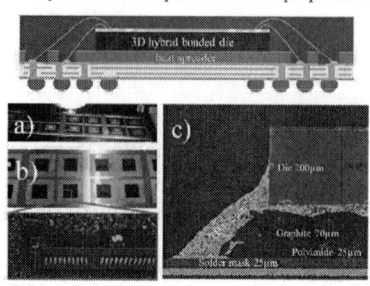

Fig. 5: Patterned heat spreader integrated between 3D die and land grid array (LGA): a) LGA on stiffener, b) pick-and-place of patterned pyrolytic graphite on LGA, zoom in red shows openings for bondpads, c) cross-section of the stack [12]

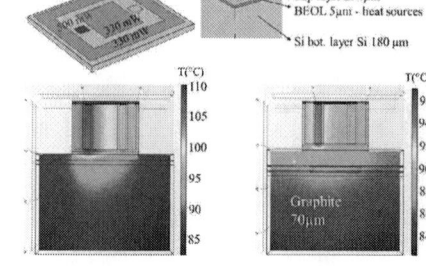

Fig. 6: FEM simulated temperature cross section calibrated with in-situ measurements on 3D circuit [12]. Hot spot mitigation is achieved with patterned graphite on LGA (right) due to heat spreading in the graphite

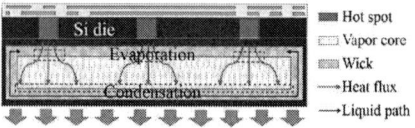

Fig. 7: Embedded vapor chamber principle. Fluid is vaporized at hot spots and goes to vapor core. Once condensed on colder areas, it goes back to hot spots by capillarity through the wick

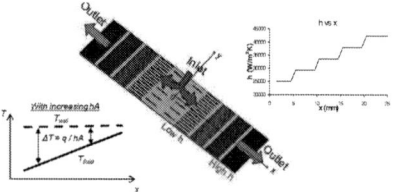

Fig. 8: Microchannel sections with stepwise varying width, tailoring thermal resistance (1/hA) along the flow path to achieve uniform wall temperature [17]

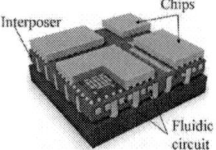

Fig. 11: STREAMS** generic autonomous smart cooling interposers with versatile microfluidic actuation, embedding thermal mapping and energy harvesting capabilities

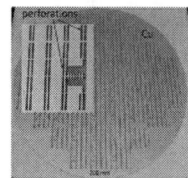

Fig. 14: 8-inch 400µm-thick bulk copper wafer with perforations. Insert: Zoom on the Cu perforation

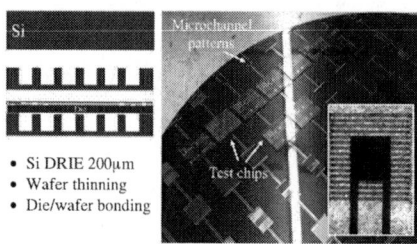
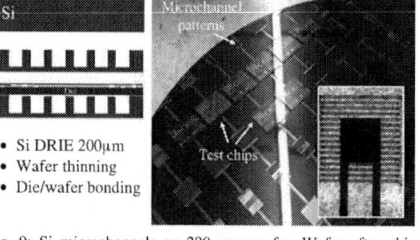

Fig. 9: Si microchannels on 200 mm wafer. Wafer after chip bonding, before dicing. In this configuration the backside of the chip is in contact with the fluid, the full stack is 660 µm thick [19]. White box shows an IR view of the bonded stack where a heater is superimposed on a localized-microchannels zone

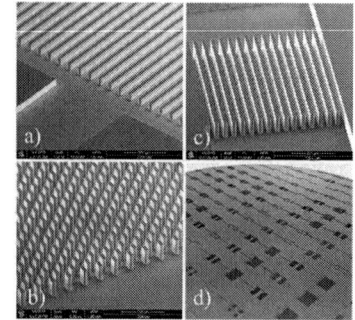

Fig. 10: Microfluidic patterns a) uniform microchannels b) pin fins c) narrow localized microchannels for hotspot d) wafer-level 1X multi-pattern trials*

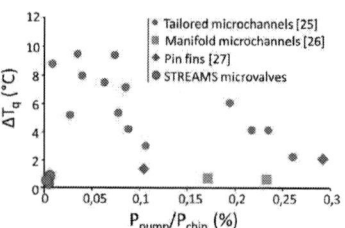

Fig. 13: Simulation with steady state CFD studies combined with temporal integration [24]. Comparison of the overall pumping performance of adaptive microvalves with previous studies [25-27]

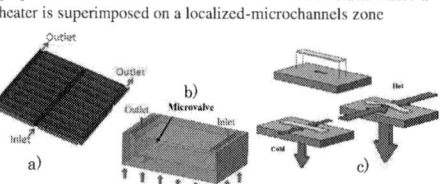

Fig. 12: STREAMS self-adaptive fluidic network: a) coolant distributor above cells array, b) microfluidic cell [22] c) adaptive microvalves cold and hot states with aperture modulation [23]

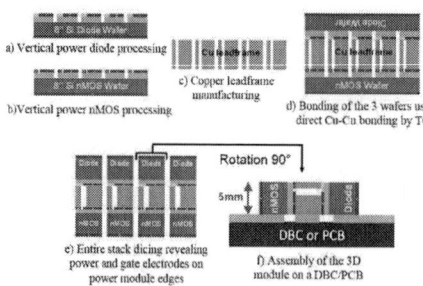

Fig. 15: Wafer-level fabrication process. Example with a 4-phase buck interleaved converter

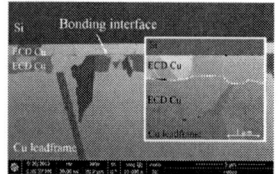

Fig. 16: Thermal interface removal. FIB SEM cross-section of Cu-Cu bonding interface [30]

References: [1] K. Triantopoulos et al., IEDM 2017, [2] C. Jeong et al., JAP 2012, [3] M. Igarashi et al., ISSCC 2014, [4] P. Vivet et al., JSSC 2016, [5] J. Michailos et al., IEDM 2015, [6] T. Brunschwiler et al., JMEP 2012, [7] P. Coudrain et al., IEEE Design & Test 2016, [8] C. Santos et al., 3DIC 2016, [9] R. Prieto et al., THERMINIC 2016, [10] C. Santos et al., 3DIC 2015, [11] R. Prieto et al., SEMI-THERM 2015, [12] J.-P. Colonna et al., THERMINIC 2017, [13] P.-H. Lee, TED 2018, [14] J. Liang et al., ITHERM 2017, [15] Q. Cai et al., J. Micromechanics Microengineering 2012, [16] S. Ryu et al., Int. J. Heat Mass Transf. 2017, [17] J. Barrau et al., ATE 2010, [18] D.B. Tuckerman et al., EDL 1981, [19] L.-M. Collin et al., ITHERM 2017, [20] L.-M. Collin et al., ICNMM 2015, [21] Y. Madhour et al., 3DIC 2013, [22] H. Azarkish et al., ITHERM 2017, [23] M. McCarthy et al., JMEMS 2008, [24] G. Laguna et al., THERMINIC 2017, [25] S. Riera et al., ATE 2015, [26] C.S. Sharma et al., Int. J. Heat Mass Transf 2015, [27] S. Riera et al., AIP 2014, [28] A. Müsing et al., PCIM Europe 2013, [29] B. Letowski et al., ISPSD 2016, [30] K. Vladimirova et al., ECTC 2018

Acknowledgements: *This work was supported by the French National Program "Programme d'Investissements d'Avenir, IRT Nanoelec" under Grant ANR-10-AIRT-05. **The research leading to these results has been performed within the STREAMS project and received funding from the European Community's Horizon 2020 program under Grant Agreement N° 688564.

978-1-5386-4219-1/18 $31.00 © 2018 IEEE

Thermal Management Research – from Power Electronics to Portables

Ki Wook Jung, Chi Zhang, Tanya Liu, Mehdi Asheghi and Kenneth E. Goodson

Mechanical Engineering Department, Stanford University, Stanford, CA 94305, U.S.A.
goodson@stanford.edu

Abstract

Thermal management is critical for electronic systems ranging from servers and smartphones to radar HEMTs and hybrid vehicle converters. Rapid research progress is being achieved both on-chip and in packaging through new materials and microfluidics. One very promising area is thermal metamaterials, which offer unusual combinations of thermal, mechanical, fluidic, and other properties by means of micro- or nanoscale heterogeneity, porosity, and/or layering. Another area is the upscaling of the performance and efficiency of fluidic systems – both capillary-based and pumped, which remove heat to an external heat rejector. This talk summarizes progress and highlights collaborations with the semiconductor industry, US defense companies and the NSF center on power electronics (POETS).

keywords: thermal management, power electronics, portables.

Introduction

Thermal management and increasingly expensive energy demands pose major challenges to the rate of increase in processor performance and the performance integration of electronic systems, from portables to power devices [1]. Energy-efficient computing schemes [2] and heterogeneous integration including 2.5, 3D and monolithic chips (Fig. 1) [3] promise substantial reduction in energy demand for emerging and growing computing needs. However, these conflicting trends have resulted in a substantial increase in heat flux and power density (W/cm^3), which makes it even more challenging to use conventional cooling technology solutions.

In the past few years, the thermal management community has made significant strides in exploring the limits of cooling for extreme heat flux applications such as data centers and power electronics [4-7]. Recently, for applied cooling of 3D monolithic circuits, we proposed nanomaterial-based thermal management for monolithic chips (Fig. 1) which integrates both conductive solutions (nano-metallics, encapsulated phase change media, high thermal conductivity polymers and 2D materials) and convective solutions (micro phase change and extreme surface contact angle engineering). These are integrated with the larger-scale chip periphery microfluidics to form an integrated monolithic thermal platform for heterogonous packaging integration.

In a contrasting effort for power electronics, we recently explored the limits of thermal management by demonstrating heat dissipation levels >1000 W/cm^2 at 10 °C superheat over 300 μm × 1 cm area using capillary fed template-fabricated copper inverse opals (CIOs), and water as working fluid [4]. More recently an exotic heat sink explored cooling limits utilizing porous copper, conformally coated in laser-etched diamond channels for removal of very large heat fluxes, ~900 W/cm^2 at ~100 °C superheat over 10^2 mm^2 (Fig. 2) [5].

Among the most promising solutions for heat removal to the package level, from power to portables, involve passive heat spreading. While vapor chambers attempt to address this need, they have reached fundamental limits in peak heat flux, thermal resistance, and thickness. The best commercial vapor chambers, thinner than 3 mm, offer nominally ~200 W/cm^2 at superheat temperature 30°C. However, recent research is pursuing even more aggressive metrics to spread 1000 W from an area of 1 cm^2 to 100 cm^2 with only 40 °C temperature drop. This is being achieved through a combination of high heat flux evaporative CIO tiles and a silicon pin fin array (or microchannels [9]) liquid delivery wick.

The progress reported in [4,5] was achieved by developing in-house fabrication strategies for CIO *metamaterials*, which offer surprising combinations of properties. We demonstrated superb control and 10× enhancement of permeability of CIOs produced via polystyrene sacrificial template and sintering followed by electrodeposition (Fig.4) [4,8]. Another example is electroplated Copper Nano Wires (CNWs) for high thermal conductivity and mechanically compliant thermal interface materials (TIMs) application. The CIOs can encapsulate phase change materials (PCMs), promising high effective heat capacity and thermal conductivity; CNWs can be also infused with PCMs [3] (Fig. 1b). Thermal metamaterials can be promising for other applications in power electronics, such as thermal switches for heat routing and transient temperature control. We recently achieved 9:1 reversible thermal resistance ratios using Li intercalation in MoS_2 multilayers, as demonstrated in [10] (Fig. 3).

Also for power electronics, we used embedded microchannels and 3D manifold liquid delivery and vapor extraction [6] (Fig. 5) to achieve heat removal levels ~700 W/cm^2 at 130°C superheat over 5^2 mm^2 area using R245fa [7]. Using single-phase water as working fluid and similar μ-cooler, we removed heat fluxes up to 250 W/cm^2 at 50 °C superheat [7]. Fluid dynamics simulations explore the feasibility of single-phase cooling at higher heat fluxes; the onset of boiling occurs at 850 W/cm^2 but must be verified experimentally. Future microfluidic research will combine ideas from [4-7] to develop a novel and innovation low-cost device capable of cooling high-power chips >1200 W/cm^2 at 10°C superheat (or 0.01 °C/W) over large silicon chip areas of 10^2-10^3 mm^2, which requires little or no fluid pumping. This approach uses a 3D manifold [6] to deliver just enough liquid to the porous copper tiles by means of capillary wicking.

In summary, here we provide examples from the rapid progress on novel materials and microfluidics that are impacting both chip-integrated and package-level thermal management and will enable further integration scaling for applications ranging from portables to power electronics.

References: [1] Fuller et al., 2011, National Academies Press, Washington, D.C. [2] Theis & Solomon, 2010, Proc. IEEE 98(12), p. 2005. [3] Mitra et al., *Computer* Vol. 48, pp. 24-33. [4] Palko et al., 2015, *Applied Physics Letters*, Vol. 107, p. 253903. [5] Palko et al., 2017, *Adv. Func. Mat.*, Vol. 27, p. 1703265. [6] Jung et al., 2017, 16th IEEE ITHERM 2017. [7] Jung et al., 2018, ASME Journal of Electronics Packaging, in press. [8] Barako et al., 2016, Nano Letters, Vol. 16, p. 2754. [9] Dusseault et al., 2014, 14th IEEE ITHERM 2014. [10] Sood et al, 2018, under review.

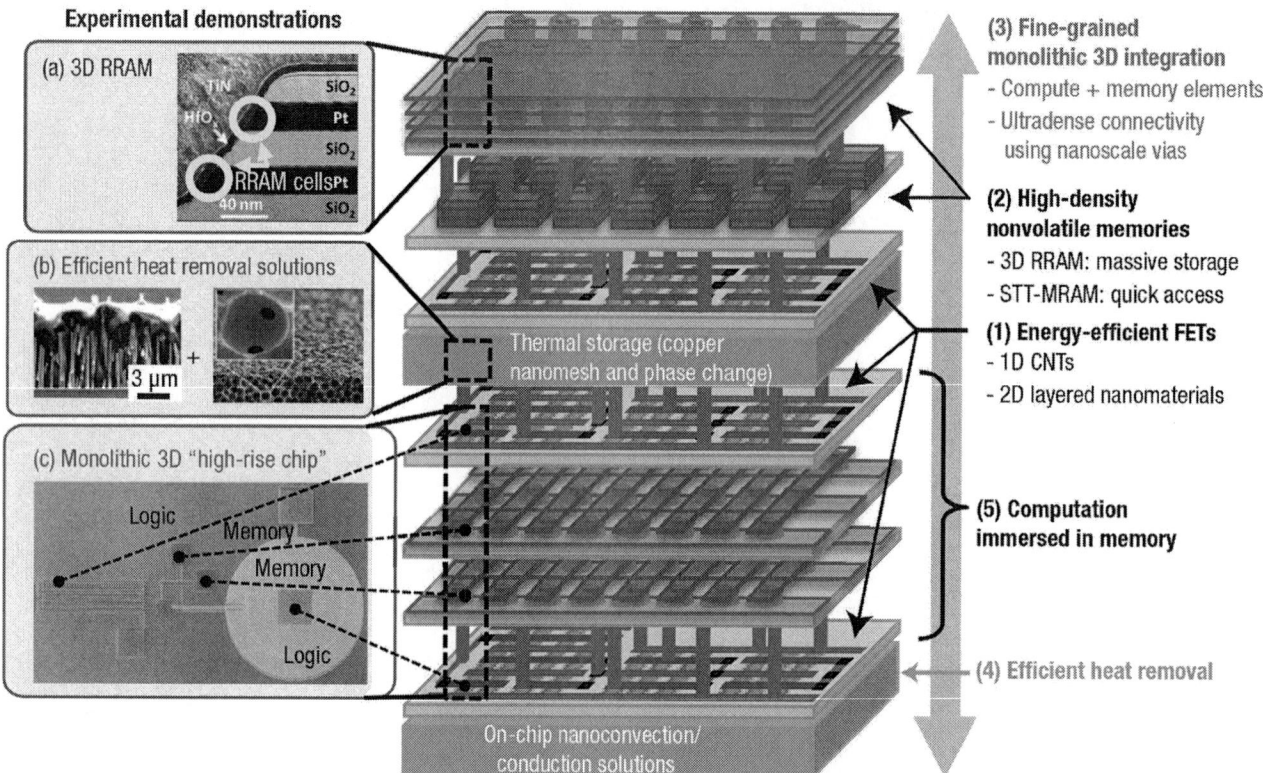

Fig. 1: Schematic of hierarchical chip-level conduction cooling strategies for managing extreme power densities for 3D integration using novel thermal conduction materials and microfluidics. These technologies promise effective internal conductivities above 5,000 W/m/K and management of local heat fluxes above 1 kW/cm² [3].

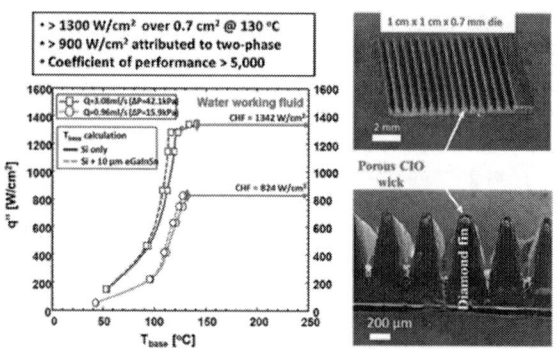

Fig. 2: Extreme heat sink uses conformal porous copper in laser-etched diamond, 900 W/cm² at 100 °C superheat over 10² mm² [5].

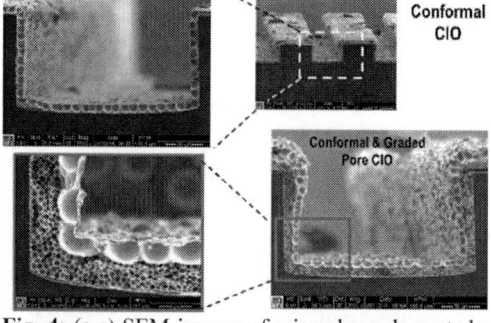

Fig. 4: (a-c) SEM images of microchannels coated with systematic porous copper for extreme capillary transport and cooling [9].

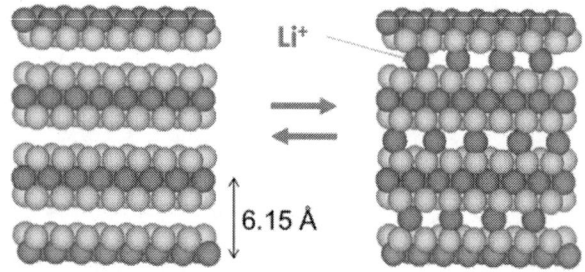

Fig. 3: Example of a thermal "metamaterial offering switching between high and low thermal conductivity states. This example achieves 9:1 reversible thermal resistance ratios using Li intercalation in MoS_2 multilayers [10].

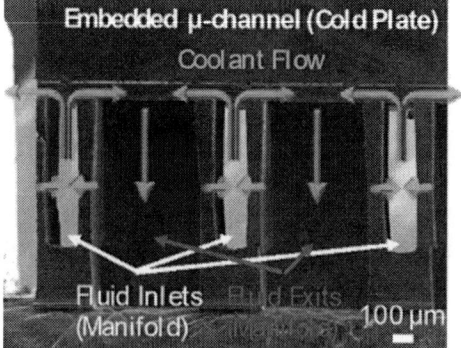

Fig. 5: Schematic and SEM of an embedded cooling microchannels bonded to a silicon-based 3D-manifold and liquid routing [6]

Electromigration Effects in Power Grids Characterized Using an On-Chip Test Structure with Poly Heaters and Voltage Tapping Points

Chen Zhou, Richard Wong*, Shi-Jie Wen*, and Chris H. Kim

Dept. of ECE, University of Minnesota, 200 Union Street SE, Minneapolis, MN 55455, USA (Email: chriskim@umn.edu)
*Cisco Systems, Inc., 285 W. Tasman Drive, San Jose, CA 95134

Abstract

A 65nm test chip to study electromigration (EM) effects in power grids was taped-out and tested. A 9x9 grid was implemented using M3 and M4 metal layers which was stressed under constant current and constant voltage modes. On-chip poly heaters were employed to raise the die temperature to 350°C without damaging the chip package. A bank of transmission gates based on IO transistors were used to tap out the M3 and M4 voltages at each intersection point of the power grid. Using the test structure, we could observe for the first time, subtle behaviors of EM such as mechanical stress dependent failure locations and self-healing due to redundant current paths.

Introduction

Electromigration (EM) effects in a single wire or a chain of wires have been studied for decades [1,2]. However, EM effects in a power grid have not been reported due to the measurement complexity. Redundant current paths in power grids make failure induced resistance jumps less obvious, and thus hard to track. In this paper, we present an on-chip test structure with poly heaters and voltage tapping points which allows efficient monitoring of subtle failures in a power grid.

Test Structure Design

The test structure is a 9x9 grid implemented in M3 and M4 metal layers, as shown in Figs. 1 and 2. Pad connection points A, B, and C are located at the two corners and the center of the grid with multiple dense vias to prevent failure. Each metal segment is 20μm in length and 0.1μm in width. 81 intersections are formed on the entire 9x9 grid. To collect EM data within an attainable stress time, single minimum size via were used to connect M4 and M3 layers at each intersection. 162 (=9x9x2) nodes are uniformly distributed on the entire grid, with half of the nodes on M4 and the other half on M3. As shown in Fig. 1, we use a M5-M4 via to tap the node voltage of M4, and a M3-M2 via to tap the node voltage of M3. This design allows us to directly measure the voltages on grid structure, without introducing any appreciable electrical or mechanical disturbance to the power grid. M5 and M2 wires were routed to the transmission gate array located at the other side of the chip to protect from the extremely high stress temperatures. The voltage drop across each 20μm M4 and M3 segment and M4-M3 via can be calculated from the measured tapping voltages.

As shown in Fig. 2, each power grid voltage is multiplexed out through individual transmission gates connected to a shared analog pad. A scan chain enables one tapping voltage at a time, and IO devices are used to suppress leakage current. The active circuits are placed more than 400μm away from the heating area to further reduce the leakage current. Three on-chip heaters [2] were used for efficient local temperature control. The heating area is 260μm x 260μm, with the power grid DUT placed in the middle. The die photo is shown in Fig. 10. The stress temperature was controlled by a software program shown in Fig. 3. The direction of the stress current was periodically reversed to prevent EM in the heaters themselves which resulted in a momentary temperature overshoot.

Power Grid Failure Locations

Fig. 4 shows all 162 tapping voltages measured from a fresh grid. The voltage drop across each segment was calculated and plotted as shown in Fig. 5. Here, the arrow indicates the magnitude and polarity of the voltage across each wire, while the circle and square markers at each intersection indicate the voltage across each via. Fig. 6 shows the first EM failure point under three different stress current configurations: A+B → C; C → A+B; and A → B. In all three cases, the first EM event happened close to the negative voltage terminal denoted V(-). This result verifies that EM failure is affected not only by current density but also by mechanical stress [3]. Fig. 7 shows how the voltages in each branch change after an EM event. The change of the voltage can be attributed to two reasons: resistance increase in the via or wire due to EM, or current increase due to EM at a nearby location.

EM Healing Effect

Fig. 8 shows the power grid resistance between A+B and C, along with the voltage drop across each adjacent node. We observed not only abrupt and progressive failures but also self-healing behavior. This is consistent with [4] where supporting data shows that wire connections can be temporarily restored. As shown in Fig. 8 (right), several voltage traces show toggling behavior for stress times from 15 to 20 hours. Based on their fluctuation magnitudes, we can find the location of the via that is undergoing self-healing. The voltage across via #39 under constant voltage stress elucidates the stress and healing cycles (Fig. 9). The time it takes for the connection to break again increases with a longer stress time.

Significance of EM Healing

To our knowledge, this is the first report of EM healing in a power grid under continuous current stress. Unlike previous works where the temperature was lowered during measurement, this work does not involve any temperature cycling, and thus thermal shrinking and expansion cannot be the reason behind the healing phenomenon. We believe EM healing is a natural process occurring in power grids due to mechanical back stress. Healing was rarely discussed in previous works because single wire test structures do not have any redundant current paths, so stress current is forced even after an EM failure, resulting in permanent damage that cannot be reversed. However, a power grid structure is different in that it has numerous redundant paths allowing the current to bypass the failure location immediately upon an EM event, providing the opportunity for healing. Since EM healing effect was overlooked in previous studies, the actual lifetime of a power grid might be significantly longer than previously thought.

References [1] S. Lee and A. S. Oates, IRPS, 2006. [2] C. Zhou, X. Wang, R. Fung, S. J. Wen, R. Wong and C.H. Kim, VLSI Technology Symposium, 2015. [3] M. H. Lin and A. S. Oates, IRPS, 2016. [4] C. Zhou, X. Wang, R. Fung, S. J. Wen, R. Wong and C. H. Kim, TDMR, 2017.

2018 Symposium on VLSI Technology Digest of Technical Papers

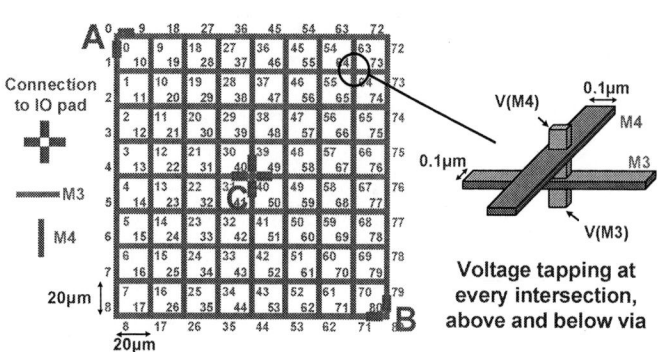

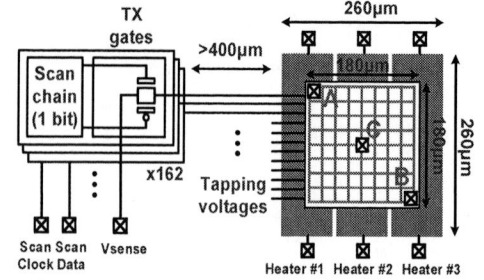

Fig. 2. EM test chip with on-chip heaters, power grid DUT, IO transmission gates and scan chain.

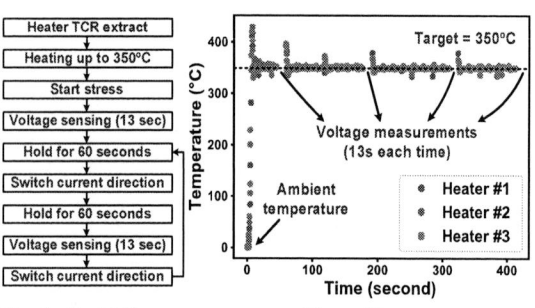

Fig. 3. (Left) Temperature control loop. Heater current direction reversed periodically to prevent EM in heaters. (Right) Temperature measured from 3 heaters.

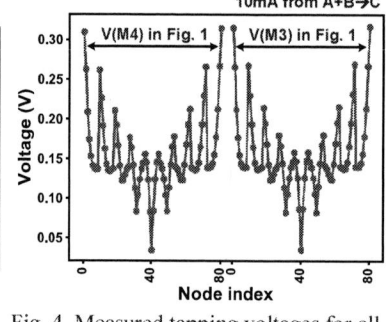

Fig. 4. Measured tapping voltages for all M3 and M4 nodes.

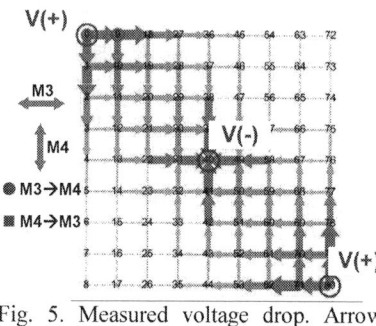

Fig. 5. Measured voltage drop. Arrow indicates the magnitude and polarity of voltage drop between adjacent nodes.

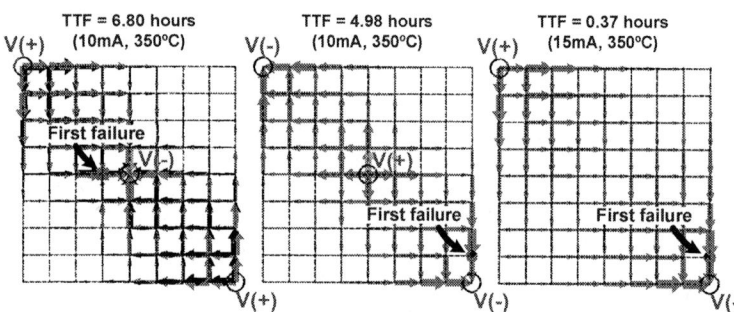

Fig. 6. First EM failure locations for different stress current configurations. Due to mechanical stress effects, the first failure occurs near the negative voltage terminal.

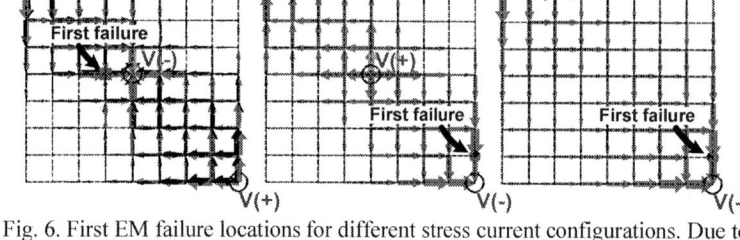

Fig. 7. Example of voltage drop traces after an EM event. The increased voltage across the via suggests a change in the via resistance or EM in nearby structures.

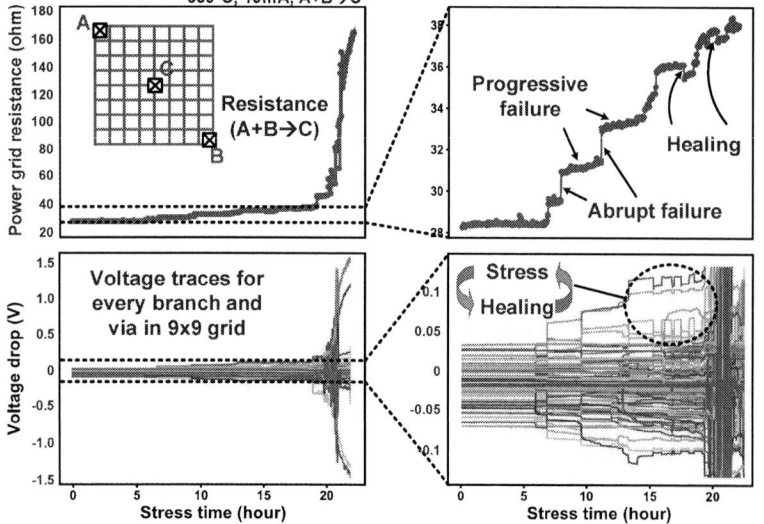

Fig. 8. (Left) Power grid resistance and voltage drop traces of all wires and vias. (Right) Zoomed in plots showing self-healing behavior.

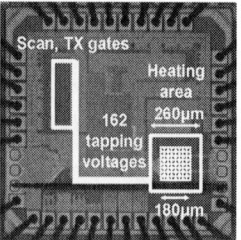

Fig. 9. Voltage across via toggles between stress mode and healing mode under constant voltage stress. Time spent in stress mode gradually increases.

Fig. 10. Die photo of 65nm test chip.

978-1-5386-4219-1/18 $31.00 © 2018 IEEE   20

Low Thermal Budget Amorphous Indium Tungsten Oxide Nano-Sheet Junctionless Transistors with Near Ideal Subthreshold Swing

Po-Yi Kuo, Chien-Min Chang, and Po-Tsun Liu*

Department of Photonics and Institute of Electro-Optical Engineering

National Chiao Tung University, Hsinchu 300, Taiwan

Tel: +886-3-5712121-52994 E-mail: kuopoyi.ee91g@gmail.com. & ptliu@mail.nctu.edu.tw*

Abstract

Amorphous indium tungsten oxide (a-IWO) nano-sheet (NS) junctionless (JL) transistors (a-IWO NS-JLTs) have been successfully fabricated and demonstrated in the category of indium oxide based thin film transistors (TFTs). We have scaled down thickness of a-IWO channel to 4nm. The proposed a-IWO NS-JLTs with low operation voltages exhibit good electrical characteristics: near ideal peak subthreshold swing (S.S.) ~ 63mV/dec., high field-effect mobility (μ_{FE}) ~ 25.3 cm^2/V-s. The novel a-IWO NS-JLTs with low temperature processes are promising candidates for monolithic three-dimensional integrated circuits (3-D ICs), vertical stacked (VS) hybrid CMOS technology, and large-scale integration (LSI) applications in the future.

Introduction

Transparent amorphous oxide semiconductor (TAOS) thin film transistors (TFTs) are of great interest for their use in mobile electronics, optoelectronics, and future displays owing to relatively high field-effect mobility (μ_{FE}), low temperature processes, and superior uniformity [1]. In recent years, c-axis aligned crystalline In-Ga-Zn-Oxide field-effect transistors (CAAC-IGZO FETs) with a nano-scaled channel exhibit extremely low off-state currents and short channel effect (SCE) immunity for large-scale integration (LSI) applications [2], [3]. However, CAAC-IGZO FETs have a lower mobility ~ 10 cm^2/V-s and poorer subthreshold swing (S.S.) with increasing channel width [2]. Moreover, the Si nanowires (NWs), poly-Si nano-sheet (NS), and poly-Si NWs junctionless (JL) FETs have been proposed and demonstrated [4]-[6]. The absence of a doping concentration gradient completely eliminates diffusion of impurities and the problem of sharp doping profile formation [4]. To improve the performance of TAOS-TFTs with μ_{FE} > 20 cm^2/V-s, amorphous tungsten oxide-doped indium oxide (a-IWO) have been developed owing to stable semiconducting films [7]. In this work, amorphous indium tungsten oxide (a-IWO) nano-sheet (NS) junctionless (JL) transistors (a-IWO NS-JLTs) have been successfully fabricated by combining JL configurations with uniform a-IWO NS channels (T_{ch} = 4nm) in the category of indium oxide based TFTs.

Device Structure

The proposed a-IWO NS-JLTs with bottom metal gate (BMG) configurations were fabricated through photo-lithography on Si wafers or glass substrates, as schematically depicted in Fig. 1. Meanwhile, a conventional Al/HfO$_2$/Si-sub (MIS) capacitor was also fabricated to study the characteristics of HfO$_2$ gate insulator (GI) on Si-substrate.

Results and Discussion

Fig.2 displays (a) cross-sectional transmission electron microscope (TEM) images of a-IWO NS-JLTs with a 4-nm-thick a-IWO NS channel and a 10-nm-thick HfO$_2$ GI and (b) cross-sectional TEM images of MIS (Al/HfO$_2$/Si-sub) capacitor. There is a negligible interfacial layer (IL) between the HfO$_2$ and the a-IWO channel in a-IWO NS-JLTs. The thickness of IL between Si-sub and HfO$_2$ in conventional MIS capacitor is about 1.9nm.The measured and normalized C-V curves of (a) conventional MIS capacitor and (b) capacitor of a-IWO NS-JTTs with 10-nm-thick HfO$_2$ GI are plotted in Fig.3. Conventional MIS capacitor has significant hysteresis and lower capacitance values compared with those in capacitor of a-IWO NS-JLTs due to thicker IL between channel and HfO$_2$ GI. Fig.4 exhibits the measured I_{DS}-V_{GS} of a-IWO NS-JLTs with 10-nm-thick HfO$_2$. The very small hysteresis (~6mV) is achieved in a-IWO NS-JLTs owing to BMG configurations.

Fig.5 shows the measured I_{DS}-V_{GS} of a-IWO NS-JLTs with different thickness of a-IWO channel. Fig.6 exhibits the measured I_{DS}-V_{GS} of a-IWO NS-JLTs with different thickness of HfO$_2$ GI. In JL configurations, there are more negative threshold voltage (V_{th}) and poorer S.S. tendencies in devices with a thicker channel thickness (T_{ch}) or under weaker gate controls [6]. The S.S. and on-currents (I_{on}) of a-IWO NS-JLTs are improved by scaling the thickness of GI owing to the good electrostatic controllability in devices with 10-nm-thick HfO$_2$. The near ideal peak S.S. ~ 63 mV/dec. can be achieved in a-IWO NS-JLTs with a-IWO = 4nm. For low voltage operation, the real I_{ON} / I_{OFF} between off-state currents and on-state currents under gate operation voltage V_{GS} = 1V is important. The measured I_{DS}-V_{GS} of a-IWO NS-JLTs with 10-nm-thick HfO$_2$ under gate operation voltage V_{GS} =1V is shown in Fig.7. The I_{ON} / I_{OFF} > 10^8 can be achieved at V_{GS} = V_{DS} = 1V owing to near ideal S.S. In addition, we also demonstrate the a-IWO NS-JLTs fabricated on a glass substrate for low temperature processes and low thermal budget applications, as shown in Fig.8 (a). Fig.8 (b) exhibits the measured I_{DS}-V_{GS} and extracted μ_{FE} of a-IWO NS-JLTs with 10-nm-thick HfO$_2$ fabricated on different substrates. The comparison of key parameters for advanced devices with nano-structure channels is summarized in Table I. The a-IWO NS-JLTs have a high μ_{FE} and near ideal peak S.S. owing to NS channels, good interface characteristics, and JL configurations.

Conclusions

The a-IWO NS-JLTs with 4-nm-thick NS channels have been successfully fabricated and demonstrated. The IL free BMG a-IWO NS-JLTs significantly improve the channel/oxide interface and the hysteresis characteristics. The electrical characteristics of a-IWO NS-JLTs with low operation voltages can be enhanced by shrinking the thickness of GI, resulting in good gate controllability, near ideal peak S.S., high μ_{FE}. The a-IWO NS-JLTs that consists of JL configurations and Ga-free & Zn-free NS channels appears great potentials for monolithic 3-D ICs, VS hybrid CMOS, and LSI applications in the future.

Acknowledgement

This work was supported by the Ministry of Science and Technology, Taiwan, under Contract: MOST 106-2221-E-009-107-MY3.

References

[1] K. Nomura *et al.*, *Nature*, vol. 432, no. 7016, pp. 488-492, 2004.

[2] Yoshiyuki *et al.*, *IEEE Electron Device Lett.*, vol. 36, no. 4, pp. 309–311, April 2015.

[3] S. H. Wu *et al.*, in *VLSI Symp. Tech. Dig.*, Jun. 2017, pp. T166–T167.

[4] J. P. Colinge *et al.*, *Nature Nanotechnology*, vol. 5, no. 3, pp. 225–229, Mar. 2010.

[5] H. B. Chen *et al.*, in *VLSI Symp. Tech. Dig.*, Jun. 2013, pp. T232–T233.

[6] P. Y. Kuo *et al.*, in *Proc. IEDM*, Dec. 2015, pp. 133–136.

[7] S. Aikawa *et al.*, *Appl. Phys. Lett.*, vol. 102, pp. 102101(1)–102101(4), Mar. 2013.

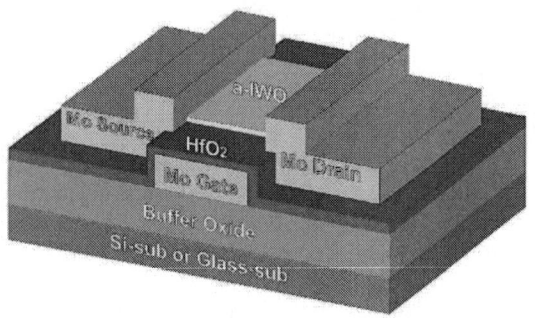

Fig.1 The schematic (not to scale) structure of a-IWO NS-JLTs. The proposed IWO NS-JLTs were fabricated in bottom metal gate (BMG) configurations.

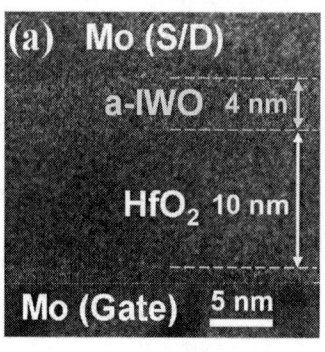

Fig.2 (a) Cross-sectional TEM images of a-IWO NS-JLTs. The thickness of a-IWO is about 4nm. The IL between HfO_2 and a-IWO is negligible. (b) Cross-sectional TEM images of conventional MIS (Al/HfO_2/Si-sub) capacitor. The thickness of IL is about 1.9nm between Si-sub and HfO_2.

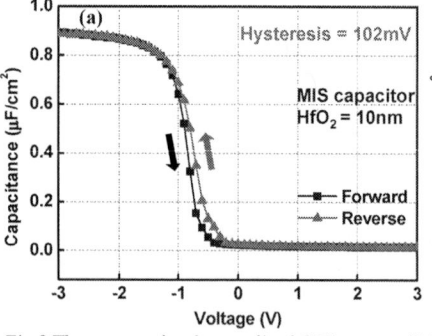

Fig.3 The measured and normalized C-V curves of (a) conventional MIS capacitor and (b) capacitor of a-IWO NS-JTTs with 10-nm-thick HfO_2. Conventional MIS capacitor has significant hysteresis and lower capacitance values compared with those in capacitor of a-IWO NS-JLTs due to thicker IL.

Fig.4 The measured I_{DS}-V_{GS} of a-IWO NS-JLTs with 10-nm-thick HfO_2. The small hysteresis ~ 6mV can be achieved.

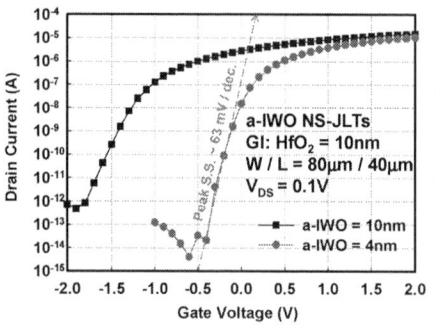

Fig.5 The measured I_{DS}-V_{GS} of a-IWO NS-JLTs with different thickness of a-IWO channel. The near ideal peak S.S. ~ 63 mV/dec. can be achieved in a-IWO NS-JLTs with a-IWO = 4nm.

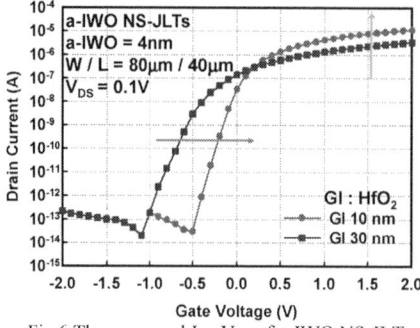

Fig.6 The measured I_{DS}-V_{GS} of a-IWO NS-JLTs with different thickness of HfO_2 GI. The S.S. and on-currents (I_{on}) of a-IWO NS-JLTs are improved by scaling the thickness of GI.

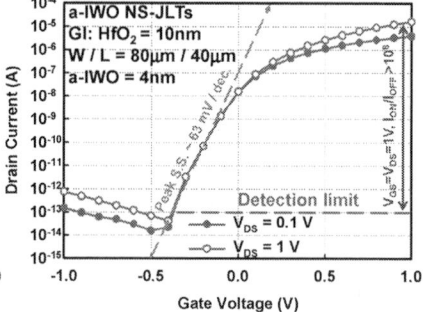

Fig.7 The measured I_{DS}-V_{GS} of IWO NS-JLTs with 10-nm-thick HfO_2 under gate operation voltage V_{GS} =1V. The I_{ON} / I_{OFF} > 10^8 can be achieved at V_{GS} = V_{DS} = 1V.

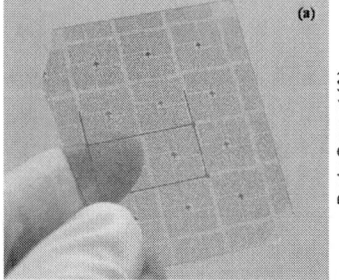

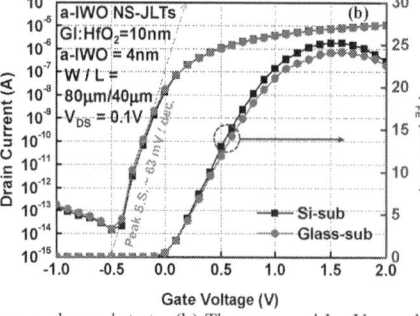

Fig.8 (a) The a-IWO NS-JLTs fabricated on a glass substrate. (b) The measured I_{DS}-V_{GS} and extracted μ_{FE} of a-IWO NS-JLTs with 10-nm-thick HfO_2 fabricated on different substrates. The electrical characteristics between Si-sub and glass-sub are almost identical.

Table I. Comparison of key parameters for advanced nano-structure devices.

Devices	a-IWO NS-JLTs	Ref.[3]	Ref.[4]	Ref.[6]
Channel Type	a-IWO Nano-Sheet (4nm)	CAAC IGZO Nano-Island	Si NW	Poly-Si NW
Peak S.S. (mV/dec.)	~ 63	~ 95	~ 63	~ 61
μ_{FE} (cm²/ V-s)	~ 25.3	~ 10	unknown	~10
Low T Processes for Glass	Yes	Yes	No	No

978-1-5386-4219-1/18 $31.00 © 2018 IEEE

Gap in pagination due to formatting issues.

Pages 23-24

Capacitor-based Cross-point Array for Analog Neural Network with Record Symmetry and Linearity

Y. Li, S. Kim, X. Sun, P. Solomon, T. Gokmen, H. Tsai, S. Koswatta, Z. Ren,
R. Mo, C. C. Yeh, W. Haensch and E. Leobandung

IBM T. J. Watson Research Center, Yorktown Heights, NY10598, USA E-mail: yulongl@us.ibm.com

Abstract

We report a capacitor-based cross-point array that can be used to train analog-based Deep Neural Networks (DNNs), fabricated with trench capacitors in 14nm technology. The fundamental DNN functionalities of multiply-accumulate and weight-update are demonstrated. We also demonstrate the best symmetry and linearity ever reported for an analog cross-point array system. For DNNs, the capacitor leakage does not impact learning accuracy even without any refresh cycle, as the weights are continuously updated during training. This makes capacitor an ideal candidate for neural network training. We also discuss the scalability of this array using optimized low-leakage DRAM technology.

Introduction

Analog-based neural network (NN) accelerators have the potential to achieve orders of magnitude improvement in training time and energy consumption compared to conventional CPU/GPU systems [1,2]. Non-volatile memory (NVM) based cross-point arrays have achieved promising results for inference tasks [3,4]. However, training NNs to high accuracy is difficult for NVM devices, since successful training depends on keeping the incremental changes in NN weight small (requiring roughly 1000 update states) and symmetric (so that positive and negative updates (or pulses) balance on average) [2, 5]. To address these shortcomings, the concept of capacitor-based cross-point array has been proposed [6] but not demonstrated. In a capacitor, charge can be added or subtracted continuously if the number of electrons is high, so analog and symmetric weight update can be achieved. In this paper, we report such a capacitor-based array using trench capacitor in 14nm technology, demonstrating new records in update symmetry and linearity. We also investigate the feasibility of scaling this to large arrays for accelerating the training of large-scale DNNs.

Array design

As shown in Fig. 1, a capacitor can serve as an analog memory, connected to the gate of a "readout" pFET. This capacitor is charged/discharged by two "current source" FETs, as controlled by two analog inverters and one digital inverter. During readout, the synaptic weight can be accessed by measuring the conductance of the readout FET. During weight update, the YW signal controls the analog inverters to drive the current sources for positive or negative updates. When these current sources are in saturation, the update currents become independent of capacitor voltage. The updated charge is determined by the length of YW pulse multiplies the current from the current source FET controlled by the XW_P/XW_N voltage. Fig. 2 shows the schematic of a 4×5 array. The layout of the array is shown in Fig. 3, and Fig. 4 shows the cross-sectional TEM image of the trench capacitor [7].

Results and Discussions

Weight update is demonstrated in Fig. 5. with multiple positive and negative update pulses applied to a single cell. Device current was measured after each pulse from the readout FET showing clear modulation by the pulses. Fig. 6(a) and (b) show the measured change in the conductance of the readout FET of a single cell, and corresponding capacitor voltage respectively, by applying ten cycles of 400 positive updates followed by 400 negative updates. 400 intermediate conductance states were achieved. This number is limited by the measurement parameters and not inherent to the circuit. Circuit simulations show the capacitor voltage (Vcap) range where expected asymmetry between positive and negative update

is better than 10% (Fig. 7). Experimental results in Fig. 6 are consistent with simulation (Fig. 8). Fig. 9 compares the experimental non-linearity-update factors [8] for our capacitor-based analog synapse against other NVM technologies. To the best of our knowledge, the capacitor-based unit cell provides the best symmetry and linearity demonstrated to date. Cell to cell variation causes extra asymmetry (Fig. 10), and better than 15% asymmetry was achieved for most cells within the 4×5 array. Multiply-accumulate operation is demonstrated on a 2×2 array (Fig.11). Randomized weights were measured individually, and then different input voltages were applied to the two columns. The current was measured on each row showing errors <0.5%. Fig. 12 demonstrates parallel weight update on a 2×2 array. Different XW_P voltages were applied on each row, with different YW pulse widths on each column. No disturb was observed during parallel update and read.

The capacitor retention time was measured by first charging the Vcap to 0.8 V, and then observing the change in readout current after turning off the current sources, which are the dominating leakage path. The corresponding Vcap change is shown in Fig. 13, with retention time in the order of seconds. To illustrate the effect of retention time on training, Fig. 14 shows test error of a simulated 784×256×128×10 fully-connected network trained on the MNIST dataset by stochastic gradient descent and backpropagation, assuming weights constantly decaying with different RC time constant τ [6]. Assuming the training cycle length per layer (forward+backward+update) is 200ns [2], the penalty in training accuracy due to capacitor charge-loss becomes negligible when τ >0.2s (10^6 × the training cycle length). Since weights are implicitly updated continuously during training, refresh cycles are not necessary. Retention requirements for a convolutional neural network (CNN) [10] are larger, due to the weight sharing (reuse) in convolutional layers (Fig. 15). The scalability of this capacitor-based array as a function of leakage is shown in Fig. 16. Since NN training can tolerate ~5-10% bad cells, the leakage spec can be significantly relaxed (2 sigma vs. worse case in DRAM). With high leakage technology such as current 14nm logic technology, large capacitor is needed. On the other hand, DRAM technology with leakages of 1 fA/cell [11] requires less than 1 fF capacitance/cell. This can be achieved with a DRAM stack capacitor, and the cell area will be dominated by FETs in control circuitry. For CNN, larger capacitances would be needed (~100 fF), requiring multiple stack capacitors and larger cell area. The scalability to larger input and more layers needs further study.

Device-to-device variation (Fig. 17) and non-ideal device characteristics may affect training accuracy (Fig. 18 (a) to (c)). While training is less sensitive to readout variation (σ=7% in this work), the variation of current sources causes mismatches between programmed charge updates (σ=6% in this work) and finite output conductance of current sources causes Vcap-depended asymmetry (Fig. 7, <10% in this work). With all these non-ideality, estimated MNIST test accuracy of 97% is achievable based on this work.

Conclusion

A capacitor-based cross-point array was demonstrated on 14nm technology using deep trench capacitors. We demonstrated record linear and symmetric weight update with 400 conductance states. This capacitor-based array, with continuous weight updates and no refresh cycle, has potential to be an ideal candidate for accelerating the training of deep neural networks.

978-1-5386-4219-1/18 $31.00 © 2018 IEEE

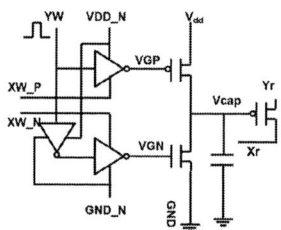

Fig. 1. Unit cell schematic of a capacitor-based cross-point array.

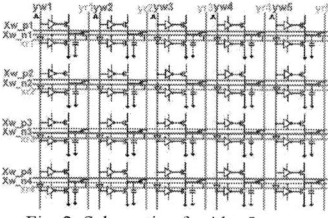

Fig. 2. Schematic of a 4 by 5 array.

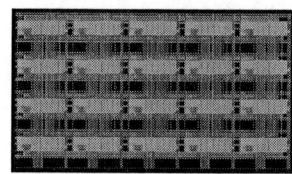

Fig. 3. Layout of the 4 by 5 array.

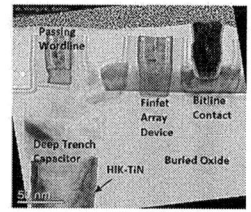

Fig. 4. TEM image of the deep trench capacitor [7].

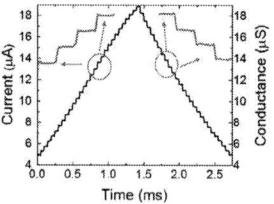

Fig. 5. Update of a single unit cell with multiple positive and negative update pulses. Pulse width: 500 ns, period: 50 μs.

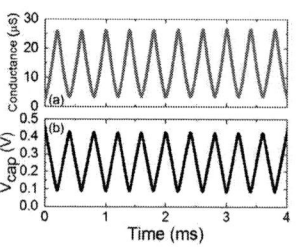

Fig. 6. (a) Experimental results for updating single-cell with 8000 pulses. (b) Corresponding capacitor voltage change. Pulse width 50 ns, period: 500 ns.

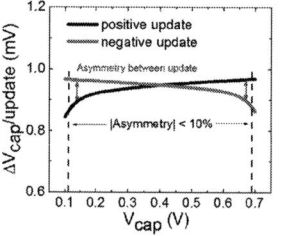

Fig. 7. Simulated asymmetry between positive and negative update as a function of Vcap. Asymmetry = (1- $\Delta Vcap_{positive}/ \Delta Vcap_{negative}$)

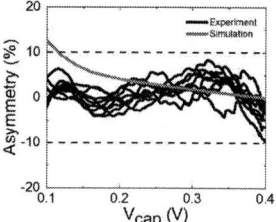

Fig. 8. Experimental update asymmetry and compared with simulation result.

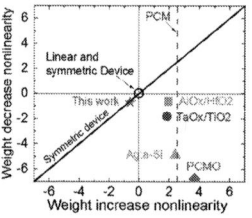

Fig. 9. Conductance non-linearity of this work compared with other NVM technologies [8]. When positive update nonlinearity equals to negative update nonlinearity, the update is symmetric. PCM (extracted from [3]) is shown as a dash line since its conductance can only be modulated gradually in one direction.

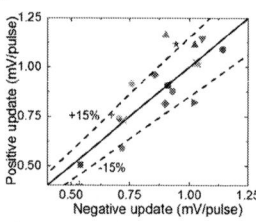

Fig. 10. Statistical results of updating single cells. Dashed line: +/-15% asymmetry. Different dots represent different unit cell in the 4×5 array.

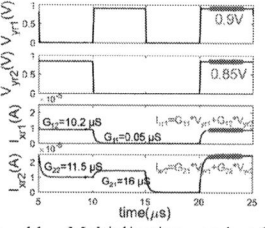

Fig. 11. Multiplication and add operation done by 2×2 array. Different input voltages were applied to different columns (yr1 and yr2). The current was measured on each row (xr1 and xr2). The measured error on each row was less than 0.5%.

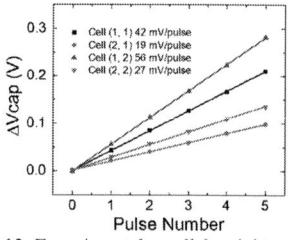

Fig. 12. Experimental parallel weight update for a 2×2 array with 5 pulses. The slope difference between cell 11 and cell 12 (also cell 21 and cell 22) is determined by different YW pulse widths in each column. The slope difference between cell 11 and cell 21 (also cell 12 and cell 22) is determined by different XW_P voltages on each row. Results match with simulation.

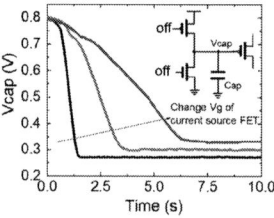

Fig. 13. Retention measurement. The retention time is in the order of seconds and can be maximized by turning the current sources deeply off, which are the dominating leakage path.

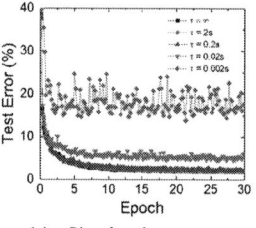

Fig. 14. Simulated test error of MNIST data set, assuming weights decay continuously with different RC time constant τ, 200ns training cycle length.

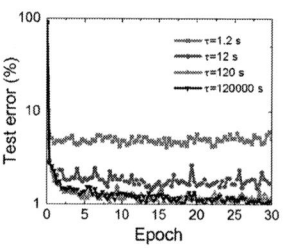

Fig. 15. Simulated retention time requirement for capacitor-based array to train convolutional neural network. 200ns training cycle length.

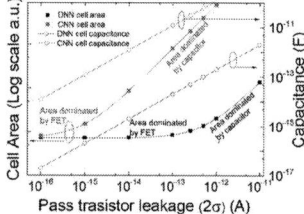

Fig. 16. Trend of required unit cell capacitance and area as a function of leakage current. When leakage is small, the area is limited by FETs in the control circuitry. Assume 200ns training cycle length, τ=0.2 s for DNN and τ=120 s for CNN. Vcap range 1V.

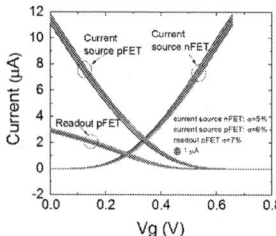

Fig. 17. Measured device variation of current source FETs and readout FETs in the 4×5 array.

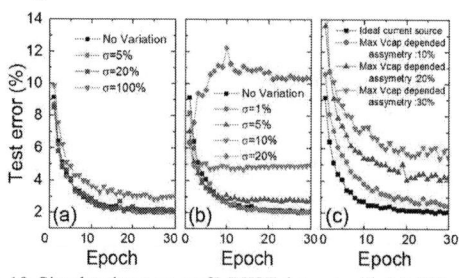

Fig. 18. Simulated test error of MNIST data set with (a) different amount of readout FET variation, σ=7% in this work, (b) different amount of current source variation σ=6% in this work, (c) non-ideal current sources, Vcap depended asymmetry <10 % in this work. τ =0.2s, 200ns training cycle length in simulation.

Reference:

[1] C. Merkel, *Computer*, vol. 49, pp. 56-64, 2016. [2] T. Gokmen, *Frontiers in neuroscience*, vol. 10, 2016. [3] G. W. Burr, *IEDM*, 2014. [4] S. Yu, *IEEE TED*, vol. 58, pp. 2729-2737, 2011. [5] S. Agarwal, *IJCNN*, pp. 929-938 2016. [6] S. Kim, *MWSCAS*, 2017. [7] G. Freeman, IEEE JSSC 2016. [8] P-Y. Chen, *ICCAD*, pp. 194-199 2015. [9] J. Liu, *ISCA*, pp. 60-71, 2013. [10] T. Gokmen, arXiv:1705.08014. [11] D. Chidambarrao, *VLSI-TSA*, 2003.

Analog Spike Processing with High Scalability and Low Energy Consumption Using Thermal Degree of Freedom in Phase Transition Materials

T. Yajima, T. Nishimura and A. Toriumi

The University of Tokyo, Bunkyo, Tokyo, Japan. yajima@adam.t.u-tokyo.ac.jp

Abstract

Spike integration and threshold processing are the basic signal processing in brain-inspired computing, such as deep learning, reservoir computing etc. In such processes, analog technology is essential for suppressing energy consumption. However, analog technology often faces problems in miniaturization due to deteriorated noise tolerance by scaling and intrinsically large analog elements such as capacitors. Here, we propose to exploit a thermal degree of freedom in phase transition materials for scalable and noise-tolerant analog spike processing. We focus on a two-terminal metal-insulator-transition VO_2 device, where quasi-adiabatic Joule heating enables efficient spike integration, and metal-insulator transition implements threshold processing. This VO_2 device is highly scalable, consuming only ~1fJ/spike (smallest so far) according to the simulation. By using this device, fully autonomous spike integration and threshold processing are also demonstrated. Exploiting the quasi-adiabatic thermal degree of freedom will facilitate scalable and energy-efficient analog implementation for a wide range of brain-inspired computing.

Introduction

Low-power brain-inspired computing often uses spike signals. In such systems, the basic signal processing consists of input spike integration and nonlinear threshold processing [**Fig. 1**], which are then interconnected on the network such as cross-bar arrays. Analog implementation of these processes has often used capacitors [1], which have hindered scaling due to their low noise tolerance and large-area occupation [**Fig. 2**]. Instead of charge integration in capacitors, we propose to integrate input Joule heat in the metal-insulator transition material VO_2 for a scalable and noise-tolerant alternative [**Fig. 2**]. Within the time scale of heat dissipation, the input Joule heat is quasi-adiabatically integrated in the insulating state of VO_2, and induces its transition to the metallic state above the transition temperature (~320K). The hysteresis of this transition mitigates input noise and stabilizes the analog operation [2]. Since the transition decreases the VO_2 resistivity by three orders of magnitude, it can easily be detected by a CMOS inverter to drive the latch circuit "R-latch" in **Fig. 3a**. This latch circuit, consisting of two resistive switching devices R_a and R_b, has only one stable solution for $R_a > R_b$, but is activated to another stable solution when $R_a < R_b$ [**Fig. 3b**]. Coupling VO_2 transition to this R-latch operation enables fully capacitor-less implementation of spike integration and threshold processing.

Experimental Demonstration

The two-terminal VO_2 device consists of an epitaxial VO_2 thin film, which are fabricated at 300°C on a single-crystalline TiO_2 (101) substrate [**Fig. 4a**]. Voltage application induces VO_2 transition from the initial insulating state to the high-temperature metallic state [**Fig. 4b**]. When the voltage is reset, the VO_2 recovers the insulating state, demonstrating threshold-switching property under DC voltage [2]. On the other hand, the dynamic property of this device under fast input spikes is defined by the time-dependent local temperature according to Joule heat accumulation and dissipation. The VO_2 local temperature, $T_{LC}(t) = C_T \int_0^t e^{-(t-p)/\tau} V(p)^2 / R \, dp$ (C_T: effective heat capacity, $V(p)$: applied voltage at time p, R: insulating VO_2 resistance), indicats Joule heat integration within the heat dissipation time scale τ. Because heat integration takes time, a considerable transition delay is observed against the applied voltage [**Fig. 5a-d**]. By using these dynamic properties of VO_2, the spike integration and threshold processing is experimentally demonstrated, by applying spike voltages on VO_2 in a setup shown in **Fig. 5a**. The VO_2 transition, which is indicated by the sudden increase in V_r, is induced only after a couple of spikes [arrows in **Fig. 6a,b**]. The estimated T_{LC} (bottom panels) confirms the input spike integration by Joule heating.

Simulation

In order to investigate the potential of the scaled VO_2 device, Joule heating inside the 10×10nm out-of-plane two-terminal structure is simulated based on the thermal diffusion equation [**Fig. 7a,b**]. In the structure, the VO_2 layer is sandwiched with two metallic W-doped VO_2 electrodes, which can minimize the contact resistance with VO_2 as well as enable Joule heat confinement by their exceptionally low thermal conductivity [3]. The simulation shows the lower energy consumption for the larger applied voltage [blue squares in **Fig. 8a**]. This is because the shorter transition delay enhances the Joule heat confinement as indicated in **Fig. 7a** (quasi-adiabatic heating). In this quasi-adiabatic regime, however, the time scales for the transition and the recovery (cooling) become too short (~ps). Therefore, the transition delay and the energy consumption are compromised at 0.7V [dashed line in **Fig. 8a**]. From a similar viewpoint, the VO_2 layer thickness is also optimized at 32nm [dashed line in **Fig. 8b**], leading to the sub-ns operation time scale and 1fJ energy consumption. This scaled VO_2 device is benchmarked with the previous reports including biological neurons [**Fig. 8c**], showing the large potential of this device due to the exploitation of the thermal degree of freedom.

Autonomous Spike Signal Processing

By using the fabricated VO_2 device, the autonomous spike signal processing is demonstrated [**Fig. 9a**]. The circuit consists of two R-latches. In the left one (brown), the input spikes are integrated by one VO_2 (R_a) and activate the output spike when R_a becomes metallic. While this left one is immediately reset by another VO_2 (R_b), it triggers the right one (purple), which generates a refractory period and cools down R_a and R_b. Based on the detailed operation of this circuit in **Fig. 9b**, the autonomous spike integration and threshold processing are experimentally demonstrated in **Fig. 9c**. Because this spike processing is fully autonomous, the simple connections of these circuits can perform various nonlinear tasks without the need of peripheral circuits. This work was supported by JST-CREST Grant Number JPMJCR14F2, Japan.

References

[1] G. Indiveri *et al.*, Frontiers Neurosci. 5, 73 (2011).
[2] T. Yajima *et al.*, IEDM (2016).
[3] S. Lee *et al.*, Science **335**, 371 (2017).
[4] G. V. Chandrashekhar *et al.*, Mat. Res. Bull. **8**, 369 (1973).

Concept

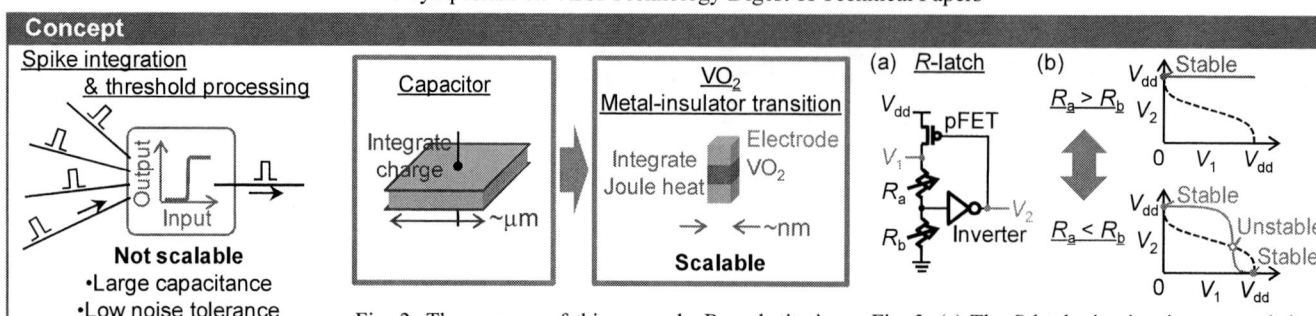

Fig. 1: The current limitation in the analog spike processing for the brain-inspired computing.

Fig. 2: The purpose of this research. By substituting the large capacitor with the metal-insulator transition VO_2, a scalable and noise tolerant spike processing becomes possible.

Fig. 3: (a) The R-latch circuit using two resistive switching devices R_a and R_b, which drives the analogue operation based on the VO_2 transition. (b) The stability diagram of the R-latch.

Experimental Demonstration

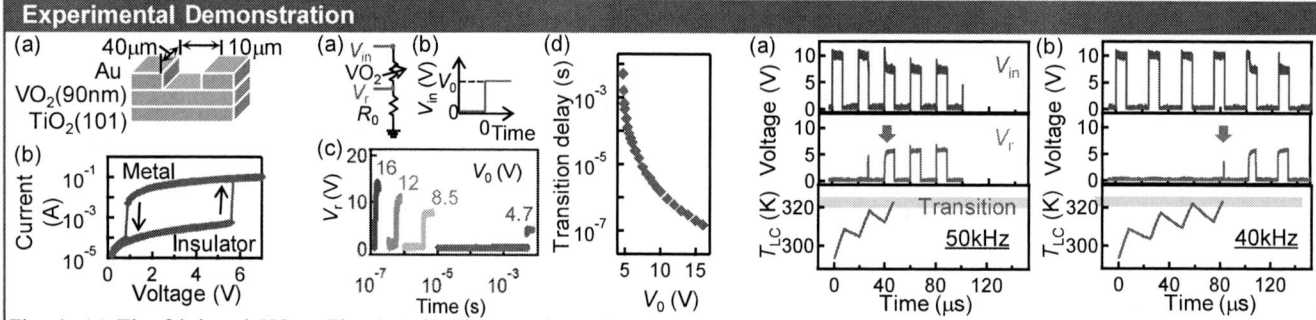

Fig. 4: (a) The fabricated VO_2 device. (b) The current-voltage characteristics, which shows the metal-insulator transition.

Fig. 5: (a,b) The experimental setting for the transition delay measurement. (c) The obtained delay curve at $R_0 = 510k\Omega$ and (d) the extracted delay.

Fig. 6: The experimental demonstration of spike integration and threshold processing, using the setup in Fig. 5a ($R_0 = 10^6\Omega$) with (a) 50kHz and (b) 40kHz spikes. The VO_2 temperature (T_{LC}) is also estimated ($\tau = 43\mu s$, $C_T = 47\mu J/K$).

Simulation

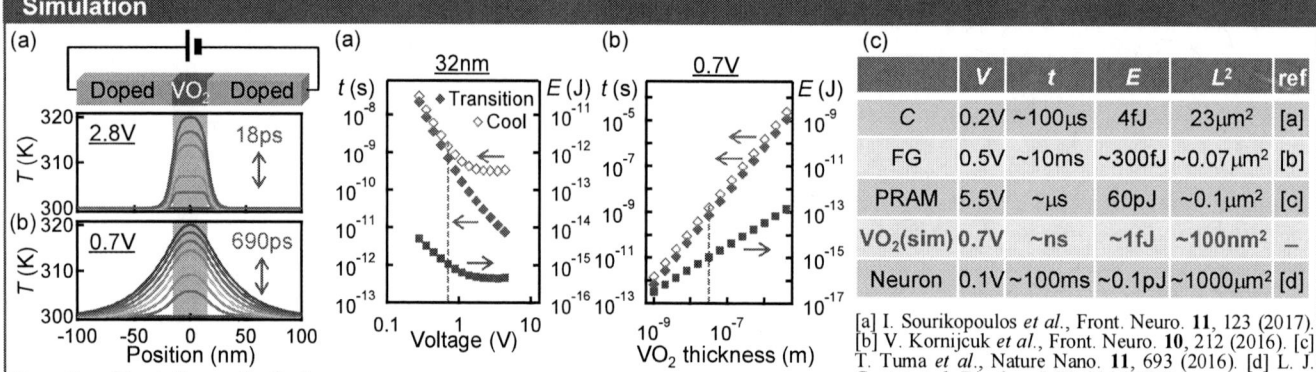

Fig. 7: Simulation of Joule heating in the W-doped VO_2 / VO_2(32nm) / W-doped VO_2 stack for two different applied voltages. Heat capacitance: 3.5J/Kcm³ [4], thermal conductivity: 0.06W/Kcm [3] for both VO_2 and doped VO_2.

Fig. 8: The simulated transition delay ("Transition"), the cooling time ("Cool"), and the energy consumption for the transition for (a) the 32nm fixed VO_2 thickness and (b) the 0.7V fixed applied voltage. The cooling time is defined by the temperature at zero position decaying to room temperature by 1/e. The optimized values are indicated by dashed lines. (c) Benchmarks of spike integration devices such as capacitor ("C"), floating gate FET ("FG"), PRAM, VO_2, and biological neuron. V: typical operation voltage, t: typical spike period, E: energy per spike, L^2: area of the integration device (not circuit).

	V	t	E	L^2	ref
C	0.2V	~100μs	4fJ	23μm²	[a]
FG	0.5V	~10ms	~300fJ	~0.07μm²	[b]
PRAM	5.5V	~μs	60pJ	~0.1μm²	[c]
VO₂(sim)	0.7V	~ns	~1fJ	~100nm²	—
Neuron	0.1V	~100ms	~0.1pJ	~1000μm²	[d]

[a] I. Sourikopoulos *et al.*, Front. Neuro. **11**, 123 (2017).
[b] V. Kornijcuk *et al.*, Front. Neuro. **10**, 212 (2016). [c] T. Tuma *et al.*, Nature Nano. **11**, 693 (2016). [d] L. J. Gentet *et al.*, Biophys. J. **79**, 314 (2000).

Autonomous Spike Processing

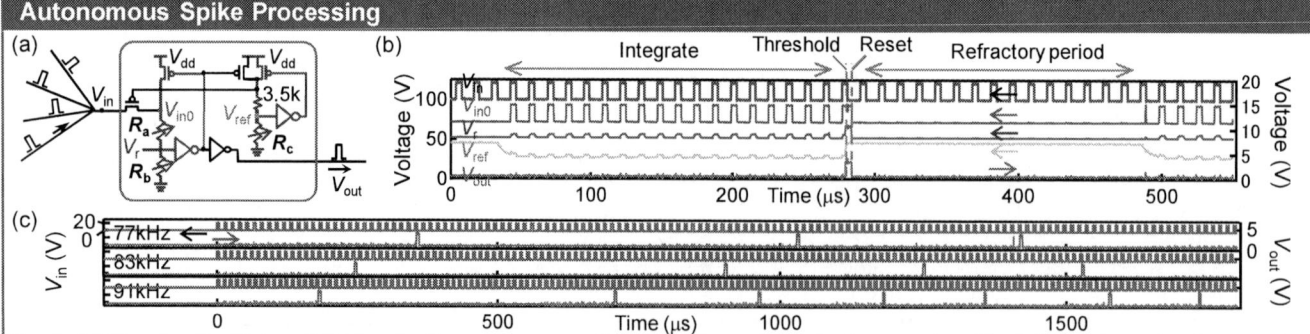

Fig. 9: (a) The circuit with two R-latches for the autonomous spike integration and threshold processing. The left one (brown) using two VO_2 devices (R_a and R_b) performs spike integration and threshold processing by R_a, and resets to the initial state by R_b. The right one (purple) using one VO_2 device (R_c) is triggered by the left R-latch and generates the refractory period in order to cool down R_a and R_b. (b) The detailed operation of this circuit, which is experimentally obtained by oscilloscope. (c) The experimental demonstration of spike integration and threshold processing for three different input spike frequencies. The circuit operates in a fully autonomous way without the need of any peripheral circuit.

An Energy Efficient FinFET-based Field Programmable Synapse Array (FPSA) Feasible for One-shot Learning on EDGE AI

J. L. Kuo[1], H. W. Chen[1], E. R. Hsieh[1], Steve S. Chung[1], T. P. Chen[2], S. A. Huang[2], T. J. Chen[2], and Osbert Cheng[2]

[1]Department of Electronics Engineering, National Chiao Tung University, Taiwan [2]United Microelectronics Corporation (UMC), Taiwan

Abstract- A pure logic 14nm FinFET with capabilities of linearly tunable V_{th} and excellent retention has been implemented as synapses in neuromorphic system. *For the first time*, a Field Programmable Synapse Array (**FPSA**) has been adopted to replace conventional R-based memory Synapse Array (**RSA**). Thanks to the wide range of V_t-tuning ability, 200X on/off ratio, and the ultra-small variability, 12%, results showed that the training power and SN ratio of FPSA are 10 times and 50 times smaller than those of the RSA, respectively. Two applications were demonstrated on FPSA array for one-shot learning applications. First, FPSA is used to detect handwritten digits of MNIST dataset. "Learned it by once" can be achieved in this task. Furthermore, FPSA has been applied to recognize goldfish in Cifar 100 dataset after learned the other 4 fish species. With the assistance from one-shot learning, results show the machine learned it faster and better on EDGE. This demonstrates the feasibility of FPSA for low-power and cost-effective synapse-based one-shot learning applications in the AIoT era.

1. Introduction

AI has been proven successful in object-classification and sound-detection by deep-learning networks. As AI encounters IoT, it will penetrate to each EDGE in society and will be a ubiquitous computing. However, deep learning claims numerous data trained on cloud, which makes it difficult to be launched to edge. In the comparison shown in Table 1, there is a huge gap of computing resource and power-draining between cloud and edge; moreover, at edge, the system is embedded and cost-sensitive, which cannot afford complex AI models with millions of weights(synapses). Nowadays, AI can only do inferring on edge [1]. Many groups have developed techniques of real training AI on edge, such as re-representation of weights in terms of few bits [2]. In Fig. 1, we propose a simple concept to realize real AI training on edge. That is, the machine only learns important features on cloud with abundant resources; after learning, the pre-trained machine is launched to edge and just needs to further learn specific and simple task based on one-shot learning [3]. By doing this, we train heavy-loading jobs on cloud and leave simple tasks on edge to reduce computing resources on edge. But what is "**one shot learning?**" Literally, **learned it by once.**" It imitates learning of our brains, that is, our brain can learn things just from simple events and pictures based on existing information stored in limited synapses of our brain. In other words, human brains do not need massive information to learn new stuffs but learn it clearly." On the design of AI hardware, people are focused on using a *high-density memory*. On the contrary, one-shot learning only requires a *small-density memory* to perform the job. We use FinFET as a low-cost, low power, and embedded hardware for one-shot learning. A typical fully-connected neural network(NN) is given in Figs. 2(a) and (b), where one can topologically map the fashion from NN style into an array style. Instead of using the resistance-based devices (RRAM [4] and PCM [5]) in Fig. 2(c), in this work, a 3-terminal FinFET is proposed, Fig. 2(d), comprising a synapse array, called *Field Programmable Synapse Array* (FPSA). We will elucidate properties of FPSA and take its advantages to build edge-AI. Then, SN ratio, power consumption of FPSA will be estimated at circuit levels using simulations.

2. Device Preparation

A 14nm logic FinFET with HKMG process has been chosen as a FPSA platform. (Fig. 3) EOT= 12Å. N- and pFinFET have been fabricated and experimentally measured to support the function of synaptic devices. Different layout sizes with various fin numbers and fin pitches have been designed to measure the V_{th} and gather the variations.

3. Results and Discussion

A. FinFET as Synapses in One-shot Learning

How can a pure logic device become a synapse? To answer this question, we first observe the learning trends of weights for synapses, Fig. 4, where convergent trends are clearly one way forward, i.e., *monotonic*. Unlike cyclic PGM/ERS of conventional memories, synapses do not necessarily return to original values.

This phenomenon matches with the V_{th} shift of FinFET after voltage stress. Therefore, it is possible to use FinFET as synapses. Fig. 5(a) shows $I_d V_{ds}$ during constant-V stress, and Figs. 5(b) and (c) are their corresponding changes of R_d. The more largely the V_{th} increases, the smaller the I_d becomes, and the higher the R_d is observed. Moreover, the ratio of R_d-change is 200 for p- and 40 for n-FinFET, which implies FinFET has the capability of continually tunable V_t. The top half of Fig. 6(a) shows const. V scheme at $V_{gs}= V_{ds}$. From Fig. 6(b), the increase of V_t follows the power law with an index, n<0.3, which means V_t becomes saturated quickly. In other words, this scheme (Fig. 6) can not realize linear-tuning capability To solve this issue, by gradually increasing voltage step by step, Fig. 7(a), a linearly tunable V_{th} can be achieved, Fig. 7(b). Furthermore, the linear tuning widow of V_t in pFinFET is larger. Fig. 8 shows the linearly varying trend of R_d during ramped V scheme. The tunable window of R_d for p is 175X. Fig. 9 shows Pelgrom plot, whose slopes, A_{vt}, represent the degree of σV_t. 12% of normalized σV_{th} in p- is larger than 10% of that in nFinFET. More importantly, the variation in FinFET is much smaller than that of resistance-based memory. Fig. 10 shows qualified test of the retention for different R_d values under one month at 85°C. As a result, pFinFET exhibits a wide range of V_{th} tunable window and larger R_d ratio.

B. The Comparison Between FPSA and Resistance-based Synapse Array (RSA)

Tables 2 and 3 show operation conditions of program(PGM) and Reading of FPSA & RSA respectively. Since 1S1R exhibits a large variation, it loses the competitiveness in terms of the power and SN ratio (Figs. 11 and 12). In comparison, by using FPSA, owing to its better σV_{th}, it shows better performance. Fig. 11 shows the comparison of reading power consumption. 10x reduction of FPSA is achieved. A huge reading power in RSA is due to sneak current, while the leakage in FinFET can be dramatically reduced. As a result, SN ratio of FPSA is wider than that of RSA (Fig. 12). Also, only 5x5 array size of RSA can maintain 10x of SN ratio. However, with benefits of low operation voltage and high PGM speed, PGM power of RSA is smaller than that of FPSA (Fig. 13). In short, FPSA has benefit of low reading power and huge SN-ratio because of ultra-low leakage and suppressed σV_t but has a drawback of relatively higher PGM power due to large operation voltage and longer PGM time.

C. FPSA as Hardware for One-shot-learning

Learn manuscripts by one-shot-learning: The machine learns handwritten by the methodology of Fig. 2. MNIST database is divided into 2 parts: one subset used to train the pre-learned CNN model on cloud; then this pre-learned model is applied to edge to learn the other subset of MNIST. The results show, with this technique, "learn it by once" has been obtained (Fig. 14)

Learn a fish species by one-shot-learning: In Fig. 15, the machine learns "goldfish" on edge from Cifar100 database after pre-learned the other 4 fish species on cloud. The results show, with pre-learning the fish species, the test error has been effectively reduced, showing great potential of using this methodology on edge.

In summary, one shot-learning on Edge by using FinFET as synapses(FPSA) shows a fast and effective training result with a low-power and low cost hardware for AI applications. pFinFET provides a wider linearly tunable window, 175x, and small σV_t. At the circuit level, SN ratio and reading power of FPSA are much better than those of RSA, with a reduction of 50x and 10x respectively. Finally, two learning tasks have been used to demonstrate this methodology on edge for simple learning tasks efficiently. This work provides us a cornerstone of how to design an embedded low-cost, energy-efficient, and simplified framework of EDGE-AI chip, based on one-shot-learning method.

Acknowledgments This work was support in part by the *Ministry of Science and Technology, Taiwan,* under contract *MOST105-2221-E009-130- MY3* and *MOST Research of Excellence program106-2233-E009-001.*

References: [1] S. Han et al., arXiv:1506.02626v3 [cs.NE] [2] J. Qiu et al., *FPGA* 2016, p. 26. [3] F. F. Li et al., *TPAMI*, vol. 28(4), p. 594. [4] M. Prezioso, *Nature*, vol. 521, p. 61 (2015). [5] S. Kim, *IEDM*, p.17.1.1 (2015).

2018 Symposium on VLSI Technology Digest of Technical Papers

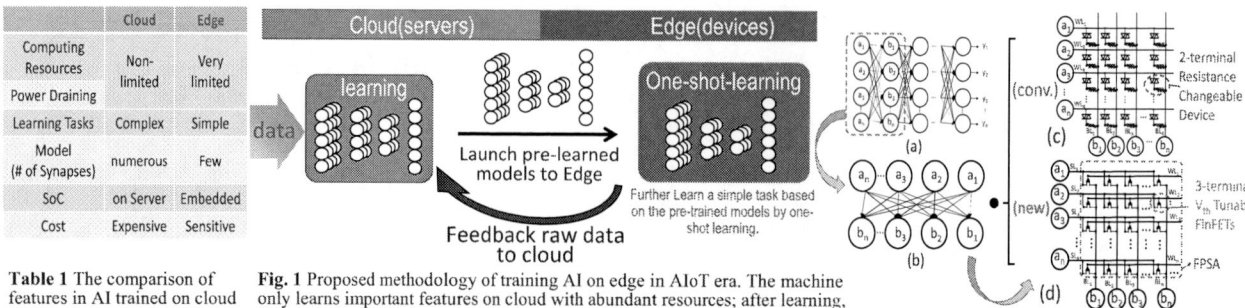

Table 1 The comparison of features in AI trained on cloud (*left*) and edge (*right*). The requirements for edge AI are much different from those in cloud AI.

Fig. 1 Proposed methodology of training AI on edge in AIoT era. The machine only learns important features on cloud with abundant resources; after learning, the pre-trained machine is launched to edge and just to further learn specific and simple task based on one-shot learning. By doing this, we train heavy-loading jobs on cloud and leave simple tasks on edge to reduce limited computing resources on edge.

Fig. 2 (a)-(b) A fully-connected NN; (c) Conventional 2-terminal resistance- based array; (d) Proposed Field Programmable Synapse Array (FPSA) constructed from FinFETs.

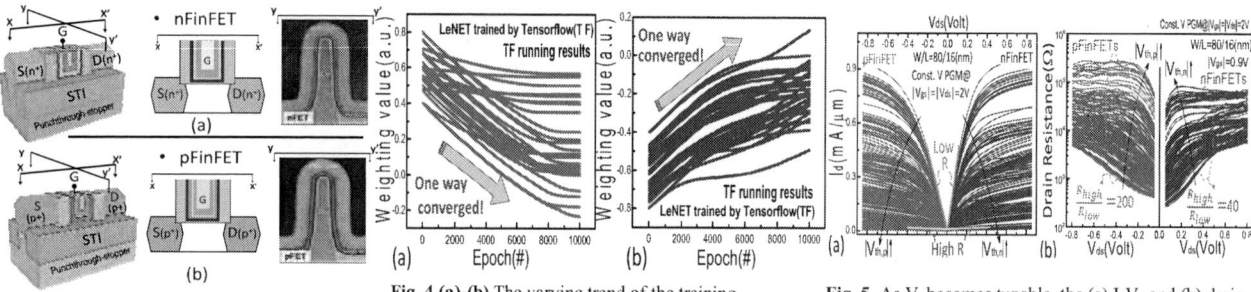

Fig. 3 (a) The schematic and TEM picture of nFinFETs; (b) pFinFETs.

Fig. 4 (a)-(b) The varying trend of the training weights shows the convergence to one direction. It gives us a hint that FinFETs actually can be the synapses of one-shot-learning applications.

Fig. 5 As V_t becomes tunable, the (a) I_dV_d and (b) drain resistance are varying as functions of tuning V_t. This property can be used to realize the weight value changes in Fig. 2.

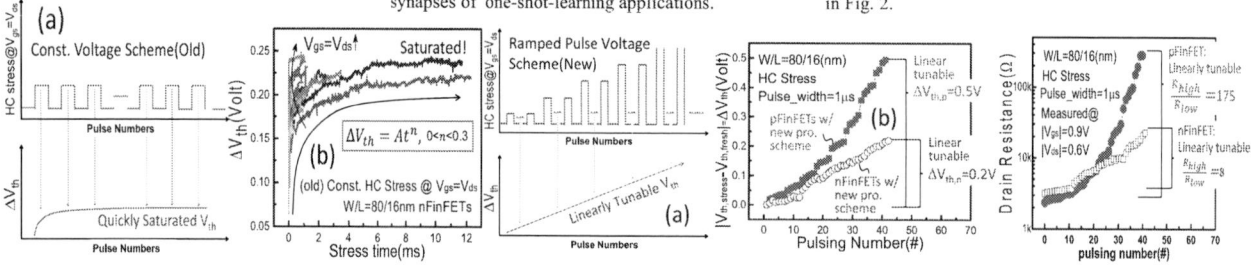

Fig. 6 (a) The conventional HC stress scheme shows the quick saturated stressed V_t as function of time. **(b)** V_t increment after const.-V stress scheme, fast saturated curves of V_t has been observed.

Fig. 7 (a) By gradually increasing the pulse voltage step by step, one can achieve linearly tunable V_t. **(b)** V_t increment after ramped V stress scheme, linear tunable V_t curves have been obtained.

Fig. 8 The linear-tuning trend of the drain resistance. The tuning ratio of p-FinFET is 175x and 8x for nFinFET.

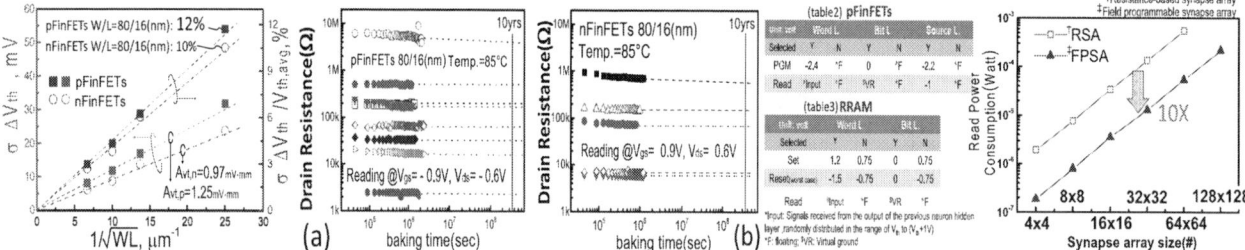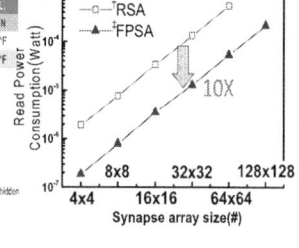

Fig. 9 The Pelgrom plot for p- and n-FinFET, whose slopes represent the standard deviation of V_t, σV_t. Both show very small σV_t.

Fig. 10 (a)-(b) the retention tests of p- and n-FinFET during 1-month at 85°C to data-keeping capability for FinFETs as synapses. Very stable data-retention for several weighting values have been observed.

Table 2 and 3 The operation conditions used in spice simulations of circuit array.

Fig. 11 The comparisons for reading-power of FPSA and RSA, showing 10x reduction for FPSA, compared to that of RSA.

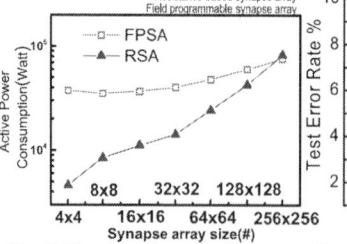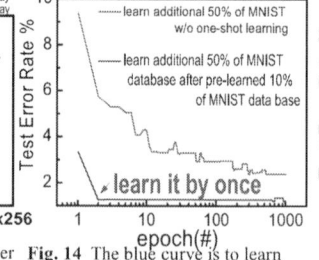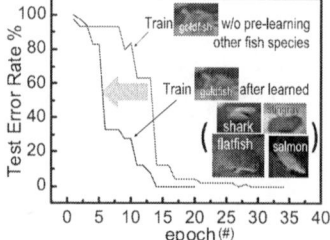

Fig. 12 The comparison of SN-ratio between FPSA and RSA. FPSA can hold a reliable SN window as macro-size expands to 173x173, w.r.t. 5x5 of RSA.

Fig. 13 The comparison of active power consumption between FPSA and RSA. FPSA exhibits higher active power at a small density. However, as density scales-up, due to sneak-path, active-power of RSA supersedes that of FPSA.

Fig. 14 The blue curve is to learn 50% MNIST dataset after pre-learned 10% of MNIST; the red curve is to learn the same 50% MNIST w/o any pre-learning. Results show the blue one can achieve one-shot-learning with much less error.

Fig. 15 In this task, the blue curve learns the goldfish after pre-learned other 4 fish species, including shark, stingray, flatfish, and salmom, showing good improved faster learning, compared to the red one, which only learns goldfish without pre-learning.

978-1-5386-4219-1/18 $31.00 © 2018 IEEE

Novel In-Memory Matrix-Matrix Multiplication with Resistive Cross-Point Arrays

Yan Liao, Huaqiang Wu*, Weier Wan[#], Wenqiang Zhang, Bin Gao, H.-S. Philip Wong[#], and He Qian

Institute of Microelectronics, Tsinghua University, Beijing, China. [#]Stanford University, USA

*E-mail: wuhq@tsinghua.edu.cn

Abstract

Resistive cross-point array can be used to implement vector-matrix multiplication in analog fashion. However, the output is in the form of analog current, and thus requires A/D conversion prior to digital storage. This paper develops and demonstrates a novel in-memory matrix-matrix multiplication method (M2M) that can compute and store the result directly inside the memory itself without requiring A/D conversion. Compared with the conventional approach, M2M provides $> 10 \times$ improvement in energy and area efficiency, and another 2 orders improvement when matrices are low-rank and sparse.

Introduction

Resistive cross-point array can implement vector-matrix multiplication (VMM) by mapping the matrix to resistance state of devices [1]. However, several challenges limit its applications as general-purpose matrix-matrix multiplication accelerator: 1) cross-point array output needs to undergo expensive A/D conversion; 2) the computed results are not directly stored in memory array.

This paper presents a novel in-memory matrix-matrix multiplication method (M2M) using resistive cross-point arrays. Multiplications are implemented in resistive cross-point array and results are directly stored in the same array without requiring A/D conversion. M2M is more energy and area efficient than conventional RRAM-based VMM for many matrix-matrix multiplication applications.

Novel In-Memory Matrix-Matrix Multiplication Method

Fig. 1(a) shows the procedures of the conventional RRAM-based vector-matrix multiplication method. Its time complexity for calculating $A \times B = C$ ($A \in R^{n \times k}$, $B \in R^{k \times m}$) is $O(m)$. A different but equivalent way to calculate the multiplication can be represented as follow:

$$A \times B = \begin{pmatrix} a_{11} \\ \vdots \\ a_{m1} \end{pmatrix} \times (b_{11} \dots b_{1n}) + \dots + \begin{pmatrix} a_{1k} \\ \vdots \\ a_{mk} \end{pmatrix} \times (b_{k1} \dots b_{kn})$$

One column of A is multiplied with one row of B to obtain one constituent matrix for C. The final result is the sum of constituent matrices (Fig. 1(b) and 1(c)) after k cycles. Resistive cross-point array that allow incremental analog resistance updates provides an efficient way to implement such computation. By applying programming pulses on the rows and columns of a cross-point array, RRAM devices (Fig. 2(a)) act simultaneously as AND gates (binary multiplier), accumulators and storage [2]. Fig. 3 demonstrates the parallel calculation scheme for a given example. Its time complexity of carrying out $A \times B = C$ ($A \in R^{n \times k}$, $B \in R^{k \times m}$) is $O(k)$.

Demonstration on RRAM Array

Fig. 2 shows the microphotograph of the chip used in measurement. The device-to-device and cycle-to-cycle variations of the devices are shown in Fig. 4 and Fig. 5. A 3×3 cross-point array is selected in the chip to carry out multiplication for different input matrix patterns. Average analog behavior for the 3×3 array is shown in Fig. 6. The measurement results for one example are demonstrated in Fig. 7. Measurements were performed for 10^2 cycles for different input patterns. We found that the measured average error (Fig. 8) decreases significantly when the dimension of matrix grows,

resulting from programming variability being averaged out by more operating cycles.

Accuracy, Sparsity, and Scalability

A compact RRAM model [3] that includes device-to-device and cycle-to-cycle variations is built for simulation. Fig. 9 shows that the average error is much smaller for larger matrices, which is consistent to the measurement results. For devices with larger variations, lower voltage pulses can be used to slow down the device resistance change per applied pulse such that the variation can be averaged out by more pulses within each quantized level (Fig. 10). Comparing two sets of input matrices with different sparsity (percentage of zero), Fig. 11 shows that the 90% sparse matrices consume $25 \times$ lower energy but have $2 \times$ higher average error than the 50% sparse matrices. Lower programming voltages can be used to recover the accuracy of sparse matrices, but will increases energy consumption by around $2 \times$ due to more operating cycles. Different from conventional cross-point based VMM, the processing time and energy of in-memory M2M scale proportionally with the rank of the matrices (Fig. 12), making it suitable for multiplication ($A \times B = C$, $A \in R^{n \times k}$, $B \in R^{k \times m}$) whose dimension $k \ll m$. Such property is found for matrix low-rank-approximation (LRA) [4], which is commonly used for data compression and recovery. Without precision loss, M2M can achieve $5 \times$ improvement in latency and energy for matrices whose rank ratio (k/m) are 0.1.

Performance comparison with RRAM-based VMM

Fig. 13 shows the system configurations of VMM and M2M that are used to estimate the latency, energy, and area. For a fair comparison, in both configurations, the final results are stored in cross-point array. The dimensions of the two input matrices are 128×32 and 32×128 respectively. Matrices with such k/m ratio can be commonly found in applications such as the recovery of LRA. The VMM system would need 128 operating cycles while the proposed in-memory M2M requires 32 cycles. Compared to VMM, in-memory M2M system could achieve savings of 75% in time, 43% in energy, and 70% in area. Besides the difference in the number of operating cycles, the benefits of M2M mainly come from eliminating A/D data conversion [5] and explicit data transfer between the array used for computation and the array used for result storage. Moreover, the energy and area benefits of M2M would be more significant for higher RRAM resistance (Fig. 15), and when matrices are sparse and low-rank.

Conclusion

In this paper, we propose and demonstrate a novel in-memory matrix-matrix multiplication approach that can store computed results directly inside a cross-point memory array without requiring A/D conversion. It enables highly efficient computing kernel that can be widely employed in applications such as data compression and image processing.

Acknowledgements: This work is supported in part by the MOST of China (2016YFA0201801), the ICFC, NSFC (61674089), NSF (1317470), and Stanford SystemX Alliance.

References: [1] M. Hu, DAC, 2016. [2] H. Li, VLSI, 2016. [2] H. Wu, IEDM, 2017. [4] Z. Zhang, Linear Algebra and its Applications, 2003. [5] B. Murmann, "ADC Survey", 2015.

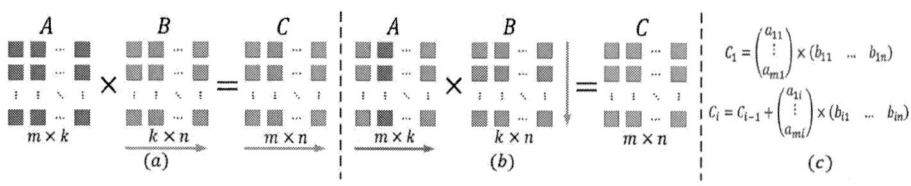
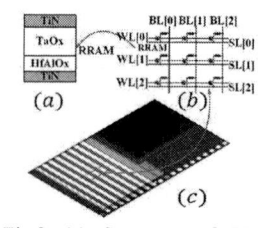

Fig. 1 (a) Procedures of conventional cross-point based vector-matrix multiplication. A is multiplied with each column of B to obtain each column of C (b) Procedures of in-memory matrix-matrix multiplication. One column of A is multiplied with one row of B to obtain one constituent matrix of C (c) During ith cycle, the multiplication of ith column vector in matrix A and ith row vector in Matrix B is implemented, then the results are accumulated with previous results. C_i is the results after i cycles.

Fig.2 (a) Structure of RRAM device. (b) Circuit diagram of a 3×3 cross-point array. (c) Microphotograph of a 1k-bit 1T1R RRAM array.

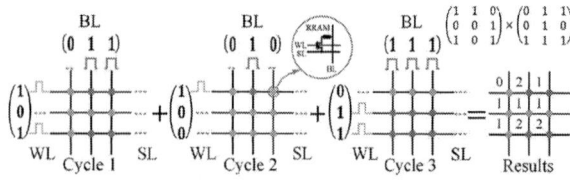
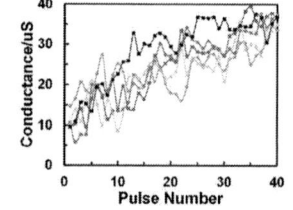

Fig.3 An example of calculating 3×3 matrix multiplication. During the ith cycle, parallel SET pulses corresponding to ith column of A and ith row of B are applied on the two sides of the cross-point array.

Fig.4 Device-to-device variation of RRAM analog behavior under consecutive SET pulses.

Fig.5 Cycle-to-cycle variation of RRAM devices' analog behavior under consecutive SET pulses

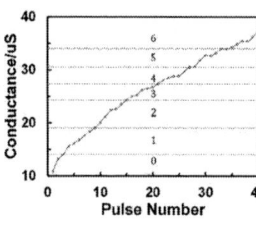
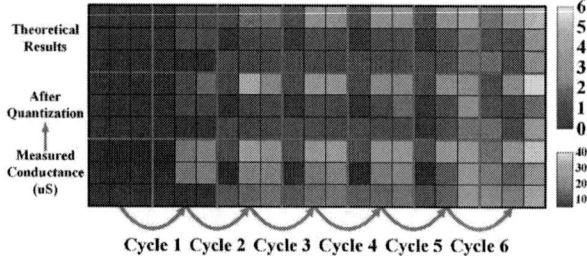
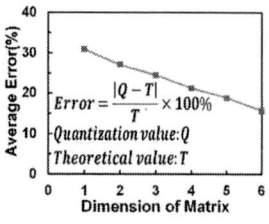

Fig.6 Average analog behavior for the 3×3 array.

Fig.7 Measurement results for one example of matrices multiplication of $A \in \mathcal{B}^{3\times6}$ and $B \in \mathcal{B}^{6\times3}$, where $\mathcal{B}$ is Boolean space.

Fig.8 Measured average error against the dimension of matrix based on different input patterns.

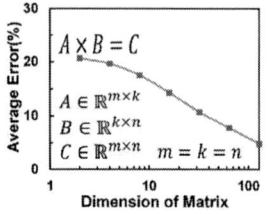
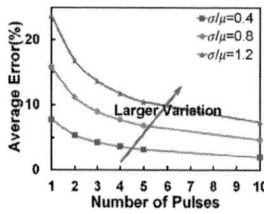
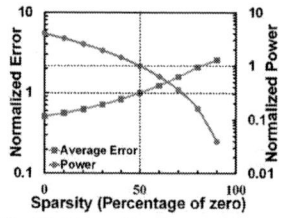
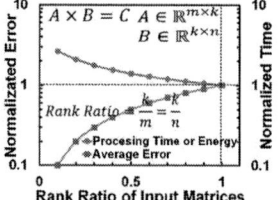

Fig.9 Simulated average error for different matrix dimensions based on stochastic RRAM model.

Fig.10 Average error against the number of pulses per quantized level for different variation.

Fig.11 Normalized average error and power consumption against the sparsity of input matrices.

Fig.12 Normalized processing time (energy) and error against the rank ratio of input matrices.

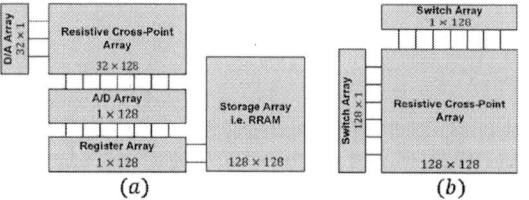
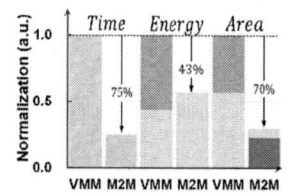
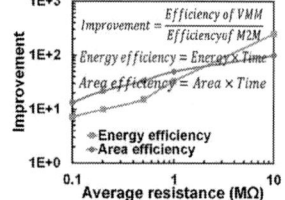

Fig.13 System configurations for calculating $A \times B = C$, $A \in \mathcal{B}^{128\times32}$, $B \in \mathcal{B}^{32\times128}$ using cross-point memory,. (a) conventional cross-point based Vector-Matrix Multiplication (VMM). (b) proposed in-memory Matrix-Matrix Multiplication (M2M).

Fig.14 Processing time, energy and area comparison between VMM and M2M. The average resistance of RRAM is 100 kΩ.

Fig.15 Energy and area efficiency benefits of M2M against the average resistance of RRAM.

978-1-5386-4219-1/18 $31.00 © 2018 IEEE

Gap in pagination due to formatting issues.

Pages 33-34

Sensors and related devices for IoT, medicine and smart-living

T. Ernst, R. Guillemaud, P. Mailley, J.P. Polizzi, A. Koenig, S. Boisseau, E. Pauliac-Vaujour, C. Plantier, G. Delapierre,
E. Saoutieff, R. Gerbelot-Barillon, E. Calvanese Strinati, S. Hentz, E. Colinet*, O. Thomas**, P. Boisseau, and P. Jallon
CEA LETI, Univ. Grenoble Alpes, Grenoble, France. Email: thomas.ernst@cea.fr
*APIX Analytics, Grenoble, France **Moovlab, Grenoble, France

Abstract

The evolutions of medicine covering genome to exposome (i.e. all types of environmental exposures) [1] opened new paths of development for electronics including low power sensors. Additionally, the frontiers for new generations of sensors between smart-living, environment and health are fading. In this paper, we will give examples based on our developments in emerging autonomous sensors and medical devices, and show how they can be included in our daily life.

Introduction

The emergence of big data is transforming our way of life, in particular in the field of personalized medicine [2], health monitoring, environment, wellness and fitness. Generic developments in those application fields for autonomous, low power interfaces and related sensors benefit from microelectronics ultimate integration capabilities. They are developed for new information processing and service paradigms within dedicated systems and software. Such fields will also require secured and sustainable solutions which is a great challenge especially at the edge of the cloud or at connected objects level.

Requirement in digitalized medicine, health and fitness

Digitalized medicine developments benefit from the huge potentialities provided by 50 years of Moore's law and sensors miniaturization. How to deploy new solutions strongly depends on cultural and economic factors as: medical doctors practices, country's wealth, public authorities, ethical culture-dependent considerations, existing regulations, which will define the more appropriate uses for a given (group of) country.

However, multiparametric distant monitoring is a major trend and can be declined in four dimensions: prevention, diagnostics, therapy and post-clinical monitoring.

It is evidenced today that miniaturized systems monitoring physiological parameters (or delivering medicine) changed life for many patients with chronicle diseases which occurrence increases in ageing populations: diabetes, heart diseases, sleep apnea, lung diseases, etc...

Physiological parameters are also relevant for wellness and fitness (in addition of usual motion sensors) mass market with the challenging specificity of monitoring on sweaty and fast moving persons (**Fig. 1**).

Emerging sensors

Typical devices for medicine use a large range of miniaturized sensors already developed for general purpose (MEMS, temperature, imagers, etc.) or specifically developed (for instance physico-chemical sensors with microfluidic and biocompatible packaging) (**Fig. 2**). New generations of MEMS are emerging for gas and particles monitoring.
Three types of contexts for monitoring and diagnostics can be defined:
- Wearable
- Mobile (including in Vitro Diagnostics)
- Implantable

Complex gas monitoring is an emerging field for exposome (air pollution) and diagnostics (for instance breath analysis). For example, volatile organic compounds (VOCs) are relevant for indoor home environment degradation or diseases (like lung cancer [3]) biomarker. It was shown recently that sensing systems using nano-mechanical resonators [4] can reach tens of ppb resolution for VOCs [5] (**Fig. 3**) or even single protein detection [6].

Wearable activity monitoring is also an emerging field that covers health and fitness. For instance we developed devices that includes multi axis accelerometers combined with machine learning. They are used to monitor physical activity and sedentary patterns in free-living conditions and to understand their impacts on health [7]. It is even also possible to include MEMs sensors within textiles which opens perspectives for ubiquitous activity monitoring (**Fig. 4**).

Wearable skin analysis is using emerging integrated photonic devices and new generations of imagers for pathology identification linked to variation of chemical or structural parameters of the skin [8]. Using this principle, we proposed recently to monitor sleep breathing disorders (considered as a chronicle disease), acute or chronic respiratory failures in using a wearable sensor measuring skin blood oxygen saturation (SpO2), and with PPG (Photoplethysmogram) analysis (**Fig. 5, 6**) [9]. This type of wearable is much more compact and comfortable than existing monitoring systems enabling their use in real sleeping conditions. Lastly, in vitro mobile diagnostic, using a combination of microfluidic, analytical protocols and optical or electrochemical sensors, is developed for fast and portable isothermal genomic diagnostic [10].

From VLSI technologies point of view, there is a general trend to integrate very closely sensing, processing, energy and communication. For instance very dense pixel sensors arrays are used for emerging genomic [11], or molecular scale sensing. In **Fig. 7** is demonstrated a crystalline NEMS resonator sensor technology with above IC monolithic integration on top of CMOS for molecular scale sensing [5]. Such an integration is also compatible with liquid-gate nanowire-FET pH-sensors arrays, used for DNA sequencing [12].

Toward autonomous connected devices

Thanks to progress in design and technologies, the power consumption of some electronic functions (sensors, RF, microcontrollers, etc.) has been strongly reduced, and tends to be compatible with ambient energy harvesting powers. Small-scale energy harvesting (below 1cm³) is indeed a great opportunity to turn medical or wellness devices into fully-autonomous sustainable systems (i.e. without battery): combined to adequate power management circuits [13], energy harvesters, such as photovoltaic cells, thermoelectric modules or mechanical energy harvesters can provide the $10\mu W$ to 10mW required by body area sensor nodes. For fitness devices, we recently proposed a harvester with a high power density ($730\ \mu W.cm^{-3}$) generating 4.95mW when worn on the arm during running [14].

For medical applications, we developed a harvester to supply pacemakers exploiting heartbeat low frequency vibrations (**Fig. 8**). Finally, for implanted devices, it is also possible to use either RF or acoustic power transfer coupled with data communication.

By 2020, 5G networks (>1Gbits/s) will create new opportunities to support highly personalized Health and Wellness. Indeed, 5G will introduce personalized services, enabled by intelligent orchestration functionalities (**Fig. 9**). During 2018 Winter Olympics, several 5G technological solutions were demonstrated [15].

Wearable systems integration: some examples

For wearables systems, integration is a key issue increasing both the measurement quality and user's comfort. These two key parameters are addressed jointly with an optimized contact with skin (through packaging and integration). We thus developed a highly integrated Electro Encephalography measurement for wellness applications, based on flexible dry electrodes, with dedicated coating to obtain low impedance contact with scalp (**Fig. 10**). Also several sensors (MEMS and optical) discussed in the previous sections were included in a wristwatch device for physiological parameters monitoring (**Fig. 11**). A flexible packaging of the device was developed to improve the mechanical contact of the sensors with the skin, which is crucial for optical measurement. This devices measures for instance pulse wave velocity (**Fig. 12**) which is an essential parameter for health monitoring (**Fig 13**).

Conclusion

The increasing medical interest for exposome data analytics will lead to knock down the frontiers between general purpose, wellness and medical applications of emerging sensors and IOT devices. This paves the way toward more preventive healthcare services.

References

[1] C.P. Wild, biomarkers & prev., vol. 14, no 8, p. 1847–1850, 2005.
[2] E. Topol, The Creative Destruction of Medicine, Basic Books, 2013.
[3] M. Hakim et al., Chem. Rev. 112, 11, 5949-5966, 2012.
[4] O. Martin et al., Sensors and actuator B, vol. 194, p. 220-228, 2014.
[5] T. Ernst et al., ESSDERC, p. 31-35, 2015, see also http://www.apixanalytics.com.
[6] M. S. Hanay et al, Nature Nanotechnology., vol. 7, pp. 602-608, 2012.
[7] T. Bastian et al, J Appl Physiol 118, pp. 716–722, 2015.
[8] A. Koenig et al., Vol. 9537 of SPIE Proc. (OSA), 95370E, 2015.
[9] A. Koenig et al., Biomedical Optics (OSA), TTu2B.4, 2016.
[10] D. Gesselin et al., Anal. Chem., vol. 89, pp. 10124–10128, 2017.
[11] J. M. Rothberg et al. Nature vol. 475, pp. 348–352, 2011.
[12] E. Accastelli et al., Biosensors 6, 9, 2016
[13] P. Gasnier et al., Journal of Solid-State Circuits 49 (7), 2014
[14] M. Geisler et al.Smart Mater. Struct. 26, 035028, 2017.
[15] E. Calvanese Strinati et. al., ETRI Journl., 40, p10-25, 2018, also www.5g-ppp.eu

2018 Symposium on VLSI Technology Digest of Technical Papers

Fig. 1. Connected wellness or fitness require monitoring on sweaty and fast moving persons and may include several physiological sensors in the future (source: moovlab®).

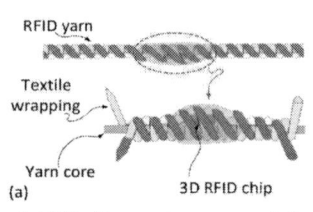 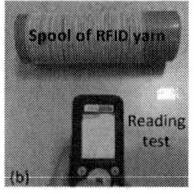  

Fig. 2. Overview of integrated sensors and estimation of their maturity (which varies depending on the precision, compactness and applications requirements).

Fig. 3. Volatile organic compounds detected thanks to emerging NEMS devices (source: apixanalytics®).

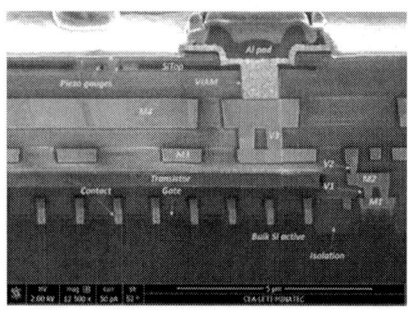

Fig. 4. (a) Textile wrapping process. (b) Spool of RFID yarn being tested with a hand held reader (source Primo1D®).

Fig. 5. Wearable sensor measuring hemoglobin oxygen saturation (SpO2) in the skin.

Fig.6. Measurement of hemoglobin oxygen saturation (SpO2) in the skin.

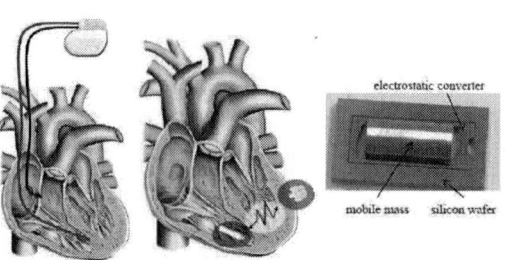 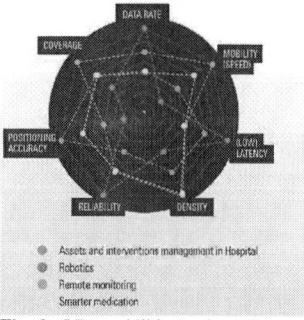

Fig. 7. Sequential monolithic integration of crystalline NEMS above CMOS for very precise molecules or particles measurements.

Fig. 8. Energy harvesters can be designed for various medical applications for instance here to supply a pacemaker.

Fig. 9. 5G capabilities and requirements for health applications (Source: 5G PPP consortium white paper).

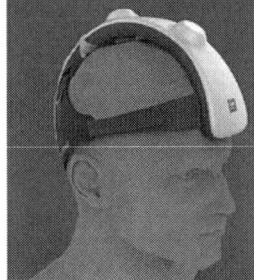 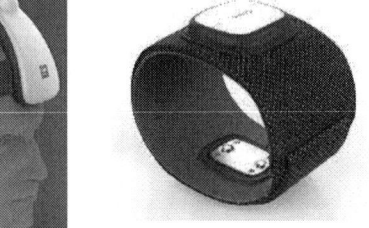 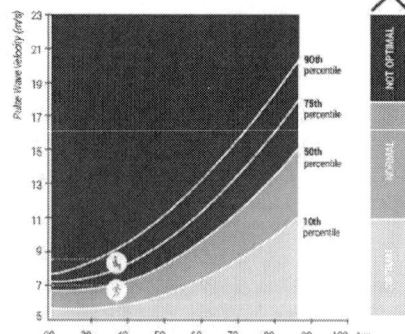

Fig. 10. Dry-electrode EEG helmet makes possible to take analysis out of lab and research centers.

Fig. 11. Wristwatch device for physiological parameters monitoring.

Acknowledgement - This work is partly supported by the French National Research Agency (ANR-15-IDEX-02) and H2020 FLAG-ERA "Convergence" project.

Fig. 12. Impact of arterial wall properties and blood pressure on pulse wave velocity. Source: Withings®

Fig. 13. Pulse wave velocity is an indicator for wellness and some cardio-vascular risks. Source: Withings®

978-1-5386-4219-1/18 $31.00 © 2018 IEEE

Development of a Multisite, Closed-loop Neuromodulator for the Theranosis of Neural Degenerative Diseases

Hsin Chen[1], Yen-Chung Chang[2], Shih-Rung Yeh[2], , Chih-Cheng Hsieh[1], Kea-Tiong Tang[1], Ping-Hsuan Hsieh[1], Yu-Te Liao[3], Ramesh Perumel[1], Ji-Feng Chuang[2], Ching-Chih Chang[2], Yu-Chieh Chen[1], Shih-Hsin Chen[1], Sung-En Hsieh[1], Yen-Peng Chen[1], Ye-Ting Chen[3], Tzu-Hao Liu[1], Yu-Ming Chang[1], Wei-Chih Lai[1], Chuang-Yi Wu[3], Yu-Hsin Chen[1], Yi-Chin Weng[1]

[1]Dept. of Electrical Eng. and [2]Dept. of Life Science, National Tsing Hua University (NTHU), TAIWAN
[3]Dept. of Electrical and Computer Eng., National Chiao Tung University (NDHU), TAIWAN
hchen@ee.nthu.edu.tw

Abstract

Stimulating specific brain regions has been found useful for treating neural disorders, such as the Parkinson's disease, epilepsy, depression, etc. However, how electrical stimulation modulates neural activities remains not fully understood. As animal models provide the advantage of recording and stimulating different disease-related regions simultaneously, this paper introduces the latest development of a multisite, closed-loop-controlled microsystem for investigating novel treatments on neural degenerative diseases with freely-moving rats. The algorithms for recognizing pathological neural activities automatically are also developed and realized in hardware, so as to control the stimulation in a closed loop and in real time. The pilot studies on the efficacy of treating the Parkinson's disease with closed-loop-controlled stimulation will be presented and discussed. Finally, the feasibility of modelling and probing how neural dynamics and connectivity are modulated by stimulation will be an important topic for future research.

Keywords: implantable microsystem, closed-loop control, neuromodulation, multisite, neural disorders

Introduction

Microelectronics has played an important role in advancing neuroscience research and improving the treatment for many neural disorders. The principle idea is exploiting micro-fabrication technologies and integrated circuits to interface with neurons at a high spatio-temporal resolution. This idea underpins the development of neural prostheses including cochlea implant, silicon retina, brain-machine interface for motor rehabilitation, etc. In addition to restoring sensory and motor functions, stimulating specific brain regions is further found useful for treating the Parkinson's disease, epilepsy, depression, etc. However, how electrical stimulation modulates neural circuits is not well understood, while the understanding is crucial for customizing stimulation patterns for individual subjects to maximize the efficacy and to minimize side-effects. To fulfill this need, we had designed a batteryless, implantable microsystem for studying the mechanisms of deep-brain stimulation [1]. This paper introduces the new design that will enable automatic detection of pathological signatures, so as to trigger stimulation only when necessary. Pilot animal studies that will be realized with the proposed system are also presented and discussed.

System Design

The architectures of the proposed microsystem and the theranostic controller are shown in Fig. 1. The microsystem consists of 16 channels of low-noise amplifiers, a 10-bit SAR ADC, a 5-bit programmable neural stimulator, an inductive-powering module, RF data transceivers, and an ASIC digital core which configure the neural recording and stimulation functions according to wirelessly-received commands. The frontend neuro-interfacing circuits and inductive-powering circuits are the same as those reported in [1], while the following aspects are improved: (1) the digital core is a customized state machine with a default operation mode. It runs more efficiently than the micro-controller in [1]. In addition, the ASIC state machine avoids the need for transmitting all micro-controller codes to initialize the chip. This ensures the microsystem operates more reliably under inductive powering. (2) Neural recordings can be wirelessly transmitted by either load-shift keying (LSK) or on-off keying (OOK) circuits. The former allows the microsystem to use only one coil for both power and bi-directional data transmission. The latter enables closed-loop control on neural stimulation upon detection of pathological signals by the theranostic controller. (3) The neural stimulator is able to generate an arbitrary waveform stored in the digital core, and the stimulation can be wirelessly triggered by the theranostic controller. As Fig. 1 illustrates, during closed-loop studies, neural recordings are transmitted to the theranostic controller continuously and classified by algorithms executed by the microcontroller or FPGA. As soon as detecting pathological signatures, controller triggers neural stimulation through the ASK channel.

The microsystem is designed and fabricated with the UMC 0.18 μm standard CMOS logic process. Fig. 2 shows the layout, and Table. I summarizes the specification and compares it with state-of-the-art designs, which also support multisite, closed-loop studies [2-4]. The main functional difference is that computationally-demanding algorithms are executed by

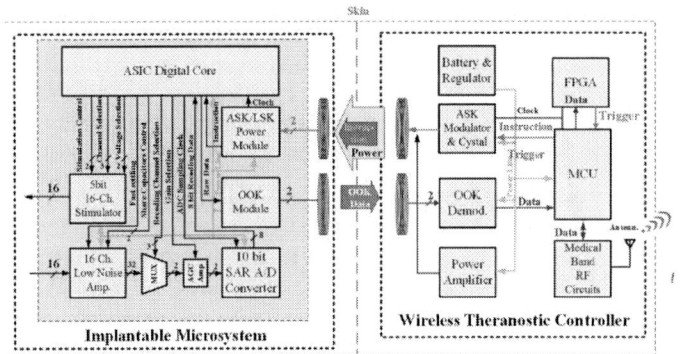

Fig. 1: The architecture of the microsystem and theranostic controller

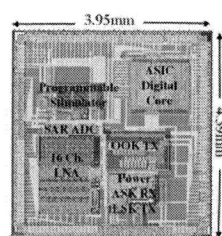

Parameter	Range
Rec. samp. rate (Hz)	10k~180 kHz/Ch.
LNA gain (V/V)	400/1000/2000
Stim. voltage (V)	2~5
Stim. pulse width	40~490 (us)
Stim frequency (Hz)	30~10k
Stim. epochs (us)	1~253
Stim. waveforms	pos., neg., biphasic pulses; arbitrary

Fig. 2: The chip layout and tunable parameter ranges

TABLE I: Specifications of the proposed design (simulation) compared to other state-of-the-art designs.

Spec.	JSSC'14 [2]	JSSC'16 [3]	VLSI'17 [4]	This work
Technology	N/A 0.18um	IBM 0.13um	TSMC 0.18um	UMC 0.18µm
Area (mm²)	4	16	25	17.3
Power (mW)	0.468 (excl. stim)	3.31~6.74	3.12~54	7~10
LNA gain (dB)	54	51.5	50/60/70	52/60/66
LNA noise (µVrms)	6.4	4.2	2.09	1.97
# of channels	4 rec./8 stim.	64	16	16
ADC resol.(bit)	8	7.1 (ENOB)	10	9.5 (ENOB)
Stim. current (uA)	116~4200	10-1000	500~3000	max. 250
Stim. waveform	biphasic pulse	biphasic pulse	biphasic pulse	biphasic pulse arbitrary
Data RX (bps)	Wireless 2.4G	ASK	BPSK, 330k	ASK, 100k
Data TX (bps)	Backscatter 800k	UWB 45M FSK 1.5M	LSK 100k	OOK 10M LSK 1M
Closed-loop function	Embed Log. DSP	Embed DSP processor	Embed BSP processor	Trig. by ext. controller

an external controller in our design, while the state-of-the-art designs have an embedded microcontroller or DSP processor for executing closed-loop-control algorithms directly. The latter could consume more power or have limited programming capacity/flexibility for the embedded algorithm.

Closed-loop-control Algorithms

Our design in [1] has been applied to investigating novel treatments for the Parkinson's disease (PD). The Parkinsonian rat models were induced in 3~4 month-old Sprague Dawley rats by unilateral injection of 6-hydroxydopamine in the medial forebrain bundle. A bipolar stimulation electrode was unilaterally implanted into the ipsilateral subthalamic nucleus (STN) of the hemi-Parkinsonian rats. Local field potentials in the cortico-basal-ganglia network were recorded with microwires. All experimental protocols has been reviewed and approved by the Institutional Animal Care and Use Committee in the NTHU, Taiwan (No.10321).

A pathological signature, called high-voltage spindle (HVSs), is observed in our experiments, as shown by the red curve in Fig. 3. HVSs are suspected to induce the pathological, beta-band synchrony observed in the Parkinsonian brain. Based on continuous wavelet transform (CWT), the HVSs are detected with latency of 0.5 s and accuracy greater than 95% [5]. However, each HVS lasts for only 2~3 s. 0.5s-long HVSs could already induce irreversible effects. Therefore, a prediction algorithm based on autoregressive modelling at interval, whose model parameters are adapted by a Kalman filter, is further proposed [6]. The prediction algorithm is proved to shorten the latency to less than 0.15s (Fig. 3b).

Animal Experiment Results

This section shows the animal experimental results with the proposed system. Fig. 4ab shows the local field potentials (LFPs) recorded from the layer 5b of primary motor cortex of both healthy and Parkinsonian rats. The corresponding time-frequency spectra show clearly that the LFPs of Parkisonian rat exhibit beta-band (20~40Hz) synchrony. On

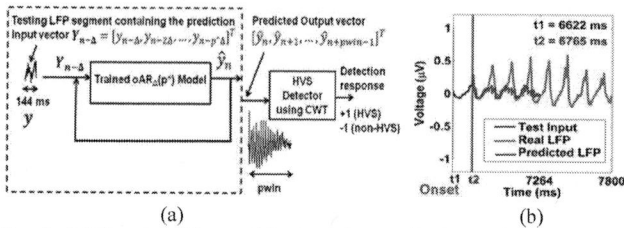

Fig. 3: (a)The algorithm architecture for predicting and detecting (b) The measured (red) and predicted (blue) HVSs.

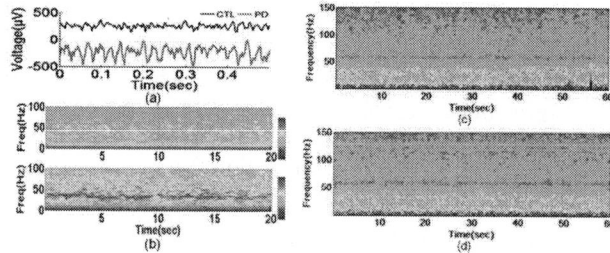

Fig. 4: (a) The LFPs recorded from a control (CTL) and a PD rats. (b) The corresponding time-freq. spectra of CTL (top) and PD (bottom) rats. (c)(d): The time-freq. spectra recorded (c)before and (d)after stimulation

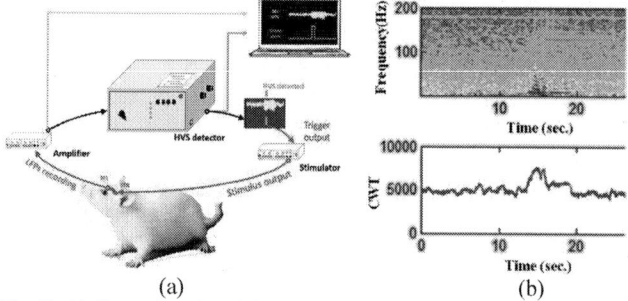

Fig. 5: (a) Setup for closed-loop experiment (b) Time-freq. spectrum of a recorded HVS (top) and corresponding CWT output (bottom).

the other hand, Fig. 4cd shows the LFPs recorded when the STN is stimulated by our microsystem. The beta-band energy reduces, while the stimulation artifact at 130Hz is visible in the time-frequency spectrum. The proposed HVS-detection algorithm is further realized in FPGA to detect HVSs and to trigger stimulation in real time. The pilot results in Fig. 5 show that the HVS can be detected and suppressed by the closed-loop stimulation effectively.

Discussion

The proposed microsystem will be fabricated in Mar. 2018 and its capability to support closed-loop studies will be explored. A miniaturized controller is being designed to allow animals to move freely during long-term experiments. In closed-loop studies, although an embedded processor help to reduce the bandwidth for data transmission, implementing the closed-loop algorithm in an external controller would not only facilitate programming flexibility but also release the constraints of power and computational capacity. This point will be discussed according to our experimental results.

References

[1] Y. P. Lin, et al., IEEE Trans. on BioCAS, p.98-112, 2016
[2] H. G. Rhew, et al., IEEE JSSCC, p.2213-2227, 2014
[3] H. Kassiri, et al., IEEE JSSCC, p.1274-1289, 2016
[4] C. H. Cheng, et al., IEEE Sym. VLSI circuits, p.c44-c45, 2017
[5] R. Perumal and H. Chen, 9th APCMBE, 2014
[6] R. Perumal, et al., J. Amer. Med. Info. Assoc., under review, 2018

High Performance High Density Gas-FET Array in Standard CMOS

Qian Yu[1], Xiaopeng Zhong[1], Farid Boussaid[2], Amine Bermak[1,3], CY Tsui[1]

[1]The Hong Kong University of Science and Technology, [2]The University of Western Australia, [3]Hamad Bin Khalifa University, Doha, E-mail: qyuac@ust.hk

Abstract

New gas sensitive FET (Gas-FET) structures are proposed to enable standard CMOS fabrication of high performance high density gas sensor arrays that are fully integrated with their associated circuitry. The performance of the proposed Gas-FET structures was investigated by fabricating an 8×8 Gas-FET array comprising 45μm×46μm sensing elements. Room temperature sub-ppm acetone sensing capability is demonstrated for non-invasive diabetes diagnosis through exhaled human breath. Performance comparison shows that the fabricated array boasts superior sensitivity while enabling ultra-low power operation at room temperature, with over 3 orders magnitude reduction in power consumption compared to previously reported standard CMOS resistive gas sensors.

Introduction

Use of a standard CMOS process to fabricate electronic noses enables miniaturization and batch fabrication of a complete electronic nose system at low costs and industrial standards [1], [2]. Previously reported standard CMOS implementations have been limited to resistive gas sensors, which exhibit low sensitivity and require a good Ohmic contact to be achieved between the sensing material and the top aluminum metal. In [3], post-fabrication of gold electrodes was investigated to implement Ohmic contacts [3]. In this paper, we propose new high performance gas-sensitive FET structures that alleviate Ohmic contact requirements. Unlike previously reported FET-type gas sensors, which rely on the use of special gate materials or structures (e.g. suspended control gate), the proposed Gas-FET structures can be directly implemented in standard CMOS process, thereby enabling the fabrication of fully integrated high performance high density gas sensor arrays. In addition, the ability to control the operating point of the proposed Gas-FETs enables a dramatic reduction of the chip power consumption.

Standard CMOS Gas-FET Array

The cross-sectional views of the proposed three standard CMOS Gas-FET structures are shown in Fig. 1(a)-(c), with their equivalent circuit schematics depicted in Fig. 1(a-1)-(c-1). Fig. 1(a) depicts an interdigitated electrode (IDE) consisting of a control gate (CG) and a floating gate. In this structure, the Gas-FET operates in the weak-inversion region through coupling of the CG voltage. Fig. 1(b) shows the second structure, which includes the IDE as floating gate and a reset switch to preset the FET operating point. Fig. 1 (c) depicts the third structure, which includes both the CG and reset switch. The architecture of the fully integrated 8×8 Gas-FET array with row decoder, column SAR ADCs, and control units is shown in Fig. 2 (a). Each column of the array was implemented using only one of the three proposed Gas-FET structures together with the readout circuitry shown in Fig. 2 (b). Each Gas-FET structure has a floating gate (FG) of exposed top metal connected to the MOSFET polysilicon gate through the other lower metal layers. Depending on the targeted application, various gas sensing materials can be deposited on top of the gate. When gas molecules are absorbed by the sensing material, the work function at the surface will change, leading to a change in the threshold voltage of the Gas-FET. Timing diagrams for the chip operation are shown in Fig. 3, with the sensing phase corresponding to the time the reset switch is off and the Gas-FET is left floating. Each Gas-FET is biased (or reset) in the subthreshold region, with its drain current (an exponential function of its gas-dependent threshold voltage) being readout for maximum sensitivity. During a fixed integration time, the drain current of each Gas-FET discharges its pre-charged integration capacitors. The global shutter ensures that the sensor array's individual responses are captured at the same time. After this global sensing current integration, voltage on integration capacitors are selected by row decoders and transferred to column SAR ADCs using source followers. Digital codes are then serially shifted out.

Measurements and Discussion

0.18μm 1P6M CMOS process was used to fabricate a fully integrated 8×8 Gas- FET array. ZnO nanorods (Fig. 1(g)) were exploited to demonstrate sub-ppm acetone sensing for non-invasive diabetes diagnosis through measurement of acetone concentration levels in human exhaled breath. Diabetes sufferers would exhibit concentration levels exceeding 1.7ppm compared to levels below 0.8ppm for healthy patients [4]. The selected ZnO sensing material was deposited using a Programmable Nanoliter Injector. The gas testing setup and chip photo are shown in Fig. 4. The chip was tested in a sealed gas chamber at room temperature. The total chip size is 1.61mm×1.28mm, with each Gas-FET occupying 45μm×46μm. For the proposed three Gas-FET structures, measured acetone responses are shown in Fig. 5 for acetone concentrations levels of interest ranging from 0.65 ppm to 1.9 ppm. The Gas-FET structure with only the control gate, is seen not to exhibit any response while the structure with reset switch only, shown in Fig. 1(b), exhibits the largest sensitivity and a response time of 24-26s. After 5572s and 3 consecutive cycles (Fig. 6), this Gas-FET with reset switch only is seen to exhibit low baseline drift levels of 0.61%-1.82% across implemented sensors of this type. Good repeatability of this Gas-FET is also observed from the error bars in Fig. 7, which shows the response level $\Delta I/I_0$ (%) as a function of acetone concentration levels ranging from 0.65ppm to 1.9ppm. The sensitivity of the Gas-FET with reset switch only (Fig. 1 (b)) is seen to be about 24 times higher than the structure with both CG and reset switch (Fig. 1 (c)). The latter suffers from the capacitor attenuation resulting from the additional control gate when ZnO surface charges change. Performance comparison (Table I) against previously reported standard CMOS gas sensors shows that the fabricated sensor array exhibits the smallest sensing area, highest sensitivity to acetone at room temperature and ultra-low power operation.

References

[1] Tzeng, Te-Hsuen, et al., JSSC 51.1 (2016): 259-272.
[2] Tang, Kea-Tiong, et al., ISSCC, 2014.
[3] Zanjani, Seyedeh Maryam Mortazavi, et al., npj 2D Materials and Applications 1.1 (2017): 36
[4] Kao, Kun-Wei, et al., Sensors 12.6 (2012): 7157-71

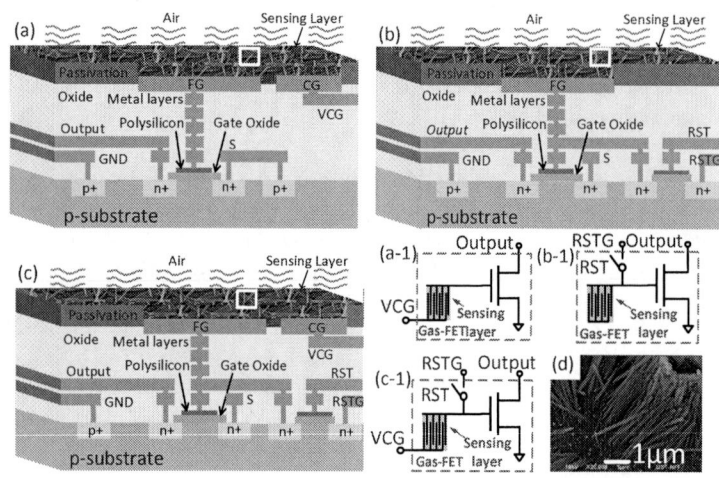

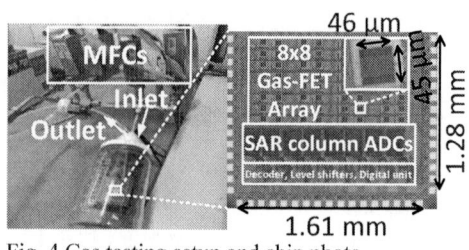

Fig. 4 Gas testing setup and chip photo.

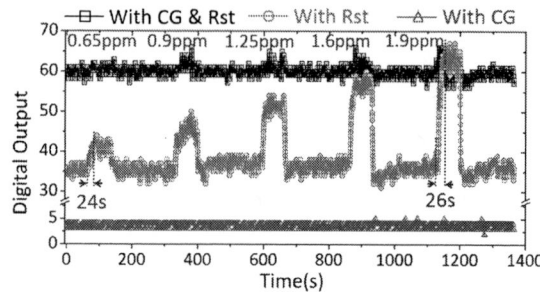

Fig. 1 Cross-sectional view of the standard CMOS Gas-FETs: (a) with CG; (b) with reset switch; (c) with both CG and reset switch; and corresponding schematic (a-1) with CG; (b-1) with reset switch; (c-1) with both CG and reset switch; (d) SEM of the ZnO nanorods deposited on top of the exposed FG (rectangular area in

Fig. 5 Acetone response of different structures in the array.

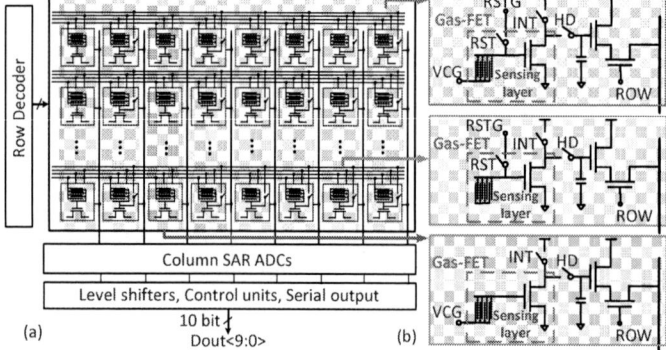

Fig. 2 (a) Top architecture of 8x8 Gas-FET array with different Gas-FET structures; (b) readout circuits in each gas sensing element with preset sub-threshold operating region by different methods.

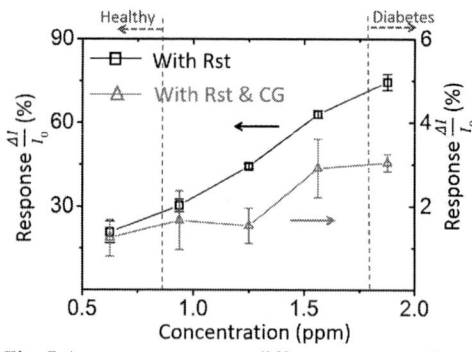

Fig. 7 Acetone response to different concentrations.

TABLE I
Comparison to previously published gas sensors

Ref.	JSSC 16 [1]	ISSCC 2014 [2]	Sensors 2012 [4]	This work
Process	0.35μm CMOS+MEMS	90 nm CMOS	In-house fabrication	0.18μm 1P6M CMOS
Total area	3.3×3.65 mm² (CMOS)	3.3×3.2 mm²	1×2.5 mm²	1.28×1.61 mm²
Type	1 Gas-Resistor	8 Gas-Resistor	1 Gas-Resistor	8×8 Gas-FET
Pixel area	1.8×1 mm²	400×400 μm²	0.25×2 mm²	45×45 μm²
ADC	10-bit SAR ADC	10-bit SAR ADC	No	10-bit SAR ADC
Sensor response	0.4% to 30 ppm Octane	30% to N/A ethanol	9% to 1 ppm Acetone/200°C	20% to 0.65 ppm Acetone
Response time	N/A	N/A	150s	26s
Power supply	3 V	0.5 V	0.5 V	1.8 V (Analog) 0.8 V (Digital)
Power	930μW	1270μW	N/A	29μW
Energy/ pixel	93nJ /Conversion	N/A	N/A	55pJ /Array conversion

Fig. 3 Timing diagram for the CMOS Gas-FET sensor readout.

Fig. 6 Repeatability and Drift characterization of the same Gas-FET structure in Fig. 1 (b) of different rows.

978-1-5386-4219-1/18 $31.00 © 2018 IEEE

High-sensitivity and low-power inertial MEMS-on-CMOS sensors using low-temperature-deposited poly-SiGe film for the IoT era

Hideyuki Tomizawa[1], Yoshihiko Kurui[1], Ippei Akita[2], Akira Fujimoto[1], Tomohiro Saito[1], Akihiro Kojima[1], and Hideki Shibata[1]

[1]Corporate Research & Development Center, Toshiba Corporation, [2]Toyohashi University of Technology

1, Komukai Toshiba-cho, Saiwai-ku, Kawasaki 212-8583, Japan

Phone: +81-44-549-2856, E-mail: hideyuki1.tomizawa@toshiba.co.jp

Abstract

In this paper, for the first time we demonstrate the material benefits of SiGe for MEMS applications based on the results of fabricated devices. To achieve SiGe inertial MEMS, we develop the deposition process for thick, low-temperature poly-SiGe film with which film stress is controlled precisely, and fabricate SiGe accelerometers having 20μm thickness. We clarify that the SiGe accelerometer shows higher sensor sensitivity and lower power consumption compared to Si one and is thus suitable for future ultra-low-power sensors.

Introduction

The market for MEMS sensors is rapidly growing in view of the ongoing emergence of the IoT era, and MEMS sensors with low-power consumption are required because of the nature of their various applications. In Table1, various types of MEMS and CMOS structure are compared[1-8]. Since SiP is unsuitable for low-power-consumption sensors owing to high parasitic capacitance induced by the wire bonding structure, utilization of SiP for IoT sensors is expected to be limited. The TSV structure has also been reported[8], but it may not make use of the merit of miniaturization because of an increase in the number of process steps and thus also the fabrication cost. To overcome above issues, the MEMS-on-CMOS structure, which is a novel MEMS structure without wire bonding and achievable by a simple fabrication process, has been proposed. With regard to MEMS material, introduction of a low-temperature process is needed so as not to degrade CMOS reliability, and thus we apply poly-SiGe film realized at below 450°C by the Plasma Enhanced Chemical Vapor Deposition(PECVD) method. Poly-SiGe is not only able to achieve the MEMS-on-CMOS structure, but has a material benefit for the inertial MEMS application as well because of the high mass density of SiGe material. In this paper, we report the SiGe film process that realizes the low temperature, thick film and controlled stress simultaneously by optimizing the condition of film deposition. We also report the sensitivity and power-consumption benefits of SiGe MEMS thanks to its high mass density compared to that of Si.

Low-temperature poly-SiGe film and MEMS fabrication

SiGe film process: Studies of poly-SiGe have been reported previously, but the thick poly-SiGe film needed for inertial MEMS had high tensile stress. Fig.1 shows the cantilever bending of a typical poly-SiGe film. Since the cantilever showed more than 7μm bending induced by the tensile stress, it would be difficult to apply the poly-SiGe film to MEMS without optimizing the film. We utilized the PECVD method by using Producer® supplied by Applied Materials Inc. to a deposition process for low-temperature poly-SiGe film at below 450°C and with a high deposition rate(3000Å/min). Fig.2 shows the stress measurement results of poly-SiGe film having 10μm thickness with various deposition recipes. A typical film (condition A)showed high tensile stress, but the film stress was widely controlled from tensile to compressive by changing the deposition recipe (condition B & C), suggesting that the PECVD tool used has a wide process window in terms of the film stress. We evaluated the cantilever bending realized by various combinations of tensile and compressive SiGe film as shown in Fig.3. Flat bending of the cantilever was achieved with 2:8 ratio of compressive and tensile film. We measured the cantilever whose design is the same as in Fig.1 by using the stress-controlled SiGe film shown in Fig.4. The amount of bending was -0.2μm, which was sufficient for the MEMS device fabrication. For further study, TEM cross-sectional and SIMS analysis were carried out as shown in Fig.5 and Fig.6, respectively.

The film thickness precisely controlled at around 20μm, and both Si and Ge content were conformal throughout the film thickness. Furthermore, boron also distributed uniformly, and the resistivity of poly-SiGe film indicated 3.4E-3ohm·cm, suggesting that the SiGe film would be applicable for MEMS devices electrically. The physical properties of the SiGe film we developed are summarized in Table 2. As a result, our thick, stress-controlled, and low-temperature poly-SiGe film is suitable for fabricating MEMS devices.

MEMS sensor fabrication: We fabricated differential capacitive type accelerometers(Acc.) using optimized poly-SiGe film as a motif of the inertial MEMS. The process flow and SEM images of our fabricated SiGe Acc. are shown in Fig.7 and Fig.8, respectively. The SiGe Acc. having 20μm thickness was successfully fabricated. SiGe film was etched with high aspect ratio whose number is around 20.

Evaluation results and discussion

Sensor sensitivity: We evaluated the device performance of the fabricated SiGe MEMS, and measured its differential capacitance by using an LCR meter while applying gravity. The SiGe Acc. exhibited good linearity(R^2=0.999), as shown in Fig.9, and worked as a MEMS sensor. Then, the differential capacitance comparison between the SiGe and the Si Acc., which were designed with the same dimensions, and the sensitivity comparison with various spring widths are shown in Fig.10 and 11, respectively. When the spring width is 2.3μm, the measured sensitivities of Acc. using SiGe and Si were 4.8fF/g and 2.2fF/g, respectively. Theoretically, the sensitivity of Acc. is proportional to mass density, and this result is attributable to about 2 times higher mass density of SiGe compared to that of Si[7]. This fact is one of the advantages of SiGe for the inertial MEMS application.

Power consumption: We evaluated the power consumption of both the SiGe and the Si Acc. to compare each performance as well as the sensitivity. A module for evaluation of Acc. is shown in Fig.12. To evaluate power consumption, the MEMS chip was mounted on a custom PCB. The AFE circuit that was also mounted on the PCB was originally designed for accelerometers[9]. The noise floor results of both the SiGe and the Si Acc. with respect to the power consumption are shown in Fig.13. The noise floor of the SiGe Acc. showed about 69% lower than that of the Si one when comparing the same power consumption. Device parameters and representative measurement results are summarized in Table3. As summarized in the table, it is clarified that the SiGe MEMS could achieve higher performance than the Si one. In other words, assuming that the noise density of the SiGe and Si Acc. are equivalent, the SiGe Acc. can achieve lower power consumption. This result is attributable to the same reason that accounts for the sensitivity benefit of SiGe material, and it was demonstrated that SiGe has the potential to reduce power consumption based on the actual device fabrication results.

Conclusion

We developed thick, low-temperature poly-SiGe film deposition technology by optimizing the film stress. We fabricated SiGe accelerometers having 20μm thickness deposited by the optimized deposition recipe, and evaluated the sensitivity and power consumption as the basic sensor performance. We firstly demonstrated the SiGe material benefit compared to Si in terms of both sensor sensitivity and power consumption by comparing the evaluations of the fabricated devices. Furthermore, since the SiGe material is applicable for MEMS-on-CMOS that has potential to decrease power consumption compared to SiP and TSV, SiGe is quite promising for realization ultra-low-power sensors for the IoT era.

Acknowledgements

This paper is partially based on results obtained from a project commissioned by the New Energy and Industrial Technology Development Organization (NEDO). The authors are grateful to Applied Materials Inc., Equipment Product Group, 200mm team for support of their study of SiGe film, and, in particular, they wish to thank H. I. Chi for technical support and fruitful discussion.

References

[1] S. Sedky and R. T. Howe, Journal of MEMS, Vol.13, No. 4, 2002.

[2] A. Scheurle et. al, MEMS 2007, pp.39-42.

[3] A. E. Franke et al, Journal of MEMS, Vol.12, No.2, pp.160-170 2003.

[4] A. R. Chaudhuri et al, TRANSDUCERS 2015, pp11-14.

[5] A. R. Chaudhuri et al, SENSORS 2016, pp532-534.

[6] C. Hirota et al, The 33rd. SENSOR SYMPO. 2016.

[7] Y. Kurui et al, SENSORS 2017, pp.181-183.

[8] BJ Woo et al, TRANSDUCERS2017, pp. 1-4.

[9] I. Akita et al, submitted to 2018 VLSI symp. on circuits.

Table1 MEMS/CMOS structure comparison table.

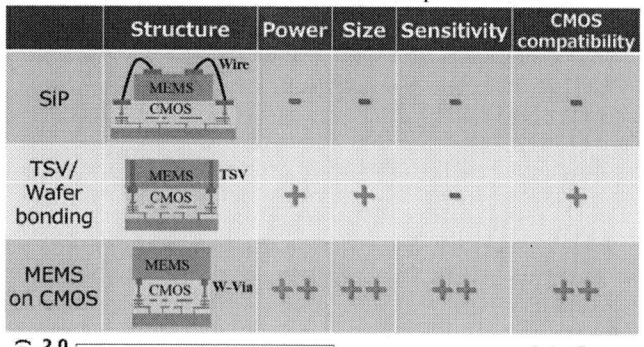

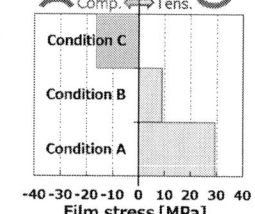

Fig.1 SiGe cantilever bending measurement without film stress control.

Fig.2 Stress measurement of SiGe film with various deposition recipes.

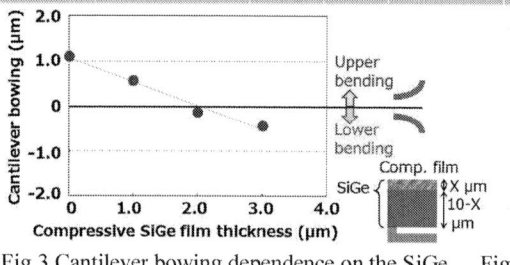

Fig.3 Cantilever bowing dependence on the SiGe compressive film. Total film thickness is 10μm.

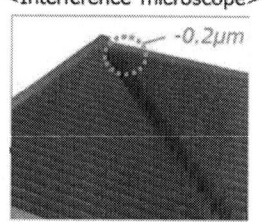

Fig.4 Stress-controlled SiGe film cantilever bending measurement.

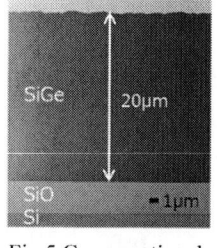

Fig.5 Cross-sectional TEM image of SiGe.

Fig.6 SIMS analysis data of the SiGe film.

Table2 Summary of SiGe film property.

SiGe	Item	Result
Film property	Stress	-70~+20MPa
	Orientation	(110)
	Resistivity	3.4E-3Ω·cm
	Ge content	70at%
	Young's modulus	140GPa

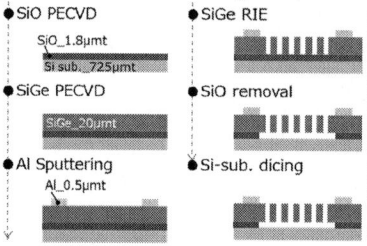

Fig.7 SiGe Acc. fabrication process.

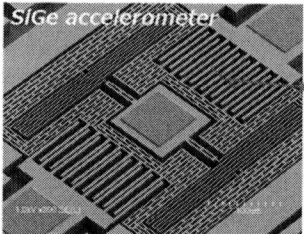

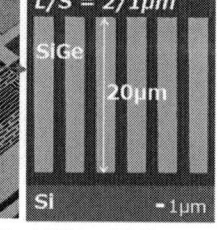

Fig.8 SEM images of SiGe ACC. and L/S.

Fig.9 Differential capacitance change of SiGe Acc. with respect to acceleration.

Fig.10 Comparison of differential capacitance of SiGe and Si Acc.. They are measured simultaneously.

Fig.11 The sensitivity for each spring width of SiGe and Si Acc..

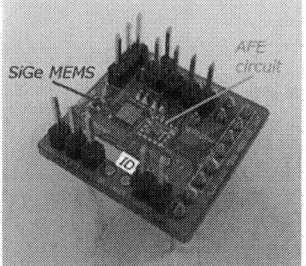

Fig.12 SiGe MEMS and AFE circuit on custom PCB.

Fig.13 Power consumption comparison between Si and SiGe Acc..

Table3 Summary of device parameters and representative measurement results.

Items		Results	
Material (Mass density, g/cm³)		Si (2.33)	SiGe (4.45)
MEMS dimension (designed)	Size(μm²)	350	
	Spring width(μm)	2.3	
	Sensing gap(μm)	1.5	
	Proof mass(μg)	4.1	7.8
Device Property (measured)	Sensor sensitivity(fF/g)	6.6	15
	Noise density(mg/√Hz) (Power consumption)	2.2 (165nW)	0.67 (169nW)

978-1-5386-4219-1/18 $31.00 © 2018 IEEE

Gap in pagination due to formatting issues.

Pages 43-44

A Comprehensive Study of Polymorphic Phase Distribution of Ferroelectric-Dielectrics and Interfacial Layer Effects on Negative Capacitance FETs for Sub-5 nm Node

Y.-T Tang[1,*], C.-J. Su[1,**], Y.-S. Wang[2], K.-H. Kao[2], T.-L. Wu[3], P.-J. Sung[1], F.-J. Hou[1], C.-J. Wang[1], M.-S. Yeh[1], Y.-J. Lee[1], W.-F. Wu[1], G.-W. Huang[1], J.-M. Shieh[1], W.-K. Yeh[1], and Y.-H. Wang[2,4]

[1]National Nano Device Laboratories; [2]Dept. of Electrical Engineering, National Cheng Kung University; Int'l College of Semiconductor Tech., National Chiao Tung University; [4]National Applied Research Laboratories, Taiwan; *Email: yttang@narlabs.org.tw, **Email: cjsu@narlabs.org.tw

Abstract

The impact of a realistic representation of gate-oxide granularity on negative-capacitance (NC) FETs at sub-5nm node is studied by a newly developed thermodynamic energy model based on the first principle calculation (FPC). For the first time, the calculation fully couples the Landau-Khalatnikov (L-K) equation with grain-size effect equation in NC-FETs. It explains the experimental results in phase transition and reveals excellent immunity against depolarization in ferroelectric (FE) layer owing to dopant concentration and stress in thin films. A sub-5nm node (L_G=10nm) NC-FET with thin FE layer (T_{FE}~2nm) is integrated to achieve low subthreshold slope (SS) of 52mV/dec via a 1.9GPa-tensor stressed interfacial layer (IL) and 12% Zr-doped HfO_2.

Introduction

Achieving higher operation speed at lower supply voltage (V_{DD}) has been the goal of Moore's law in the past decades. The power dissipation scales with V_{DD} quadratically, and V_{DD} scaling is constrained by the minimum SS. The NCFETs showing steep SS has been attracting much attention owing to the ferroelectricity in CMOS-compatible Hf-based oxide. Many groups have demonstrated NCFETs exhibiting minimum SS < 60mV/dec.[1-2] To realize sub-5nm scaling technology, the gate stack should be thinner than 2nm. Conventional DC performance simulations by TCAD for NCFETs usually assume a single-crystalline $Hf_{0.5}Zr_{0.5}O_2$ (HZO) (space group: Pca21) without considering the interface and granularity (size effects) of the gate oxide. However, due to the polycrystalline structures of HZO, the impact of grain distribution may play a role by reducing L_g, illustrated schematically by **Fig. 1**. The discrepancy between the theoretical predictions (with a single-crystalline ferro-dielectric) and experiment results can be foreseen. Different from the ordinary NCFETs with single orthorhombic state, in this work, we discuss the instability from multiphase-transition for NCFETs for the first time, and the importance of phase-stabling factor from the IL layer is also discussed.

Modeling and Experimental

In our NCFETs simulation, the electric field dependency of the ferroelectric polarization (P) is considered within Landau-Khalatnikov and Kittel equations

$$\rho \frac{dP}{dt} = -\frac{dU}{dt} + E_{field} \quad (1)$$

$$C_{NC}^{-1} = \upsilon_{FE} C_{FE}^{-1} + \upsilon_{AFE} C_{AFE}^{-1} + \upsilon_{DE} C_{DE}^{-1}$$

$$C_{FE}^{-1} = \left[2\alpha_{FE} + 12\beta_{FE}P^2 + 30\gamma_{FE}P^4 \right] T_{FE} ; C_{DE}^{-1} = T_{FE} / \varepsilon$$

$$C_{AFE}^{-1} = \left[\left(2\alpha_p + \alpha_n + 24\beta P_a^2 \right) - 48\beta P_a P + 24\beta P^2 \right] T_{FE}$$

, where α_{FE}, β_{FE}, γ_{FE} are parameters for a ferroelectric material and $\alpha_{P(N)}$ β, P_a are parameters for anti-ferroelectric (AFE) material [1]. T_{FE} denotes the thickness of the NC material and υ represents the phase fraction. To clarify the stability of ferroelectric phase of the 2nm-thick NC, a thermodynamical model [4][5] is employed to simulate the stabilization condition of metastable ferroelectric phase. According to this model, the phase within grains can be computed based on bulk free-energy (U), entropy (S), surface energy(γ), dimension of grains, and stress from doping and ILs, receptively. The free energy formula of a grain, G=U-TS+γA+PV, is used. Note these parameters are also dependent on the film composition of $Hf_{1-x}Zr_xO_2$.

Results and Discussion

A. Doping and Thickness Effects on HZO Phases and Grain Sizes

Fig. 2 shows phase map produced from the FPC. Tetragonal (t), orthorhombic (o), and monoclinic (m) indicate preferred phases for various Zr compositions with varying film thicknesses. Obviously a very small m-phase fraction (1.5%) is found in 9nm HZO. Decreasing Zr ratio makes grain distribution function departure from the o-phase window, leading to a smaller remanent polarization (P_r). These FPC results agree very well with the experimental observation [3] as shown in **Fig. 3**. Based on the discussion above, for the 2nm thin film, the expected maximal o-phase fraction of Zr/(Zr+Hf) is 12%, and which is lower than that of conventional $Hf_{0.5}Zr_{0.5}O_2$. **Fig. 5** shows the polarization versus electric field (P-E) and inverse capacitance versus polarization for o-, t-, and m-phase respectively. As reported in Ref [1], the minimal SS is obtained as $1/C_{MOS}$ tangent to $(-1/C_{NC})$. **Fig. 6** shows the polarization dependence of $1/C_{MOS}$ and $(-1/C_{NC})$ with different IL and FE-layer thickness. **Fig. 6**(a) shows 9nm FE-layer with 0.4nm IL leading to the minimum SS. On the other hand, **Fig. 6**(b) reveals IL thickness can be also used to tune $1/C_{MOS}$ to match $(-1/C_{NC})$ for the minimal SS.

B. IL Effects on HZO Stress and Grain Sizes

Fig. 7 shows the phase diagram of 2 and 7nm HZO thin films with various grain size and stress. The circle-line refers to the phase-transition boundary. In the case without stress, 2nm and 7nm HZO shows the AFE and FE-like (t/o-ratios=3/7) behaviors, respectively. With increasing the tensile stress, the transition from o- to t-phase became more pronounced, and this is consistent with experimental P-E data of the IL layer of Al_2O_3 (**Fig. 7** inset). The compressive stress presents a tendency of t→o transition. According to the molecular dynamics of FPC, ILs of SiO_2 and GeO_2 characteristically provide 1.8 and 2.4GPa compressive stress on HZO. It can be seen that, at 7nm HZO, the volumetric ratio of o-phase between GeO_2 (91%)/SiO_2 (78%) is 1.2, and at 2nm, ratio increases to 20 (87% vs 4.2%). Strong P_r behavior has been evidently shown in Ge FinFETs [2]. Dopant concentration is an alternative way to enlarge o-phase window. **Fig. 8** shows phase diagram for 2 and 7nm HfO_2, compared with HZO, a maximum polarization (71%) in 2nm HfO_2 is achieved with a tensile stress of 1.9GPa.

C. Phase Distribution on Electrical Characteristics

Fig. 9 shows I_D-V_G characteristics of NCFETs with T_{FE}=2 and 7nm and various o-/t-phase ratio. Here we focus o-/t- ratio because the HZO electrical characteristics are dominated by o- and t-phases in this size. Obviously, the phase ratio of 1/0 shows the maximum I_{ON}/I_{OFF} with SS_{min}=52mV/dec in 2nm and SS_{min}=13mV/dec in 7nm. The more t-phase results in higher SS_{min}. **Fig. 10** shows charge density of the NCFETs with T_{FE}=2 and 7nm. The optimal o/t ratio to achieve SS_{min}<60mV/dec is 0.9 and 0.85 for T_{FE}=2 and 7nm, respectively.

Conclusions

A design guideline of 2nm FE-$Hf_{1-x}Zr_xO_2$ NCFETs with strained ILs has been reported based on the developed thermodynamic energy model. The stress analysis reveals stable o-phase in $Hf_{0.88}Zr_{0.12}O_2$ with tensor stress while $Hf_{0.50}Zr_{0.50}O_2$ with compressive stress, which cannot be simulated by conventional TCAD. With careful design of the doping concentration and strained ILs, a novel NCFETs is expected to operate at a quarter-volt for the application of low-power IoT.

Acknowledgment

The authors would like to thank Prof. M. A. Alam, and Prof. C. Hu for the valuable comments. This work was support in part by the Ministry of Science and Technology, Taiwan (MOST-106- 2633-E-009 -001 and MOST 106-2221-E-492-034).

References

[1] K. Karda et al., APL, vol. 106, p.163501, 2015. [2] C.J. Su et al., VLSI2017 p.152. [3] Nano Lett. , 2011, 11 (9), pp 3983, 2012. [4] M. H. Park et al., Nanoscale, vol.9, p.9973, 2017. [5] R. Materlik et al., JAP, vol. 117, p. 134109, 2015.

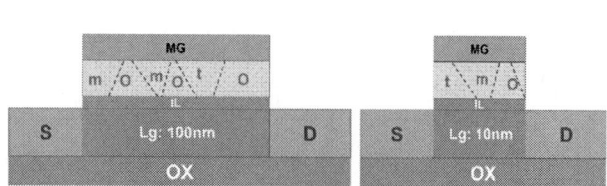

Fig. 1 Schematics of NC-FETs with polymorphic FE gate stack in long and nano-scaled L_g.

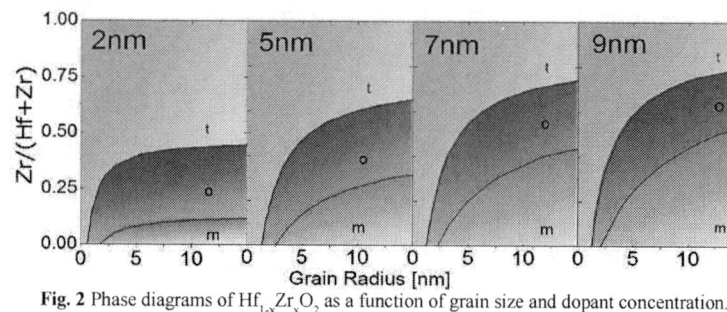

Fig. 2 Phase diagrams of $Hf_{1-x}Zr_xO_2$ as a function of grain size and dopant concentration.

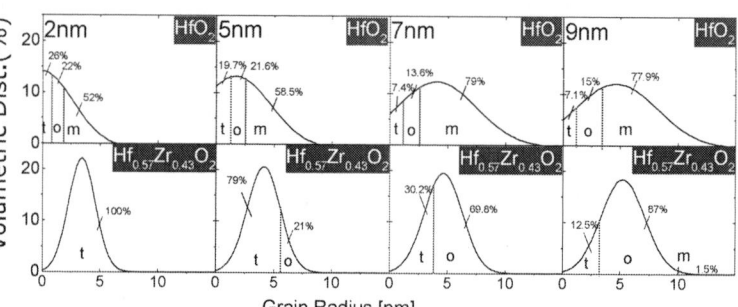

Fig. 3 Grain size distribution with critical grain size (vertical lines) at various thickness of HfO_2 and $Hf_{0.57}Zr_{0.43}O_2$, respectively. A very small m-phase fraction (1.5%) of HZO is found in 9nm, agreeing well with the experimental data.

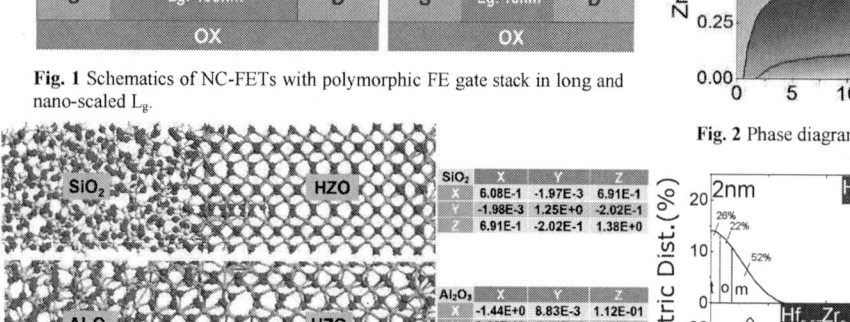

Fig. 4 Relaxed unit cell of HZO on ILs (a) SiO_2 (b) GeO_2 and (c) Al_2O_3. The tables show the stresses induced from SiO_2 is about 1.08 GPa (compressive), Al_2O_3 is about 2.46 GPa (tensile), and GeO_2 is about 2.44 GPa (compressive).

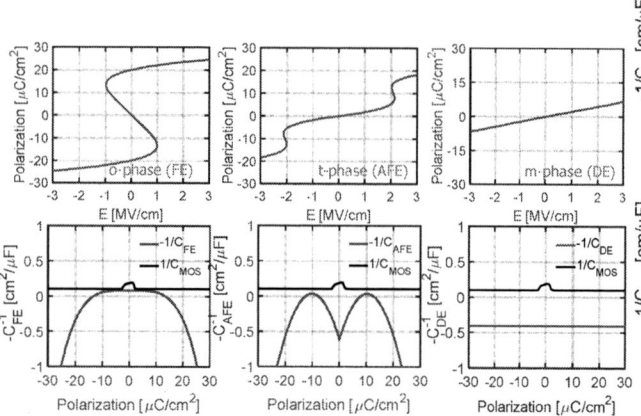

Fig. 5 L-K curves of measured polarization vs. electric field for 9nm HZO film with IL=0.4nm (SiO_2).

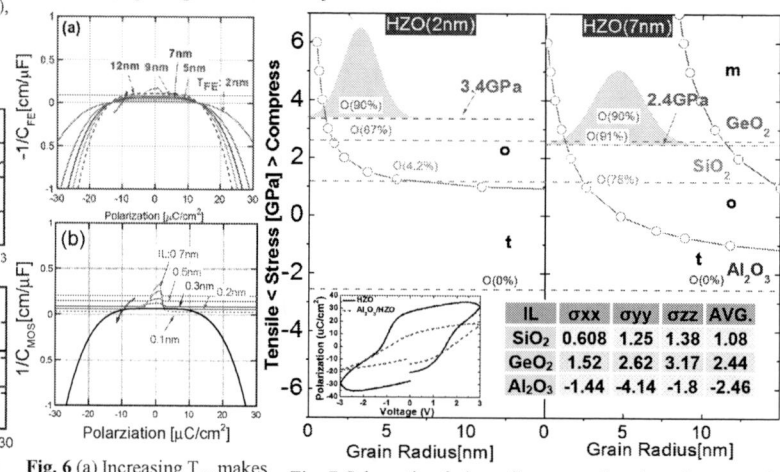

Fig. 6 (a) Increasing T_{FE} makes $1/C_{FE}$ tangent to $1/C_{MOS}$ as at $T_{FE}=9nm$ (min. SS). In contrast, lowering IL in (b) does as well.

Fig. 7 Schematic of phase diagram as function of stress and grain size distribution in 2nm and 7nm HZO. Dashed lines refer to the stress of ILs, as computed by MD-FPC. Inset shows less P-V of Al_2O_3/HZO than HZO experimentally. The unit in stress is GPa.

IL	σ_{xx}	σ_{yy}	σ_{zz}	AVG.
SiO_2	0.608	1.25	1.38	1.08
GeO_2	1.52	2.62	3.17	2.44
Al_2O_3	-1.44	-4.14	-1.8	-2.46

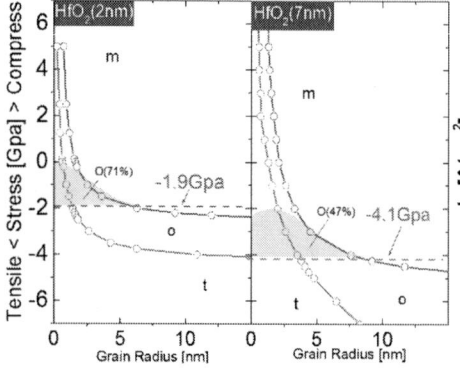

Fig. 8 Phase diagrams as function of stress and grain size distribution in 2nm and 7nm HfO_2. Dashed lines give the maximum fraction of o-phase and the required stress.

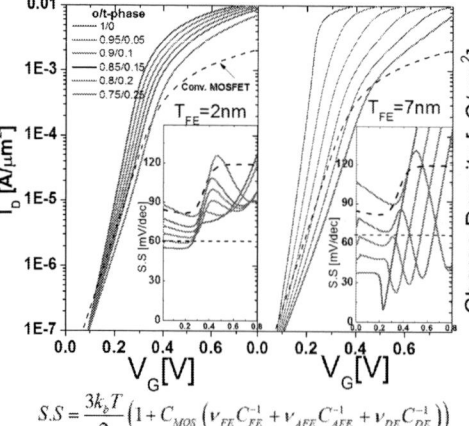

$$S.S = \frac{3k_bT}{2}\left(1 + C_{MOS}\left(\nu_{FE}C_{FE}^{-1} + \nu_{AFE}C_{AFE}^{-1} + \nu_{DE}C_{DE}^{-1}\right)\right)$$

Fig. 9 I_D-V_G characteristics of T_{FE}=2nm and 7nm in the NCFETs and the related SS equation.

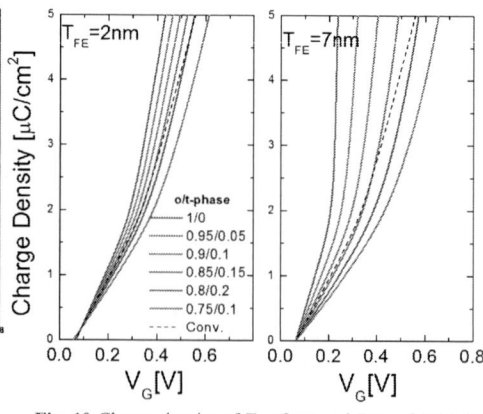

Fig. 10 Charge density of T_{FE}=2nm and 7nm of HZO in the NCFETs.

First Experimental Demonstration of Negative Capacitance InGaAs MOSFETs With Hf$_{0.5}$Zr$_{0.5}$O$_2$ Ferroelectric Gate Stack

Q. H. Luc[1], C. C. Fan-Chiang[1], S. H. Huynh[1], P. Huang[1], H. B. Do[1], M. T. H. Ha[1], Y. D. Jin[1], T. A. Nguyen[1], K. Y. Zhang[1], H. C. Wang[1], Y. K. Lin[1], Y. C. Lin[1], C. Hu[1,2], H. Iwai[1,3], and E. Y. Chang[1]

[1] National Chiao Tung University, Hsinchu, Taiwan. E-mail: edc@mail.nctu.edu.tw
[2] University of California at Berkeley, Berkeley, USA. [3] Tokyo Institute of Technology, Yokohama, Japan.

Abstract

We demonstrate, for the first time, the negative capacitance (NC) In$_{0.53}$Ga$_{0.47}$As nMOSFET with 8-nm Hf$_{0.5}$Zr$_{0.5}$O$_2$ (HZO) as ferroelectric (FE) dielectric for sub-60 mV/dec subthreshold swing (SS). The impact of annealing treatments on the FE properties and electrical characteristics of NC InGaAs nMOSFETs are investigated. Optimized annealing condition results in NC effects at the HZO/Al$_2$O$_3$/InGaAs nMOSFETs with steep SS property (~ 11 mV/dec).

Introduction

Ferroelectric (FE) materials with negative capacitance (NC) effects have been proposed to satisfy the quests of steep subthreshold swing (SS) and low supply voltage operation for enabling high-performance low-power sub-7 nm technology node [1]. NC field-effect transistors (FETs) have been reported on not only conventional Si but also on high carrier mobility Ge-based devices with SS < 60 mV/dec [2]-[12]. In$_x$Ga$_{1-x}$As materials are considered to replace Si as channel materials in nFETs due to their extraordinarily high electron mobility. There is an effort to realize NC InGaAs nMOSFET by using La$_2$O$_3$ gate oxide; unfortunately, steep SS can be observed only at low temperature due to inadequate quality of high-k/InGaAs interface [13]. In this work, we fabricate and examine ferroelectric Hf$_{0.5}$Zr$_{0.5}$O$_2$ (HZO) NC InGaAs nMOSFETs under different annealing conditions. The NC In$_{0.53}$Ga$_{0.47}$As nMOSFET exhibits sub-20 mV/dec SS which is the first report for III-V-based nMOSFETs.

Device Fabrication

Fig. 1(a) presents the key fabrication steps of NC InGaAs nMOSFETs using a gate last process. For ferroelectric gate oxide, an 8-nm HZO (Hf/Zr = 1/1) film was deposited using an Ultratech Fiji G2 ALD system with an Al$_2$O$_3$ (~1 nm) interfacial layer. Post deposition annealing was performed at 500 °C in forming gas ambient to achieve high interface quality of high-k/III-V MOS structure [14]. The 100-nm TiN gate electrode was formed and the post metallization annealing (PMA) was applied with different annealing temperatures and durations in N$_2$ ambient. The TiN/HZO/Al$_2$O$_3$/In$_{0.53}$Ga$_{0.47}$As MOSCAPs were processed using identical PMA conditions as the MOSFETs to investigate the FE properties of HZO film as seen in Fig. 1(b). For comparison, MOSFETs and MOSCAPs were also fabricated without the PMA as the control samples.

Results and Discussion

The polarization-voltage (P-V) hysteresis and the current response to a triangular voltage excitation measurements on the HZO/Al$_2$O$_3$/InGaAs MOSCAPs for various PMA conditions are shown in Fig. 2(a) – (d) and (e) – (h), respectively. The FE properties and polarization switching behaviors were observed for all samples. Although the control sample exhibited the FE behaviors after the post deposition annealing process, the PMA conditions showed a great impact on the FE properties. Comparing to the control sample, the remnant polarization

(2P$_r$) increased by more than 4 times as PMA at 600 °C for 60 sec (Fig. 3). The P$_r$ to coercive field (E$_c$) ratio is similar for different PMA conditions, attributing to the large E$_c$ obtained with high annealing temperature for long duration (Fig. 4). This may be related to the dielectric leakage issues. The HRTEM and the fast Fourier transformation (FFT) (insets) images seen in Fig. 5 reveal the polycrystalline nature of HZO films with clear interfaces in TiN/HZO/Al$_2$O$_3$/InGaAs structures. In Fig. 6, the GIXRD spectra confirm the polycrystalline structure with a mixture of monoclinic and orthorhombic phases for all samples. Noting that in the control sample the large grain size can be the reason regarded to the weak FE properties (Fig. 2(a)).

The transfer and point SS characteristics of the In$_{0.53}$Ga$_{0.47}$As MOSFETs are shown in Fig. 7(a) – (d) and (e) – (h), respectively. Despite good I$_{DS}$-V$_{GS}$ behavior, due to the poor FE characteristics, NC effects cannot be observed for the control sample (Fig. 7(a) and (e)). The observed I$_{DS}$-V$_{GS}$ hysteresis (positive ΔV_{th}) in the control sample is caused by the charge trapping/detrapping effect due to insufficient annealing of the HZO thin film without PMA process. The PMA process, on the other hand, helps to not only enable the NC behaviors evidenced by steep SS and counter-clockwise ΔV_{th} in I$_{DS}$-V$_{GS}$ characteristics but also increase the drive current by more than three times. As depicted in Fig. 8, abrupt SS accompanies with the large ΔV_{th} for higher temperature and longer annealing time which is caused by the unmatched C$_{FE}$ and C$_{MOS}$ [15]. This is the first experiment demonstrates InGaAs NC MOSFET and the ΔV_{th} can be improved by the optimization of the HZO's thickness and annealing conditions. PMA at 500 °C-30 sec resulted in similar forward and reverse SS along with 5 times reduction in ΔV_{th} (Fig. 9). It has to be mentioned that in this work the NC effect was prohibited for PMA's temperatures below 500 °C (data not shown). Fig. 10 presents the electrical characteristics of NC InGaAs nMOSFET with PMA at 500 °C for 30 seconds in N$_2$ ambient. Very steep SS exhibited over 3 orders of I$_{DS}$ magnitudes at V$_{DS}$ of 0.05 and 1 V with $|\Delta V_{th}|$ of 0.7 V. Peak polarization (P) transition was also observed in G$_M$ characteristics at V$_{DS}$ of 0.05 and 1 V for either forward or reserve sweeps. The P transition locates at the same V$_{GS}$ where the sudden switching appears. This is an evidence for the NC effects observed at the transistor with ferroelectric layer. Fig. 11 and Table 1 benchmark device parameters with the state-of-the-art reports which demonstrate steep SS negative capacitance FETs at room temperature.

Conclusion

NC InGaAs nMOSFETs with FE Hf$_{0.5}$Zr$_{0.5}$O$_2$ gate dielectric for steep SS characteristics (SS$_{for}$ = 30mV/dec and SS$_{rev}$ = 11 mV/dec) are experimentally demonstrated, for the first time. Sub-60 mV/dec property achieved by annealing treatment in InGaAs nMOSFETs is beneficial for enabling III-V NC FETs for future high-speed low-power CMOS logic applications.

978-1-5386-4219-1/18 $31.00 © 2018 IEEE

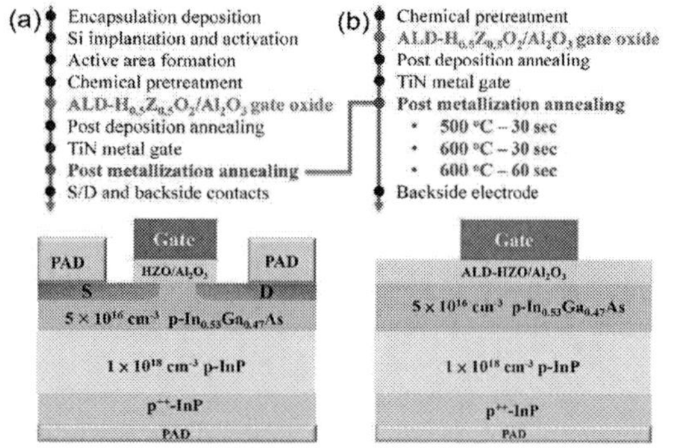

Fig. 1. Process flow and schematic of the FE InGaAs (a) NC nMOSFETs and (b) MOSCAPs.

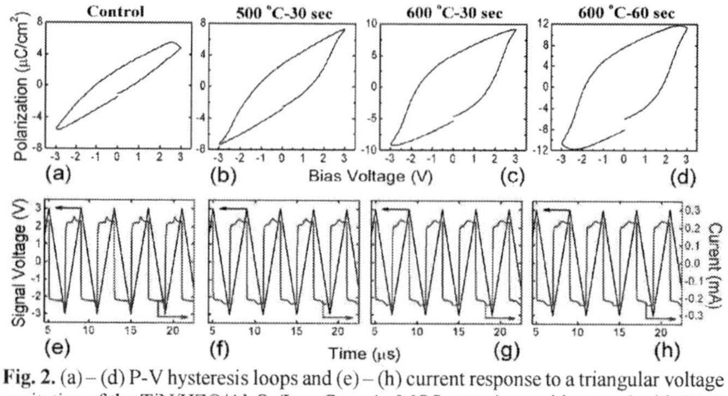

Fig. 2. (a)–(d) P-V hysteresis loops and (e)–(h) current response to a triangular voltage excitation of the TiN/HZO/Al$_2$O$_3$/In$_{0.53}$Ga$_{0.47}$As MOS capacitors without and with PMA treatments.

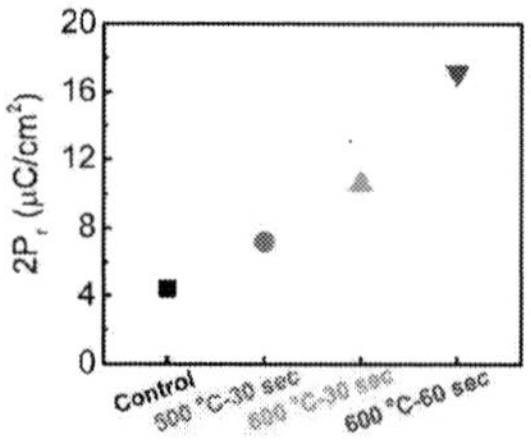

Fig. 3. Remnant polarization for samples without and with PMA treatments.

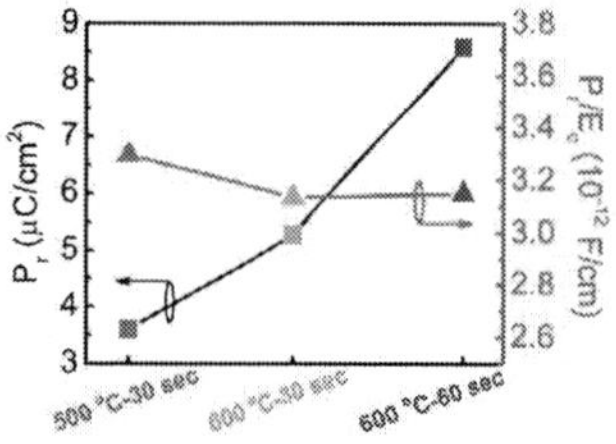

Fig. 4. P$_r$ and P$_r$/E$_c$ ratio for different PMA conditions.

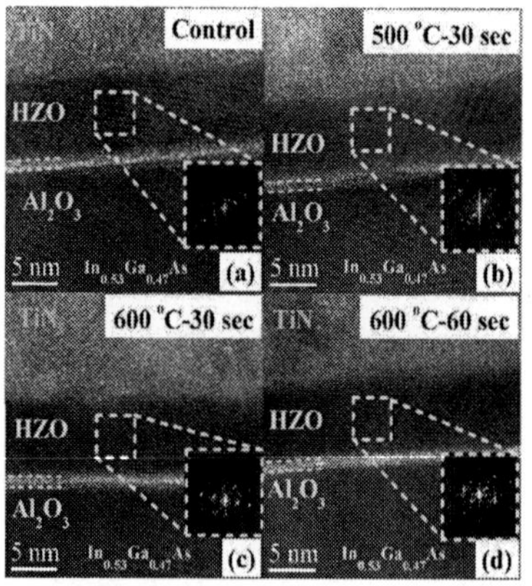

Fig. 5. HRTEM and FFT (insets) images of the MOS structures with ferroelectric ALD-HZO (~ 8 nm)/Al$_2$O$_3$ (~ 1 nm).

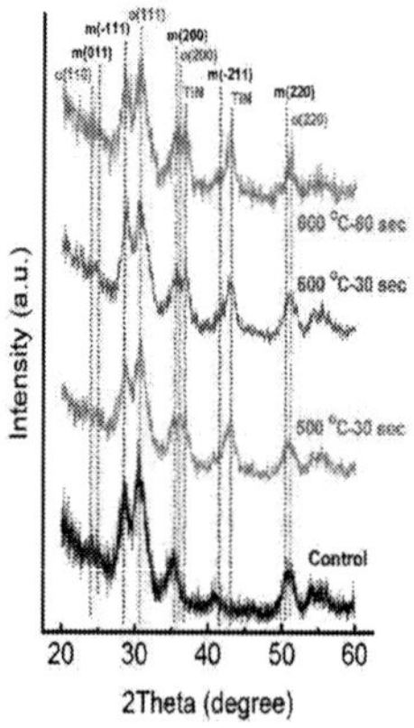

Fig. 6. GIXRD of HZO thin films without and with PMA treatments.

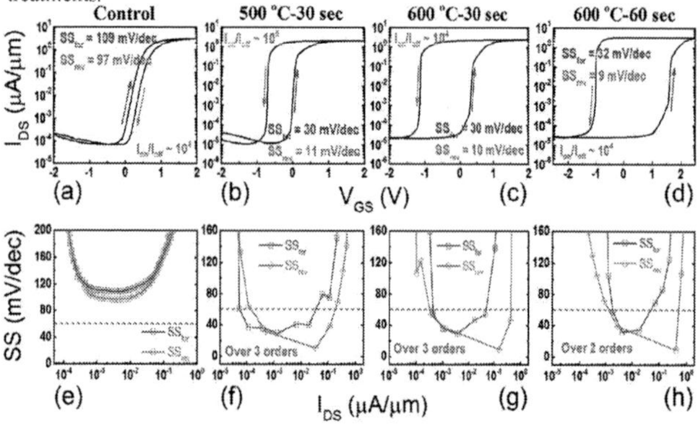

Fig. 7. (a)–(d) Transfer and (e)–(h) point SS characteristics of the TiN/HZO/Al$_2$O$_3$/In$_{0.53}$Ga$_{0.47}$As control and NC MOSFETs at V$_{DS}$ = 0.05 V. The fabricated nMOSFETs feature gate length (L$_G$) and gate width (W$_G$) of 6 and 100 μm, respectively.

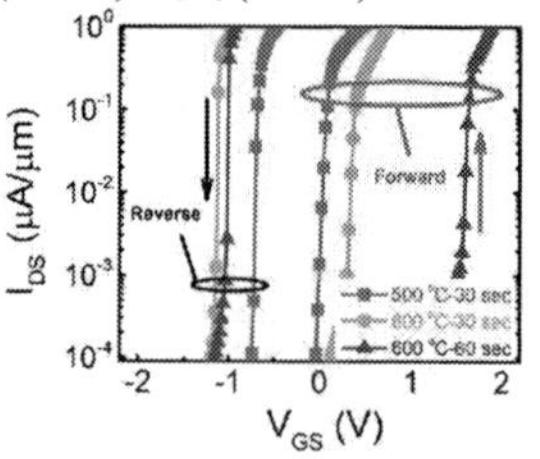

Fig. 8. Transfer characteristics of PMA samples.

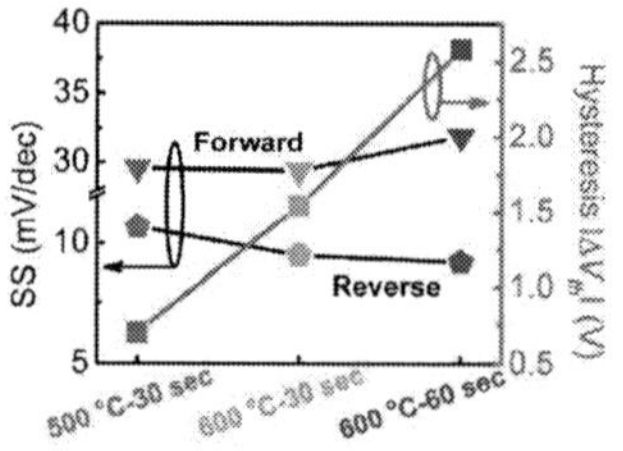

Fig. 9. Minimum SS and |ΔV$_{th}$| versus annealing conditions.

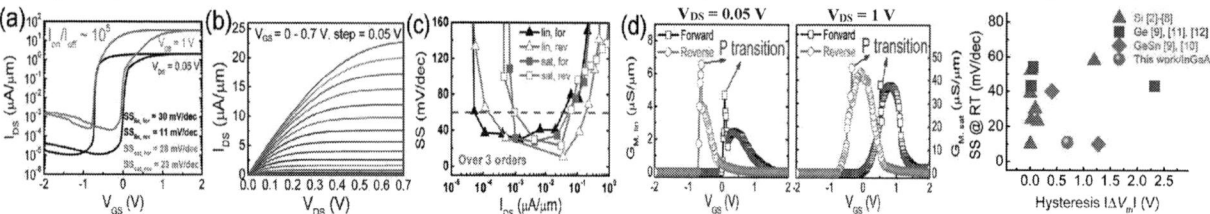

Fig. 10. (a) Transfer, (b) output, (c) point SS and (d) G$_M$ characteristics of FE In$_{0.53}$Ga$_{0.47}$As NC nMOSFET underwent PMA at 500 °C for 30 second.

Fig. 11. Sub-60 mV/dec SS at room temperature versus |ΔV$_{th}$|.

TABLE I. Benchmark of devices parameters for the FE InGaAs NC MOSFET achieved in this work and other literature reports.

	Substrate	Material	L$_G$ (μm)	|ΔV$_{th}$| (V)	SS$_{min}$*2 (mV/dec)	I$_{on}$/I$_{off}$	I$_{on}$*3 (μA/μm)	Range*4
pMOS	Si [3]	HZO/30 nm	10	1.2	58	8	N/A	N/A
	Ge [9]	HZO/6.5 nm	5	2.34	43	3	22	1 order
	GeSn [10]	HZO/6 nm	3	1.28	10	3	15	2 orders
	GeOI [11]	HZO/10 nm	0.105	4.3	7	5	20	1 order
nMOS	Si [5]	HZO/9.5 nm	10	0.14	23	8	N/A	4 orders
	Si [8]	HAO/7 nm*1	30	0.004	39	5	N/A	2 orders
	GeOI [11]	HZO/2 nm	0.302	0.017	43	5	N/A	1 order
	This work	HZO/8 nm	6	0.7	11	5	41	3 orders

(*1) HAO/Hf$_x$Al$_{1-x}$O$_2$; (*2) SS$_{min}$ at V$_{DS}$ = 0.05 V; (*3) I$_{on}$ at V$_{DS}$ = V$_{OV}$ = ± 1 V; (*4) Sub-60 mV/dec over the range of I$_{DS}$ magnitude.

References: [1] S. Salahuddin *et al., Nano Lett.*, vol.8, 2008, p.405; [2] C. H. Cheng *et al., IEEE Electron Device Lett.*, vol.35, 2014, p.274; [3] K. S. Li *et al., IEDM2015*, p.620; [4] Y. C. Chiu *et al., Phys. Status Solidi RRL*, vol.11, 2017, p.1600368; [5] M. H. Lee *et al., IEDM2015*, p.616; [6] M. H. Lee *et al., IEEE Electron Device Lett.*, vol.36, 2015, p.294; [7] M. H. Lee *et al., IEDM2016*, p.306; [8] C. C. Fan *et al., IEDM2017*, p.561; [9] M. H. Lee *et al., IEDM2017*, p.565; [10] J. Zhou *et al., IEDM2016*, p.310; [11] J. Zhou *et al., IEEE Electron Device Lett.*, vol.38, 2017, p.1157; [12] W. Chung *et al., IEDM2017*, p.365; [13] C. Y. Chang *et al., IEDM2016*, p.322; [14] Q. H. Luc *et al., Jpn. J. Appl. Phys*, vol.53, 2014, p. 04EF04; [15] A. I. Khan *et al., IEDM2011*, p.255.

978-1-5386-4219-1/18 $31.00 © 2018 IEEE

Response Speed of Negative Capacitance FinFETs

Daewoong Kwon[1*], Yu-Hung Liao[1*], Yen-Kai Lin[1], Juan Pablo Duarte[1], Korok Chatterjee[1], Ava J. Tan[1],
Ajay K. Yadav[1], Chenming Hu[1], Zoran Krivokapic[2] and Sayeef Salahuddin[1]

[1]Department of Electrical Engineering and Computer Sciences, University of California, Berkeley, CA 94720, U.S.A.
[2]GlobalFoundries Inc; * these authors contributed equally to this work; E-mail: sayeef@berkeley.edu

Abstract

We report on the measurement of a 101-stage ring oscillator (RO) consisting of state-of-the-art 14 nm FinFET devices with a ferroelectric gate layer that exhibits negative capacitance. We show that the gate stage delay as a function of applied voltage can be directly modeled from DC characteristics of the individual NC-nFET and NC-pFET devices that constitute the RO, thereby demonstrating that there is no slowdown of the NC effect at the highest speed tested - per-stage delay as small as 7.2 ps.

Introduction

Negative Capacitance (NC) [1]-[5] promises to significantly reduce supply voltage requirement in CMOS technology. One question is often raised as to the limit of speed of NC effect in NCFETs. Although theoretical analysis [4] has shown that the intrinsic response time is expected to be <0.5 ps, it is only very recently that ROs have been demonstrated with transistors that show clear NC effect [6]. We note here that ROs are the most appropriate platform to study the speed of NCFETs as it directly probes the large signal response of the devices. In this context, we study the fabricated ROs at high speed (delay<10ns). The postulate is that if the NC gate stack cannot respond to these frequencies, one should observe an extra slowdown beyond the usual circuit simulation. We follow a simple procedure to test this hypothesis. We have first developed a compact model based on the BSIM CMG framework that fits the DC transfer and output characteristics of the NCFETs. Next, this compact model is fed into a RO simulator to extract the capacitances in the RO by matching the measured speed and I_{DDA} (active current drawn by the RO). Subsequently, the measured delay, I_{DDA}, R_{eff}, and C_{eff} reaching up to delay<10ns are compared with simulation. We find that the DC models accurately predict these parameters, thereby demonstrating that the circuit behavior at these high frequencies is reflective of the DC characteristics. This in turn indicates that there is no discernible delay that can be attributed to the speed limit of the NC effect down to the shortest measured delay, which was ~7.2 ps in our case.

Experiments, Results and Discussions

State-of-the-art replacement gate 14 nm FinFET devices with doped HfO_2 ferroelectric layer of 3 nm thicknesses were fabricated in a GlobalFoundries process. 101 stage FO3 ring oscillators were used to study RO performance. The highest voltage to be applied (lowest delay measured) was chosen so that it does not damage the devices. Representative DC characteristics (V_{th}-matched) of an NC-nFET and NC-pFET with that of a control (having much thinner dielectric layer) are shown in Fig. 1(a). The steepening of the subthreshold swing is clearly evident for both cases (despite a thicker dielectric). Note that, for comparison, simulation of a nFET and pFET with 3 nm gate dielectric are also included. Fig. 1(b) shows the regions of operation in terms of the ferroelectric 'S' curve. At small V_G, the load line is still in the NC region, leading to steeper swing. At large V_G, the load line goes beyond the NC region, leading to a smaller ON current; here the larger thickness of the dielectric limits the ON current. In [6], to distinguish this behavior from those devices that are still in the NC region in the ON condition, such devices were termed 'PCFET'. Nonetheless, the important point here is that as the voltage is swept up and down, one traverses in the 'S' curve following ABC-CBA (Fig. 1(b)). This path dictates the amount of I_{DDA} at a given voltage. We also note that the RO is made

with only n-type work function gate metal. As a result, the V_{th} is not matched and V_{th} for the NC-pFET is significantly larger than the NC-pFET.

We start with a BSIM-CMG model development for the devices. Fig. 2 and 3 show the model calibration to data. Excellent agreement between the model and the measurement is obtained. Next, 101 stage FO3 ring oscillators are simulated. Fig. 4 shows simulated and measured I_{DDA} for a voltage range, where the maximum value of the voltage (V_{DD}) is low enough that the p-FET is still in the NC region but n-FET is beyond the NC region. This means that as the voltage goes up and down the RO will always be in the NC region for at least one type of devices. The I_{DDA} simulated at such V_{DD} closely matches the measured values (Fig. 4a). This clearly shows the efficacy of the DC model in this low voltage region. We next proceed to higher voltages. Fig. 5 shows the comparison between measured and simulated I_{DDA} as a function of V_{DD}. The agreement, observed in Fig. 5, between simulation and measured data is a critical result as I_{DDA} is strongly related to characteristics such as I-V and C-V; the agreement shows that I-V and C-V characteristics remain the same at these high frequencies as DC. Moreover, the simulated C_{eff} and R_{eff} (Fig.5) match well with the measured data. The measured R_{eff} is extracted by the relation of $R_{eff} = V_{DD}/(I_{DDA}-I_{DDQ})$ and the C_{eff} is calculated as C_{eff} = (delay time)/R_{eff}. Finally, gate stage delay is plotted in Fig. 6. The measured speed can be well predicted by simulation. One important observation is that the C_{eff} increases linearly with V_{DD} (Fig. 5(b)). In conventional ROs, this is a signature of the fact that the devices can respond to input signal changes at the oscillation frequency. As mentioned before, the nFET and pFET did not have matched V_{th}. To see the effect of this mismatch, we have simulated the ROs with the same device parameters but by artificially matching their V_{th}. This significantly improves delay in the low V_{DD} region (Fig. 7). At the very high speeds, the effect is smaller, indicating that it is the series resistance and parasitic capacitance that limits the highest speed (8.5 ps) rather than the intrinsic device. Finally, we measured ROs made of 4 fin devices that provide larger ON current. The ROs behave similarly (Fig. 8) to 2 fin ROs and a gate delay as small as 7.2 ps can be measured (Fig.9).

Conclusions

We have demonstrated that the behavior of the RO based on 14 nm NCFETs can be predicted from DC characteristic of the individual devices, just like conventional CMOS technology, thereby demonstrating that there is no extra slowdown attributable to NC effect even at the very high frequencies used in this study. Stage delay time up to 7.2 ps has been measured for FO3 inverters and limited by resistance and capacitances of the circuit as usual. Therefore, well optimized NCFETs should operate at very high frequencies at reduced power dissipation.

References: [1]. S. Salahuddin et al., *Nano Lett.*, p.405, 2008. [2] K. S. Li et al., *IEDM*, p. 22.6.1, 2015. [3] A. I. Khan et al., *EDL*, p. 111, 2016. [4] D. Kwon et al., *EDL*, p. 1, 2018. [5] Chatterjee et al, 38.9 (2017): 1328-1330. [6] Z. Krivokapic et al., *IEDM*, p. 15.1.1, 2017.

2018 Symposium on VLSI Technology Digest of Technical Papers

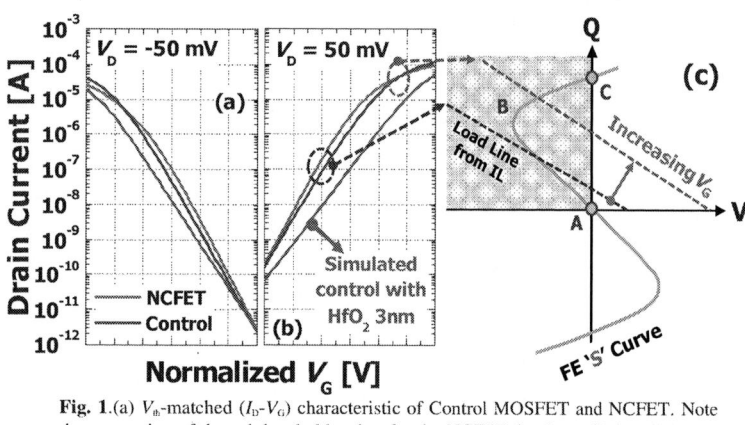

Fig. 1. (a) V_{th}-matched (I_D-V_G) characteristic of Control MOSFET and NCFET. Note the steepening of the subthreshold swing for the NCFET despite a thicker dielectric layer. For comparison, simulated characteristics of devices with 3 nm DE are included (b) The 'S' curve of the Ferroelectric showing the region of operation. At low V_G, the load line intersects in the negative capacitance region leading to steeper swing. At higher V_G, the load line intersects in a region where capacitance is positive. As the voltage is swept up, the device traverses following the ABC path of the S curve and when it is swept down, it follows the CBA.

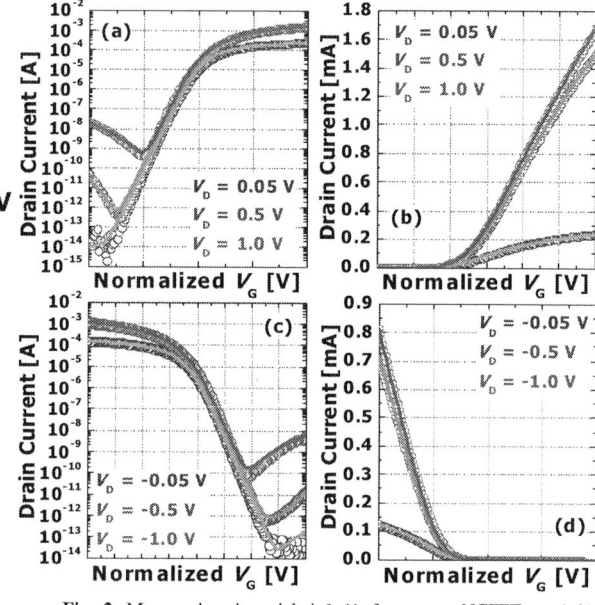

Fig. 2. Measured and modeled I_D-V_G for n-type NCFET on (a,b) semilog, linear scale and (c,d) for a p-type NCFET for semilog, linear scale. Dotted and solid lines represent measurements and modeling, respectively.

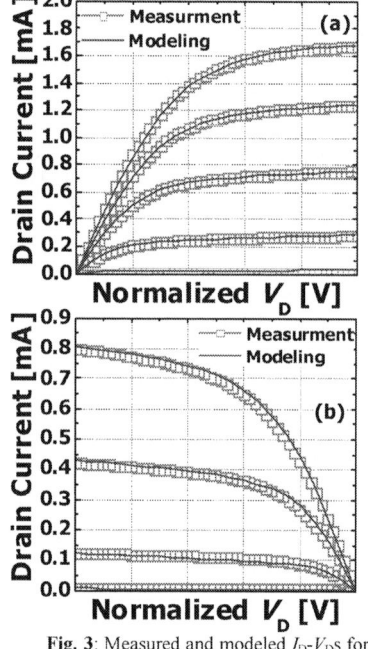

Fig. 3: Measured and modeled I_D-V_{DS} for (a) n-type and (b) p-type NCFETs at various V_Gs.

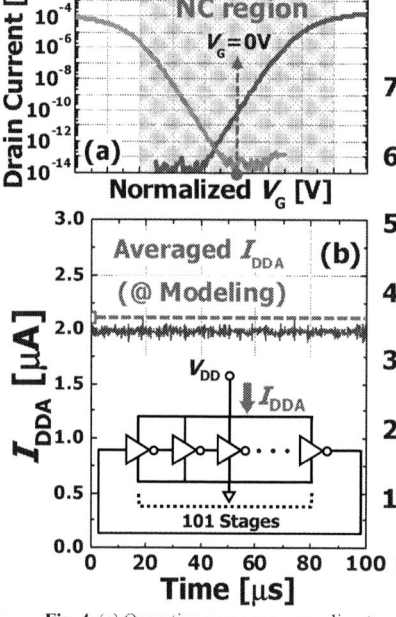

Fig. 4: (a) Operation range corresponding to a V_{DD} where pFET remains in NC regime. (b) Simulated and measured I_{DDA} at V_{DD} indicated in (a).

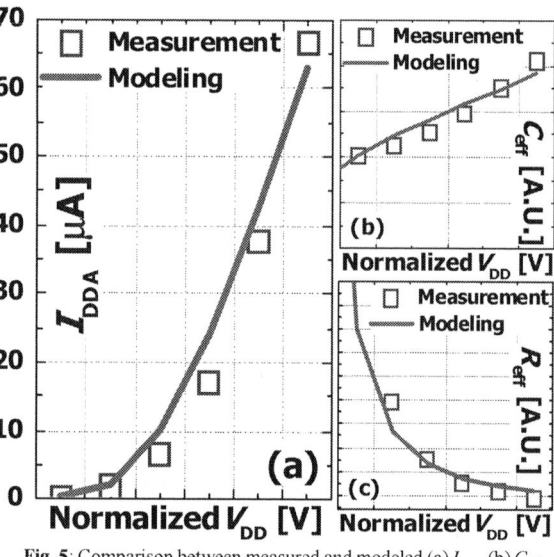

Fig. 5: Comparison between measured and modeled (a) I_{DDA}, (b) C_{effs}, and (c) R_{effs} as a function of V_{DD}. I_{DDA} is average value of measured current while oscillator is ON.

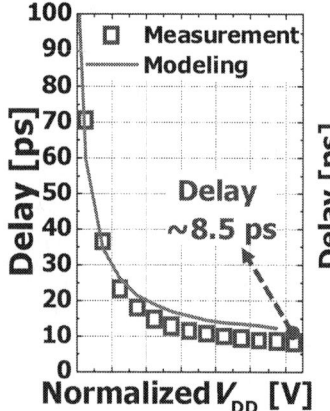

Fig. 6: Measured and modeled delays Vs. V_{DD} curves for 2 fin NCFET ring oscillator.

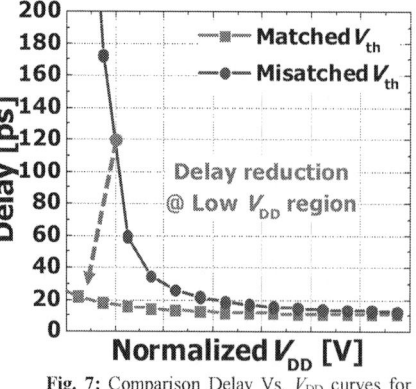

Fig. 7: Comparison Delay Vs. V_{DD} curves for the measured oscillator which has mismatched V_{th} between n and pFETs with simulated behavior for the same oscillator at matched V_{th}.

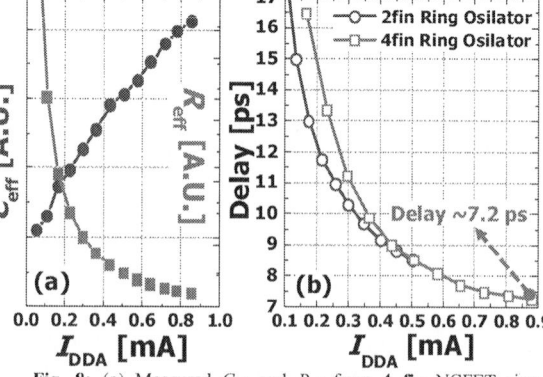

Fig. 8: (a) Measured C_{eff} and R_{eff} from **4 fin** NCFET ring oscillator. Behavior of C_{eff} and R_{eff} follow expected trends. (b) Delay Vs. I_{DDA} curves for 2 fin and 4 fin NC-FinFET ring oscillators. Stage delay as small as 7.2 ps was measured.

978-1-5386-4219-1/18 $31.00 © 2018 IEEE

Ferroelectric Switching Delay as Cause of Negative Capacitance and the Implications to NCFETs

B. Obradovic, T. Rakshit, R. Hatcher, J. A. Kittl, M. S. Rodder

Samsung Advanced Logic Lab, Austin, TX 78754

b.obradovic@samsung.com

Abstract

We report on measurements and modeling of FE HfZrO/SiO$_2$ Ferroelectric-Dielectric (FE-DE) FETs which indicate that phenomena attributed to Negative Capacitance can be explained by a delayed response of ferroelectric domain switching. No traversal of the stabilized negative capacitance branch is required. Modeling is used to correlate the hysteretic properties of the ferroelectric material to the measured transient and subthreshold slope (SS) behavior. It is found that steep SS can be understood as a transient phenomenon, present only when significant polarization changes occur. The technological implications of this finding are investigated, and it is found that NCFETs are most likely not suitable for high-performance CMOS logic, due to voltage, frequency, and voltage polarity limitations.

Introduction

Since Negative Capacitance (NC) was proposed [1], numerous groups have reported FE-DE-FETs with sub-60 mV/dec SS and inductive-like transient pulse responses of FE-DE cap stacks [2-4]. However, the theoretical explanation of the observed behavior is not fully settled. The prevailing theory is that of the stabilized P-V S-curve, a negative capacitance path available to ferroelectrics only with suitable stabilization, and devoid of domain switching [1]. The S-curve theory rests on a number of untested assumptions, but most troubling is the lack of a microscopic model of the stabilization (Fig. 1). More recently, it has been proposed that domain switching is in fact required for NC to be observed, and that the S-curve is never traversed [3, 5]. In this paper, we propose that NC phenomena are caused by the delay in polarization switching [6], without the traversal of the S-curve (Fig. 2), based on analysis and modeling of prior [3] and new data for FE-DE (HfZrO/SiO$_2$) devices.

Measurements and Analysis

We analyze two categories of data: transient pulsing of FE-DE Cap stacks (Fig. 3), and the SS of FE-DE-FETs as a function of peak voltage (V_{peak}) and time. Transient pulsing of FE-DE Caps produces the well-known spike behavior [3]; instead of the expected monotonic voltage increase, an initial voltage pulse is formed, followed by a more gradual rise to full voltage (Fig. 4). This behavior is qualitatively straightforward to explain using our delay model. The initial sharp rise in the voltage takes place prior to ferroelectric switching, while only the non-ferro portion of the polarization is active. After a brief delay, the ferroelectric domains begin to switch, leading to a large increase in polarization. This is balanced by an increased current into the FE-DE Cap, resulting in a voltage drop across the stack. Once the initial delay is over, the FE-DE Cap stack more gradually charges to the final voltage. During the onset of ferroelectric switching, apparent negative capacitance is observed, but only for sweeps with sufficiently high V_{peak} (Fig. 4). The reason is clear from the associated hysteresis (Fig. 5): at low V_{peak}, the loop becomes too tight for a meaningful negative capacitance region to form (in samples with wide V_C distributions/hysteresis loops, and/or longer timescales, discrete domain switching may be seen at lower V_{peak}). It can also be seen that bipolar and unipolar input waveforms result in very different behavior: only bipolar inputs produce spikes (Fig. 6). After the initial bipolar to unipolar transition, no additional spikes are observed. This can also be understood from the associated hysteresis (Fig. 7); only the bipolar loop produces a large change in polarization. The unipolar loop is very tight and no appreciable ΔP occurs. Without a large and delayed increase in polarization, no spike is produced. While the unusual behavior of pulse trains is interesting, of more immediate technological relevance is the SS behavior of FE-DE FETs. As shown in Figs. 8, 9, and 10, the polarization delay effect can produce sub-60 mV/dec SS. Due to the delay in FE-DE Cap polarization, a large ΔQ can take place even with minimal (or no) ΔV_g, forcing the surface potential ψ to change more rapidly than the applied V_g. Since the phenomenon is ultimately driven by polarization switching, it is V_{peak} dependent in the same way as the pulse train response. This is verified experimentally in Fig. 11, in which it is apparent that sub-60 mV/dec SS is observed only for sweeps with V_{peak} above the FE-DE Cap coercive voltage. Due to delay, sub-60 SS may manifest at low voltages, as long as V_{peak} is sufficiently high. For $V_{peak} < \sim 2V$ the SS response follows standard electrostatics. Observed discrete sub-60 events are caused by discrete domain switching (Fig. 11 inset), contrary to the prediction of the S-curve model. The transient nature of the effect suggests that it only manifests in a range of input switching frequencies: if too fast, there is no ferroelectric response; if too slow, the behavior is quasistatic. The rate of the ferroelectric response determines the available frequency range for NC. As seen in Fig. 12, the polarization response is sensitive to timescale and V_{peak}. At the relatively low voltages compatible with modern CMOS, the polarization response times are measured in ms or slower. Thus, sub-60 mV/dec SS is practically attainable only at very low switching frequencies. This is evidenced in Fig. 13, which shows the measured SS on a single FE-DE FET as a function of the sample time. It can be seen that sub-60 behavior is observed only well below kHz frequencies for HfZrO FE materials. Furthermore, at even lower frequencies (e.g. DC), the response once again loses the NC signature; this is the quasistatic regime in which the FE delay is negligible compared to the input frequency. Finally, it should be noted that the hysteretic behavior of FE-DE FETs is complicated by the presence of traps in the FE layer (Fig. 14); the latter tend to produce clockwise hysteresis, partially or wholly cancelling the counter-clockwise FE hysteresis (lack of hysteresis suggested to indicate stabilized S-curve behavior [2]). The effect is V_{peak} and time dependent, and at high V_{peak}, the FE hysteresis is seen to dominate (large ΔP). The presence of traps likewise tends to diminish the NC effect on rising pulses, where the effective Vt is increasing due to electron trapping (Fig. 15).

Conclusions

Pulse and SS responses (V_{peak} and time-dependent) of FE-DE Caps and FETs are consistent with a delayed polarization switching model. A stabilized S-curve model is not needed to and cannot explain the data. The distinction is of great technological significance; while the "stabilized S-Curve FE-DE FET" is straightforwardly applicable to CMOS circuits (with the proviso of sufficiently rapid response), the "switching delay FE-DE FET" is not usable as a sub-60 SS logic FET. The requirement of polarization switching makes steep slope low-VDD operation unlikely, the frequencies at which steep slopes are available are many orders of magnitude slower than needed. Furthermore, the lack of unipolar NC response is challenging. Additional work is needed to confirm that indeed the NC effect is driven by polarization switching (Fig. 16). If so, useful applications of FE-DE FETs will be as NVMs, but likely not logic FETs.

[1] S. Salahuddin et al, *Nano Lett. 8*, (405) 2008. [2] J. Jo et al, *IEEE Elec Dev Lett 37*, (245) 2016. [3] P. Sharma et al., *IEEE Elec Dev Lett 39*, (272), 2018. [4] Z. Krivokapic et al., *IEEE IEDM* 2017. [5] C.M. Krowne et al., *Nano Lett. 11*, (988) 2011. [6] B. Obradovic *et al.*, https://arxiv.org/abs/1801.01842 (2018).

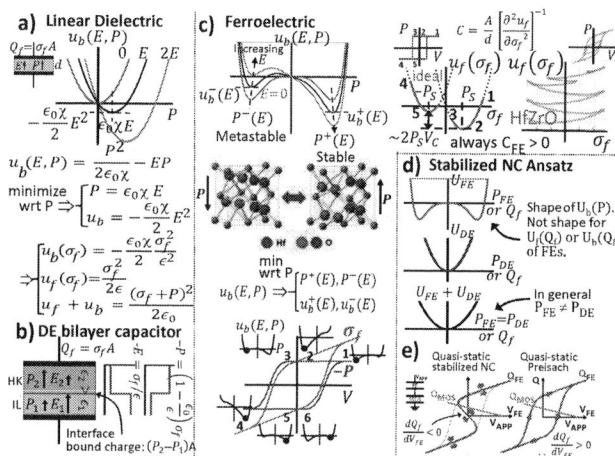

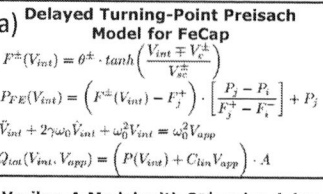

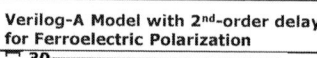

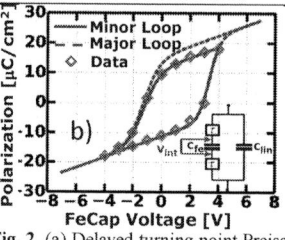

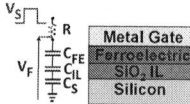

Fig. 1. Both linear DEs (a) and FEs (c) minimize their internal energy u_b ($\neq u_f$ energy of free charge config.) adopting their stable or meta-stable P, under external fields or free charges Q_f. The energy dependence on P is different than on Q_f. (b) In bilayers (shown for IL/HK), P varies from layer to layer, with discontinuity at interface (in epitaxial thin perovskite FE-DE multilayers, strong coupling can lead to $P_1=P_2$, and dielectric like behavior, but in this case $u=u_1(P_1)+u_2(P_2)+u_{coupl}(P_1,P_2)$, $\partial^2 u/\partial P_1^2$ (fixed P_2) >0, $\partial^2 u/\partial P_2^2$ (fixed P_1) >0 and total cap positive) . (d) Neg cap Ansatz assumes either same polarization for two layers or functional form for $U(Q_f)$ not applicable to FEs. (e) Quasi-static neg. cap trajectory ("S-curve") and physical hysteretic behavior.

Fig. 2. (a) Delayed turning point Preisach model used in simulations to match and explain experimental data. (b) Example of P on Q HfZrO MIM caps. The dynamic params. (natural frequency ω_0 and damping ratio γ) are calibrated with FE-DE stack transient data (Figs. 3, 4).

Fig. 3. Experimental setup for transient pulse measurements. FE-DE Cap stack: FE layer (HfZrO), non-FE dielectric (IL), and Si substrate. Voltage pulses (V_S) are applied to the stack through an access resistor while stack voltage (V_F) is monitored.

Fig. 4. Measured and simulated transient response for the capacitor stack subject to a bipolar square-pulse excitation. "Anomalous" spikes in the early part of the response to each pulse are clearly visible in data and sim.

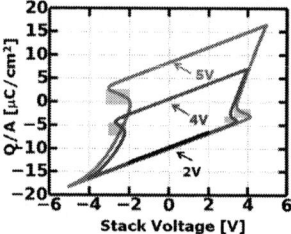

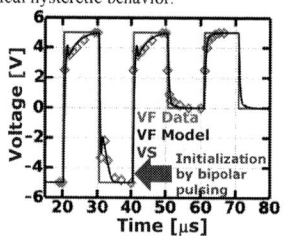

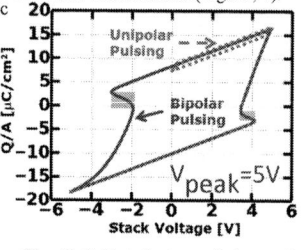

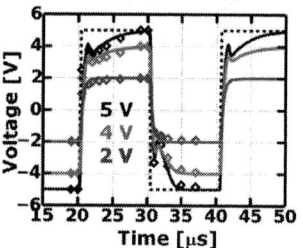

Fig. 5. The simulated hysteresis for bipolar switching for several V_{peak} values is shown. Regions of apparent negative capacitance ($dQ/dV < 0$) are shown shaded. At 5V and 4V, there are clear NC regions (though transient in nature), while the 2V hysteresis loop is too tight to produce NC.

Fig. 6. Measured and simulated transient response for the capacitor stack subject to unipolar square-wave pulses. The stack is initialized by a sequence of bipolar pulses (shaded blue region), followed by a sequence of unipolar pulses. On the first unipolar pulse, the "anomalous" spike is observed, but it is absent on subsequent pulses.

Fig. 7. P-V trajectory of the stack during bipolar and unipolar pulsing. Initial pulses are bipolar (blue solid line), establishing a wide loop with a large ΔP. Subsequent pulses are unipolar (red dashed line), resulting in a tight minor loop with minimal FE ΔP. No apparent NC is exhibited on the minor loop.

Fig. 8. Surface potential ψ and gate voltage V_G for a series of bipolar switching events. As each switching event starts, ψ initially tracks V_G slowly, since the FE domains are not yet switching. When the domains start to switch, ψ changes rapidly. Average slope during FE switching is indicated by the black dashed line. Quasistatic ψ ($\omega_0 \to \infty$) is shown (red dashed line) for reference.

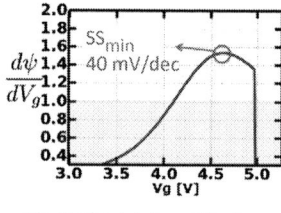

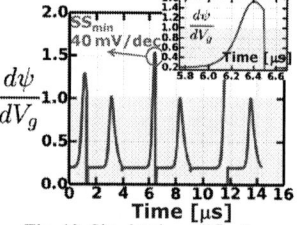

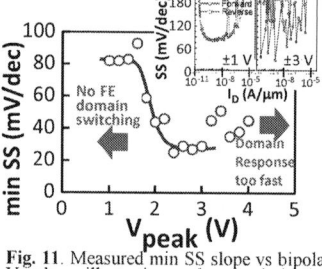

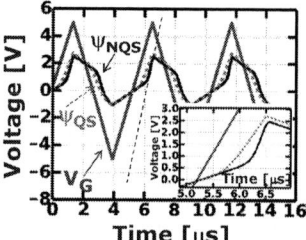

Fig. 9. Simulated amplification as a function of Vg, showing the small voltage interval where sub-60 is achievable on a µs timescale at a high bipolar $V_{peak}= +/-5V$.

Fig. 10. Simulated amplification as a function of time showing brief periods where sub-60 is achievable on a µs timescale at a high bipolar $V_{peak} = +/-5V$.

Fig. 11. Measured min SS slope vs bipolar V_{peak}, illustrating the polarization switching driven sub-60 behavior. Full bipolar Id-Vg shown in Fig. 15. Sub-60 data points (inset) attributed to discrete domain switching events.

Fig. 12. Measured polarization vs pulse width (FE HfZrO MIM) as a function of V_{peak} illustrates slow dynamics of polarization switching, clearly detriment for logic devices.

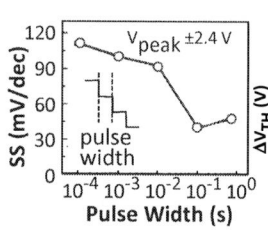

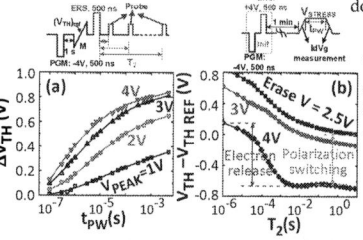

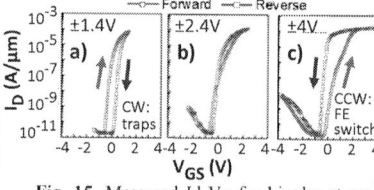

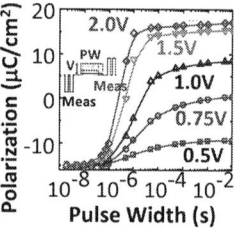

Fig. 13. Measured SS slope as a function of pulse-width duration showing timescale for sub-60 behavior only achievable at a ms or slower timescale, precluding fast logic applications.

Figs. 14. a) Measured Vt shift of FET vs pulse width, as a function of V_{peak}. b) Decoupling of charge-trap-detrap effects and polarization switching due to different time scales of these effects, also highlighting the detrimental slow dynamics of polarization switching and sub-60.

Fig. 15. Measured Id-Vg for bipolar stress, a) at low (+/-1.2V) V_{GS} range there is no sub-60 behavior and hysteresis is dominated by trapping-detrapping (clockwise), b) at +/- 2.4V, sub-60 events observed and cancellation of hysteresis due to overlap of trapping-detrapping and FE polarization switching hysteresis, c) at +/- 4V, stronger sub-60 due to FE polarization switching which now dominates hysteresis (CCW).

	Sub-60 conditions		
	Switching	S-Curve	Exper.
Peak $V_{FE} \ll V_c$	No	Yes	No
Peak $V_{FE} > V_c$	Yes	Yes	Yes
$f < f_{FE}$	Yes	Yes	Yes
$f \gg f_{FE}$	No	Yes	No
Unipolar	No	Yes	No

(note: f_{FE} ~ kHz to MHz for HfZrO)

Fig. 16. Conditions for observing sub-60 SS behavior in the switching model and the S-curve model, with experimental data (this work). V_{FE} is the voltage across the FE layer. V_c is the coercive voltage. f_{FE} is the response frequency at the given V_{FE}.

Negative Capacitance, n-Channel, Si FinFETs: Bi-directional Sub-60 mV/dec, Negative DIBL, Negative Differential Resistance and Improved Short Channel Effect

Hong Zhou, Daewoong Kwon, Angada B. Sachid, Yuhung Liao, Korok Chatterjee, Ava J. Tan,
Ajay K. Yadav, Chenming Hu, and Sayeef Salahuddin

Department of Electrical Engineering and Computer Sciences, University of California, Berkeley, CA 94720, U.S.A.
E-mail: sayeef@eecs.berkeley.edu

Abstract

We report on negative capacitance (NC) FinFETs with ferroelectric $Hf_{0.5}Zr_{0.5}O_2$ (HZO) as gate dielectric on fully depleted silicon on insulator (FDSOI) substrate with various channel length (L_{CH}) of 450 nm to 30 nm and multiple fin widths (W_{FIN}) of 200 nm to 30 nm. We demonstrate all signature characteristics expected from NCFET: nearly hysteresis free operation (~3 mV), <60 mV/decade subthreshold swing (SS) with an average SS of 54.5 mV/dec for ~2 orders of I_D and to the best of our knowledge, for the first time in Si MOSFETs, negative Drain Induced Barrier Lowering (DIBL) and Negative Differential Resistance (NDR). Remarkably, we observe significant improvement in the short channel effect compared to control FinFETs: both SS and DIBL are substantially lower for the NCFET for the same L_{ch}/W_{Fin} ratio. Importantly, these benefits become increasingly larger for shorter channel lengths.

Introduction

Negative capacitance (NC) effect promises to provide a boost in the surface potential of a MOSFET that can steepen the subthreshold swing, even below the limit of 60 mV/decade and thereby make it possible to reduce supply voltage requirements [1]-[7]. NC effect relies on the matching of the negative differential capacitance of the Ferroelectric 'S' curve with other positive capacitances of the FET. This means that as the channel is scaled down, the increasing drain capacitance could help capacitance-matching as the drain capacitance softens the otherwise strongly voltage-dependence capacitance of the FET. Consequently, NC effect can effectively counteract the degradation of SS with scaling. In addition, the coupling through the drain capacitance also leads to the effect of negative DIBL that counteract the increase of DIBL with scaling. Thus, NC effect can significantly improve the short channel effect (SCE) in a MOSFET. While a lot of the work on NCFET has focused on the achievement of <60 mV/dec in the SS, systematic study of improved short channel effect in scaled transistor is lacking. In this work, we explore NC FinFETs as a function of L_{CH}/W_{Fin} ratio down to a channel length of 30 nm. We find that for all values of L_{CH}/W_{Fin}, NC FinFETs show improved SS and DIBL compared to control FinFETs.

Experiments

The key fabrication processes for HZO NC-FinFET and HfO_2 FinFET are shown as Fig. 1(a). Device fabrication starts with mesa isolation, source/drain As^+ ion implantation, Fin definition, atomic layer deposition (ALD) dielectrics and source/drain and gate metallization. L_{CH} and W_{Fin} are scaled from 450 nm to 30 nm and 200 nm to 30 nm, respectively. Both devices share the same fabrication process except the ALD dielectric steps, namely NC-FinFET has 4 nm of $Hf_{0.5}Zr_{0.5}O_2$ and FinFET has 4 nm of HfO_2 as the reference device. Fig. 1(d) is a false-colored SEM image of a fabricated HZO NC-FinFET with L_{CH}/W_{Fin}=450 nm/30 nm. Note that for all transistors, NCFET or control, the FIN is lightly doped.

Results and Discussions

Fig. 2(a) shows the I_D-V_{GS} of a HfO_2 FinFET and HZO NC-FinFET with high L_{CH}/W_{Fin}=450 nm/30 nm at V_{DS}=0.05 V. NC-FinFET shows a steeper slope and higher on-current compared with HfO_2 FinFET. The extracted SS~I_D for two devices is shown in Fig. 2(b), confirming the reduced SS of NC FinFET. Meanwhile, an average SS=54.5 mV/dec is extracted for 2 decades of I_D at a I_D level of 10^{-7} ~ 10^{-5} μA/Fin, which is 1~3 orders of magnitudes higher than the gate leakage current (I_G). Fig. 3(a) describes the I_D-V_{GS} plots of the NC-FinFET with L_{CH}/W_{Fin}=450 nm/50 nm, showing sub-60 mV/dec SS and negative DIBL effect. The corresponding NDR effect of the same device at V_{GS}=0.2 V is shown in Fig. 3(b). Both negative DIBL and NDR are observed. Fig. 4 shows output characteristics. Shown in Fig. 5 is a representative measurement of double sweep from a device with dimension L_{CH}/W_{Fin}=150 nm/40 nm. Very small anti-clockwise Hysteresis (3 mv) hysteresis is observed. Other devices also possess similarly small anti clockwise hysteresis. Fig. 6(a) and 6(b) show the I_D-V_{GS} plots of HfO_2 FinFET and HZO NC-FinFET with short channel lengths (L_{CH}/W_{Fin}= 30/60 nm) at V_{DS}=0.5 V. Again, a smaller SS for NCFETs is observed. Importantly the improvement in SS exists for a large voltage range, even beyond the threshold voltage, such that the increase in ON current is visible even in the log scale. This, in conjunction with steeper SS gives ~100X increase in the ON/OFF ratio at V_{DS}=0.5 V and V_{GS}=-0.6 V for the NCFET at L_{CH}=30 nm. Fig. 7 shows the L_{CH} and W_{Fin} dependence of the average SS at V_{DS}=0.05 V. Many devices show average SS < 60 mV/dec including down to 60 nm channel length.

Fig. 8(a) and 8(b) show the SS and DIBL as a function of channel lengths for W_{Fin}=60 nm. For all channel lengths a significant improvement in SS and DIBL are seen for the NCFETs. Fig. 9(a) and 9(b) show SS and DIBL as a function of L_{CH}/W_{Fin} for varying W_{fin}. Note that the SS hovers around close to 60 mV/dec value for L_{CH}/W_{Fin}~1. For a L_{CH}/W_{fin}= 30nm/60nm =0.5, which for a conventional FinFET will be considered a really bad electrostatic design, NC FinFET still has an average SS~85 mV/dec. Indeed, a corresponding control FinFET shows SS~175 mV/dec.

Conclusions

In conclusion, we demonstrated NC-FinFETs that show <60 mV/decade, negative DIBL and NDR, increased ON current and ~100X increase in ON/OFF ratio. Importantly, for short channel length transistors (~L_{CH}=30 nm), we show that NC FinFETs provide significantly improved SS and DIBL at the same L_{CH}/W_{FIN}. This improvement in the short channel effect indicates that it should be possible to relax the stringent requirement on the aspect ratio (L_{ch}/W_{FIN}) needed for scaling at the ultra scaled nodes. Ability to use a larger W_{FIN} could also alleviate series resistance problem. In addition to reducing the supply voltage requirement, NC effect is therefore expected to help scaling continue beyond the end of the current roadmap.

References: [1]. S. Salahuddin et al., *Nano Lett.*, 2008. [2] K. S. Li et al., *IEDM*, 2015. [3] A. I. Khan et al., *EDL*, 2016. [4] M. Si et al., *Nature Nanotech.*, 2017. [5] D Kwon et al., *EDL* 2018. [6] C. Hu et al., *DRC*, 2015. [7] V. Hu et al., IEDM, 2017.

2018 Symposium on VLSI Technology Digest of Technical Papers

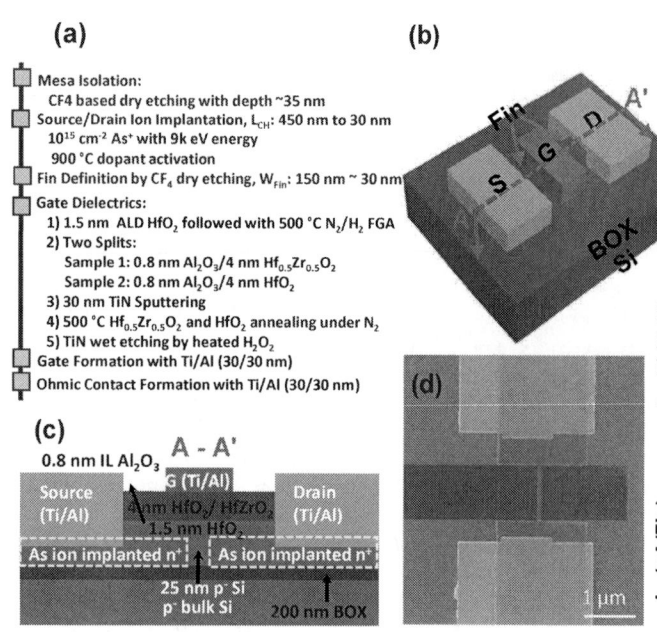

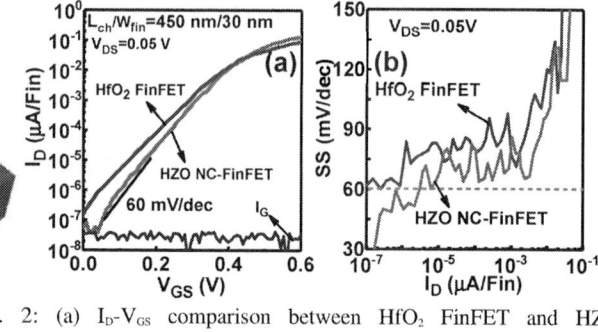

Fig. 2: (a) I_D-V_{GS} comparison between HfO$_2$ FinFET and HZO NC-FinFET with high L_{CH}/W_{Fin}=450 nm/30 nm at V_{DS}=0.05 V and (b) Extracted SS~I_D of two devices, showing the reduced SS of NC FinFET. An average SS=54.5 mV/dec for around 2 decades of I_D.

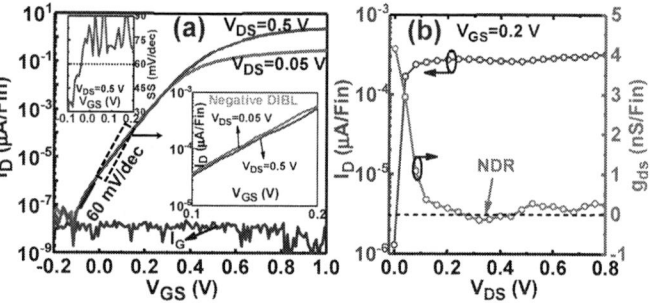

Fig. 1: (a) Description of fabrication processes of HfO$_2$ FinFET and HZO NC-FinFET, (b) Tilted top view of the device schematic on fully depleted SOI (FDSOI) substrate, (c) Cross-section schematic view of the devices with each layer thickness marked, and (d) False-colored scanning electron microscopy (SEM) image of a fabricated device with W_{fin}=30 nm.

Fig. 3: (a) I_D-V_{GS} plot of HZO NC-FinFET with L_{CH}/W_{Fin}= 450 nm/60 nm, showing SS<60 mV/dec and negative DIBL and (b) I_D-g_{ds}-V_{DS} plot of the same device. Negative differential resistance (NDR) is observed at V_{GS}=0.2 V.

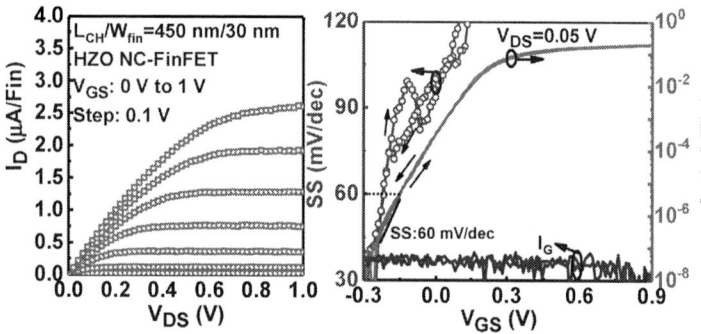

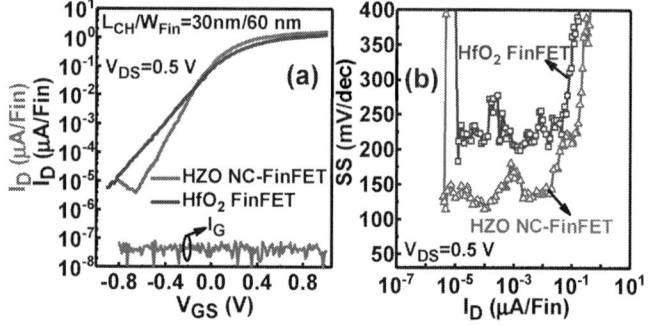

Fig. 4: I_D-V_{DS} characteristics of HZO NC-FinFET with L_{CH}/W_{fin}= 450 nm/30 nm.

Fig. 5: Anti clockwise, small (3 mV) hysteresis NC-FinFET I_D-V_{GS} with dual sweep SS<60 mV/dec and L_{CH}/W_{fin} = 150 nm/40 nm.

Fig. 6: (a) I_D-V_{GS} plots of short channel HfO$_2$ FinFET and HZO NC-FinFET with low L_{CH}/W_{fin}= 30/60 nm and (b) Extracted SS at V_{DS}=0.5 V, showing less SS for HZO NC-FinFET, verifying that NC can minimize short channel effect (SCE) at small L_{CH}.

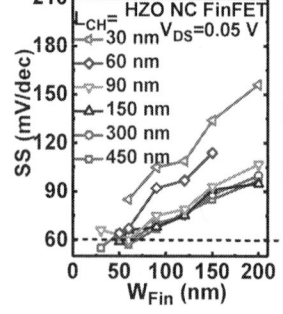

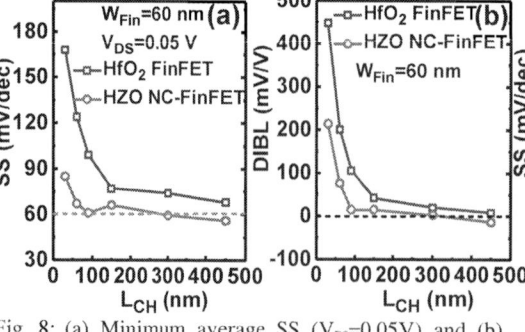

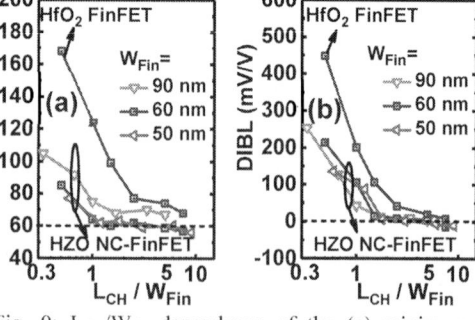

Fig. 7: W_{Fin} dependence of the minimal average SS (V_{DS}= 0.05V) of NC-FinFET for various L_{CH}. Many devices show SS < 60 mV/dec.

Fig. 8: (a) Minimum average SS (V_{DS}=0.05V) and (b) DIBL scaling metrics for HfO$_2$ FinFET and HZO NC-FinFET with W_{Fin}=60 nm, verifying that HZO NC-FinFETs suffer from less SCE. **Minimal average SS is extracted from averaging SS among one decade of I_D.**

Fig. 9: L_{CH}/W_{fin} dependence of the (a) minimum average SS at V_{DS}=0.05V and (b) DIBL comparisons between HZO NC-FinFETs with various W_{Fin}. For comparison data from control FinFET with HfO$_2$ dielectric is shown (W_{Fin}=60 nm).

978-1-5386-4219-1/18 $31.00 © 2018 IEEE

Gap in pagination due to formatting issues.

Pages 55-58

True 7nm Platform Technology featuring Smallest FinFET and Smallest SRAM cell by EUV, Special Constructs and 3rd Generation Single Diffusion Break

WC Jeong, S. Maeda, HJ Lee, KW Lee, TJ Lee, DW Park, BS Kim, JH Do*, T Fukai, DJ Kwon, KJ Nam, WJ Rim*, MS Jang*, HT Kim*, YW Lee*, JS Park, EC Lee*, DW Ha, C.H. Park, H.-J. Cho, S.-M. Jung and H.K. Kang

Semiconductor R&D Center, *Foundry Business, Samsung Electronics, San#16, Banweol-Dong, Hwasung-City,
Gyeonggi-Do, 445-701, Republic of Korea, email: wcjeong@samsung.com

Abstract

7nm platform technology that takes full advantage of EUV lithography was developed, where EUV was straightforwardly used for single patterning of MOL and BEOL, not just as a means for cutting of SADP/SAQP. The combination of 27nm fin pitch (FP) and 54nm contacted poly pitch (CPP) as well as the high density SRAM cell size of 0.0262 um^2 is the smallest in the reported FinFET platform. Further scaling is secured with special constructs and the 3rd generation single diffusion break. Full working of 256M bit SRAM and large-scale logic test chip was demonstrated with guaranteed reliability.

Introduction

FinFET has been the main stream technology for logic platform [1-6]. However, manufacturing cost of cutting-edge logic technology has not been reduced along with scaling since around 28nm due to the introduction of multiple patterning, such as LE^n (Litho-Etch^n), SADP (Self-Aligned Double Patterning), SAQP (Self-Aligned Quadruple Patterning), and cut technology. To overcome this, EUV lithography was developed. In this paper, we describe our true EUV 7nm platform technology, where the advantages of EUV were fully utilized in single patterning for MOL and BEOL [5]. By further aggressive FEOL scaling from the previous technology, this technology achieved smallest transistor size and smallest SRAM cell among the reported technologies[1-5]. In addition, special constructs and the 3rd generation single diffusion break (SDB) were enabled for further area scaling. Through this 7nm platform technology, cost-effectiveness of cutting-edge technology will revive in the semiconductor industry.

Process Key Features

Fig. 1 shows the scaling trend based on CPP and Mx pitch with reported technologies. Fig. 2 indicates that this work offers the smallest transistor size among reported FinFET platform technologies. Table I shows technology comparison with previous technologies [1-5].

Result and Discussion

AC performance will be 20-30% better in speed and 30-50% better in power, demonstrated with various library sets as shown in Fig. 3. By applying EUV single patterning, BEOL uniformity dramatically improved as shown in Fig. 4. EUV single patterning realizes superior 2D fidelity as shown in Fig. 5, suggesting strong immunity to weak points in random layout. Unlike low-k$_1$ lithography, EUV does not use strong resolution enhancement technology, but rather uses small wavelength light to resolve small patterns. This is a big advantage for yield ramp up of random logic products.

Special constructs that connect two contacts locally have been enabled for vertical scaling down. They make 10% scaling down by lowering track height, possible only through EUV. Fig. 6 shows allowed patterns which are designated for wider yield window. Single diffusion break was also developed for further scaling. The 3rd generation SDB has reached the ultimate form of SDB with stable layout effect. As shown in Fig. 7, it offers 15% horizontal scaling down by reducing CPP grid. Fig. 8 indicates more stable layout dependence over various layouts compared with the 2nd generation [6].

Fig. 9 is a butterfly curve of the SRAM cell and SRAM cell sizes in the reported FinFET platform technologies, showing that the SRAM cell size of this work is the smallest in the world [1-5]. Fig. 10 is the shmoo plot from a 256M bit SRAM array and a wide operation margin was confirmed. Fig. 11 shows the pass/fail dies of SRAM 256Mb. Vmin/Vmax margin are also confirmed to be wide.

Fig. 12 is a micrograph of a logic test chip, including CPU and GPU cores. Operation was fully confirmed. Fig. 13 shows the shmoo plot of the GPU core, indicating a wide operation margin. Reliability is also well guaranteed. Although the sidewall surface orientation of {110} is not a good orientation for NBTI [7], device lifetime is successfully guaranteed, as shown in Fig. 14.

Conclusion

True EUV 7nm platform technology was developed. EUV improves pattern uniformity and is inherently immune against weak points in random layout. It offers the smallest transistor and SRAM cell sizes. Special constructs and single diffusion break makes it more scalable than ever. This true EUV 7nm platform technology will return cost effectiveness back to the semiconductor industry.

Reference

[1] H.-J. Cho, VLSI Tech., p.12 (2016), [2] S. -Y. Wu. IEDM, p.43 (2016), [3] C. Auth, IEDM, p.673 (2017), [4] S. Narasimha, IEDM, p.689 (2017), [5] D. Ha, VLSI Tech., p.T68 (2017), [6] S. Yang, VLSI Tech., p.T70 (2017), [7] S. Maeda, IRPS, p.8 (2004)

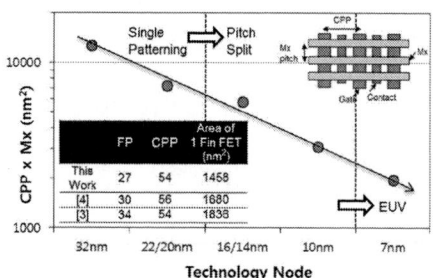

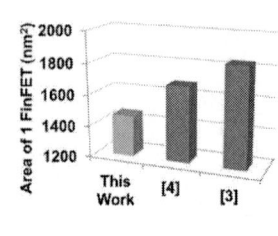

	2G 7nm [This Work]	1G 7nm [5]	10nm [1]
Fin	5G	4G	3G
Gate	3G	2G	1G
SD	5G	4G	3G
SDB	3G	-	2G
Gate Stack	3G	2G	1G
Contact	EUV	EUV	MPT
BEOL	EUV	EUV	MPT

Fig. 1. Scaling trend based on CPP and Mx pitch. EUV is indispensable to true 7nm patterning technology.

Fig. 2. Transistor size comparison. This work achieved smallest area.

Table I. Technology comparison. Here "G" means generation. This work was established on many of innovations.

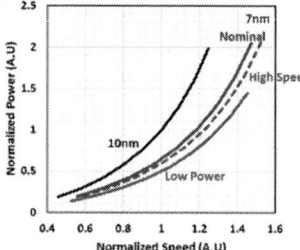

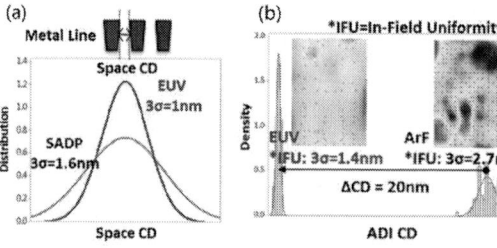

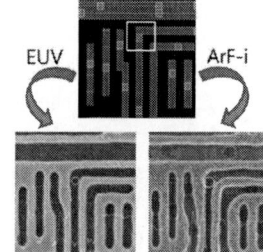

Fig. 3. Power and speed of various 7nm libraries in comparison with 10nm[1].

Fig. 4. EUV is superior to ArF in uniformity on (a) Space CD (~1.6X) and (b) hole (~2X).

Fig. 5. EUV 2D fidelity is 70% better than ArF.

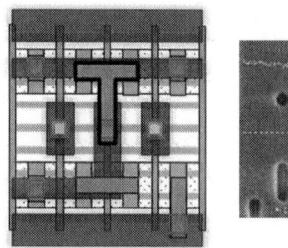

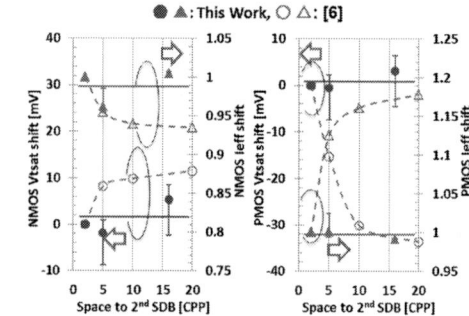

Fig. 6. Special constructs could make 10% vertical scaling by lowering track height

Fig. 7. Single diffusion break

Fig. 8. Suppression of local layout effect from 2nd diffusion break

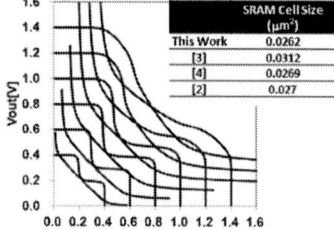

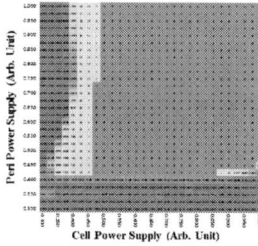

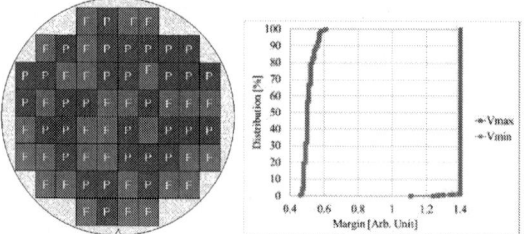

Fig. 9. Butterfly curve of SRAM cell. Inlet shows the SRAM cell size

Fig. 10. Shmoo plot of 256M bit SRAM

Fig. 11. 256Mb SRAM yield and Vmin/Vmax distribution

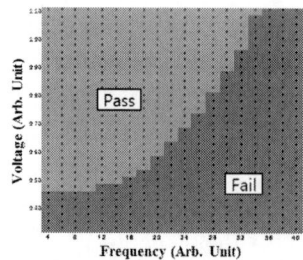

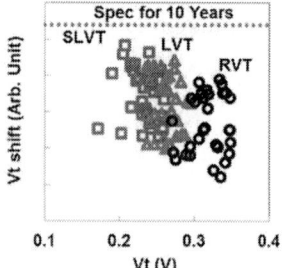

Fig. 12. Chip micrograph of logic test chip

Fig. 13. GPU Shmoo plot in logic test chip

Fig. 14. NBTI lifetime plot

Nanosecond Laser Anneal for BEOL Performance Boost in Advanced FinFETs

Rinus T.P. Lee, N. Petrov, J. Kassim, M. Gribelyuk, *J. Yang, L. Cao, K.B. Yeap, T. Shen, A.N. Zainuddin, A. Chandrashekar, S. Ray
E. Ramanathan, A.S. Mahalingam, R. Chaudhuri, J. Mody, D. Damjanovic, Z. Sun, R. Sporer, T.J. Tang, H. Liu, J. Liu, B. Krishnan
GLOBALFOUNDRIES, 400 Stone Break Road Extension, Malta 12020, New York, USA, email: rinus.lee@globalfoundries.com

Abstract

Nanosecond laser-induced grain growth in Cu interconnects is demonstrated for the first time using 14nm FinFET technology. We achieved a 35% reduction in Cu interconnect resistance, which delivers a 15% improvement in RC and a gain of 2 – 5% in I_{Dsat}. Additionally, reliability was enhanced with an improvement in dielectric V_{BD} and Cu EM performance without impacting the ULK mechanical integrity. Our results demonstrate a path to extending Cu interconnects for performance boost in 14nm FinFETs and beyond.

Introduction

BEOL interconnect resistance is expected to dominate product performance at advanced technology nodes [1]. Additionally, Cu interconnect resistance is projected to increase abruptly as minimum metal line-widths becomes less than the electron mean free path in Cu for these advanced nodes. This is due to increase electron scattering from grain boundaries, surfaces and interfaces [2, 3]. Alternative metals in development to replace Cu are also not ready for high volume implementation. Hence, we need to continue to develop processes to extend the use of Cu in advanced nodes. In this paper, an ultra-fast anneal for Cu using Nanosecond (nS) laser is reported. A reduction in Cu line resistance with improved RC, I_{Dsat}, V_{BD} and EM performance was realized. Our results show a clear correlation of these improvement to Cu grain size modification with nS laser anneal.

Concept and Process Flow

The concept of laser anneal for Cu grain size modification was first proposed by T. Nogami and co-workers [4]. The ultra-fast heating/quenching process and rapid crystallization of Cu in confined structures was predicted to enhance the growth of large Cu grains. This improves reliability [4] and reduces Cu resistance [5].

In this work, 11 levels of Cu metallization starting with 64nm pitch at M1 were used. nS laser was selected as it allows us to evaluate the full melt option without structural integrity issues compared to a mS laser anneal. Details on the use of nS laser anneal in our flow is shown in Fig 1. We start with post Cu electroplating [Fig 1(a)]. Typically, the Cu overburden is left intact during the anneal step. However, in our nS laser flow [Fig. 1(b)], a CMP step was added to remove the Cu overburden prior to nS laser annealing. Low κ is then deposited and confines Cu in the structure [Fig. 1(c)]. This allows a full reflow of Cu (sub-melt/melt) with nS laser anneal to enhance Cu grain growth [Fig, 1(d)]. Control devices in this work received a Cu anneal of 100°C, 60mins. Process knobs used in this work were substrate chuck temperature (T_{chuck}) and nS laser peak temperature (T_{peak}). nS laser anneal was applied at M1/V0. nS laser temperature splits were performed on each quadrant of the wafer (inset of Fig 2).

Results and Discussion

Fig. 2 shows the Cu line resistance values for devices with/without nS laser anneal. A reduction of 35% in line resistance with nS laser anneal was obtained at T_{peak2}. However, a slight capacitance (~9%) increase was found (Fig. 3) in devices with nS laser anneal. This is surprising as there is < 3% change in dielectric constant for blanket ULK films with/without nS laser anneal (see Fig. 4). This agrees with published data [6, 7]. We ascribed this to differences in heat dissipation and laser absorption coefficients for blanket ULK and patterned ULK between Cu lines. To address the increase in capacitance, we could increase the ULK thickness and/or optimize T_{chuck}, as Fig 3 shows that Δcapacitance is dependent on ΔT_{chuck}. The mechanical integrity of ULK films subjected to nS laser anneals were also evaluated as shown in Fig. 5. Inset shows the thickness difference of < 0.3nm pre/post nS laser anneal indicating there is no shrinkage. By optimizing T_{chuck}, we obtain up to 12% improvement in Young's modulus compared to the control. This provides a more robust ULK integration capability. The comparable optical extinction coefficients also inferred that dielectric constant remains unchanged. Hence, the nS laser condition of ($T_{peak2} + T_{chuck2}$) was used for device processing.

Fig. 6 compares the RC performances of devices with/without nS laser anneal. It shows a significant 15% RC improvement for devices with nS laser anneal. Additionally, HOL macros in Fig. 7 and 8 shows comparable (or lower) contact-to-via and via-chain resistances for devices with nS laser anneal. This indicates that nS laser anneal is compatible with downstream processing and free from stress-induced voids typically associated with thermal-based processes for Cu grain modification.

TEM analysis with orientation and phase analysis was used to compare Cu grain size for devices with/without nS laser anneal to explain the line resistance improvement in this work. Cross-sectional orientation and phase maps for two sets of Cu interconnect along the y-direction is shown in Fig. 9. It is evident that devices with nS laser anneal have larger Cu grains with a more bamboo-like structure than control devices. A total of 927 grains were measured from control devices and 578 grains from devices with nS laser anneal using the same interconnect length of ~500nm. Details are summarized in Fig. 10, which reveals a trend of larger Cu grains (~2.7× larger) with nS laser anneal.

The impact of a high temperature BEOL nS laser anneal on FEOL FinFETs is a potential concern and was also evaluated. Remarkably, we see a 2 – 5% improvement in I_{Dsat} for N/PMOS (Fig. 11 & 12), which was retained when processed to M11. This is credited to the reduced line resistance at M1 with nS laser anneal. Our results suggest that we could implement nS laser anneal at only the critical Cu layers to maximize performance without a trade-off in throughput/cost.

Potential reliability concerns with high temperature laser anneal were also addressed. Fig. 13 shows the Weibull distribution fit for the dielectric V_{BD} of devices with/without nS laser anneal. The lower bound of 95% confidence interval of nS laser anneal devices does not overlap with the upper bound of 95% confidence interval of the control devices. This indicates that nS laser anneal improves V_{BD} (+10). This is proposed to be related to a higher breakdown strength ($E_{BD} = V_{BD}$/Spacing) of the dielectric material due to the nS laser anneal process. Higher E_{BD} can be explained by the dielectric having less terminal bonds (e.g. -CH3, -OH) and more interconnecting bonds (e.g. -Si-O-Si-) post laser annealing. EM lifetime for devices with/without nS laser anneal are also shown in Fig. 14 and 15. For upstream V0-M1 and downstream V1-M1 measurements, devices with nS laser anneal show an improvement in EM lifetime compared to control devices. This is attributed to the larger bamboo-like Cu grains with nS laser anneal.

Summary

We demonstrate a novel nS laser anneal process to extend Cu interconnect technology for 14nm FinFETs and beyond. Cu interconnect resistance is reduced significantly with nS laser anneal leading to improvements in RC and I_{Dsat} performance. Additionally, V_{BD} and EM lifetime was also improved due to advantageous structural changes of the dielectric and Cu with nS laser anneal. Table 1 summarizes key benefits with the use of nS laser anneal for Cu.

Acknowledgement

The authors gratefully acknowledge Ultratech Inc. for nS laser processing. We would also like to thank S. Metha, W. Taylor and R. Fox for discussions. *J. Yang is an IBM assignee at GLOBALFOUNDRIES.

References

[1] G. Yeric, "Circuit Application Requirements", *IEEE IEDM Short Course*, 2014.
[2] K. Fuchs, "The conductivity of thin metallic films according to the electron theory of metals", *Mathematical Proc. of the Cambridge Phil. Society*, pp. 100 – 108, 1938.
[3] W. Steinhogl, *et al.*, "Tungsten interconnects in the nano-scale regime", *Micro.Eng.*, vol. 82, pp. 266 – 272, 2005.
[4] T. Nogami, *et al.*, "Method of improving Cu damascene interconnect reliability by laser anneal before barrier polish", *USA Patent #6,103,624*, Aug 2000.
[5] O. Gluschenkov. *et al.*, "Constrained nanosecond laser anneal of metal interconnect structures", *USA Patent #9,412,658*, Aug 2016.
[6] C. Fenouillet-Beranger, *et al.*, "nS laser annealing for junction activation preserving inter-tier interconnections stability within a 3D sequential integration", *IEEE S3S*, 2016.
[7] M. Redzheb, *et al.*, "Laser anneal of oxycarbosilane low-k film", *IEEE IITC/AMC*, pp. 156 – 158, 2016.

2018 Symposium on VLSI Technology Digest of Technical Papers

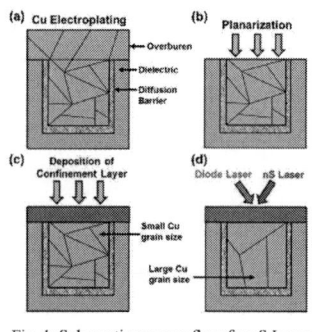

Fig. 1: Schematic process flow for nS Laser anneal used in this work. Control devices received a 100°C, 60mins anneal before planarization.

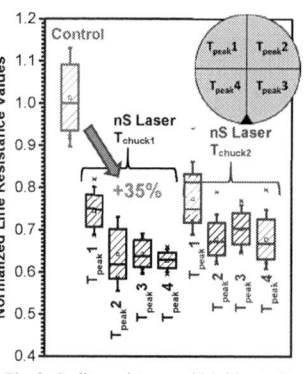

Fig. 2: Cu line resistance with/without nS laser anneal at different T_{peak} and T_{chuck}.

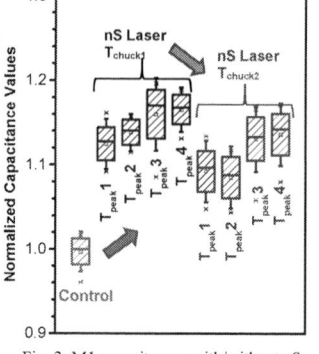

Fig. 3: M1 capacitances with/without nS laser anneal at different T_{peak} and T_{chuck}.

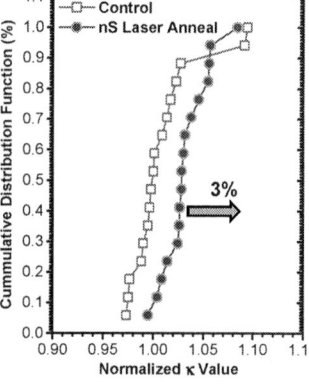

Fig. 4: CDF plot for CV extracted κ values.

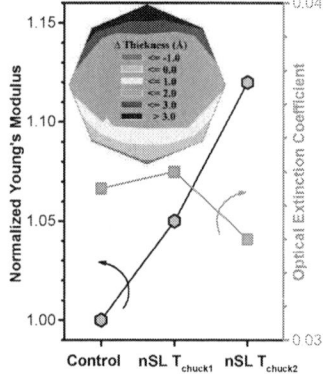

Fig. 5: ULK properties with/without nS laser anneal. Inset shows the thickness variation across 300mm pre/post nS laser.

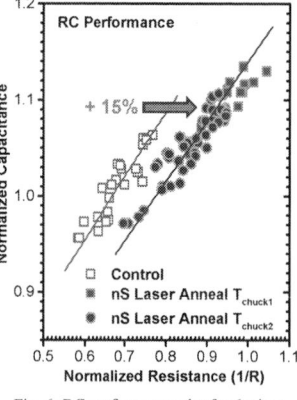

Fig. 6: RC performance plot for devices with/without nS laser anneal.

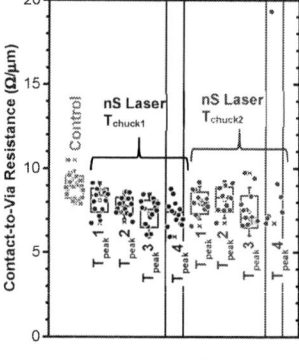

Fig. 7: Contact-to-via resistance values are similar for devices with/without nS laser anneal. Suggests that nS laser has no impact to HOL structures (→ yield).

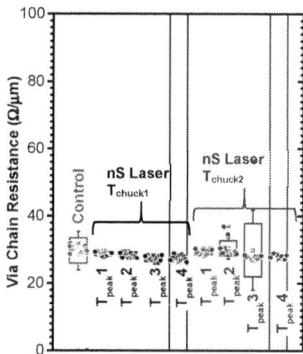

Fig. 8: Via resistance values are similar for devices with/without nS laser anneal. Suggests that nS laser has no impact to HOL structures (→ yield).

Fig. 9: Orientation, Phase and TEM maps of Cu lines for (a) control and (b) devices with nS laser anneal.

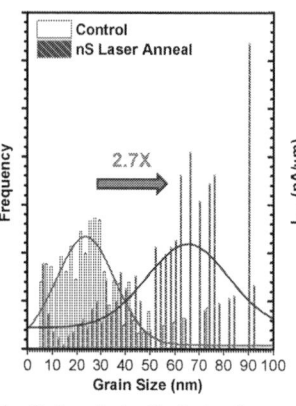

Fig. 10: Cu grain size distributions for devices with/without nS laser anneal.

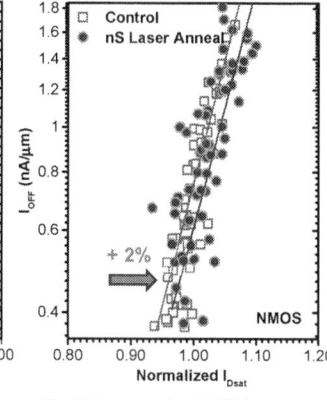

Fig. 11: I_{OFF}-I_{Dsat} for NMOS for devices with/without nS laser anneal.

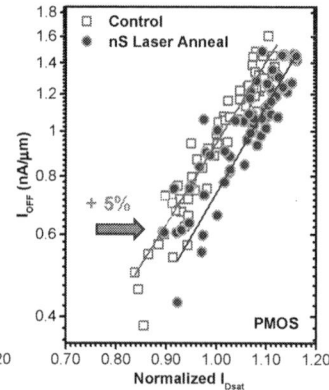

Fig. 12: I_{OFF}-I_{Dsat} for PMOS for devices with/without nS laser anneal.

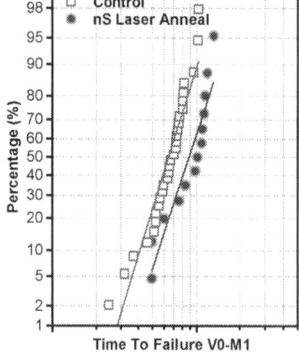

Fig. 13: M1-M1 dielectric V_{BD} for devices with/without nS laser anneal.

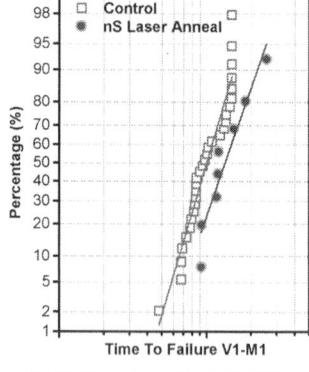

Fig. 14: Upstream V0-M1 EM lifetime for devices with/without nS laser anneal.

Fig. 15: Downstream V0-M1 EM lifetime for devices with/without nS laser anneal.

Summary: nS Laser Anneal vs Conv. Anneal	
Parameters	Performance
Line Resistance	+ 35%
Capacitance	- 9% (tunable)
RC Performance	+ 15%
Median Cu Grain Size	+ 2.7X
N/PMOS I_{Dsat}	+ 2%, +5%
M1-M1 Dielectric V_{BD}	+ 10%
V0-M1 EM Lifetime	+ 27%
V1-M1 EM Lifetime	+ 36%

Table 1: Summary of performance benefits demonstrated in this work with/without nS laser anneal.

978-1-5386-4219-1/18 $31.00 © 2018 IEEE

From Memory to Sensor: ultra-Low Power and High Selectivity Hydrogen Sensor Based on ReRAM Technology

Zhiqiang Wei[1], Kazunari Homma[1], Koji Katayama[1], Ken Kawai[1], Satoru Fujii[1], Yasuhisa Naitoh[2], Hisashi Shima[2], Hiroyuki Akinaga[2], Satoru Ito[1] and Shinichi Yoneda[1]

[1]Panasonic Semiconductor Solutions Co., Ltd., 1 Kotari-yakemachi, Nagaokakyo, Kyoto 617-8520, Japan
[2]National Institute of Advanced Industrial Science and Technology, 1-1-1 Higashi, Tsukuba, Ibaraki 305-8565, Japan

Abstract

We have fabricated a novel hydrogen sensor using optimized 0.18-μm ReRAM process. Our ReHsensor (Resistive Hydrogen Sensor) conforms with the ISO26142 standard in that it exhibits exceptional sensing capabilities, including high sensitivity, wide hydrogen concentration range (up to 4 vol.%) in air and N_2 ambient, high gas selectivity (no reaction with CH_4, CO, CO_2, CH_3OH, and CH_3COCH_3) and is immune to poisoning by SO_2 and hexamethyl disiloxane (HMDS). As it does not require a heater, the power consumption of the ReHsensor is very low, at 0.35 mW. We used this hydrogen sensor device to develop a battery-powered all-in-one wireless hydrogen sensor unit for IoT applications.

Introduction

There is growing interest in building a hydrogen-based economy. Fuel cell vehicles and hydrogen refueling stations are already in the marketplace. Hydrogen sensors will be increasingly important to guarantee the safety of hydrogen usage. However, conventional hydrogen sensors [1] are energy-draining, since they require a heater. They will also operate in an oxygen-containing atmosphere.

In this study, we present a new type of hydrogen sensor (ReHsensor) that is formed in a conductive filament based on ReRAM technology. It does not require a heater, and has the advantages of ultra-low power consumption, high gas selectivity, and sensitivity in atmospheres that lack oxygen.

Fabrication of ReHsensor

The item being tested was fully integrated using 0.18-μm CMOS technology based on the ReRAM process [2]. In Fig. 1a, the 0.5 μm × 0.5 μm ReRAM cell is placed between two metal layers that consist of $Ir/Ta_2O_5/TaO_x/TaN$ stacked layers. Three improvements were developed to enhance its hydrogen sensitivity (Fig. 1b). First, instead of the Ir, we selected 20 nm-thick Pt as the top electrode because of its catalytic properties, and made it much thinner than Ir (50 nm). Second, we enlarged the device to 5 μm × 5 μm to increase the exposure area. Finally, the via on the top electrode was shifted to the edge, and a window was opened by etching the passivation and insulation layer on the Pt electrode. A top view and cross section of the ReHsensor are shown in Fig. 1c and 1d. To enhance the gas reaction by exposing the interface between Pt and filament, lateral devices with nanogap [3] (Fig. 2) were also fabricated for selectivity and poisoning testing.

Characteristic of ReHsensor

The tests were based on ISO 26142, which defines the performance requirements and test methods of hydrogen detection apparatus. The ReHsensor shows excellent sensing capability over a large range of hydrogen concentrations as well as resistive switching (Fig. 3). As shown in Fig. 4, the hydrogen sensitivity increases with increasing hydrogen concentration. The response time decreases with increasing hydrogen concentration (Fig. 5). The power consumption of ReHsensor at 4% is as low as 0.35 mW. The repetition property of ReHsensor exposure to 100% H_2 is shown in Fig. 6. The hydrogen response in N_2 atmosphere is shown in Fig. 7. The current change is much greater than that in an air atmosphere. Gas selectivity was also (Fig. 8). There is no response to CH_4, CO, CO_2, CH_3OH, and CH_3COCH_3. Siloxanes deposit silicide on a conventional hydrogen sensor's catalyst surface and inhibit the reaction with hydrogen. To

ensure a maintenance-free hydrogen sensor, it needs to be resistant to siloxane contamination. The hydrogen sensitivity of the ReHsensor before and after a 60-min exposure test to 10 ppm HMDS is shown in Fig. 9a. The results confirm that the ReHsensor is immune to siloxane vapor. ReHsensor also shows high resistance to SO_2 poisoning (Fig. 9b).

To investigate the mechanism of the hydrogen response of the ReHsensor, a virgin ReHsensor was exposed to two cycles of air and 100% H_2 at room temperature (Fig. 10). In the first exposure, the resistance of the ReHsensor in air was about 11.9 kΩ, and saturated at 2.0 kΩ in 100% H_2. The resistance did not recover to its initial value after the removal of hydrogen ambient, and saturated at 7.6 kΩ. The same pattern was seen for the second exposure. Current-voltage data at the three levels and fitting curves based on the hopping conductive mechanism are shown in Fig. 11. The current J of the hopping conductive mechanism is expressed by the following Equation,

$$J = S_m N(E_F) \frac{kT}{q\gamma} sinh\left(\frac{q\gamma}{kT}(V_a - JR_{ext})\right) \quad (1)$$

where q is the unit of electronic charge, V_a is the applied voltage, and R_{ext} is the external resistance. $N(E_F)$ is the density of states for electrons. The value of $N(E_F)$ of level II is about 9.1-fold that of level I (Fig. 12, HRS). Since $N(E_F)$ of level II (Fig. 12 in air after H_2) does not return to that in level I after removal of hydrogen, it appears that O-H complexes are created by hydrogen atoms in the immediate vicinity of the oxygen vacancies and remain after the removal of the hydrogen gas. With O-H complex formation, new defect states are introduced and the incorporation of hydrogen atoms is likely to increase the conductivity of level II. On the other hand, the density of states for electrons of level III (Fig. 12 in H_2) is about 3.6-fold that of level II. This means that the states for electrons corresponding to the difference between level II and III vanish after removal of hydrogen gas. With high activity oxygen ions in filament, ReHsensor shows high hydrogen sensitivity at room temperature.

Using the above high-performance hydrogen sensor device, we successfully developed an all-in-one wireless ReHsensor unit (Fig. 13). The unit consists of the ReHsensor plus a ReRAM-embedded MCU, battery, and wireless modules.

Summary

A hydrogen sensor using $Pt/Ta_2O_5/TaO_x/TaN$ was fabricated employing enhanced ReRAM technology. It was demonstrated to sense hydrogen gas at room temperature, with high selectivity and low power consumption of less than 0.1% of that required by a conventional hydrogen sensor. The ReHsensor passed all ISO261242 standard tests, indicating that it has potential for ultra-low power hydrogen-sensing applications.

Acknowledgements

The authors thank Y. Konishi, T. Yamada, H. Miyatake, K. Suzuki, and T. Iwami (New-Cosmos Electric Co. Ltd.,) for hydrogen testing.

References

[1] T.Usagawa and Y. Kikuchi IEEE Sensors Journal **12**, 2243 (2012)

[2] Z. Wei and K. Eriguchi, IEEE TED, **64**, 2201-2206 (2017)

[3] Y. Naitoh, et al., 78th JSAP Autumn Meeting, 2017

[4] T. Hübert, et al. Sensors and Actuators B, **157**, 329– 352 (2011)

[5] Y. Sasago, et al. VLSI Tech.Dig.,T106- 107 (2017)

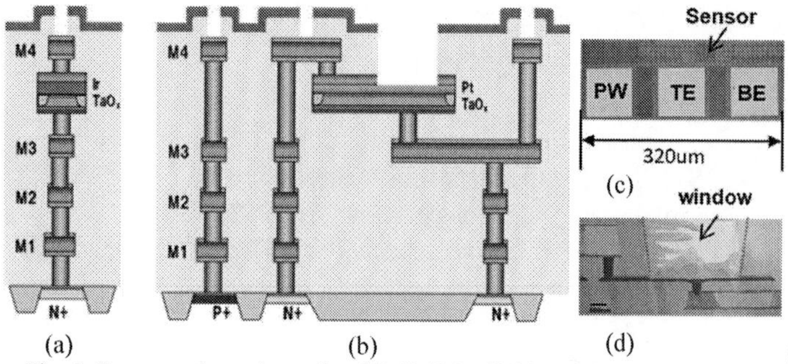

2018 Symposium on VLSI Technology Digest of Technical Papers

Fig. 1 Cross section schematic of ReRAM cell (a) and ReHsensor cell (b), ReHsensor chip (c) and TEM cross section of window (d).

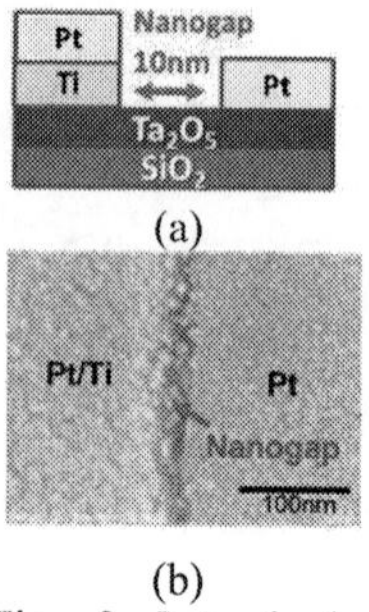

Fig. 2 Lateral device with nanogap, schematic (a) and top view (b).

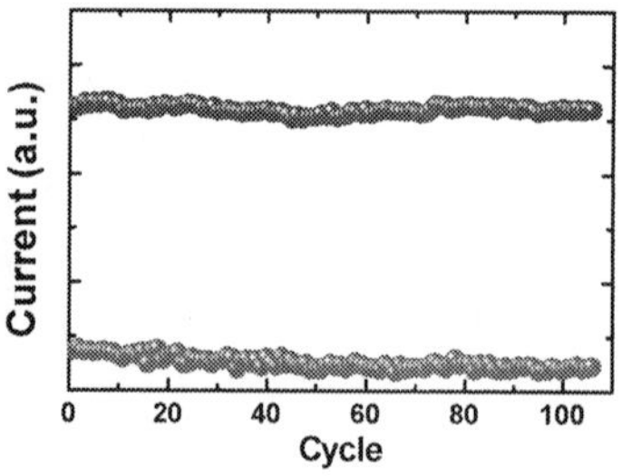

Fig. 3 Resistive switching of ReHsensor.

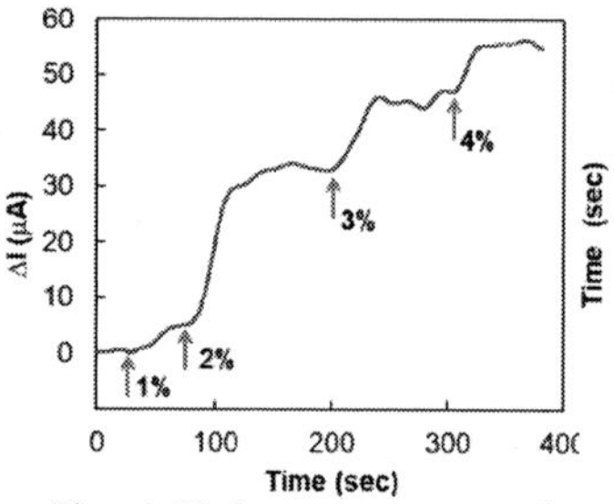

Fig. 4 Hydrogen response of ReHsensor.

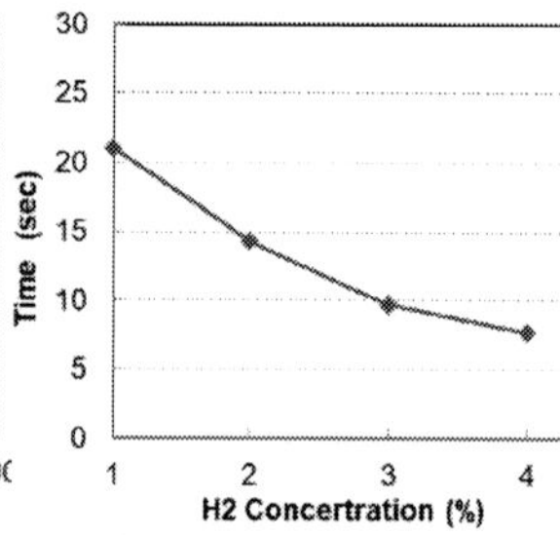

Fig. 5 Hydrogen concentration dependence of response time.

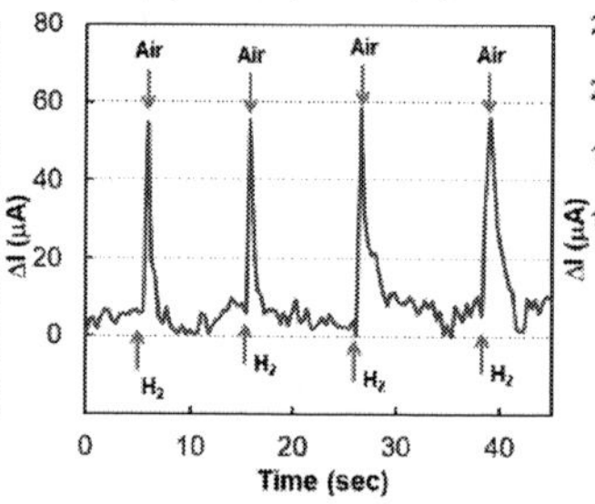

Fig. 6 Repetition property of 100% hydrogen response.

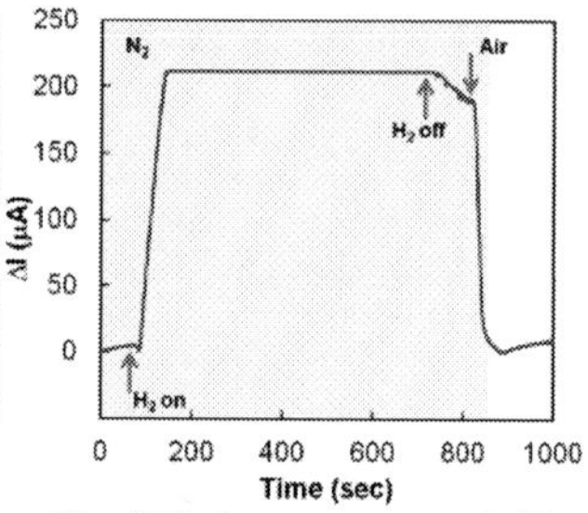

Fig. 7 Hydrogen response in N_2 atmosphere (0.5 vol.% H_2).

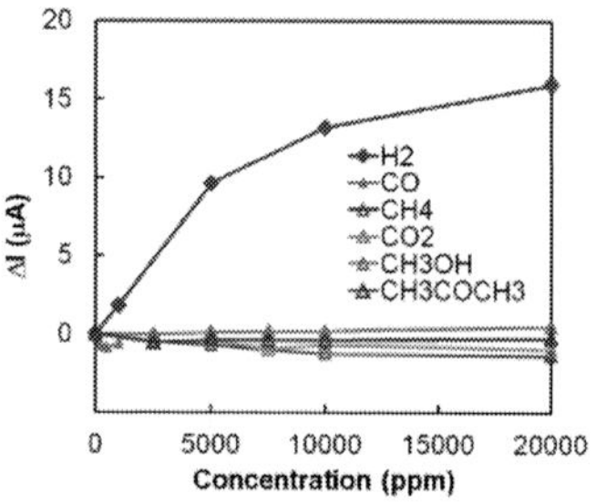

Fig. 8 Selectivity of ReHsensor.

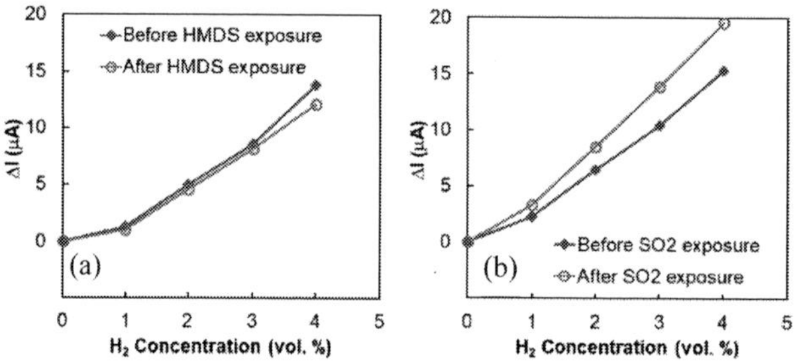

Fig. 9 Test results for HMDS (a) and SO_2 (b) poisoning.

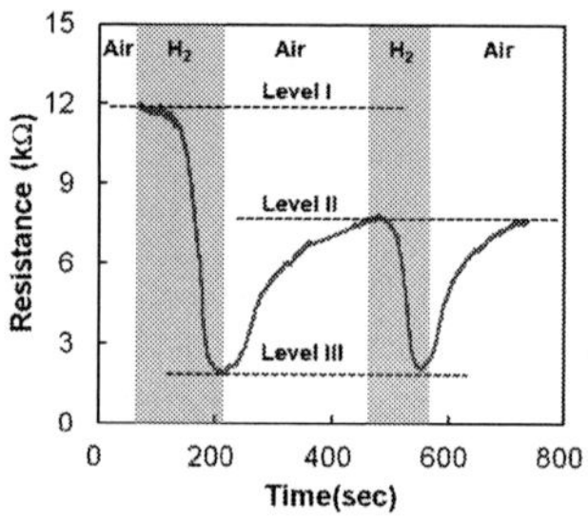

Fig. 10 Relative response of the ReHsensor exposed to air and 100% H_2

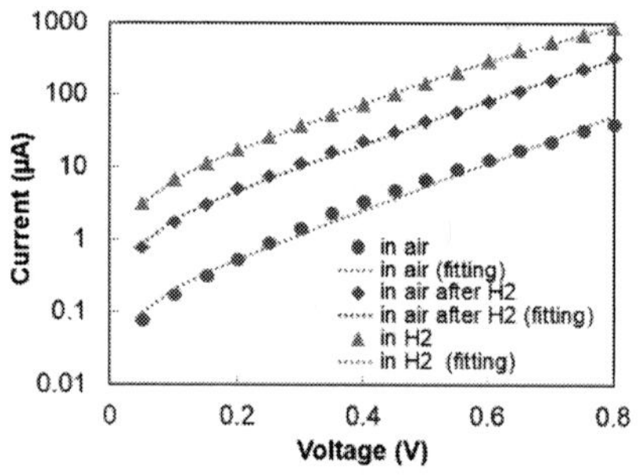

Fig. 11 Current-voltage measured data and fitting curves based on the hopping conductive mechanism of the ReHsensor.

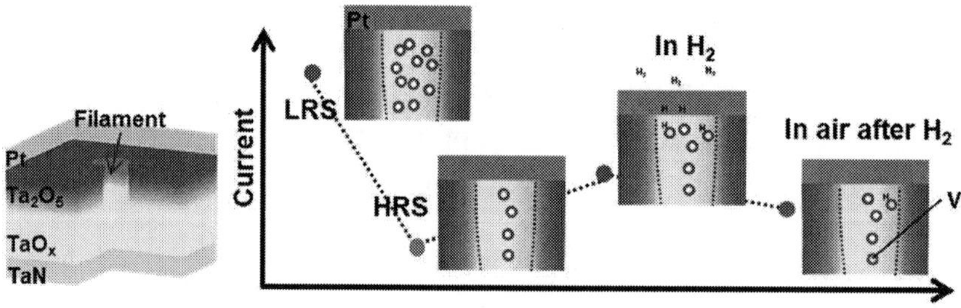

Fig. 12 Mechanism of hydrogen response of ReHsensor (Vo: Oxygen Vacancy, HRS: High Resistance State, LRS: Low Resistance State).

Table 1 Benchmarks for hydrogen sensor

Sensor type	ReRAM (This work)	Catalytic combustion[4]	FET [5]
Sense environment	Does not require O_2	Require 5-10% O_2	-
Temperature	R. T.	400 °C	115 °C
Heater	Does not require heater	Require heater	Require Heater
Power	0.35 mW	1000 mW	10 mW
	Does not require heater	With heater	Without heater
Device size	0.3 mm x 0.12 mm	2.4 mm x 2.4 mm	2 mm x 2mm
Response time (1%)	21 sec	<30 sec	0.8 sec
Warm-up time	0 (No need)	<1sec	-
Selectivity CH_4, CO, CO_2, CH_3OH CH_3COCH_3	High selectivity	Not H_2 selective	-
Poisoning HMDS, SO_2	Resistant to poisoning	Susceptible to poisoning	-

Fig. 13 All-in-one wireless ReHsensor unit.

Demonstration of Ultra-Low Voltage and Ultra Low Power STT-MRAM designed for compatibility with 0x node embedded LLC applications

Guenole Jan, Luc Thomas, Son Le, Yuan-Jen Lee, Huanlong Liu, Jian Zhu, Jodi Iwata-Harms, Sahil Patel, Ru-Ying Tong, Vignesh Sundar, Santiago Serrano-Guisan, Dongna Shen, Renren He, Jesmin Haq, Zhongjian Jeffrey Teng, Vinh Lam, Yi Yang, Yu-Jen Wang, Tom Zhong, Hideaki Fukuzawa, and Po-Kang Wang

TDK-Headway Technologies, Inc., Milpitas, CA, USA

E-mail: guenole.jan@headway.com

Abstract

We present for the first time STT-MRAM devices with ultra low operating voltage and power compatible with next generation 0x node logic voltages. By engineering the tunnel barrier and improving the efficiency of the devices we report a record low writing voltage of 0.17V for a 1ppm error rate, which has been achieved for a 20ns write operation using a writing current of only 35uA. We further demonstrate error rates below 10^{-9} at voltage and current at 0.25V and 50uA using 10ns writing pulses on the same 30nm devices with extended 400C thermal budget while preserving functionality and data retention at 85ºC. Finally, TDDB studies confirm the almost unlimited endurance of these devices at the operating voltage.

Keywords: Low Power, STT-MRAM, Backhopping, TDDB

Introduction

Scaling of SRAM cell size is becoming a major concern for advanced nodes with an average cell size reduction of 0.6x per node as shown on *Fig 1*. Among alternative memories, STT-MRAM has shown great promises for embedded LLC applications thanks to its demonstrated speed [1], endurance and compatibility with BEOL integration. One last area of concern for using MRAM at advanced nodes is the relatively high operating voltage required for reliably switching the devices (500 to 700mV typ. as shown on *Fig 2.*). This concern arises from two main factors. First, due to loading of the access transistor, the total available voltage for the typical Write "1" (P to AP switching) polarity is reduced by at least V_t before even considering other contributions such as transistor and contact resistance as shown in *Fig 3*. Secondly, as technology nodes shrink, the core voltage of the chip is reduced as well, as plotted on *Fig 4*, reaching voltages close 0.5V for 0x applications. We estimate that an operating switching voltage of the MTJ devices well below 300mV is required for 0x nodes applications based on the available V_{DD} and V_t data. In this paper, we report on a MTJ stack comfortably meeting the 300mV writing voltage requirement. Low writing current is also reported making this paper the first low voltage, low power demonstration of embeddedable STT-MRAM.

Switching Voltage reduction and Reliability

The resistance area product (RA) of the MTJ film can in theory be reduced to lower the switching voltage of an MTJ device without impacting other properties. The result of such RA scaling study is presented on *Fig 5-6* validating the proportional relationship between switching voltage and RA implied by the constant current density needed to reverse the magnetization. Breakdown voltage is found to vary logarithmically with RA, thus widely opening the writing window for lower RA films as shown in *Fig 6*. For a target RA around 3 to 4 ohm.um^2, and using a 300mV operating voltage and 10^{15} cycles the expected error rate of a device array is to be well below 1ppb (10^{-9}).

Device size reduction

For 0x node applications, a device diameter size at or below 30nm is needed to yield a reasonable cell size. We integrated low RA film stacks into devices down to 15nm and subsequently annealed them for more than 3.5hours at 400°C to simulate integration at the M1 or M2 level. *Fig 9* shows the TEM micrograph of an integrated 35nm device after such anneal. Thanks to improvements in the device fabrication process, the properties (switching voltage, TMR) of different size devices are kept almost constant as shown on *Fig 8*.

Smaller devices will require much lower writing current but will see a reduction of the energy barrier between the two storage states impacting data retention. The anisotropy field (H_K) of the device needs to be dramatically increased to maintain data retention. On *Fig 10*, H_K and its variation vs. temperature for two different stack flavors demonstrates that very high H_K (>11kOe) and reduced temperature dependence can be achieved simultaneously. The coercivity (H_C) of a 30nm device is shown at different temperatures up to 125°C on *Fig 11*. H_C of this device remains still close to 1kOe at 125°C.

STT switching data

Smaller devices are also more prone to the backhopping (BH) phenomenon [3] which is understood as the effect of the back torque applied on the pinned layer after the device switching. As shown on *Fig 13*, the BH effect is enhanced for smaller pulse width where more current is needed to switch the free layer as well as higher operating temperature. Pinned layer improvements have been implemented to mitigate the occurrences of BH in our devices.

The results of all those breakthroughs are most notably displayed on *Fig 14* showing the switching error rate (BER) of a 30nm device at 10ns write pulse. An error rate below 1e-9 at a write "1" voltage of 240mV (50uA/30nm) with a wide write margin verified up to 360mV is achieved. No read disturb has been observed on the lower voltage end after more than 10^9 pulses. Finally, on *Fig 15* the BER vs. pulse width SHMOO is presented for different temperatures using 10^6 write cycles. The device shows very wide write and read margins at all temperatures down to a few ns. While BH is still observed on our devices on the write 1 polarity, especially at higher temperatures, its ppm level onset occurs at double the operating voltage (BER=1ppm). *Fig 16* summarizes the critical device properties needed for 0x node LLC integration and the result from the devices described in this work.

Conclusions

We have shown for the first time ultralow voltage and power operation of an STT-MRAM which can fulfill the most stringent requirements for 0x node LLC applications.

References

[1] G. Jan *et al.*, Symp VLSI Tech symp. IEEE,. pp 50-51. (2016)
[2] G. Jan *et al.*, Symp VLSI Tech symp. IEEE,. pp 50-51. (2015)
[3] T.Min *et al.* J. App. Phys.,105, 07D126 (2009)

2018 Symposium on VLSI Technology Digest of Technical Papers

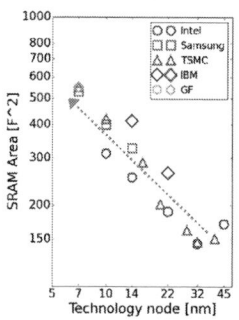

Fig 1 Scaling of SRAM cell area vs. technology node normalized in f².

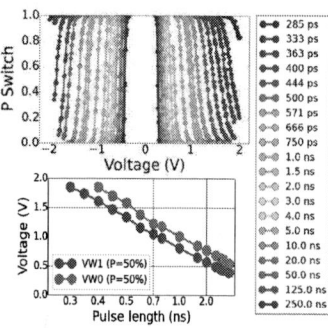

Fig 2 [top] Switching properties of the devices presented in [1]. Sub-ns switching is achieved [bottom] linear dependence of the 50% switching voltage vs. write pulse width.

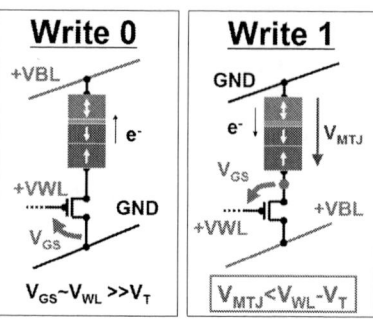

Fig 3 STT_MRAM Cell diagram showing biasing conditions for Write "0" and"1". Loading on the "W1" polarity leads to very low MTJ voltage available.

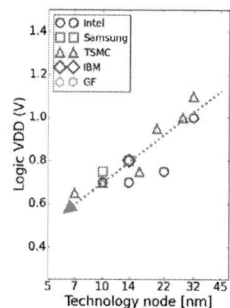

Fig 4 Evolution of Logic V_{DD} voltage with technology node.

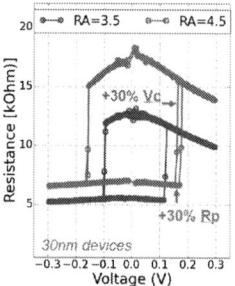

Fig 5 Resistance vs. voltage curves for similar size devices with different MgO barrier resistance area product (RA). V_{Switch} and resistance scale linearly with RA.

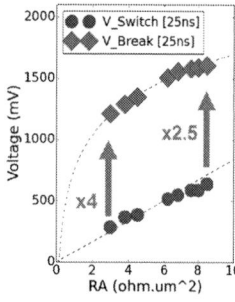

Fig 6. Correlation between RA and switching voltage plotted for different tunnel barriers. As RA drops the switching window improves.

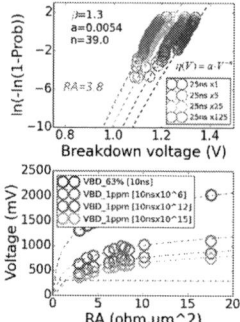

Fig 7. [Top] ramped voltage source TDDB data from RA=3.8 devices using the same method described in [2]. [Bottom] extrapolated error rate curves vs. RA

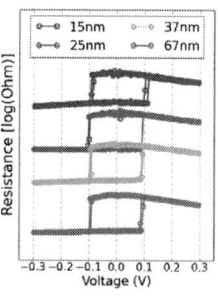

Fig 8 Quasi-static resistance vs. voltage loops for a range of device size. The switching voltage stays constant with device size

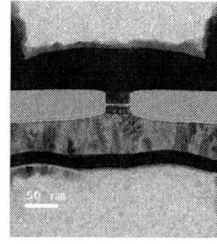

Fig 9 TEM picture of a fabricated 35nm device after 400C annealing for 3.5h.

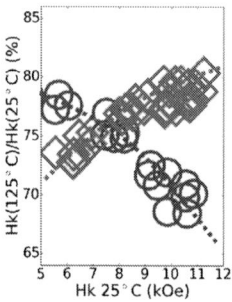

Fig 10 Temperature dependence of the HK for two different MTJ stack types. The anisotropy field drops by less than 20% between RT and 125ºC.

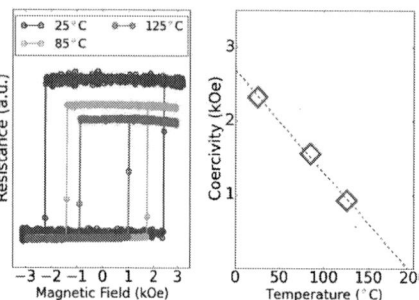

Fig 11 Resistance vs. magnetic field curves of a 30nm device taken at different temperatures .The coercivity of the device drops with increasing temperature but remains just under 1kOe at 125ºC.

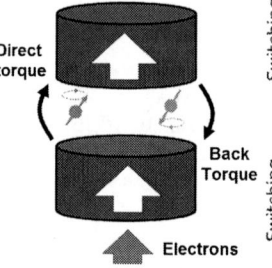

Fig 12 Under STT, both layers experience torque. While most models assume the Pinned Layer (PL) to be fixed, the backhopping issue arises from the PL reduced stability under torque.

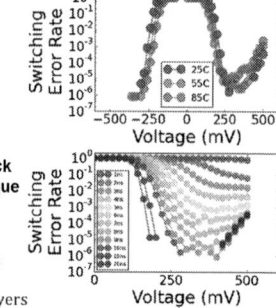

Fig 13 Effect of backhopping on the switching. The backhopping prevents achieving reasonable error rate especially at high temperature and short pulses.

Fig 14 Bit Error rate of a 30nm device taken at 25ºC using 10ns pulses. The cross symbols indicate that no error occurred during 10^9 write cycles at those voltages.

Fig 15 Switching SHMOO for the same device as shown on fig 14 taken at different temperatures, voltages and pulse length. The color bitmap scale is using -|sigma| of the switching probability. The numbers displayed represent the number of errors at a given voltage and pulse length condition during a 1.2x10⁶ BER cycle test. Clear read and write margin are observed throughout the temperature range. Some very low probability backhopping can be observed at high write "1" voltages.

	Target for 0x nm LLC	This Work	
Device Diameter (electrical CD)	<30nm	30nm	
Switching Voltage W1 @1ppm	<300mV	170mV	240mV
Switching current @1ppm	<50uA	35uA	50uA
Writing pulse length	<20ns	20ns	10ns
Switching Energy (pJ) @1ppm	<1ppm	0.12pJ	0.12pJ
Thermal budget @400ºC	3hours (>10ML)	3.5h	
Operating temp. range	>85ºC	85ºC	

Fig 16 Summary table of the properties of the MTJ devices presented in this paper

978-1-5386-4219-1/18 $31.00 © 2018 IEEE

Gap in pagination due to formatting issues.

Pages 67-68

3D sequential stacked planar devices on 300 mm wafers featuring replacement metal gate junction-less top devices processed at 525°C with improved reliability

A. Vandooren, J. Franco, B. Parvais**, Z. Wu, L.Witters, A. Walke, W. Li, L. Peng, V. Desphande, F.M. Bufler***, N. Rassoul, G. Hellings, G. Jamieson, F. Inoue, G. Verbinnen, K. Devriendt, L. Teugels, N. Heylen, E. Vecchio, T. Zheng, E. Rosseel, W. Vanherle, A. Hikavyy, B. T. Chan, R. Ritzenthaler, G. Besnard*, W. Schwarzenbach*, G. Gaudin*, I. Radu*, B.-Y. Nguyen*, N. Waldron, V. De Heyn, D. Mocuta and N. Collaert.
IMEC, Kapeldreef 75, 3001 Leuven, Belgium *SOITEC, Parc Technologique des Fontaines, 38190 Bernin, France ** also with VUB, Dept. of Electronics and Informatics, Pleinlaan 2, 1050 Brussels, Belgium ***also with ETH Zürich, Switzerland.

Abstract

3D sequential integration requires top MOSFETs processed at low thermal budget, which can impair the device reliability. In this work, top junction-less device are fabricated with a maximum processing temperature of 525°C. The devices feature high k /metal replacement gate and low temperature Si:P and SiGe:B 60% raised SD for NMOS and PMOS respectively. Device matching, analog and RF performance of the top tier devices are in-line with state-of-the-art Si technology processed at high temperature (>1000°C). The top Si layer is transferred on CMOS planar bulk wafers with W metal-1 interconnects, using a SiCN to SiCN direct wafer bonding.

Introduction 3D sequential integration is a very attractive path to continue increasing the circuit functionality per area for next technology nodes, without requiring further reduction of the device dimensions. Because of the high alignment accuracy dependent only on the lithography stepper performance, high device density per chip area can be reached when stacking devices on top of another in the same front-end process flow. However, extremely dense 3D inter-tier connections need to be integrated and proper routing of circuits is challenging. Other potential benefits of 3D sequential integration are the reduced interconnect wire length, the simplified co-integration of heterogenous channel materials (such as Ge or III-V) and hybrid technologies such as RF, memory or optical IO on logic. The most critical challenge of 3D sequential integration is the management of the process thermal budget. The top tier thermal budget needs to be reduced to avoid degrading what is below, namely the bottom devices, the interconnects and the bonding interface. Much work has been done towards lowering the thermal budget of the top devices [1-4]. In this work, the top devices are processed at a temperature as low as 525°C thanks to the low temperature offset spacers, cyclic deposition/etch selective epitaxy and extension-less integration using junction-less devices with channel doping set prior to top silicon layer transfer. This enables NMOS on PMOS (or PMOS on NMOS) transistors stacking, as well as potential applications such as RF/analog on logic. Low thermal budget of the top device processing precludes however any "reliability anneal", which impacts the quality of the gate dielectric [5]. Reliability of the top devices is therefore one of the major issues when decreasing the process temperatures below 800°C. Previous gate stack studies showed that a LaSiOx-based interface dipole [6] can provide sufficient PBTI reliability for NMOS devices, but a solution for PMOS devices is not readily available. In this work, junction-less top devices are proposed to enable improved reliability at low thermal budget thanks to their lower electric field and inherent better reliability in comparison to inversion-mode devices [7,8].

Device fabrication The 3D sequential process flow is summarized in Fig.1. The bottom tier devices consist of high temperature NMOS and PMOS bulk planar transistors with a HfO2/TiN replacement metal gate stack fabricated on 300 mm wafers. Contacts to the source and drain regions are made of a Ti/TiN barrier and W metal in direct contact with silicon, enabling improved thermal stability up to 550°C [9] over standardly used silicide and suppressing any contamination of FEOL tools with silicide-based species such as Ni or Pt. After W metal-1 interconnects, an oxide is deposited and planarized followed by SiCN layer deposition. A thin Si layer of good uniformity is then transferred by direct SiCN wafer bonding [10] of a blanket SOI wafer to the bottom tier device wafer. The SOI wafer is beforehand implanted and annealed at high temperature to activate the dopants, followed by oxide and SiCN deposition. Before bonding, an anneal at a temperature corresponding to the maximum temperature of the top device processing is performed on the carrier and donor wafers to ensure full degassing of the bonding dielectrics and preclude any voids

formation after bonding during further processing (Fig.2). The wafer bonding is followed by an anneal at 350°C to ensure proper SiCN bonding strength. The SOI substrate is then removed by silicon grinding and wet and dry etch processes, followed by the top device processing. This one features an RMG process using a dummy gate stack deposited at 500°C, a TiN/HfO2 gate stack deposition at 430°C, a nitride offset spacer deposited at 480°C and low temperature raised source/drain selective epitaxy at 500°C and 525°C for SiGe60%:B PMOS and Si:P NMOS respectively, grown on the ASM Intrepid™ XP epi cluster in combination with Previum® pre-clean,. Dopant activation of $2x10^{20}$ at/cm^3 is obtained in the epi layers from 4-point-probe sheet resistance and Hall mobility measurements. Fig.3 shows the TEM cross-section of the final stacked devices and the 3D contacting to the bottom devices.

Top device DC electrical results The I_D-V_G characteristics of the NMOS and PMOS top transistors are presented in Fig.4. In junction-less devices, the threshold voltage depends on channel thickness and channel doping, in agreement with simulations (Fig.5). The short channel effects, subthreshold slope and DIBL (Fig.6) of the devices are well controlled down to a physical gate length of 40nm, for a film thickness of 10nm. Extracted CET of the devices is 14Å. Fig.7 shows that, for an optimal performance and complete turn off, junction-less devices require a proper combination of the channel active doping and channel thickness. Comparison of the top stacked devices and equivalent SOI devices fabricated using the same low temperature process flow (Fig.8) results in similar performance hinting that the silicon layer transfer has no significant impact on the layer integrity. TCAD simulations [11] properly reproduce the threshold voltage decrease with channel doping. NBTI and PBTI measurements (Fig.9) show that low thermal budget inversion mode MOSFETs compatible with top tier integration have a ΔV_{th} ~20× larger than required for a 10 year reliable operation spec. In contrast, the junction-less top devices show significantly improved reliability thanks to inherently reduced oxide electric field at operating conditions [7]. BTI reliability specifications are met by increasing the doping in the channel.

Top device RF, analog and matching performance The device matching coefficient (A_{VT}) ranges between 1.7 and 3.1 mV.μm for NMOS devices with a channel doping varying from $1x10^{18}$ to $9x10^{18}$at/cm^3, while slightly higher values are obtained for PMOS. These values are obtained thanks to the good silicon film uniformity and are in line with doped planar bulk technologies at the same inversion thickness (Fig. 10). Compared to advanced finFET devices, the planar devices offer a higher f_T/f_{max} due to reduced parasitics. RF characterization leads to f_T of ~80 GHz for 80 nm gate length devices, in line with state-of-the art Si data [12-17], despite the low thermal budget (Fig.11). Voltage gain (A_v) and g_m/I_D (Fig.12) show overall good performance with improved g_m for thicker film at the cost of slightly reduced voltage gain.

Bottom device reliability I_D-V_G and C-V curves show functional bottom devices (Fig.13). Due to the thermal steps required for the top tier fabrication, the bottom tier inversion mode NMOS with a HfO2/TiN/W gate stack meet BTI specs, despite no dedicated reliability anneal (Fig.9).

Conclusions This work demonstrates that junction-less devices are suitable for top tier sequential 3D stacking thanks to the low complexity of their fabrication, which does not require junction formation, and the channel doping and activation performed prior to layer transfer. The top tier junction-less devices processed at T≤525°C show good device performance without impact from layer transfer. Junction-less devices present the additional advantage of a reduced electric field which enables NBTI/PBTI reliability, despite the large oxide defect densities related to the low budget thermal process.

2018 Symposium on VLSI Technology Digest of Technical Papers

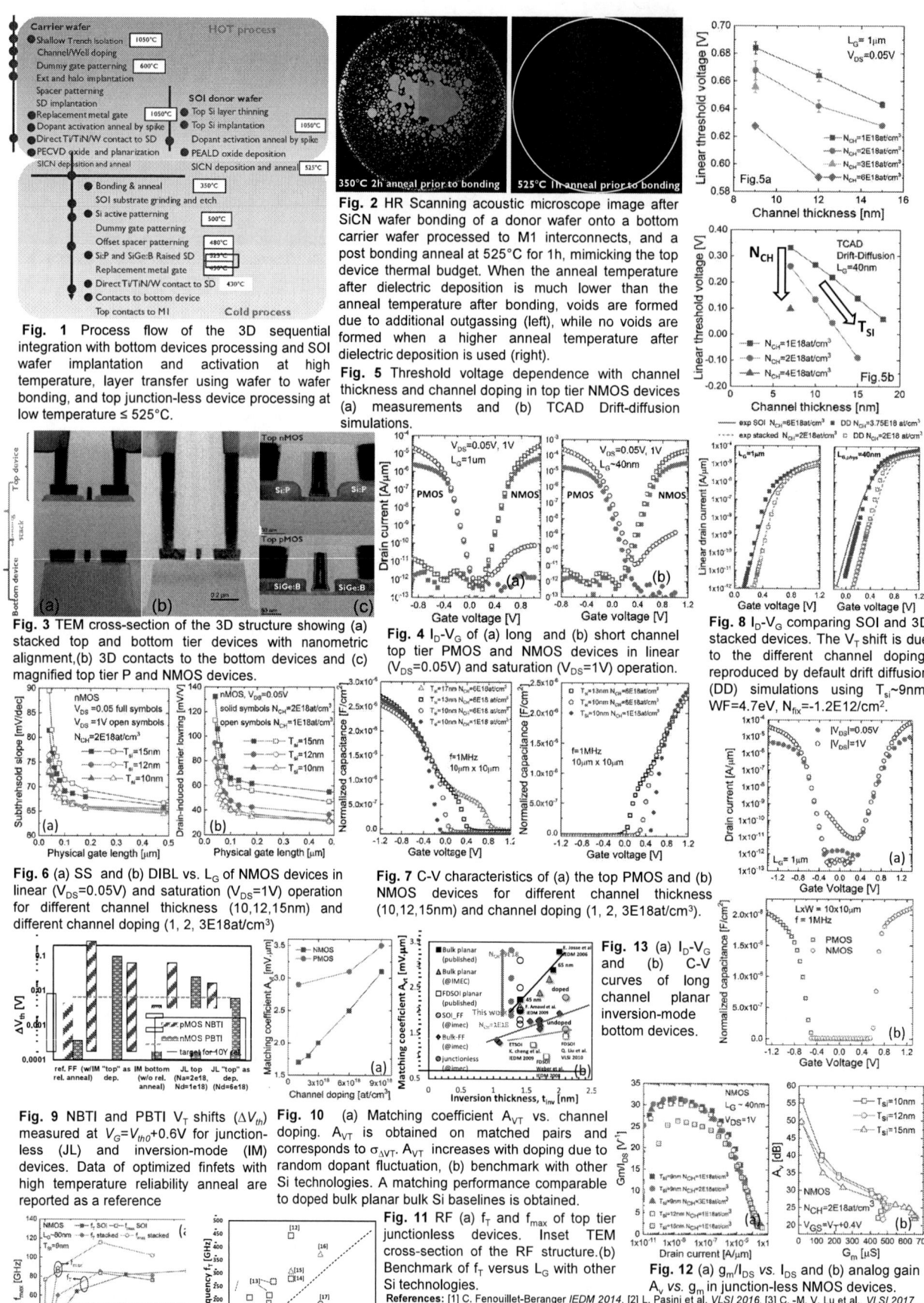

Fig. 1 Process flow of the 3D sequential integration with bottom devices processing and SOI wafer implantation and activation at high temperature, layer transfer using wafer to wafer bonding, and top junction-less device processing at low temperature ≤ 525°C.

Fig. 2 HR Scanning acoustic microscope image after SiCN wafer bonding of a donor wafer onto a bottom carrier wafer processed to M1 interconnects, and a post bonding anneal at 525°C for 1h, mimicking the top device thermal budget. When the anneal temperature after dielectric deposition is much lower than the anneal temperature after bonding, voids are formed due to additional outgassing (left), while no voids are formed when a higher anneal temperature after dielectric deposition is used (right).

Fig. 5 Threshold voltage dependence with channel thickness and channel doping in top tier NMOS devices (a) measurements and (b) TCAD Drift-diffusion simulations.

Fig. 3 TEM cross-section of the 3D structure showing (a) stacked top and bottom tier devices with nanometric alignment,(b) 3D contacts to the bottom devices and (c) magnified top tier P and NMOS devices.

Fig. 4 I_D-V_G of (a) long and (b) short channel top tier PMOS and NMOS devices in linear (V_{DS}=0.05V) and saturation (V_{DS}=1V) operation.

Fig. 8 I_D-V_G comparing SOI and 3D stacked devices. The V_T shift is due to the different channel doping, reproduced by default drift diffusion (DD) simulations using T_{si}~9nm, WF=4.7eV, N_{fix}=-1.2E12/cm².

Fig. 6 (a) SS and (b) DIBL vs. L_G of NMOS devices in linear (V_{DS}=0.05V) and saturation (V_{DS}=1V) operation for different channel thickness (10,12,15nm) and different channel doping (1, 2, 3E18at/cm³)

Fig. 7 C-V characteristics of (a) the top PMOS and (b) NMOS devices for different channel thickness (10,12,15nm) and channel doping (1, 2, 3E18at/cm³).

Fig. 13 (a) I_D-V_G and (b) C-V curves of long channel planar inversion-mode bottom devices.

Fig. 9 NBTI and PBTI V_T shifts (ΔV_{th}) measured at V_G=V_{th0}+0.6V for junction-less (JL) and inversion-mode (IM) devices. Data of optimized finfets with high temperature reliability anneal are reported as a reference

Fig. 10 (a) Matching coefficient A_{VT} vs. channel doping. A_{VT} is obtained on matched pairs and corresponds to $\sigma_{\Delta VT}$. A_{VT} increases with doping due to random dopant fluctuation, (b) benchmark with other Si technologies. A matching performance comparable to doped bulk planar bulk Si baselines is obtained.

Fig. 11 RF (a) f_T and f_{max} of top tier junctionless devices. Inset TEM cross-section of the RF structure.(b) Benchmark of f_T versus L_G with other Si technologies.

Fig. 12 (a) g_m/I_{DS} vs. I_{DS} and (b) analog gain A_v vs. g_m in junction-less NMOS devices.

References: [1] C. Fenouillet-Beranger *IEDM 2014*, [2] L. Pasini et al. *VLSI 2016*, [3] C. -M. V. Lu et al., *VLSI 2017*, [4] L. Brunet et al., *VLSI 2016* [5] B. Linder, *IRPS 2016*, [6] J. Franco et al., *IRPS 2017*, [7] M. Toledano-Luque et al., *EDL* 35(12), 2014, [8] A. Veloso et al., *VLSI 2016* [9] A. Mallik et al., *IEDM2017*, [10] S.-W. Kim et al., *3D System Intl. Conf.* 2015, [11] Sentaurus Device, M-2016.12 [12] C.H. Jan et al., *IEDM* 2010, [13] P. Van Der Voorn et al., *VLSI* 2010, [14] https://electronics360.globalspec.com/article/4078/samsung-foundry-adds-rf-to-28-nm-cmos [15] https://www.globalfoundries.com/technology-solutions/cmos/fdx/28nm-hkmg-technologies [16] R. Carter et al., *IEDM* 2016 [17] S.-Y. Wu et al., *IEDM* 2014.

978-1-5386-4219-1/18 $31.00 © 2018 IEEE

An over 120 dB wide-dynamic-range 3.0 μm pixel image sensor with in-pixel capacitor of 41.7 fF/um² and high reliability enabled by BEOL 3D capacitor process

M. Takase, S. Isono, Y. Tomekawa, T. Koyanagi, T. Tokuhara, M. Harada and Y. Inoue

Technology Innovation Division, Panasonic Corporation, Osaka, Japan, e-mail: takase.masayuki@jp.panasonic.com

Abstract

We realized a simultaneous-capture wide-dynamic-range image sensor with 3.0 μm pixels using novel in-pixel 3D capacitors located in BEOL. We achieved high capacitance density of 41.7 fF/μm² and low leakage current density of 3.6×10^{-10} A/cm² at 1 V by applying 3D structure and optimizing dielectric deposition process. TDDB investigations showed that estimated failure time at 125°C is more than 10 years. We demonstrated over 120 dB dynamic range image sensor with 3.0 μm pixels, which is enabled by this BEOL 3D capacitor process technology.

(Keywords: image sensor, capacitor, 3D, BEOL, high reliability and wide dynamic range)

Introduction

Recently, image processing technology has been rapidly growing. Especially, functions of camera have been expanding from just capturing images to sensing, object-recognition and decision-making. Thus, the demands for high performance and high resolution have been increasing.

Several pixel structures with DMOS capacitors have been reported in order to enhance sensor performance [1]. We have presented a simultaneous-capture wide-dynamic-range (SC-WDR) organic photoconductive film (OPF) image sensor with dual-sensitivity pixels (DS-pixels) that has two types of cells, a high-sensitivity cell and a high-saturation cell, as shown in Fig. 1 [2]. Each high-saturation cell has a DMOS capacitor to increase full well capacity. However, this structure doesn't allow us to shrink the pixel size because there is no space to put DMOS capacitors and transistors on a Si substrate. In order to overcome this shrinking limitation, we propose a novel pixel structure with in-pixel 3D capacitor in BEOL.

Pixel structure

Fig. 2 is comparison of pixel structures between a 6.0 μm pixel with DMOS capacitors and a 3.0 μm pixel with 3D capacitors. In a 6.0 μm pixel, both DMOS capacitors and pixel transistors can be designed on a Si substrate. On the other hand, in a 3.0 μm pixel, DMOS capacitors cannot be designed on a Si substrate due to the limitation of space. We propose in-pixel capacitors in BEOL instead of DMOS capacitors on a Si substrate. The implementation of capacitors in BEOL enables the pixel design to be easier and more flexible.

In-pixel capacitors require high capacitance density, low leakage, high breakdown voltage, and long TDDB lifetime in order to keep image quality with small pixel size. In order to realize both high capacitance and low leakage, we designed the 3D capacitors in which trenches are located in BEOL layer in each pixel. 3D structure enables us to use a thick dielectric film.

Experimental results

A. 3D capacitor configuration and characteristics

The electrical properties and reliabilities of capacitors were investigated using a test module on a 300 mm wafer. In this test module, we fabricated two kinds of structures, a simple planer structure and a 3D structure with trenches. Fig. 3 shows a cross sectional image of 3D capacitor. Each trench has 100 nm width and 400 nm depth. The 3D capacitor consists of a TiN/HfO₂/TiN stack. TiN and HfO₂ were deposited by atomic layer deposition process in order to ensure film uniformity inside the trenches. Following top electrode deposition, post annealing was performed in an H₂ ambient at 400°C for 30 min.

Fig. 4 shows the relation between capacitance density and leakage current density as a function of HfO₂ thickness. A 3D capacitor overcomes the trade-off between capacitance density and leakage current density in a planer capacitor. In case of 3D capacitor with 18-nm-thick HfO₂, high capacitance density of 41.7 fF/um² and low leakage current density of 3.6×10^{-10} A/cm² at 1 V were realized. Compared with previous reports [2-4], this 3D capacitor has relatively high capacitance density with low leakage current density as shown in Fig. 5.

In addition, we successfully improved breakdown voltage by optimizing deposition process of HfO₂. Fig. 6 shows I-V characteristics with 18-nm-thick HfO₂ before and after process optimization. It indicates that high voltage can be applied to this capacitor. TDDB characteristics at 125°C was also investigated. Fig. 7 shows the Weibull distribution of capacitors with 18-nm-thick HfO₂, which shows β of 1.33, 1.65 and 1.65 at applied voltage of 6.4 V, 6.2 V and 6.0 V, respectively. We projected the use voltage (V_{use}) at 10 year / 0.5 % / 32 mm² / 125°C on the assumption that an image sensor has 32M pixels and each pixel has a 1 μm² capacitor. As shown in Fig. 8, we achieved V_{use} = 2.33 V at 41.7 fF/um², which can be applied to analog circuit.

Thus, we realized a 3D capacitor with sufficient performance required for 3.0 μm pixel image sensor.

B. Image sensor characteristics

We fabricated an image sensor using a 65 nm 1PS 3Cu 1Al CMOS technology. As shown in Fig. 9, we successfully implemented in-pixel 3D capacitors located in BEOL in each 3.0 μm pixel. Table 1 lists the performance summary of this sensor, along with other reports. In this work, full well capacity of 489ke⁻ was realized using this 3D capacitor with 1 μm² per pixel applied up to 2.0 V. Despite of its small pixel size, we have realized over 120 dB dynamic range with 3.0 μm DS-pixel, almost the same as 6.0 μm DS-pixel with DMOS capacitors. We verified that in-pixel 3D capacitors in BEOL are able to enhance image sensor performance.

Conclusion

We have successfully developed a novel pixel structure with in-pixel capacitor located in BEOL. The in-pixel capacitor with high capacitance density of 41.7 fF/um², low leakage current density of 3.6×10^{-10} A/cm² at 1V, and high reliability was realized by applying 3D structure and optimizing high-k dielectric deposition process. Using this 3D capacitors, we demonstrated over 120 dB dynamic range in 3.0 μm pixel. This technology is a promising candidate for future image sensing.

References

[1] L. Stark, J. M. Raynor, F. Lalanne, and R. K. Henderson, Symp. VLSI Technol., pp. 242-243, 2016.

[2] K. Nishimura et al., ISSCC, pp. 110-112, 2016.

[3] T. Ando et al., IEDM, pp. 236-239, 2016.

[4] Y. Koda et al., ECS Trans. 2016, vol.72, pp. 91-100, 2016.

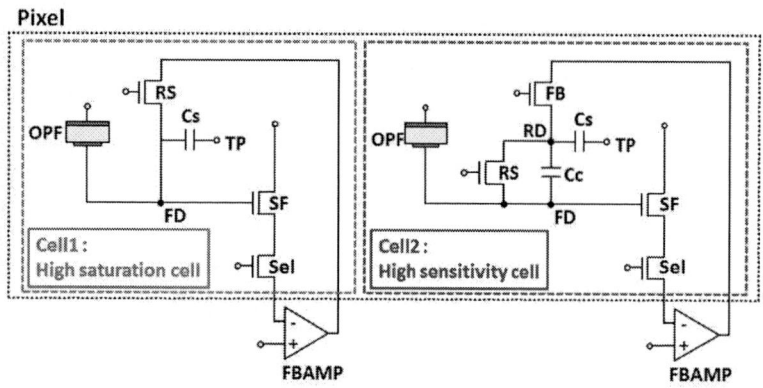

Fig. 1 Pixel circuit of dual sensitivity pixel that has two types of cells, a high-sensitivity cell and a high-saturation cell

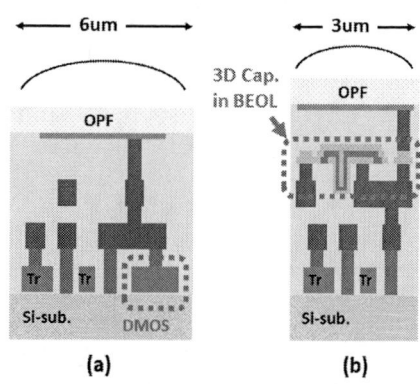

Fig. 2 Comparison of pixel structures of (a) 6.0 um pixel with DMOS capacitor and (b) 3.0 um pixel with in-pixel capacitor in BEOL

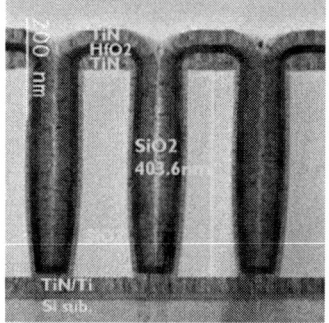

Fig. 3 Cross-section of test module

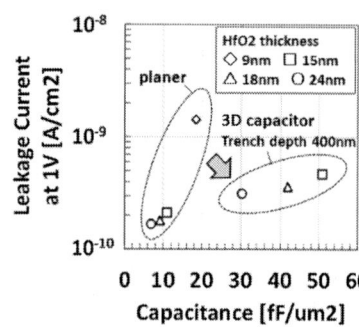

Fig. 4 Leakage current vs capacitance density as a function of HfO$_2$ thickness

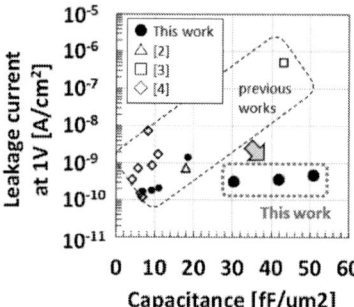

Fig. 5 Leakage current vs capacitance density comparison with previous papers

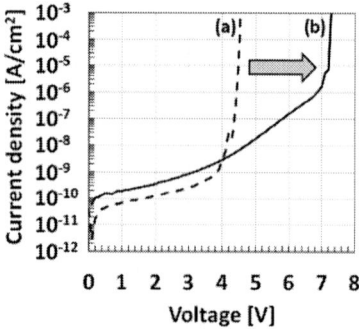

Fig. 6 Current density vs applied voltage (a) before and (b) after optimization

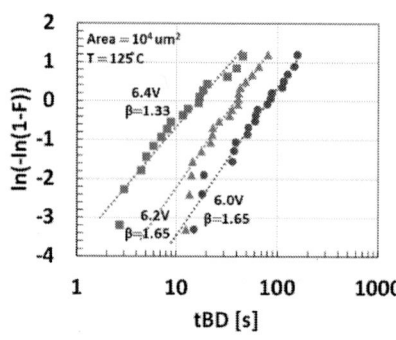

Fig. 7 Weibull distribution of capacitors with 18-nm-thick HfO$_2$ at 125°C

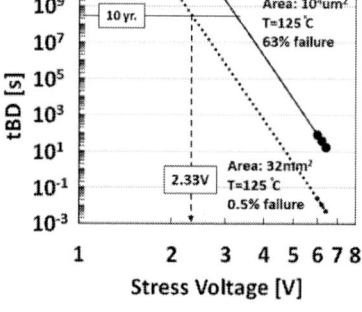

Fig. 8 TDDB lifetime projections at 10 year / 0.5 % / 32 mm^2 / 125°C

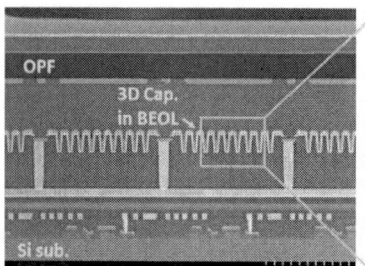

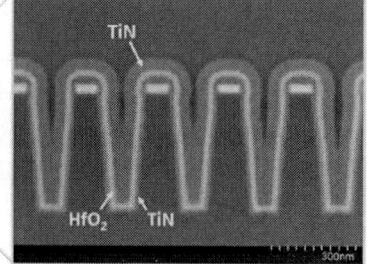

Fig. 9 Cross-sectional view of pixel structure with in-pixel 3D capacitor in BEOL

Table 1 Chip characteristics

	[1]	[2]	This work
Pixel size [um]	3.75 x 3.75	6.0 x 6.0	3.0 x 3.0
Pixel topology	Single pixel	Dual-sensitivity	Dual-sensitivity
Capacitor type	DMOS	DMOS	Capacitor in BEOL
Sensitivity Ratio	N.A.	14	14
Full-well capacity [ele]	8.1k	600k	489k
Random noise [ele]	2.7	5.0	6.2
SC WDR [dB]	102	123	121

Selective Pore-Sealing of Highly Porous Ultralow-k dielectrics for ULSI Interconnects by Cyclic Initiated Chemical Vapor Deposition Process

Seong Jun Yoon[1], Kwanyong Pak[2], Hyun Jun Ahn[1], Alexander Yoon[3], Sung Gap Im[2], and Byung Jin Cho[1]

[1]Dept. of Electrical Engineering, KAIST, Daejeon 34141, Republic of Korea
Phone: +82-42-350-3485 Email: bjcho@kaist.edu
[2]Dept. of Chemical and Biomolecular Engineering, KAIST, Daejeon 34141, Republic of Korea
[3]Lam Research Corporation, San Jose, California 95134, USA

Abstract: A selective pore-sealing of highly porous ultralow-k (pULK) dielectrics by a cyclic initiated CVD (iCVD) process has been successfully developed. A negligible increase of the pULK thickness and the k value was achieved even after the hermetic pore-sealing. The pore-sealed pULK films show low leakage current and excellent dielectric reliability, comparable to the commercialized low-k dielectric. The selective pore-sealing process does not deposit the pore-sealing layer on Cu surface. The porosity difference between pULK and Cu surfaces is attributed to the origin of the selectivity in the cyclic iCVD process.

Introduction: As a porosity of inter/intra-layer dielectrics increases in modern ULSI interconnects, severe process-related issues have arisen, *e.g.* a penetration of CVD/ALD barrier metal precursors into porous dielectrics [1]. To mitigate the problems, several pore-sealing methods have been developed, however, they have several drawbacks, *e.g.* plasma-induced damages [2-3]. In this work, we present a damage-free and selective pore-sealing method by the cyclic iCVD process.

Experiments: As a base pULK dielectric, a PECVD porous SiCOH (pSiCOH) film with k = 2.0 (porosity ~ 47%) was selected. We had reported a basic concept of the cyclic iCVD process to form a polymeric pore-sealing layer at pSiCOH surfaces (Fig. 1) [4]. The iCVD process injects monomers & initiators into a chamber and converts the initiators into radicals by heating filaments to induce a free-radical polymerization of the adsorbed monomers at the pSiCOH surfaces. After that, the pSiCOH film is post-annealed to remove unreacted monomers underneath the surface polymers. By repeating the iCVD & post-annealing cycles, the pSiCOH surfaces are gradually sealed. We found that the sealing becomes hermetic after the 3rd cycle, where the k_{eff} is only 2.16 (Fig. 2). For the each iCVD process, 1,3,5-trimethyl-1,3,5-trivinyl cyclotrisiloxane (V3D3) and di-tert-butyl peroxide (TBPO) were selected as a monomer and an initiator, respectively. The final form of the pore-sealing layer is poly-V3D3 (pV3D3).

Results and Discussion: Since iCVD does not require any plasma ambient and the substrate temperature is maintained at near room temperature, there are neither thermal nor plasma-induced damages to pSiCOH films, which is extremely important to keep the low-k nature during pore-sealing [4]. Furthermore, we have found that the cyclic iCVD pore-sealing reduces the leakage current and enhances the TDDB lifetime of pSiCOH as the iCVD & pore-annealing cycle is repeated (Fig. 3) [4]. The hermetically sealed pSiCOH shows an excellent long-term reliability, comparable to a typical dense low-k dielectric. Another important advantage of the cyclic iCVD process is the selective deposition of pV3D3 only on the pSiCOH surface, not on the Cu surface. To confirm the pore-sealing capability and selectivity, both of Cu & pSiCOH surfaces were formed on the same wafers (Fig. 4). After that, some wafers underwent the cyclic iCVD process (3 cycles) and the other wafers not. Then, ALD TiN was deposited on all the wafers. While a severe penetration of ALD TiN precursors into the pSiCOH layer is observed for the case of no cyclic iCVD cycle, a complete blocking against the ALD TiN precursors is observed for the case of 3 cycled samples, which indicates the polymerization occurred at the pSiCOH surface and the surface was hermetically sealed (Fig. 5). Meanwhile, no deposition of pV3D3 layer is observed on the Cu surface (Fig. 6). The possible mechanisms for the selectivity are either from the difference in material itself or the difference in porosity. To identify the dominant mechanism of the selectivity, a dense SiCOH film (k = 2.7, porosity = 7.2 %) is prepared and compared. Formation of pV3D3 was inhibited on both Cu and dense SiCOH surfaces during the cyclic iCVD process and there was no remarkable difference between the two surfaces. When porous and dense SiCOH surfaces are compared, however, the amount of pV3D3 formed on the both surfaces was very different (Fig. 7). Fig. 8 shows that V3D3 monomer adsorption on dense surfaces is much less than that on porous surfaces, which suggests that the polymerization rarely occurs on dense surfaces. Fig. 9 shows that the use of lower pressure during the each iCVD process increases the incubation time and slows down the thickening of the pore-sealing layer because the monomers are hardly adsorbed on the outer pSiCOH surfaces. This indicates that lowering the process pressure will help increase the selectivity. Moreover, at the early cycles (3-4) of the cyclic iCVD process, both of hermetic pore-sealing and virtually zero-thickness increase can be achieved simultaneously. Since the iCVD process utilizes the adsorption-first-and-polymerization-last mechanism, which is similar to a typical ALD mechanism, the conformal pore-sealing on narrow pSiCOH trench/vias is feasible. Fig. 10 shows EDS Ti signal mapping results on patterned pSiCOH surfaces after ALD TiN deposition. The cyclic iCVD process achieved a successful pore-sealing of the patterned pSiCOH (gap between the patterns < 20 nm), showing no ALD precursor penetration on all sides of the patterned pSiCOH.

Conclusion: It has been shown that the pV3D3 pore-sealing layer deposited by the cyclic iCVD process can seal the pores at highly porous dielectric surfaces with minimum increases of the k value. Furthermore, a selective pore-sealing only at the porous dielectric surfaces has been successfully demonstrated. The results in this work can pave the way for the use of highly porous dielectrics with no worry of the process-induced damage and degradation of the dielectric reliability.

References: [1] P. Verdonck, et al., *ECS J. Solid State Sci. Technol.*, vol. 2, pp. N103, 2013. [2] A. Kobayashi, et al., *IITC*, 2013. [3] Y. Sun, et al., *Microelectron. Eng.*, vol. 137, pp. 70, 2015. [4] S. J. Yoon, et al., *ACS Nano*, vol. 11, pp. 7841, 2017.

2018 Symposium on VLSI Technology Digest of Technical Papers

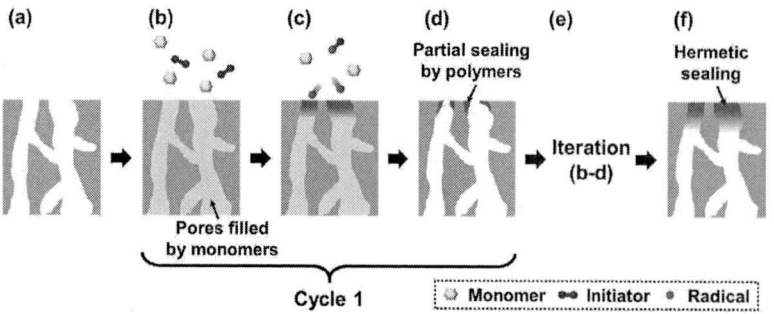

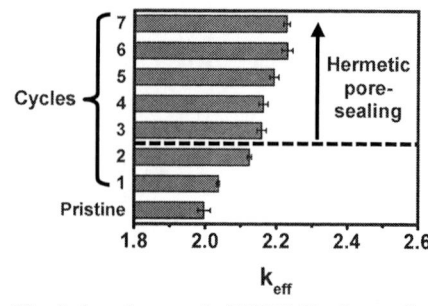

Fig. 1. Schematic illustrations of pore-sealing by cyclic iCVD process. (a) A pristine pSiCOH film. (b) Injection of vaporized monomers & initiators. The monomers are condensed in the pores. (c) The initiators are converted into radicals. Polymerization starts from the surface. (d) Post-anneal to desorb unreacted monomers passing through the surface polymers. Still some polymer fragments remain. (e) By repeating b-d, the amount of the condensed monomers decreases and the remaining polymer fragments are reinforced. (f) After a certain cycle, the pSiCOH surfaces are hermetically sealed. [4]

Fig. 2. k_{eff} change of pSiCOH film by cyclic iCVD process. After the 6th cycle, k_{eff} saturates to 2.23 (k of pure pV3D3 = 2.2). From ref. 4, the pore-sealing is already effective from the 3rd cycle, where k_{eff} = 2.16.

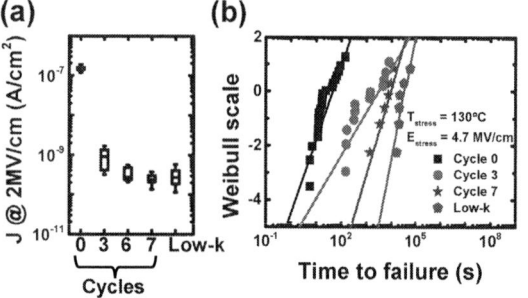

Fig. 3. (a) Leakage current densities of pristine (cycle 0), pore-sealed (cycle 3, 6, 7) pSiCOH and low-k dielectric at 2 MV/cm. (b) TDDB lifetimes of pristine, pore sealed (cycle 3, 7) pSiCOH and low-k dielectric. The k & porosity of the low-k dielectric are 2.7 & 7.2% respectively.

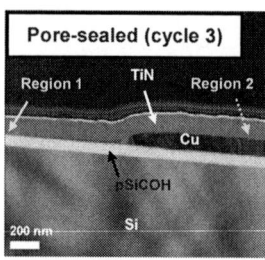

Fig. 4. Bright field TEM image of pore-sealed (cycle 3) pSiCOH (region 1) & Cu (region 2) surfaces in a single wafer. ALD TiN was deposited after the pore-sealing.

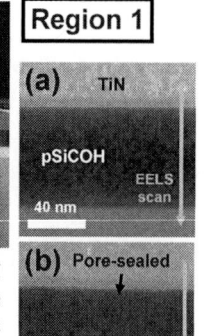

Fig. 5. Scanning TEM images of (a) pristine and (b) pore-sealed pSiCOH layers of region 1 in Fig. 4. (c) Related EELS Ti signals in pristine & pore-sealed pSiCOH layers.

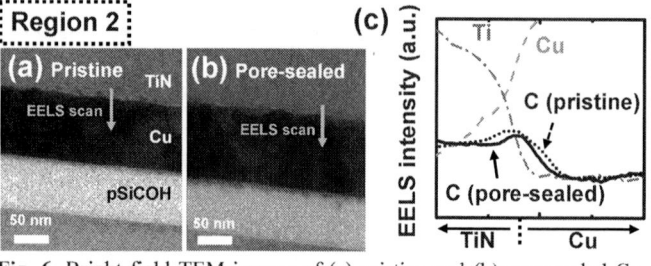

Fig. 6. Bright field TEM images of (a) pristine and (b) pore-sealed Cu surfaces of region 2 in Fig. 4. (c) EELS Cu, Ti, C signals at Cu-TiN interfaces of pristine & pore-sealed samples. Similar C line profiles are seen for both cases.

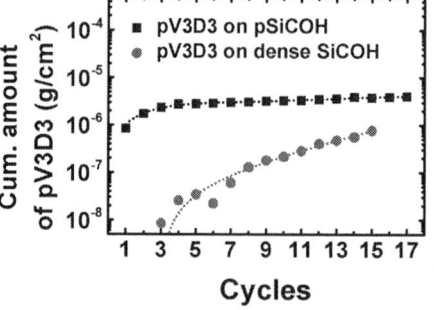

Fig. 7. Cumulative amount of pV3D3 deposited on porous & dense SiCOH films in each cycle of the cyclic iCVD process.

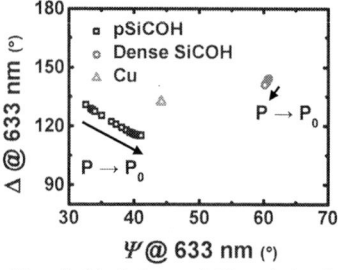

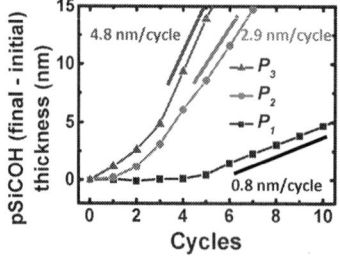

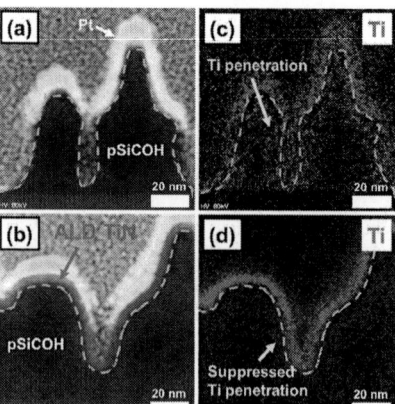

Fig. 10. Scanning TEM images of patterned (a) pristine and (b) pore-sealed pSiCOH films. EDS Ti signals in (c) pristine and (d) pore-sealed pSiCOH. Conformal pore-sealing at pattern sidewalls is confirmed.

Fig. 8. Variations of Ψ and Δ of porous, dense SiCOH and Cu films during V3D3 adsorption measured by ellipsometric porosiemtry. Large variation means large adsorption of V3D3. P & P0: partial & saturation vapor pressures of V3D3.

Fig. 9. Thickness changes of pSiCOH films during cyclic iCVD process with various iCVD process pressures. P1 is the standard condition in this study. P3 > P2 > P1.

978-1-5386-4219-1/18 $31.00 © 2018 IEEE

Performance and Reliability of a Fully Integrated 3D Sequential Technology

A.Tsiara[1,2,3], X.Garros[1,3], L.Brunet[1,3], P.Batude[1,3], C.Fenouillet-Béranger[1,3], K.Triantopoulos[1,3], M.Cassé[1,3], M.Vinet[1,3], F.Gaillard[1,3] and G.Ghibaudo[2,3]

[1]CEA-LETI, MINATEC, Grenoble, France [2]IMEP-LAHC, France [3]Université Grenoble Alpes, France. Email: artemisia.tsiara@cea.fr

Abstract

We investigate in detail, for the first time, *both performance and reliability* of a 3D sequential integration process. It is clearly demonstrated that the *top level transistor can be successfully processed at 630°C* with almost *no impact on the performance and reliability of the bottom level*. It is also highlighted that top level devices *meet the P&NBTI reliability requirements*. Finally an example of successful and robust 3D logic integration is proposed based on a 3D inverter combining a top-level PMOS with a bottom-level NMOS.

Introduction

3D sequential integration, with its unique 3D contact characteristics offers a large set of applications. It is an alternative to the traditional scaling for computing application. It also offers opportunities for smart and scaled sensors [1]. But the fabrication of the two level transistors faces process integration challenges [2-4]. Indeed the top MOSFET must be processed at low temperature ($\approx 500°C$) in order to preserve the integrity of bottom level MOSFET and BEOL levels. However, using a low thermal budget (TB) could reduce the top device performance and could degrade its reliability [5-6]. For the first time, an in depth analysis of the performance and reliability of a fully integrated 3D sequential technology is presented. The first part shows the performances of the two device levels; focusing on how the top level processing impacts the performance of the bottom one. Then an extensive study of reliability (LFN, BTI and HC) of both levels is reported. Finally the last part presents an example of this monolithic 3D integration, based on a study of a two-level 3D inverter.

Process fabrication

The process flow is illustrated in Fig. 2. Gate-first CMOS FDSOI route serves as 2D reference (REF) in our study. For the "Bottom Level" (BL), the Si channel is 7nm thick and a HfSiON/TiN/Poly gate stack is used. Next, we developed a low-temperature sequential integration process to fabricate NMOS & PMOS transistors on a top tier as depicted on the cross-sectional TEM image (Figs. 1-2). The Top Level (TL) devices also feature a 7nm thick Si channel thickness but uses an HfO2/TiN stack. The maximum temperature of the TL process is fixed by the formation of the SiN spacers at 630°C. More details about the device fabrication can be found in [7]. Note that the TB reduction of the top MOSFET down to 500°C, is feasible as shown in [7], but has not been done here.

Bottom & Top transistor performance and t₀-reliability

Figs. 3&4 first summarize the electrical performance of long channel devices (L=10µm) of the bottom level. The key feature is that, the transport as well as the gate stack properties are not affected by the processing of the top level. Indeed no difference of Equivalent Oxide thickness (EOT) and mobility is seen between the reference wafer (REF) and the one which has seen the processing of the TL (3D seq.). Fig. 3 also shows that the EOT & work function extracted from CVs of TL N&PMOS matches the value of the BL level *i.e.* 1.12nm & 4.5eV, respectively. Looking at the short-channel device performance (Fig. 5), we notice a small degradation of the saturated current I_{ON} of bottom level transistors after processing the top level. Fig. 6 demonstrates that this is due to a small increase of the sheet resistance of the S/D contacts. This is consistent with previous observations and can be related to a structural change of the silicide layers for temperatures above 550°C [4, 8]. Matching results are also reported in Fig. 7. Excellent A_{VT} value of $\approx 1mV.\mu m$ is measured for both planar REF transistors and BT devices of 3D sequential wafers. This confirms the benefit of undoped FDSOI films to reduce variability, as well as that top processing has only minor impact on BL performance. Charge Pumping (CP) measurements are then presented in Fig. 8. Interface state density from CP is found very low $\approx 2.10^{10}/cm^2/eV$ for both REF & 3D seq. wafers. This proves that the excellent Si interface passivation achieved for REF FDSOI technology can be preserved after TL processing. For Low Frequency Noise (LFN), the noise level is found even higher for the REF transistors (Fig. 9). By fitting the normalized Power Spectral Density at 10Hz with the CNF/CMF model [9], we extract the bulk oxide trap density (N_t) and the correlated mobility fluctuations factor (Ω). Ω is almost the same since it mainly depends on the centroid of the carriers with the channel. However N_t is divided by 3 after the TL processing. This suggests that additional anneal due to TL process favors defect curing in the BL gate oxide. In contrast, mobility of TL devices is degraded (Fig. 10). This can be ascribed here to enhanced Coulomb scattering resulting from a higher

interface trap density D_{it} compared to BL (Fig. 11). Several process knobs, as high pressure Deuterium anneal, are under study to improve the Si interface passivation and, by turn, improving mobility. Inserting HPD2 into the process route enabled us to achieve the D_{it} values found on high temperature bottom level devices without degrading the leakage current (Fig. 12) [10, 11]. I_{on}/I_{off} Figure Of Merit for short gate lengths top level PMOS, with SiGe$_{27\%}$ raised S/D is also reported in Fig. 13 [1].

Bottom & Top Level reliability

We first address how the processing of the top level impacts the reliability of bottom level transistors, by always comparing 2D REF and BL 3D sequential wafers. BTI degradation at 125°C is analyzed in Figs. 14&15. Clearly no difference is seen on the BTI amplitudes and on the $V_{G,stress}$ dependences for REF devices and bottom transistors of the 3D integration scheme. Hot Carrier (HC) reliability is also evaluated in 30nm NMOS. Stresses are performed under worst case conditions $V_G=V_D$ at 125°C (Fig. 16). The I_{Dsat} drift is almost the same whatever the stress level. Furthermore, negligible recovery is observed during the relaxation ($V_{stress}=0V$) for both variants, confirming that interface state generation is responsible for HC degradation, in both cases. This last result, also, suggests that the integrity of dielectrics over the channel/drain junction is preserved after the TL processing and that Self Heating which can enhance HC degradation is not higher in the 3D integration scheme in spite of the presence of the top level (Fig. 19) [12]. HC lifetime of BL devices is finally extracted in Fig. 17. V_D for a 10% drive current degradation and one-year operation is estimated to 0.995V, that satisfies the technology requirements ($V_D > 0.9V$). To conclude, it is obvious that *the top level process integration does not affect the gate oxide reliability of the bottom level CMOS transistors*. Regarding top level transistors, N&PBTI reliability is studied in Fig. 18. Minimum Vg for a 5-year lifetime is very close to their BL counterparts and over 1V for both N&PMOS. This is therefore the first demonstration that *Top Level of a 3D seq. integration meet the BTI requirements*. A complete benchmark of performance and reliability, of published results [13-17], is presented at the end of the abstract.

Circuit operation – 3D inverter performance & reliability

We then study the circuit operation of a two-level inverter, fabricated using this 3D sequential integration scheme. Fig. 20 shows the $V_{in}-V_{out}$ characteristics of an inverter made of a PMOS fabricated on the top on a BL NMOS with two different configurations, independent or shared crossing drain contacts. Good operation of the 3D inverter is evidenced for all values of V_{DD}. Figs. 21-22 also prove that experimental data can be perfectly reproduced by a homemade SPICE-like model. The model, calibrated against characteristics of individual transistors of the two levels, is very suitable to predict the inverter consumption at any V_{DD}. Fig. 23 shows that it is also very useful to predict how V_T of both top PMOS and bottom NMOS must be tuned to achieve the right inverter performance in terms of switching bias $V_{1\to0}$ and dissipated power. The same inverter is then stressed under NBTI condition using a fast BTI methodology. Experimental V_{out} drifts of Fig. 24 (left) are perfectly captured by BTI modeling (right) calibrated against device data of Fig. 18. This already validates our approach combining SPICE like and NBTI modeling to predict aging of 3D sequential circuits.

Conclusion

Performance and reliability of a fully integrated 3D sequential integration scheme have been in depth analyzed for the first time. It is demonstrated that a 3D integration with a processing of a top level transistor at 630°C has only minor impact on the performance and reliability of the bottom level. Top level devices show already good performance and great enough N&PBTI reliability. Reliability of 3D inverters combining a top-level PMOS with a bottom-level NMOS is also measured and modeled for the first time.

References

[1] P. Batude, IEDM'17 [2] Y.-J. Lee, IEDM'09 [2] A. Kumar, VLSI'17 [3] C. Fenouillet-Beranger, ESSDERC'14 [4] C. Fenouillet-Beranger, IEDM'14 [5] K. C. Saraswat, SOI Conf.'99 [6] E.-K Lai, IEDM'06 [7] L. Brunet, VLSI'16 [8] P. Batude, VLSI'11 [9] G. Ghibaudo, Phys. Stat. Sol. 1991 [10] C.-M.V.Lu, VLSI'17 [11] A. Tsiara, S3S'17 [12] K. Triantopoulos, IEDM'17 [13] C-C Yang, IEDM'16 [14] C- Yang, IEDM'15 [15] K. Usuda, IEDM'14 [16] C-C Yang, IEDM'13 [17] J. Franco, IRPS'17

2018 Symposium on VLSI Technology Digest of Technical Papers

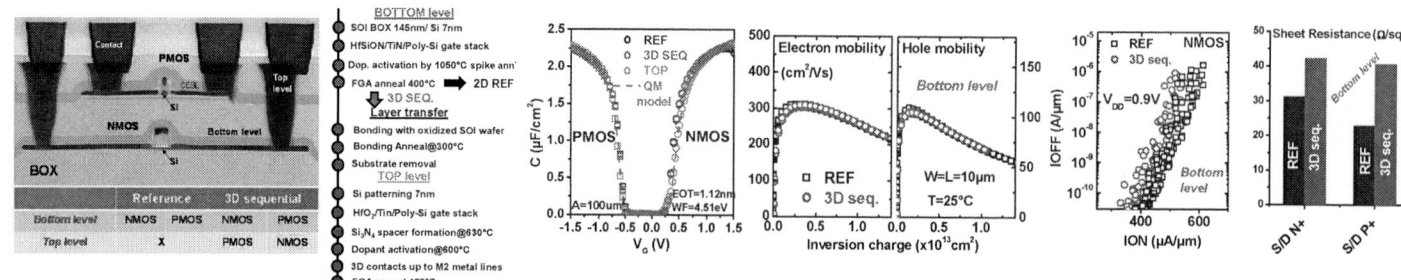

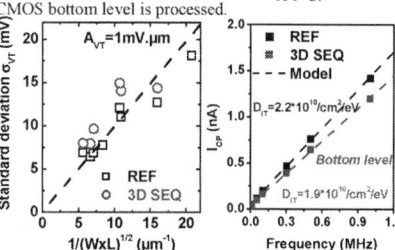

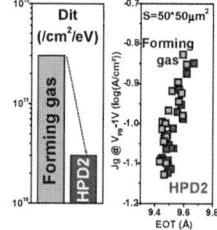

Figure 1 TEM image of 3D integration scheme. N&PFETs are fabricated on BOTTOM and TOP layers. For 2D planar references, only CMOS bottom level is processed.

Figure 2 Process flow of 3D sequential scheme. Max temperature of the top level is 630°C.

Figure 3 CV characteristics measured on long channel N&PMOS of BL transistors of 3D seq. tech., 2D REF and TL. EOT is 1.12nm in all cases.

Figure 4 Electron and hole mobility for 2D reference and bottom transistors of a 3D seq. technology. Mobility is unchanged after top level processing.

Figure 5 I_{on}/I_{off} for NMOS bottom level devices. Small I_{on} degradation is observed after the top level process.

Figure 6 Sheet resistance of N+ & P+ Source/Drain regions. Increased resistance after the TL process explains I_{on}/I_{off}.

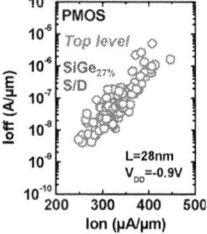

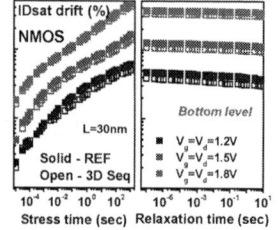

Figure 7 Pelgrom FOM for REF and BL 3D seq. technology. Excellent A_{vt} is measured even after the TL fabrication.

Figure 8 Charge pumping current vs frequency. Very low D_{it} value is extracted for both 2D REF and BL 3Dseq transistors.

Figure 9 Normalized PSD@10Hz fitted with CNF/CMF model. Bulk trap density N_t is strongly reduced after 3D integration.

Figure 10 Electron & hole mobility for TOP level extracted on long channel N&PMOS from CV split.

Figure 11 Conductance-Voltage GV characteristics for TOP level N&PMOS. D_{it} is extracted to ~$5.10^{11}/cm^2/eV$ by GV peak modeling.

Figure 12 Investigation of Deuterium as back end forming gas seems to improve D_{it} values and does not degrade the leakage current.

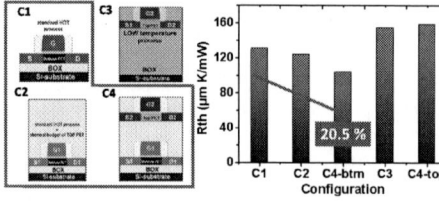
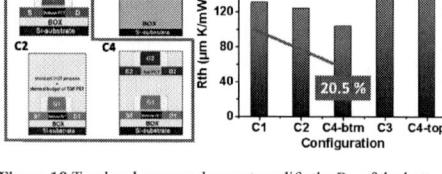

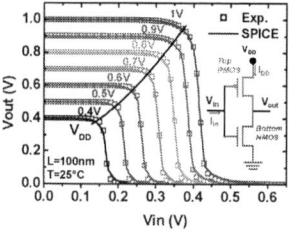

Figure 13 I_{on}/I_{off} FOM for short gate length PMOS TOP level devices fabricated at low temperature. Use of $SiGe_{27\%}$ for raised S/D.

Figure 14 NBTI shift vs time measured on 2D REF and BL PMOS of 3D seq. using fast methodology (1µs range) at T=125°C.

Figure 15 BTI shifts for N&PMOS transistors showing no additional degradation due to the Top Level process.

Figure 16 HC drift in L=30nm BL NMOS for various stress conditions of VG=VD. Again no difference is seen between 2D REF and BL 3Dseq.

Figure 17 HC extrapolation, of short channel BL NMOS, of Time-To-Failure for different stress voltage. Pass of the criterion $V_D > 0.9V$.

Figure 18 NBTI/PBTI Time-To-Failure extrapolation for TL N&PMOS compared to the BL counterparts. Both levels meet the lifetime requirements at 5 and 10 years (see benchmark).

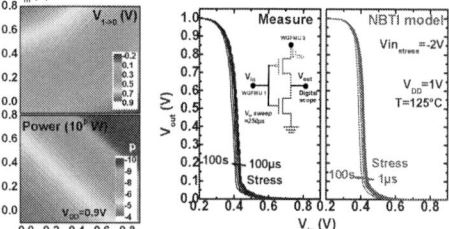
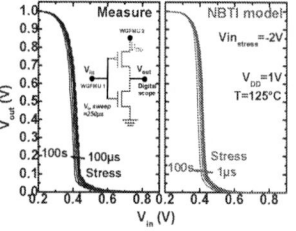

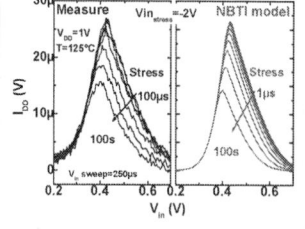

Figure 19 Top level process does not modify the R_{th} of the bottom level transistor (C1&C2) [12]. Self-Heating is not enhancing HC degradation.

Figure 20 V_{in}-V_{out} characteristic of an inverter integrated at 3D sequential scheme for different V_{DD} and two configurations.

Figure 21 Fit of the experimental data with a SPICE-like model for the common drain setup at different V_{DD}.

Figure 22 Very good agreement between simulated and measured I_{DD} current for different V_{DD} (left). Fit of the calculated consumption as well (right).

Figure 23 Simulated switching voltage (top) and dissipated power (bottom) of the inverter w.r.t. N&PMOS $V_{t,n}$ & $V_{t,p}$.

Figure 24 Measured (left) and simulated (right) drifts of V_{in}-V_{out} characteristic of the two-level inverter under NBTI stress at $V_{in,stress}$=-2V and 125°C

Figure 25 Current drifts at V_{DD}=1V, under NBTI stress of an inverter integrated at 3DSI (left) in very good agreement with the simulated characteristics (right).

		This work	IEDM'16[13]	IEDM'15[14]	IEDM'14[15]	IEDM'13[16]	IRPS'17[17]
Performance	Channel	Si mono	Epi-like Si	Epi-like Si	Poly-Ge	Si LC	Si
	Gate	TiN	TaN	TaN	TaN	TiN	TiN
	Structure 3D seq.	FDSOI on FDSOI	FinFET (TL&BL)	NWFET on bulk FinFETs	FinFET on bulk CMOS	UTB on bulk MOSFETs	Planar capacitors
	T_{si} (nm)	7	53.4	16	47	14	-
	EOT (nm)	1.12	-	-	-	2.1	1
	TOP level I_{on}/W_{eff}	P: 250 (I_{off}=40nA/µm V_{DD}=-0.9V)	P: 352 (I_{off} unknown V_{DD}=-1V)	P: 220 (I_{off} unknown V_{DD}=-1V)	P: 311 (I_{off} unknown V_{DD}=-1V)	P: 62 (I_{off} unknown V_{DD}=-1V)	-
Reliability	BTI@T=125°C Crit=50mV	VG@10y/3σ, PASS if VG>0.9V					V_{ov}@10y/ 30mV, T=25°C
	PMOS (TL/BL)	0.98V/1.01V OK		X			TL: V_{ov}=0.7V BL: V_{ov}=1.05V
	NMOS (TL/BL)	1.26V/1.37V OK					
	HCI@T=125°C ΔId_{sat}=10%	VD_{max}@1y, PASS if VD>0.9V		X			X
	NMOS	BL: 0.91V OK					

Benchmarking performance and reliability of published results for 3D sequential technology

Acknowledgement - This work was supported by the French Public Authorities through NANO2017, EQUIPEX FDSOI11, LabEx Minos ANR-10-LABX-55-01 programs.

Metal/P-type GeSn Contacts with Specific Contact Resistivity down to 4.4×10^{-10} Ω-cm^2

Ying Wu,[1] Wei Wang,[1] Saeid Masudy-Panah,[1] Yang Li,[1] Kaizhen Han,[1] Liuhuiquan He,[1] Zheng Zhang,[2]
Dian Lei,[1] Shengqiang Xu,[1] Yuye Kang,[1] Xiao Gong,[1,*] and Yee-Chia Yeo[1,*,†]

[1] Department of Electrical and Computer Engineering, National University of Singapore (NUS), 117576 Singapore;
[2] Institute of Material Research and Engineering, A*STAR (Agency for Science, Technology and Research), 138634 Singapore
*Phone: +65 6516-7871, E-mail: elegong@nus.edu.sg, yeo@ieee.org † Currently with TSMC.

ABSTRACT

Ga and Sn surface-segregated p$^+$-GeSn (Seg. p$^+$-GeSn) was grown by molecular beam epitaxy (MBE) to achieve an average active Ga doping concentration of 3.4×10^{20} cm^{-3} and surface Sn composition of more than 8%. This enables the realization of record-low specific contact resistivity ρ_c down to 4.4×10^{-10} Ω-cm^2. The average ρ_c extracted from 14 sets of Ti/Seg. p$^+$-GeSn Nano-TLM test structures, a collection of more than 90 devices is 6.5×10^{-10} Ω-cm^2. This is also the lowest ρ_c for non-laser-annealed contacts. Ti contacts to p$^+$-GeSn films with and without Ga and Sn surface segregation were fabricated. It is shown that the segregation of Ga and Sn at the Ti/p$^+$-GeSn interface leads to 50% reduction in ρ_c as compared with a sample without segregation.

I. INTRODUCTION

Ge$_x$Sn$_{1-x}$ has higher hole mobility than Si, Si$_{1-y}$Ge$_y$, and Ge, and is a promising channel material for future p-channel FETs (PFETs) [1,2]. Ge$_x$Sn$_{1-x}$ can also be used in source/drain (S/D) contacts. For the aggressively scaled FETs, S/D contact resistance (R_c) is one of the key factors that limits the drive current or speed performance of transistors. Sub-10^{-9} Ω-cm^2 specific contact resistivity (ρ_c) is desired for future advanced technology node [3]. Ga is a promising doping species to achieve high active doping concentration for p-type Ge (Ge$_x$Sn$_{1-x}$) because of its higher solid solubility limit in Ge than boron [4]. Incorporating Sn into Ge can provide additional benefits for ρ_c reduction by reducing the hole effective mass, shifting up the valence band edge of Ge, and boosting the hole tunneling between metal and p-type Ge$_{1-x}$Sn$_x$. 50% reduction of ρ_c has been obtained for Ti/p$^+$-Ge$_{0.95}$Sn$_{0.05}$ as compared to that of Ti/p$^+$-Ge [3].

In this work, we realized a Ti/p$^+$-Ge$_{0.95}$Sn$_{0.05}$ contacts with ρ_c down to 4.4×10^{-10} Ω-cm^2 (average ρ_c of 6.5×10^{-10} Ω-cm^2) by using three key technology enablers: (1) a heavily doped p$^+$-GeSn with an average active Ga doping concentration as high as 3.4×10^{20} cm^{-3}, (2) Ga surface segregation to further enhance the doping concentration at the contact interface, and (3) Sn surface segregation to decrease the hole effective mass and lower tunneling barrier (Fig. 1). In addition, it is experimentally verified that use of Sn surface segregated p$^+$-GeSn (Seg. p$^+$-GeSn) film achieves a 50% lower ρ_c as compared to the p$^+$-GeSn control without Ga and Sn surface segregation.

II. GROWTH OF GESN WITH GA AND SN SURFACE SEGREGATION

Ga and Sn surface segregated Ge$_{1-x}$Sn$_x$ layer was grown by MBE on Ge (100) substrate. Fig. 2 shows the Ga SIMS depth profiles of the Seg. p$^+$-GeSn and p$^+$-GeSn without segregation (control). As compared to the p$^+$-GeSn control, obvious Ga surface segregation with a surface Ga doping concentration of ~6×10^{21} cm^{-3} and a surface-to-bulk ratio of ~10 was observed in the Seg. p$^+$-GeSn sample. The active Ga doping concentration N_A was evaluated by infrared Ellipsometry Spectroscopy (IR-ES) based on Drude model [5]. The real (ε_1) and imaginary (ε_2) parts of the infrared dielectric function of Seg. p$^+$-GeSn and p$^+$-GeSn control are shown in Fig. 3. The measured active doping concentrations of Seg. p$^+$-GeSn and p$^+$-GeSn control are 3.4×10^{20} cm^{-3} and 1.6×10^{20} cm^{-3}, respectively. It should be noted that the IR-ES measurement gives an average carrier concentration. A higher N_A at the surface of Seg. p$^+$-GeSn film is expected due to Ga surface segregation.

Fig. 4 shows the Sn SIMS profiles of the Seg. p$^+$-GeSn and p$^+$-GeSn control samples. A uniform Sn profile was obtained in the p$^+$-GeSn control while Sn segregation was observed for the Seg. p$^+$-GeSn sample. The Sn composition near the surface (>8%) is 4 times higher than that in the bulk. Fig. 5 shows the Sn composition at the sample surface measured by Angle-resolved XPS with various take-off angles. The increase of Sn near the surface was clearly observed which is consistent with SIMS results.

II. NANO-TLM TEST STRUCTURE AND CHARACTERIZATIONS

The ρ_c was extracted based on the Nano-TLM structures [6]. The 3D schematic of Nano-TLM structure is shown in Fig. 6(a). The zoom-in image of the mesa region of Nano-TLM is shown in Fig. 6(b) indicating the mesa width W, the metal line width L_c, and the contact distance L_d. The cross-section view of Nano-TLM along the line EE' in Fig. 6(a) is shown in Fig.

6(c). The metal lies above the SiO$_2$ insulation layer except for that in the mesa region in order to suppress the leakage current flowing from metal into p$^+$-GeSn outside the mesa region. Key process steps for fabricating the Nano-TLM structure are listed in Fig. 7 and the SEM image of fabricated Nano-TLM structure is shown in Fig. 8.

The TEM image of the Ti/Seg. p$^+$-GeSn Nano-TLM along the line FF' in Fig. 6(a) is shown in Fig. 9 (a). The calibrated L_c of the Ti line is 206 nm [Fig. 9(b)]. High-resolution TEM (HRTEM) image in Fig. 9(c) shows a good and reaction-free interface formed between Ti and Seg. p$^+$-GeSn.

III. RESULTS AND DISCUSSION

The method of ρ_c extraction using Nano-TLM structure is well established, and details are given in Ref. 3. In the Nano-TLM structure, two measurement schemes, parallel and cross measurements, are used. The measured parallel terminal resistance R_p and cross terminal resistance R_x (symbols) of Ti/Seg. p$^+$-GeSn Nano-TLM structures with L_c of 206 nm, 156 nm, and 138 nm are shown in Fig. 10 (a), (b), and (c), respectively. The R_p and R_x in Fig. 10 (a) to Fig. 10 (c) were collected from four sets of Nano-TLM structures with the same L_c. Good consistency is observed for different sets of Nano-TLM structures with the same L_c value. The best-fit curves of R_p and R_x are also shown in Fig. 10. The contact resistance (R_c), semiconductor sheet resistance (R_{sh}), and metal sheet resistance (R_{shm}) can be obtained by the fitting of R_p and R_x, shown in the lower right corner of Fig. 10 (a), (b), and (c), respectively.

The sub-10^{-9} Ω-cm^2 ρ_c for Ti/Seg. p$^+$-GeSn extracted from 14 sets of Nano-TLM structures on two Seg. p$^+$-GeSn samples, a collection of more than 90 devices, are shown in Fig. 11. A ρ_c as low as 4.4×10^{-10} Ω-cm^2 is achieved for the Ti/Seg. p$^+$-GeSn contact. The average ρ_c for all extracted devices is 6.5×10^{-10} Ω-cm^2. The ρ_c of 1.4×10^{-9} Ω-cm^2 was also extracted for Ti/p$^+$-GeSn control. More than 50% reduction in ρ_c is obtained for Ti/Seg. p$^+$-GeSn as compared to that of Ti/p$^+$-GeSn control (Fig. 12). The significant reduction of ρ_c results from Ga surface segregation which enhances the doping concentration at the contact interface, decreasing the tunneling width for holes, and Sn surface segregation which lowers the Schottky barrier and decreases the hole effective mass, boosting hole tunneling for Ti/p$^+$-GeSn contact [3]. The combination of Ga surface segregation and Sn surface segregation provides an attractive way to achieve extremely low ρ_c in future advanced FETs.

Fig. 13 benchmarks the ρ_c of metal/p-Ge or Ge$_x$Sn$_{1-x}$ contacts reported in the literature [3,4,7-11]. The surface segregation of Sn and Ga leads to an average ρ_c of 6.5×10^{-10} Ω-cm^2, which is one of the lowest ρ_c for metal/p-type Ge(GeSn) contacts even though advanced implantation and annealing techniques are not used. The sub-10^{-9} Ω-cm^2 ρ_c is also the lowest value for non-laser-annealed contacts for any metal/p-type Group-IV semiconductor contacts (Fig. 14) [4,12,13]. Further reduction of ρ_c could be possible by boosting the active Ga doping concentration using advanced implantation techniques and laser annealing [10].

IV. CONCLUSION

The Ga and Sn surface segregated p$^+$-GeSn film with an average active Ga doping concentration of 3.4×10^{20} cm^{-3} and surface Sn composition of more than 8% was grown by MBE. ρ_c as low as 4.4×10^{-10} Ω-cm^2 (average ρ_c is 6.5×10^{-10} Ω-cm^2) was achieved by forming Ti contact on Seg. p$^+$-GeSn. This makes the Ti/Ge$_{1-x}$Sn$_x$ contact very attractive, not only for GeSn channel transistors, but also SiGe or Ge p-channel transistors with GeSn S/D regions.

Acknowledgement. The authors acknowledge support from NUS Trailblazer Grant (R-263-000-B43-733), and Ministry of Education (MOE) Academic Research Fund (R-263-000-B50-112).

References:
[1] Y.-C. Yeo *et al.*, *IEDM* 2015, pp. 28. [2] D. Lei *et al.*, *VLSI* 2017, pp. 198. [3] Y. Wu *et al.*, *JAP* 122, 224503, 2017. [4] L.-L. Wang *et al.*, *IEDM* 2017, pp. 549. [5] W. Wang *et al.*, *JAP* 119, 155704, 2016. [6] W. Liu *et al.*, *EDL* 35, pp.178, 2014. [7] H. Miyoshi *et al.*, *JJAP* 53, 04EA05, 2014. [8] P. Bhatt *et al.*, *EDL* 35, pp. 717, 2014. [9] L. Hutin *et al.*, *JES* 156, pp. H522, 2009. [10] O. Gluschenkov *et al.*, *IEDM* 2016, pp. 448. [11] H. Miyoshi *et al.*, *VLSI* 2014, pp. 978. [12] N. Stavitski, *et al.*, *EDL* 29, pp. 378, 2008. [13] Z. Zhang, *et al.*, *EDL* 34, pp. 723, 2013.

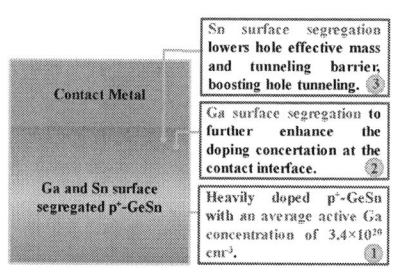
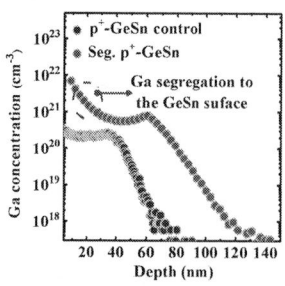
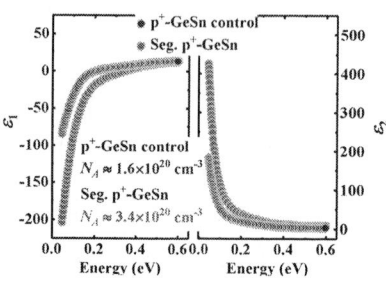
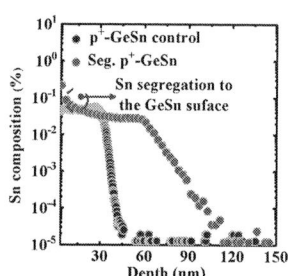

Fig. 1. Schematic of metal contact on Ga-doped p⁺-GeSn with Ga and Sn surface segregation to enhance the doping concentration at contact interface and boost hole tunneling, achieving a lower ρ_c.

Fig. 2. Ga SIMS profiles of the p⁺-GeSn control and surface segregated p⁺-GeSn (Seg. p⁺-GeSn). Ga surface segregation with surface-to-bulk ratio of ~10 was observed on Seg. p⁺-GeSn film.

Fig. 3. Real (ε_1) and imaginary (ε_2) parts of the infrared dielectric function of the grown p⁺-GeSn films, showing the average active Ga doping concentration (N_A) of 1.6×10^{20} cm⁻³ and 3.4×10^{20} cm⁻³ for the p⁺-GeSn control and Seg. p⁺-GeSn, respectively.

Fig. 4. Sn SIMS profiles of the p⁺-GeSn control and Seg. p⁺-GeSn. The Sn composition near the surface is about 4 times higher than that bulk in Seg. p⁺-GeSn due to the Sn surface segregation.

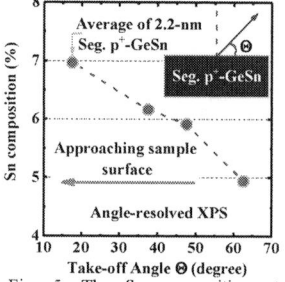
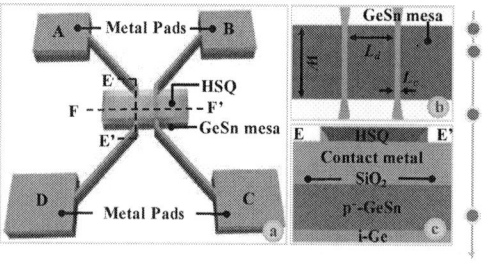
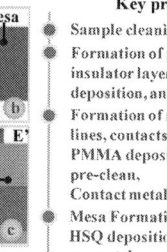
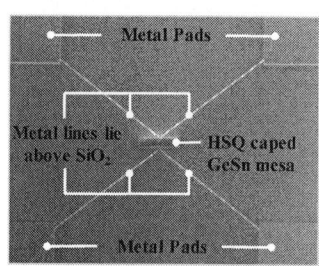

Fig. 5. The Sn composition at surface region of Seg. p⁺-GeSn as a function of photoelectron take-off angle Θ, as determined by angle-resolved XPS. The Sn composition increases with decreasing Θ, indicating Sn surface segregation.

Fig. 6. (a) 3D schematic of the Nano-TLM structure showing the metal pads, lines, and HSQ caped GeSn mesa. (b) The zoom-in image of GeSn mesa without the HSQ cap showing the key parameters (i.e. the mesa with, W, the metal line width, L_c, and the contact distance, L_d) of Nano-TLM structure. (c) The cross-section along the line EE' in (a), showing the contact metal lies above the SiO₂ insulation layer except for that in the contact region.

Key process steps
● Sample cleaning
● Formation of patterned SiO₂ insulator layer: Lithography, SiO₂ deposition, and lift-off
● Formation of metal structure (pads, lines, contacts):
   PMMA deposition and lithography, pre-clean,
   Contact metal deposition, lift-off
● Mesa Formation:
   HSQ deposition and lithography, mesa etch

Fig. 7. The key process steps of Ti/Seg. p⁺-GeSn nano-TLM structure fabrication, showing the SiO₂ insulation layer formation, metal structure (metal pads, and lines) deposition, and GeSn mesa formation.

Fig. 8. Tilted-view SEM image of the fabricated Ti/Seg. p⁺-GeSn Nano-TLM structure. The metal pads, lines, and HSQ capped GeSn mesa are shown. The mesa width (W) of the fabricated Ti/Seg. p⁺-GeSn Nano-TLM structure is 2.5 µm.

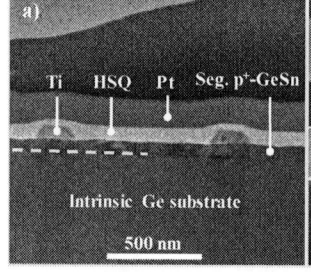
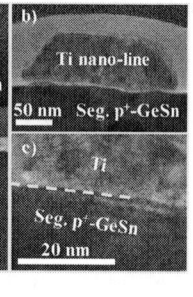
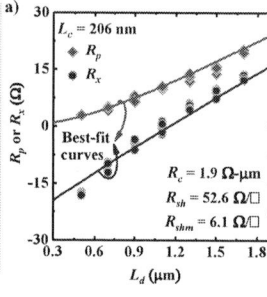
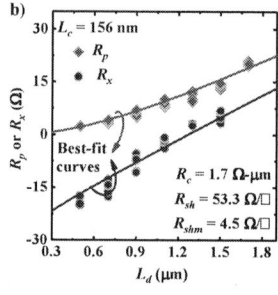
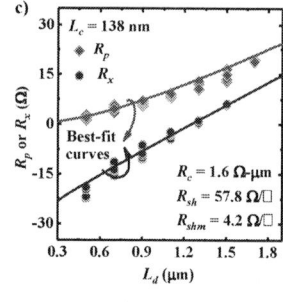

Fig. 9. (a) Cross-section TEM image of the fabricated the Ti/Seg. p⁺-GeSn Nano-TLM structure along line FF' in Fig. 6(a). (b) Zoomed-in image of Ti contact to Seg. p⁺-GeSn. The calibrated value of L_c of the Ti nano-line is 206 nm. (c) High-resolution TEM image shows a flat, reaction-free interface between Ti and Seg. p⁺-GeSn.

Fig. 10. Extracted R_p and R_x (symbols) at different L_d and best fitting curves of Ti/Seg. p⁺-GeSn Nano-TLM structures with L_c of a) 206 nm, b) 156 nm, and c) 138 nm. R_p is measured when the current is forced to flow from port A to port B and the voltage drop is measured between ports C and D. R_x is measured when the current is forced to flow from port A to port C and the voltage drop is measured between ports B and D. The R_p and R_x were collected from four sets of Nano-TLM structures with the same L_c. Good consistency is observed for different sets of Nano-TLM structures with the same L_c value.

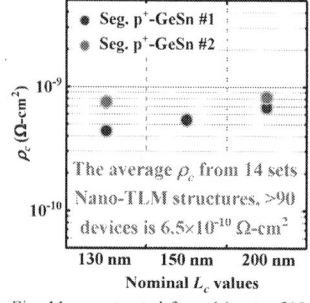
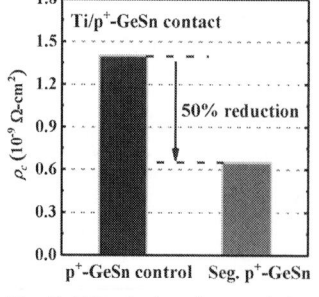
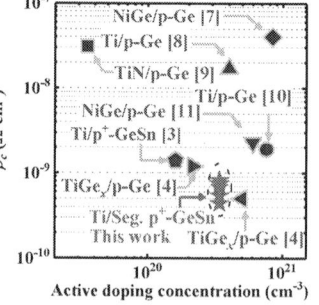
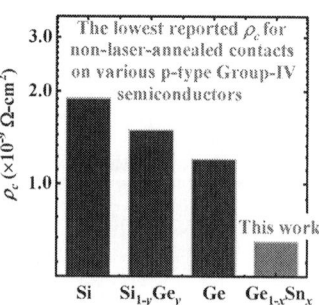

Fig. 11. ρ_c extracted from 14 sets of Nano-TLM structures, a collection of >90 devices, on two Seg. p⁺-GeSn samples, showing an average ρ_c of 6.5×10^{-10} Ω-cm². A ρ_c as low as 4.4×10^{-10} Ω-cm² is also achieved.

Fig. 12. 50% reduction of ρ_c was obtained from Ti/Seg. p⁺-GeSn as compared to that of Ti/p⁺-GeSn control without Ga and Sn surface segregation.

Fig. 13. Benchmarking of reported ρ_c values on p-type Ge and GeSn. Even though advanced implantation and annealing techniques are not used, ρ_c down to 4.4×10^{-10} Ω-cm² (average ρ_c is 6.5×10^{-10} Ω-cm²) is achieved for Seg. p⁺-GeSn, the lowest for the metal/p-type Ge-based contacts.

Fig. 14. Benchmarking of the lowest ρ_c for non-laser annealed on various p-type Group-IV semiconductors. Average ρ_c of 6.5×10^{-10} Ω-cm² achieved on Seg. p⁺-GeSn is the lowest ρ_c for any non-laser annealed metal/p-type Group-IV semiconductor contacts.

978-1-5386-4219-1/18 $31.00 © 2018 IEEE

Gap in pagination due to formatting issues.

Pages 79-80

Multiple Workfunction High Performance FinFETs for Ultra-low Voltage Operation

M. Togo, R. Asra, P. Balasubramaniam, X. Zhang, H. Yu, S. Yamaguchi, E. Geiss, H. S. Yang, B. Cohen, H-C. Lo, O. Hu, H. Lazar
O. Kwon, D. Burnett, J. Versaggi, E. Banghart, M. K. Hassan, E. Bazizi. L. Pantisano, J. G. Lee, S. B. Samavedam, and D. K. Sohn
GLOBALFOUNDRIES, Malta, NY, USA, email: Mitsuhiro.Togo@globalfoundries.com

Abstract—A multiple workfunction (multi-WF) integration technology was developed for ultra-low voltage operation in high performance FinFETs. It is essential to solve three key issues in the multi-WF process, a) short channel effect (SCE) degradation due to removing halo implants b) gate resistance increase due to multi-WF stack, and c) gate dielectric reliability degradation due to additional patterning. In this study, we resolve these issues through the combination of junction engineering and workfunction metal (WFM) boolean engineering in long channel (LC) and short channel (SC) devices for SCE, WFM stack optimization for gate resistance, and HK interface optimization for reliability. In logic devices, 15/13% N/PFET DC and 14% AC performance were improved without SCE or reliability degradation. In SRAM devices, 43% Vt mismatch (Vtmm) improvement resulted in record Vmin yield down to 0.4V on 128Mb 0.064μm² SRAM array.

INTRODUCTION

Mobile electronics are a major driver for improving device performance, reducing power, and improving SRAM Vmin yield in low power SOC applications [1], [2]. Multi-WF is a key technology to enable lower voltage operation with the flexibility to optimize threshold voltage (Vt) and achieves low voltage SRAM Vmin yield [3], [4]. While prior publications have focused on Vt optimization in multi-WF processes, this paper describes the key difficulties to implement multi-WF integration: how to improve SCE and mitigate multi-WF integration process impact on device reliability. In this study, we developed a novel multi-WF integration technology to fully leverage multi-WF benefits to improve logic device performance, reliability, and to achieve high SRAM Vmin yield on a 14nm FinFET baseline.

EXPERIMENT

In a replacement metal gate (RMG) module, after dummy gate removal, a thin $HfTiO_x$ interfacial layer (IL) was formed by pretreatment to protect the HK/IL-SiO_2 gate dielectric during WFM patterning without consuming any gate space as seen in Fig. 1. Multi-WFM deposition and patterning were implemented to control multiple Vt flavors. In this flow, the top WFM was deposited by in-situ WFM stack followed by a low resistance contact metal. The Vt was fully controlled by gate workfunction without halo implants in logic FETs or SRAM, which saves halo implant masks and process steps. In order to reduce SC and LC Vt differences without halo implants, WFM boolean change in LC and SC devices with the same Vt flavor was applied to PFETs. This multiple workfunction integration technology was demonstrated on 14nm logic and SRAM FinFETs.

RESULTS and DISCUSSIONS

A. Short channel effect suppression in multi-WF

In order to fully utilize the multi-WF benefit to control Vt without halo implants, SCE immunity is essential. Devices without halo implants have degraded SCE by 50-60mV in both NFET and PFET as seen in Fig. 2. In order to improve SCE, junction architecture was optimized by using TCAD simulation. Fig. 3(a) shows on-current and off-current density for optimized junctions in N/PFET and Fig. 3(b) shows the process conditions e.g. spacer thickness impact on Vt roll-off (Vt difference between LC and SC devices) and DC performance (Ieff-Idoff). In the NFET, an optimized V-shape source/drain (S/D) edge structure separates the main path of on-current and off-current, so that off-current can be controlled without performance degradation. On the other hand, the PFET S/D structure remains close to each other to increase channel mobility from embedded SiGe stressor near the S/D bottom. Thus Vt roll-off improvement by process control causes DC performance degradation in no halo implant. Fig. 2 shows Vt roll-off Si data for NFET and PFET after improving SCE. In Fig. 2(a), NFET SCE was suppressed by the optimized V-shape S/D structure. In PFET, a different WFM is applied between SC and LC devices by WFM

boolean change. SC devices used WFM1 and the LC device used WFM2, which combination successfully suppressed PFET Vt roll-off.

Fig. 4 (a) shows a multi-WF device without halo implants has improved N/PFET effective channel mobility by 19/16% by eliminating halo dopant scattering. In Fig. 4 (b), on-state resistance breakdown analysis shows NFET channel resistance (Rch) was decreased 14% keeping low S/D resistance (Rsd). Fig. 5 shows N/PFET performance was improved by 15/13%. Fig. 6 shows the ring oscillator (RO) frequency and quiescent current (IDDQ). By improving DC performance using multi-WF without halo implants, AC performance was improved by 14%.

B. Gate resistance reduction in multi-WF

Fig. 7 shows gate resistance as a function of WFM thickness for multi-WF, multi-WF with in-situ top WFM, and single-WF as a reference. In single-WF, thicker WFM does not increase gate resistance. However, in multi-WF, thick WFM with multiple WFM stacks increases gate resistance about two times. In-situ top WFM reduces multi-WF gate resistance down to the same gate resistance as single-WF at thick WFM by suppressing oxygen diffusion into the underlying WFM [5].

C. Gate dielectric reliability improvement in multi-WF

Fig. 8 shows N/PFET inversion capacitance (Cinv) and inversion gate leakage (Iginv) in multi-WF, multi-WF with pretreatment, and single-WF as a reference. In Fig 8(b), all PFET WF1 (multi-WF1, multi-WF1 with pretreatment, and single-WF1) do not have WFM patterning. The others (all NFET WF1/2 and PFET WF2 in Fig. 8(a), (b)) have WFM patterning. In Fig 8(a), the NFET multi-WF1/2 process decreases both Cinv and Iginv compared with single-WF1/2. In Fig 8(b), the PFET multi-WF1 has comparable Cinv and Iginv values to that of single-WF1 because of no WFM patterning damage. On the other hand, the PFET multi-WF2 increases Iginv despite the Cinv decrease. In order to solve this multi-WF integration problem, the interfacial layer on HK/IL-SiO_2 is formed by pretreatment before the first WFM deposition [6]. Fig. 9 indicates the interfacial layer on HK/IL-SiO_2 is $HfTiO_x$ and its thickness is only 0.3Å thicker than without pretreatment. In Fig. 8(a), (b), the IL-$HfTiO_x$ formed by pretreatment does not change Cinv or Iginv in the NFET multi-WF1/2 or the PFET multi-WF1 since there is no WFM patterning degradation. However, in Fig. 8(b), the PFET multi-WF2 with IL-$HfTiO_x$ by pretreatment reduced Iginv to the same value as single-WF2 with a consistent Cinv with multi-WF2. Fig. 10 shows a Quantile-Quantile (Q-Q) plot of the gate dielectric breakdown voltage (Vbd) and device gate leakage (Iginv). In Fig 10(b), the PFET multi-WF2 shows leakage increase and more than 500mV Vbd degradation compared to single-WF2 due to HK/IL-SiO_2 damage by WFM patterning. The IL-$HfTiO_x$ by pretreatment significantly improves Iginv and Vbd of the PFET multi-WF2 as well as retains the same reliability as the single-WF2. In Fig. 10(a), the NFET leakage was reduced and NFET Vbd was improved 100mV through multi-WF processing in NFETs. The sub-angstrom thick IL-$HfTiO_x$ formed by pretreatment protects the HK/IL-SiO_2 from the multi-WF process damage and maintains the high on-current drivability and low gate resistance.

D. SRAM Vmin yield improvement with multi-WF

Fig. 11 shows that multi-WF devices without halo implants significantly improved Vtmm in Pull down/Pass gate/Pull up (PD/PG/PU) devices by 44/42/41/% by eliminating random dopant fluctuations (RDF). Fig. 12 shows SRAM static noise margin (SNM) and write margin (WRM) for multi-WF device without halo implants and single-WF with halo implants. Eliminating halo dopant scattering by multi-WF control significantly improved SNM and WRM, and resulted in record 128Mb 0.064μm² SRAM Vmin yield down to 0.4V as seen in Fig. 13.

CONCLUSIONS

A multiple workfunction integration technology was developed to improve logic performance and SRAM Vmin yield. Three multi-WF integration issues were overcome, and higher logic performance and SRAM Vmin yield down to 0.4V were achieved. The combination of optimized S/D structure and WFM boolean in LC and SC devices suppressed SCE caused by removing halo implants. The in-situ top WFM suppressed gate resistance increase. Thin IL-HfTiO$_x$ on HK formed by WFM pretreatment protected HK/IL-SiO$_2$ gate dielectric during WFM patterning and improved gate dielectric reliability.

ACKNOWLEDGEMENT

The authors would like to thank the GLOBALFOUNDRIES Fab8 team in Malta NY for their support.

REFERENCES

[1] D. Ha et al, VLSI Tech. Symp. Dig., p. 68, 2017
[2] S. Yang et al, VLSI Tech. Symp. Dig., p. 70, 2017
[3] M. Togo et al, VLSI Tech. Symp. Dig., p. 196, 2013
[4] S.-Y. Wu et al, IEDM Tech. Dig., p. 43, 2016
[5] A. Veloso et al, VLSI Tech. Symp. Dig., p. 34, 2011
[6] S. Yamaguchi et al, Microelectronics Reliability, p. 49, 2017

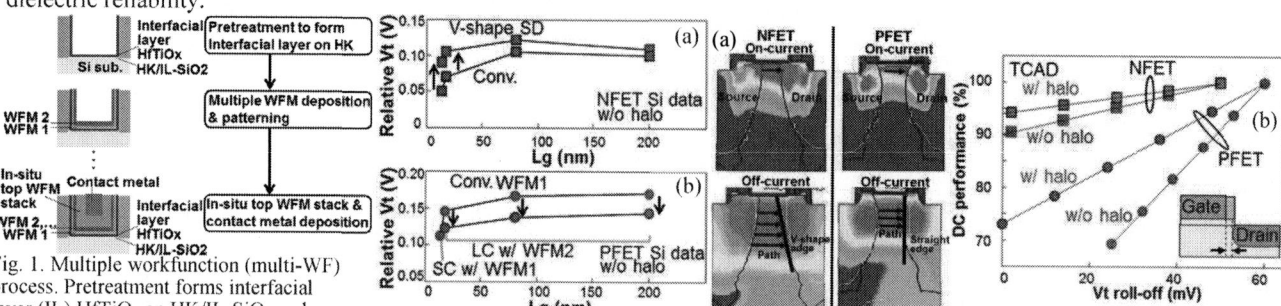

Fig. 1. Multiple workfunction (multi-WF) process. Pretreatment forms interfacial layer (IL) HfTiO$_x$ on HK/IL-SiO$_2$ and protect HK/IL-SiO$_2$ during workfunction metal (WFM) patterning. Top WFM is in-situ WFM stack to decrease gate resistance.

Fig. 2. Vt roll-off Si data of (a) NFET with/without optimized S/D structure and (b) PFET with/without multiple WFM. Both N/PFET have no halo.

Fig. 3. (a) N/PFET on/off-current density in optimized S/D structure. NFET device structure is optimized to suppress SCE. PFET device structure is optimized to increase mobility by eSiGe. (b) Process control impact (e.g. spacer width) on Vt roll-off and DC performance (Ieff-Idoff) in N/PFET with optimized S/D structure as shown in (a) for with/without halos. TCAD simulation.

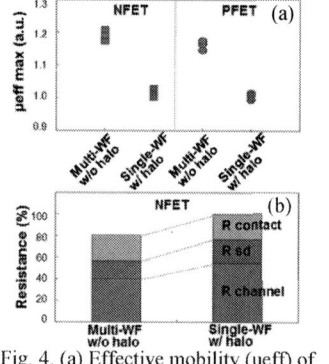

Fig. 4. (a) Effective mobility (ueff) of N/PFET in multi-WF without halos and single-WF with halos and (b) NFET resistance breakdown analysis.

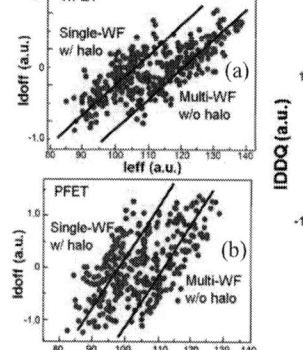

Fig. 5. N/PFET Ieff and Idoff in multi-WF without halos and single-WF with halos.

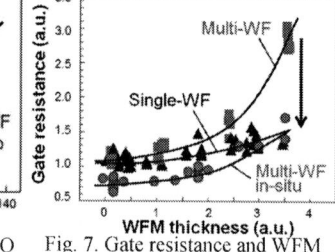

Fig. 6. RO frequency and IDDQ in multi-WF without halos and single-WF with halos.

Fig. 7. Gate resistance and WFM thickness in multi-WF, multi-WF with in-situ top WFM, and single-WF.

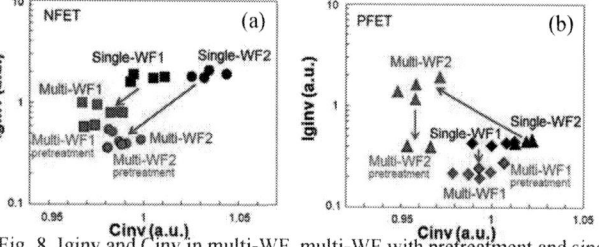

Fig. 8. Iginv and Cinv in multi-WF, multi-WF with pretreatment and single-WF for (a) NFET and (b) PFET. All PFET WF1 have no WFM patterning. The others (all NFET WF1,2 and PFET WF2) have WFM patterning.

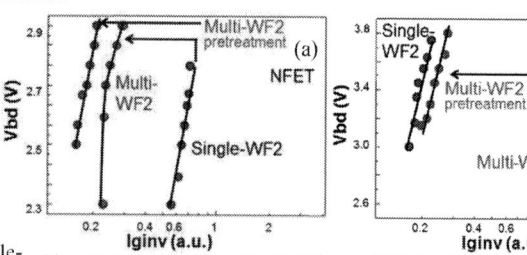

Fig. 10. Quantile-Quantile (Q-Q) plot of Vbd and Iginv in multi-WF, multi-WF with pretreatment, and single-WF for (a) NFET WF2 and (b) PFET WF2.

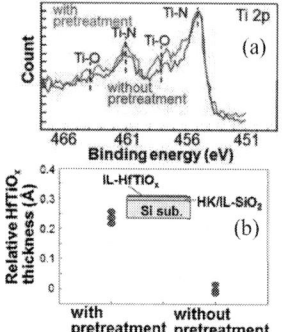

Fig. 9. (a) IL on HK with and without pretreatment by XPS. (b) IL HfTiO$_x$ thickness with and without pretreatment.

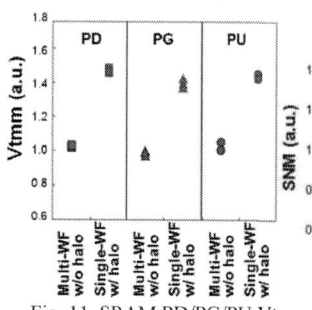

Fig. 11. SRAM PD/PG/PU Vt mismatch (Vtmm) in multi-WF without halos and single-WF with halos.

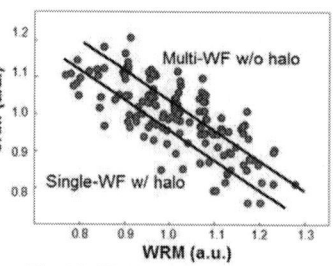

Fig. 12. SRAM SNM and WRM in multi-WF without halos and single-WF with halos.

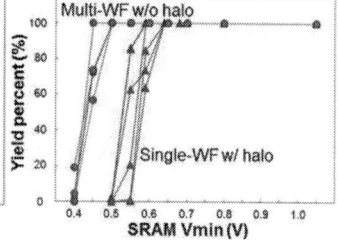

Fig. 13. 128Mb 0.064µm² SRAM Vmin yield with multi-WF without halos and single-WF with halos.

978-1-5386-4219-1/18 $31.00 © 2018 IEEE

An In-depth Study of High-Performing Strained Germanium Nanowires pFETs

J. Mitard, D. Jang, G. Eneman, H. Arimura, B. Parvais*, O. Richard, P. Van Marcke, L. Witters, E. Capogreco, H. Bender,
R. Ritzenthaler, H. Mertens, A. Hikavyy, R. Loo, H. Dekkers, F. Sebaai, A. Milenin, N. Horiguchi, A. Mocuta, D. Mocuta, N. Collaert

IMEC, Kapeldreef 75, 3001 Leuven, Belgium E-mail: Jerome.Mitard@imec.be *Also with Vrije Universiteit Brussel, Dep ETRO

Abstract:

An in-depth study of scaled nanowire Ge pFETs for digital and analog applications is proposed. Improved device characteristics are first obtained after gaining a good understanding of the HPA on device performance. Up to 45% higher $I_{D,SAT}$ is obtained at I_{OFF}=3nA/fin when comparing to best Si GAA nFET and similar $I_{D,SAT}$ is found when benchmarking to mature 14/16nm pFinFET technology at -0.5 V_{DD}. The temperature dependent study of $I_{D,SAT}$ highlights that the mechanism limiting the transport in Ge at short channel are neither purely diffusive nor fully ballistic.

Introduction

With the orders of magnitude increase in connected devices, Internet-Of-Things (IoT) is going to trigger a massive influx of data which requires devices satisfying a myriad of performance, form-factor and cost needs. Thanks to scaling boosters, the FinFET technology is seen to be extended to the 7-5nm node and for now, the Gate-All-Around (GAA) architecture appears to be the most practical solution beyond 5nm since most of the process steps can be re-used and short gates can be integrated without degrading electrostatics. One important challenge of these lateral nanowires (NWs) FETs is the amount of surface channel which is significantly decreasing as compared to FinFET. Stacking them (i.e. >3NWs) is required to indeed improve the drive per footprint but at a serious expense of parasitic capacitance and resistance increase. Replacing the silicon NWs by a high mobility channel material is seen to provide the necessary current boost per footprint limiting the use of tall structures with the risk of AC device performance loss [1].

We propose here an in-depth study of the electrical properties of high performing stacked Ge GAA FETs and benchmark them to their Si counterparts using a common 14/16nm technology platform and identical device design. A transport study is also carried out to get an insight into the mechanisms driving the current at short L_G.

The key role of the HPA in boosting the device performance

Fig.1 details the fabrication of the Ge GAA devices used in this work. More information can be found in [2,3] but the major difference lies in the use of source and drain (S/D) material consisting of $Si_{0.35}Ge_{0.65}$ grown on $Si_{0.3}Ge_{0.7}$ SRB. This means that there is almost no additional stress coming from the embeeded S/D which is typically the case for tight Contact Poly Pitch (CPP) technology (sub-40nm CPP at 5nm node). In this fabrication process, one step remains critical to provide high performance Ge-channel GAA devices: High Pressure Anneal (HPA). Indeed, its benefit can be observed at different modules. -Contact and junctions- In Fig. 2, the reduction of contact and sheet resistance is clearly seen after applying HPA-450°C on the samples. This is attributed to a better Boron activation in the $Si_{0.35}Ge_{0.65}$ S/D underneath the Ti/TiN direct contact. Lower reverse current after HPA-450°C is also obtained (Fig. 3) most likely because of a reduced electrical field at the junction sides potentially lowering I_{OFF}. -Gate stack- A statistical analysis is carried out to isolate the behavior of each transistor against annealing. As in [4], a boost of the drain current is also reported (Fig. 4) but it is worthwhile noticing that only the highest performing Ge GAA FETs are improved. In other words, our optimum HPA step is not really effective for devices with excessive amount of initial interface traps density (D_{IT}). This is typically the case for the Si-cap-passivated highly-doped $SiGe_{70\%}$ FinFET which is present at the bottom of the device [5]. The V_{TH} controllability of Ge GAA FETs also benefits from the HPA step leading to a much tighter distribution of this parameter over the entire wafer (Fig. 5). A Pelgrom plot (Fig. 6) first shows a decent A_{VT} slope for GAA/FINFET [6] although an increased maturity of the process is mandatory to give physical meaning of the variation source seen at short L_G (28nm-Lgmin).

In-Depth Study of the High Performance GAA FETs: Si vs Ge

In the previous section, we have quantified the specific impact of HPA on the V_{TH} variability. However, the sources of intrinsic or extrinsic variation in Ge GAA FETs are multifold and result mostly from the cross-interaction between the process choices and the design of the target device. To progress and get early feedback toward Ge-GAA specifications for the industry, the best performing devices were carefully selected for the analysis. In Fig.7, the cross-TEM images of two excellent Ge GAA pFETs are provided. I_D-V_G characteristics of 28nm-$L_{G,min}$ and 70nm-L_G Ge 2NWs FETs are shown in Fig. 8 and compared to the state-of-the-art 2NWs Si GAA FETs [7]. Note that the comparison is made for NWs FETs with same device design and sharing the same 14/16nm technology platform at imec. The transfer characteristics shown in Fig. 9 confirm that competitive performance of the scaled Ge NWs pFETs can be demonstrated when compared to a relevant nFET counterpart namely 2NWs Si nFETs. Indeed, ~45% higher I_{ON} per footprint is obtained after work function tuning to set the devices at I_{OFF}=3nA/fin (Fig. 8-inset). The Ge devices were also compared to industry standard Si pFinFET using a BSIM-CMG model card calibrated on the 14/16nm FinFET technology. It is found that our 28nm-L_G Ge GAA pFET have matched performance when compared to "mature" FinFET device. Further gain in favor of Ge GAA FETs is expected towards future nodes. Indeed, since the stressor is embeeded into the substrate, no strain loss is expected when reducing the contact size as foreseen at 7nm node and beyond for Si FinFET with embeeded stressors in S/D (Fig. 11).

Analog figures of merit are also provided since there is an increasing interest in CMOS mixed-mode (analog-digital) systems with the IoT market. The transconductance efficiency (Fig. 12) of Ge GAA pFETs reaches values ~34V^{-1} (ideal=38V^{-1}) at relatively low I_D as well as a good intrinsic gain (>30dB) (Fig. 13). These results are aligned with current Si FinFET technology and are obtained thanks to the high mobility of the Ge channel maintained at short channel and good electrostatic control from the GAA architecture.

Temperature dependence analysis has been carried out to get physical insights into the mechanisms driving the carrier transport in Ge GAA pFETs. Fig. 14 shows a strong impact of defects playing a role in most of the transport regimes. Indeed, it is interesting to notice that the SS parameter is found to have an higher sensitivity to temperature than that of the Si counterparts (Fig. 15). An improved gate stack (C_{IT}/C_{OX} ratio) is seen to further boost the performance of our Ge GAA pFETs. Carrier transport at operating conditions (high lateral field) remains dominated by an optical phonon scattering mechanism as expected from relatively long channel devices. Carrier velocity is extracted [8] from the shortest Ge GAA (28nm-$L_{G,min}$) and its temperature dependency is found to be almost temperature-independent with V_{DS} (Fig. 16). This result can be explained by competing mechanisms which have opposite temperature effect on carrier velocity [9-10]. At the shortest L_G, our Ge GAA FET seems to have a higher contribution of ballisticity than at longer channels dominated by V_{SAT} (diffusive). The ballistic ratio which can always be extracted, is assessed by comparing to values from simulations (Fig. 17) and found to be around 33% assuming a stress level of -1.7GPa is kept until end of the process.

Conclusions and Summary

In this work, we have provided an early understanding of what Ge nanowire pFETs can offer in term of device performance for high-end analog and HP digital applications. We have shown that an optimization of key process steps such as the thermal budget (HPA) or gate stack (C_{IT}/C_{OX}) can make Ge GAA pFETs a serious contender for a GAA FET technology beyond the 7nm node.

978-1-5386-4219-1/18 $31.00 © 2018 IEEE

2018 Symposium on VLSI Technology Digest of Technical Papers

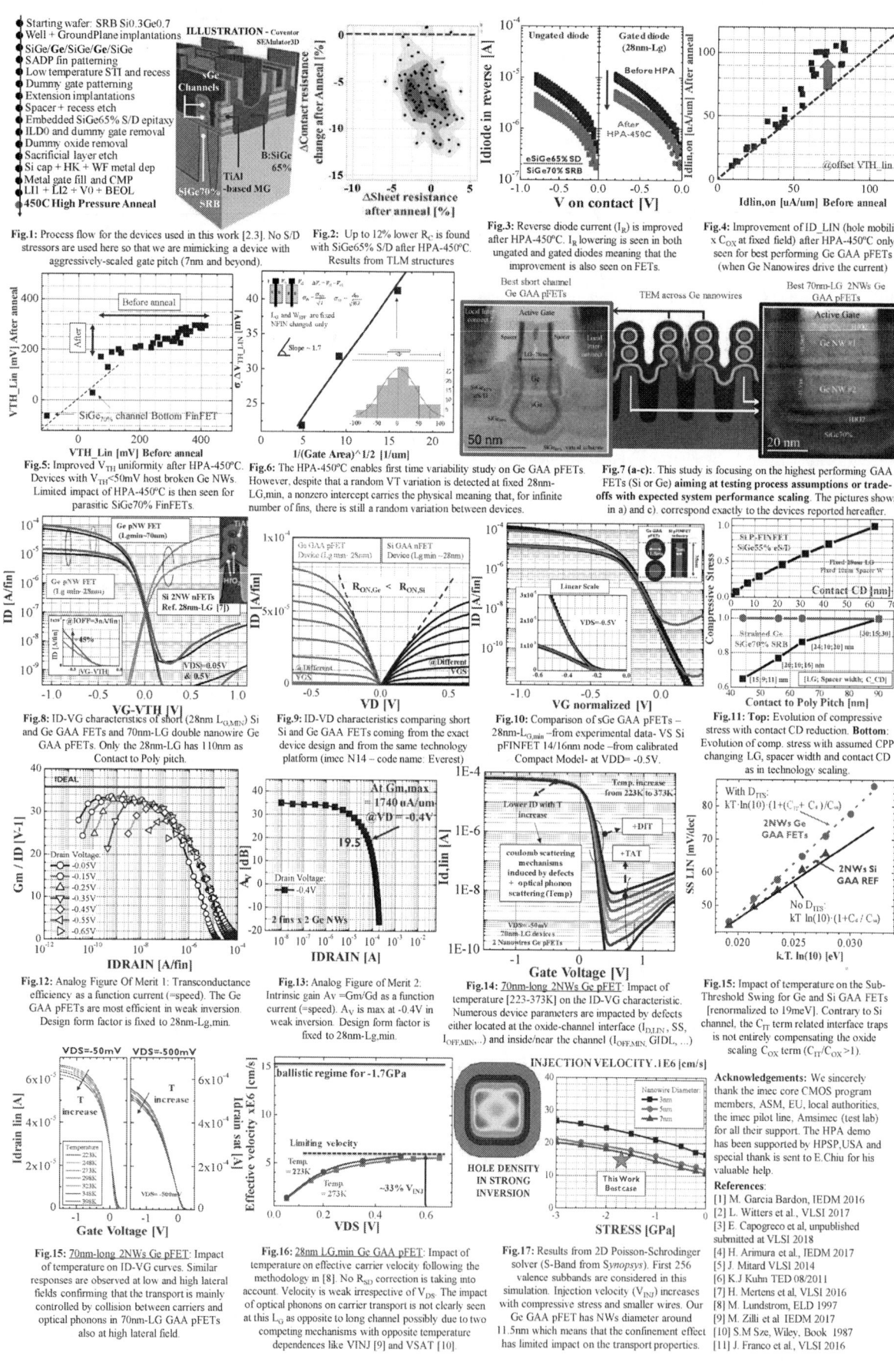

- Starting wafer: SRB Si0.3Ge0.7
- Well + GroundPlane implantations
- SiGe/**Ge**/SiGe/**Ge**/SiGe
- SADP fin patterning
- Low temperature STI and recess
- Dummy gate patterning
- Extension implantations
- Spacer + recess etch
- Embedded SiGe65% S/D epitaxy
- ILD0 and dummy gate removal
- Dummy oxide removal
- Sacrificial layer etch
- Si cap + HK + WF metal dep
- Metal gate fill and CMP
- LI1 + LI2 + V0 + BEOL
- 450C High Pressure Anneal

Fig.1: Process flow for the devices used in this work [2,3]. No S/D stressors are used here so that we are mimicking a device with aggressively-scaled gate pitch (7nm and beyond).

Fig.2: Up to 12% lower R_C is found with SiGe65% S/D after HPA-450ºC. Results from TLM structures.

Fig.3: Reverse diode current (I_R) is improved after HPA-450ºC. I_R lowering is seen in both ungated and gated diodes meaning that the improvement is also seen on FETs.

Fig.4: Improvement of ID_LIN (hole mobility x C_{OX} at fixed field) after HPA-450ºC only seen for best performing Ge GAA pFETs (when Ge Nanowires drive the current)

Fig.5: Improved V_{TH} uniformity after HPA-450ºC. Devices with V_{TH}<50mV host broken Ge NWs. Limited impact of HPA-450ºC is then seen for parasitic SiGe70% FinFETs.

Fig.6: The HPA-450ºC enables first time variability study on Ge GAA pFETs. However, despite that a random VT variation is detected at fixed 28nm-LG,min, a nonzero intercept carries the physical meaning that, for infinite number of fins, there is still a random variation between devices.

Fig.7 (a-c): This study is focusing on the highest performing GAA FETs (Si or Ge) aiming at testing process assumptions or trade-offs with expected system performance scaling. The pictures shown in a) and c). correspond exactly to the devices reported hereafter.

Fig.8: ID-VG characteristics of short (28nm $L_{G,min}$) Si and Ge GAA FETs and 70nm-LG double nanowire Ge GAA pFETs. Only the 28nm-LG has 110nm as Contact to Poly pitch.

Fig.9: ID-VD characteristics comparing short Si and Ge GAA FETs coming from the exact device design and from the same technology platform (imec N14 – code name: Everest)

Fig.10: Comparison of sGe GAA pFETs – 28nm-$L_{G,min}$ –from experimental data- VS Si pFINFET 14/16nm node –from calibrated Compact Model- at VDD= -0.5V.

Fig.11: Top: Evolution of compressive stress with contact CD reduction. **Bottom:** Evolution of comp. stress with assumed CPP changing LG, spacer width and contact CD as in technology scaling.

Fig.12: Analog Figure Of Merit 1: Transconductance efficiency as a function current (=speed). The Ge GAA pFETs are most efficient in weak inversion. Design form factor is fixed to 28nm-Lg,min.

Fig.13: Analog Figure of Merit 2: Intrinsic gain Av =Gm/Gd as a function current (=speed). Av is max at -0.4V in weak inversion. Design form factor is fixed to 28nm-Lg,min.

Fig.14: 70nm-long 2NWs Ge pFET: Impact of temperature [223-373K] on the ID-VG characteristic. Numerous device parameters are impacted by defects either located at the oxide-channel interface ($I_{D,LIN}$, SS, $I_{OFF,MIN}$,..) and inside/near the channel ($I_{OFF,MIN}$, GIDL, ...)

Fig.15: Impact of temperature on the Sub-Threshold Swing for Ge and Si GAA FETs [renormalized to 19meV]. Contrary to Si channel, the C_{IT} term related interface traps is not entirely compensating the oxide scaling C_{OX} term (C_{IT}/C_{OX}>1).

Fig.15: 70nm-long 2NWs Ge pFET: Impact of temperature on ID-VG curves. Similar responses are observed at low and high lateral fields confirming that the transport is mainly controlled by collision between carriers and optical phonons in 70nm-LG pFETs also at high lateral field.

Fig.16: 28nm LG,min Ge GAA pFET: Impact of temperature on effective carrier velocity following the methodology in [8]. No R_{SD} correction is taking into account. Velocity is weak irrespective of V_{DS}. The impact of optical phonons on carrier transport is not clearly seen at this L_G as opposite to long channel possibly due to two competing mechanisms with opposite temperature dependences like VINJ [9] and VSAT [10].

Fig.17: Results from 2D Poisson-Schrodinger solver (S-Band from *Synopsys*). First 256 valence subbands are considered in this simulation. Injection velocity (V_{INJ}) increases with compressive stress and smaller wires. Our Ge GAA pFET has NWs diameter around 11.5nm which means that the confinement effect has limited impact on the transport properties.

Acknowledgements: We sincerely thank the imec core CMOS program members, ASM, EU, local authorities, the imec pilot line, Amsimec (test lab) for all their support. The HPA demo has been supported by HPSP,USA and special thank is sent to E.Chiu for his valuable help.

References:
[1] M. Garcia Bardon, IEDM 2016
[2] L. Witters et al., VLSI 2017
[3] E. Capogreco et al, unpublished submitted at VLSI 2018
[4] H. Arimura et al., IEDM 2017
[5] J. Mitard VLSI 2014
[6] K.J Kuhn TED 08/2011
[7] H. Mertens et al, VLSI 2016
[8] M. Lundstrom, ELD 1997
[9] M. Zilli et al IEDM 2017
[10] S.M Sze, Wiley, Book 1987
[11] J. Franco et al., VLSI 2016

Si/SiGe superlattice I/O finFETs
in a vertically-stacked Gate-All-Around horizontal Nanowire Technology

G. Hellings[1,*], H. Mertens[1], A. Subirats[1], E. Simoen[1], T. Schram[1], L.-A. Ragnarsson[1], M. Simicic[1], S.-H. Chen[1], B. Parvais[1,5], D. Boudier[2], B. Cretu[2], J. Machillot[3], V. Pena[3], S. Sun[4], N. Yoshida[4], N. Kim[4], A. Mocuta[1], D. Linten[1] and N. Horiguchi[1]

[1]imec, kapeldreef 75, Leuven, Belgium *Geert.Hellings@imec.be
[2]Normandie University, UNICAEN, ENSICAEN, CNRS, GREYC, 14000 Caen, France
[3]Applied Materials, Leuven, Belgium [5]VUB, Brussels, Belgium
[4]Applied Materials, 3050 Bowers Av., Santa Clara, CA 95053, USA

Abstract

This work presents Si/SiGe superlattice finFETs (FF) for 1.8V/2.5V I/O applications in vertically-stacked Gate-All-Around horizontal nanowire technology (hNW) technology. Superlattice FF have a higher I_{ON} than I/O hNW reference devices and can be more easily integrated into a GAA hNW technology than Si I/O FF. These novel I/O FET structures exhibit competitive analog performance and are superior as ESD protection devices.

Introduction

Vertically stacked horizontal gate-all-around nanowire FETs (hNW) are promising candidates to replace finFETs (FF) in future technology node, because they allow for more aggressive gate length scaling [1]. While much research has focused on CORE devices with $V_{DD}<0.7V$, many SoC applications require also I/O components operating at e.g. 1.8 or 2.5V to be integrated on-chip in a cost-effective manner.

The integration of thick-oxide Si finFETs (Si I/O FF) in a hNW technology would potentially become quite complex, as it would require one area to have a Si/SiGe/Si/SiGe/Si epi stack (for CORE hNW) and another to have Si epi (for Si I/O FF). These two substrate areas would have to be dealt with throughout the entire FEOL integration flow. On the other hand, the vertical space between two stacked hNWs does not accommodate the thick I/O gate stack, causing comparatively low I_{ON} when hNW spacing is smaller than twice the gate stack thickness (hNW performance gap, fig.1).

Therefore, the goal of this paper is to investigate the feasibility of a Si/SiGe superlattice finFET for I/O applications (I/O superlattice FF) in a hNW technology: a finFET-like structure is made using the available epi stack in hNW technology, thus avoiding two different substrate areas.

First, the fabrication of superlattice I/O finFETs, as well as I/O hNW is discussed. Second, their electrical performance is compared to Si I/O FF, including TCAD simulations. BTI reliability and analog performance metrics are discussed. Finally ESD robustness is evaluated.

Superlattice I/O FF fabrication

The process flow for the fabrication of superlattice I/O FFs and I/O hNW is shown in fig. 2. After well implants the $Si_{0.70}Ge_{0.30}$/Si active films were grown and patterned using SADP. After STI process, dummy gates and spacers were defined, before the SiGe/Si fins were replaced by Si:P epi in the s/d areas for nFET (SiGe:B for pFET). Then the dummy gate is removed, exposing the Si/SiGe active fin. To fabricate I/O hNW ("B" in fig. 2), the sacrificial SiGe is etched using HCl vapor etching (also required for CORE hNW – "A"), followed by I/O gate stack deposition. To fabricate superlattice I/O finFET, the SiGe etching is skipped and I/O gate stack is deposited on the Si/SiGe superlattice fin ("C"). In this work, the I/O gate stack targeting 1.8V V_{DD} consists of 3nm ALD-SiO_2, 2nm HfO_2 and p-type WF metal deposition. The ALD-SiO_2 layer thickness is increased to 5nm for 2.5V operation. The fabrication process is further illustrated with FET TEM images in fig. 3. Note the 2.5V gate dielectric fully occupies the room between the stacked hNWs. The superlattice I/O FF is shown in fig.3(e). Note that the fin profile can be enhanced using the SiN liner discussed in [2].

Electrical Characterization

The superlattice I/O FFs exhibit good electrostatic control, as shown by the I_{DSB}-V_G curves in fig.4 with $I_{ON}>1200\mu A/\mu m$ for the 1.8V superlattice FF. Particularly the pFET shows excellent SS=69mV/dec

and DIBL=30mV/V, at L_G=70nm. Note that the bulk leakage can still be improved by using optimized layout and dedicated I/O junctions. The I_{ON}-I_{OFF} plot (fig.5) clearly shows the benefit of the superlattice I/O FF, compared to I/O hNW, which is impacted by the geometric problem shown in fig.1. Comparing to Si I/O FF with a similar gate stack, both superlattice I/O n/pFF show high I_{ON}, with best devices outperforming Si I/O FF by ~25%. V_T-L_G is plotted in fig. 6 and shows a very limited roll-off for both the I/O hNW and the superlattice I/O FF down to 70nm (1.8V gate stack) and 90nm (2.5V gate stack). In absolute terms, the nFET V_T is rather high due to the p-type WF metal used in this work, while the pFET V_T is lower because of the Si/SiGe E_V offset.

TCAD simulations

Using the actual geometries from the TEM, 3D process/device TCAD simulations were performed to benchmark I/O hNW, Si I/O FF and superlattice I/O FF. The extracted I_{ON}-I_{OFF} values (fig. 7(a)) confirm the smaller I_{ON} for hNW I/O devices, especially when Si-loss occurs through over-etching. The superlattice I/O FF can yield higher I_{ON} depending on the Ge%. V_T (fig. 7(b)) for nFET is shown to be similar ($E_{C,Si}=E_{C,SiGe}$), while the pFET V_T shows a dependency on the E_V offset. Finally, a cross-section of the I/O hNW and superlattice I/O FF in fig. 8 illustrates the larger volume available for conduction in the latter. Observe also that both Si and SiGe contribute to the current conduction in the s/d regions, while the SiGe forms as somewhat of a Quantum Well in the channel.

BTI Reliability

BTI reliability was evaluated on the ALD-deposited 3nm SiO_2 I/O gate stacks (fig.9). In the superlattice pFF, NBTI is very well under control, with V_{Gmax} (10y lifetime, 125°C)>V_{DD}=1.8V. The nFF PBTI fell short although this cannot be attributed to SiGe-presence in the channel, since it was also observed in the Si-only hNW, but rather to the high-k + p-type WF metal being used without PDA. Based on previous work [3,4,5], this can probably be improved by using n-type WF metal for the nFF, simultaneously lowering nFF V_T.

Analog and ESD performance

As a metric for analog performance, the g_m/g_{ds} is plotted in fig. 10 for superlattice FF with L_G=70-250-1000nm, showing voltage gains in the 40-60dB. Finally, the input-referred noise power in these devices is plotted in fig.11 and shown to be comparable (leading edge FETs) to the Si FF reference with similar I/O gate dielectric.

Finally, proper ESD protection circuitry is also essential to any SoC. Consequently, the 1.8/2.5V superlattice I/O technology was evaluated as ESD protection using Transmission-Line-Pulsing (TLP) system (fig. 12), using gated diode structures. Clearly, the superlattice FF gated diode can withstand higher (+50%) ESD current levels than hNW I/O, comparable to Si FF [6, 7].

Conclusions

Si/SiGe superlattice I/O FFs were fabricated in a horizontal NW technology, targeting 1.8V and 2.5V applications. Good electrostatic control (SS=69mV/dec) and high $I_{ON}>1200\mu A/\mu m$ were reached in prototype L_G=70nm FETs, superior to hNW-based I/O reference FETs, due to increased cross-sectional area as shown with TCAD. BTI reliability and analog metrics and ESD performance were shown to be competitive with Si I/O FF technology, which would be significantly more complex to fabricate in advanced GAA hNW technology.

978-1-5386-4219-1/18 $31.00 © 2018 IEEE

References: [1] Mertens et al. IEDM2016 [2] Mertens et al. IEDM2017 , [3] Rzeepa et al. IRPS2017 [4] Hellings et al., IRPS2017 [5] Franco et al. IRPS2017 [6] Chen et al., IEDM2014 [7] Chen et al., IEDM2016

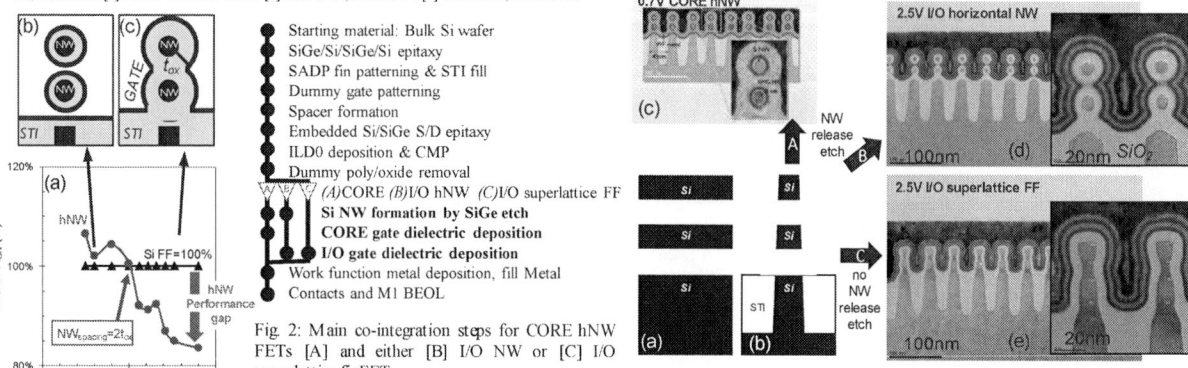

Starting material: Bulk Si wafer
SiGe/Si/SiGe/Si epitaxy
SADP fin patterning & STI fill
Dummy gate patterning
Spacer formation
Embedded Si/SiGe S/D epitaxy
ILD0 deposition & CMP
Dummy poly/oxide removal
*(A)*CORE *(B)*I/O hNW *(C)*I/O superlattice FF
Si NW formation by SiGe etch
CORE gate dielectric deposition
I/O gate dielectric deposition
Work function metal deposition, fill Metal
Contacts and M1 BEOL

Fig. 2: Main co-integration steps for CORE hNW FETs [A] and either [B] I/O NW or [C] I/O superlattice finFET.

Fig. 1: (a) normalized I_{ON}, as function of physical dielectric thickness for stacked hNW and Si finFET. hNW I_{ON} is reduced significantly for thicker I/O dielectrics ($NW_{spacing} < 2t_{ox}$, gate no longer wraps around NW) (b-c).

Fig. 3: Starting from a Si/SiGe/Si/SiGe/Si superlattice (a) on which fins are formed (b), hNW is fabricated by NW release (SiGe) etch (c-d). Skipping this etch yields a superlattice I/O finFET (e). Observe coalescent dielectrics in-between the hNW in (d), **preventing metal gate between hNW.**

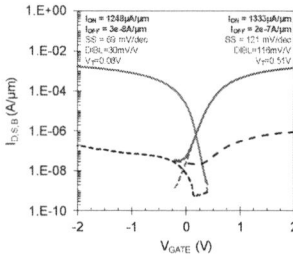

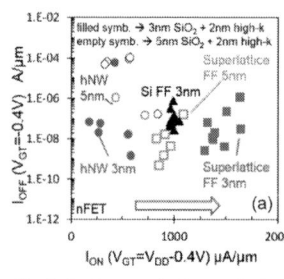

Fig. 4: I_{DSB}-V_G for I/O superlattice FF (nFET and pFET), with important DC performance metrics (L_G=70nm, 1.8V gate stack with 3nm SiO2 IL).

Fig. 5: I_{ON}-I_{OFF} metric for superlattice I/O FF, I/O hNW and ref. Si I/O FF with the same gate dielectric (V_{DD}=1.8V/2.5V for 3nm/5nm IL resp.), showing improved I_{ON} for superlattice I/O FF at 1.8V vs. others (I/O hNW pFETs not available).

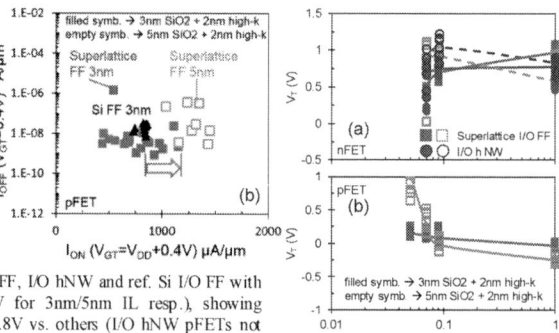

Fig. 6: (a) nFET and (b) pFET V_T, as function of L_G. Limited V_T-rolloff observed down to L_G=70nm (3nm SiO2 IL stack for 1.8V operation)

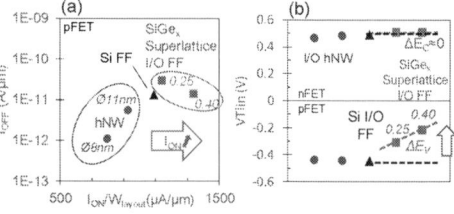

Fig. 7: (a) TCAD simulated I_{ON}-I_{OFF} for Si FF reference, hNW with different NW diameter and SiGe superlattice FF with 1.8V-target gate stack. (V_{DD}=1.8V, V_{GT}=0.4/1.4V). (b) TCAD simulated nFET/pFET V_T for these structures, showing V_T-lowering in superlattice FF pFET, due to E_V offset. (WF$_{GATE}$=4.6eV)

Fig. 8: 3D TCAD simulation of ON-state current density in hNW (a) and superlattice FF (b) (cut through middle of fin). hNW shows current crowding in the extension region, while superlattice FF has more volume available for current conduction.

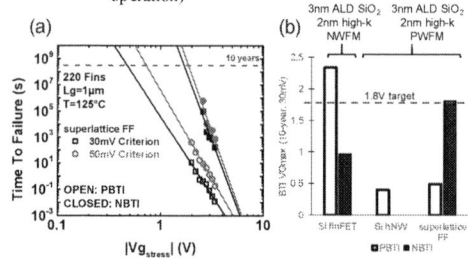

Fig. 9: (a) BTI analysis for the 1.8V superlattice FF at 125°C, showing pFET NBTI exceeding the required V_{Gmax}. (b) The reduced nFET PBTI reliability can not be attributed to SiGe presence in the channel, rather to the PWFM gate stack (Si hNW also shows reduced PBTI reliability).

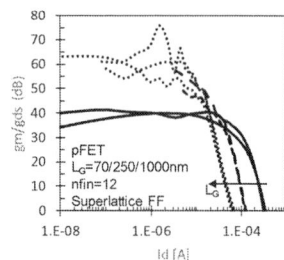

Fig. 10: Voltage gain g_m/g_{ds} for 70nm superlattice pfinFET with 1.8V gate stack (3nm SiO2 IL+2nm high-k).

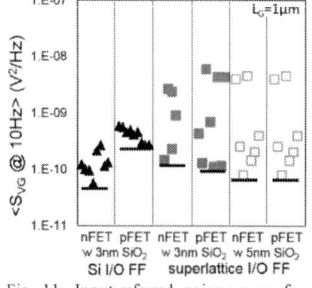

Fig. 11: Input-referred noise power for 1.8V Si I/O FF and 1.8V/2.5V superlattice FF. Comparing to Si I/O FF yields a small degradation for the sFF nFET, while best sFF pFET have lower <S_{VG}>.

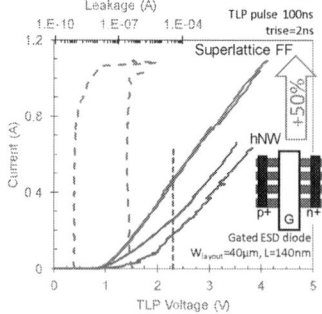

Fig. 12: Gated diode in hNW and superlattice FF I/O technology (ESD protection device), measured with the TLP tool, showing improved ESD robustness for the superlattice FF gated diode (~30mA/μm), comparable to Si FF gated diodes.

Leakage aware Si/SiGe CMOS FinFET for low power applications

Gen Tsutsui, *Curtis Durfee, Miaomiao Wang, *Aniruddha Konar, Heng Wu, Shogo Mochizuki, Ruqiang Bao, Stephen Bedell, Juntao Li, Huimei Zhou, *Daniel Schmidt, *Chun Ju Yang, James Kelly, Koji Watanabe, Theodore Levin, *Walter Kleemeier, Dechao Guo, Devendra Sadana, Dinesh Gupta, *Andreas Knorr, Huiming Bu

IBM Research, 257 Fuller Rd, Albany NY 12203, USA, *GLOBALFOUNDRIES Inc., NY, USA, email: gtsutsu@us.ibm.com

Abstract

Leakage in Si/SiGe CMOS FinFET is examined. Si cap passivation effectively improves SiGe pFET D_{it}, subthreshold slope, and mobility, which improves pFET DC performance by 20%. SiGe GIDL is higher than Si by a factor of 9, though GIDL is limited to 50pA/um. SiGe GIDL reduction knobs to meet Si counterpart is demonstrated. The results open the door to the next stage of Si/SiGe CMOS FinFET such as low power and low leakage applications.

Introduction

Leakage control is one of the key factors for low power applications. Replacing conventional Si with SiGe as a channel material modulates interface trap density (D_{it}) and bandgap, which are crucial for the two major transistor leakage components, subthreshold leakage and gate induced drain leakage (GIDL). Si/SiGe CMOS on (100) planar has been manufactured [1-3] with reasonably controlled pFET D_{it}. D_{it} is higher with (110) surface orientation compared to (100) as shown in Fig. 1 and refs [4-6], which makes D_{it} control more challenging with SiGe channel pFinFET [7] with (110) fin sidewall. In this paper, D_{it} and GIDL in low Ge content Si/SiGe CMOS FinFET [8-10] are examined. D_{it} is degraded with increased integration such as CMOS enablement (nFET implementation), multi-V_t, and fin pitch/height scaling. Si cap passivation [11,12] is identified as a major knob to improve D_{it}. 80% pFET D_{it} reduction is demonstrated, which improves both subthreshold leakage and channel hole mobility resulting in 20% pFET DC performance improvement. GIDL in SiGe pFET is higher than Si by a factor of 9, though GIDL is limited to 50pA/um. GIDL reduction knobs are explored by both TCAD and HW. It is demonstrated that SiGe GIDL can be reduced to a level similar to Si by carefully designing junction gradient and location along with physical gate length (L_{gate}).

D_{it} in SiGe channel pFinFET

Fig. 2 shows pFET D_{it} transition in Si/SiGe CMOS FinFET. D_{it} of the initial process flow, unipolar SiGe pFinFET with 10nm ground rule, was 3e11cm^{-2}.eV^{-1}. D_{it} is degraded by a factor of 2 by enabling CMOS flow (nFET implementation) along with multi-V_t implementation. Other process changes in fin module to enable fin pitch and height scaling resulted in a further D_{it} increase by a factor of 1.5. Consequently, D_{it} became 9e11cm^{-2}.eV^{-1} with the final integration flow. Interfacial layer (IL) and high-k processes are common across all the cases as well as Ge content of SiGe fin which is 20%. Fig. 3 shows pFET D_{it} – Ge content in SiGe fin. D_{it} increases as Ge content increases, and Dit is 9e11cm^{-2}.eV^{-1} at a Ge content of 20%. D_{it} is successfully reduced to 2e11cm^{-2}.eV^{-1} by introducing Si cap passivation process as described in the next section. Fig. 4 shows long channel SS and hole mobility dependence on D_{it} where a strong correlation is observed.

Si cap passivation

Among the trials we have made to improve pFET D_{it} such as IL formation, high-k and metal gate stack, thermal budget in RMG, H_2 anneal and so on, Si cap passivation worked the most effectively. Si cap passivation is to form thin layer Si epi layer (~5A) on SiGe fin. Fig. 5(a) and (b) compares fin cross sectional images between with and without Si cap passivation. Images are taken post high-k deposition process. No major structural difference is observed indicating that Si cap layer is consumed during IL formation. It is also confirmed by elemental analysis that excess Si cap layer does not exist (Fig. 5(c)). Fig. 6 shows cumulative distribution function (CDF) plot of long channel linear SS comparing with and without Si cap. Si cap passivation improves not only SS, but also within wafer SS

variation. Fig. 7 shows hole mobility – N_{inv}. Hole mobility gain of 60% is observed with SiGe fin with Si cap compared to conventional Si fin. Fig. 8 shows I_{off} – I_{eff} of both n- and pFET where pFET DC performance improvement is observed with no nFET penalty. Fig. 9 shows pFET I_{eff} at I_{off}=240pA/um as a function of D_{it}. 20% pFET DC performance improvement is achieved by Si cap passivation, which reduces D_{it} from 9e11 cm^{-2}.eV^{-1} to 2e11cm^{-2}.eV^{-1}. The performance gain is due both mobility and short channel effect improvement. RO delay is also improved with Si cap passivation as shown in Fig. 10. The NBTI benefit with SiGe compared to Si [8] is retained with Si cap passivation as shown in Fig. 11.

GIDL in SiGe pFinFET

GIDL in SiGe pFET is one of the concerns due to its narrower bandgap compared to Si [13,14]. Fig. 12 shows I_{doff} – I_{soff} of SiGe pFinFET with Ge content of 20%. The trend deviates from 1:1 trend line at around I_{doff} of 100pA/um, and an I_{doff} floor of around 50pA/um is observed. Fig. 13(a) shows typical I_{ds} – V_{gs} with different V_{ds} bias conditions. In this paper, pFET GIDL current(I_{gidl}) is defined as

$$I_{gidl}=I_{sub}@V_{gs}=V_{tsat}+0.3V, V_{ds}=-0.75V, V_s=V_{sub}=0V. \quad (1)$$

Fig. 13(b) shows I_{gidl} – V_{ds} where I_{gidl} shows a trend in $V_{ds}>0.7V$. Note that gate leakage is ~1/100x of GIDL. Fig. 14 compares I_{gidl} between SiGe and Si pFinFET, where SiGe I_{gidl} is 9x of Si. Figs. 15 and 16 are TCAD results exploring knobs to improve I_{gidl}. I_{gidl} depends on junction gradient and boundary location [15]. I_{gidl} reduction as increasing RTA temperature is observed in Fig. 16, which is associated with poorer short channel control due to a broader junction. Short channel control is recovered by increasing L_{gate} enabling SiGe pFET leakage to meet a similar level as Si. Fig. 17 shows HW demonstration of I_{gidl} reduction by increasing RTA temperature.

Conclusion

Leakage in Si/SiGe CMOS FinFET is examined. SiGe pFET D_{it} is degraded by a factor of 3 through the implementation of key modules such as CMOS enablement (nFET implementation), multi-V_t, and fin pitch/height scaling. SiGe pFET D_{it} is recovered by Si cap passivation, which effectively reduces D_{it} from 9e11 to 2e11cm^{-2}.eV^{-1}, satisfying D_{it} criteria. D_{it} reduction contributes to improve SS and its variability as well as hole mobility. Consequently, 20% pFET DC performance improvement along with 10% AC performance improvement is demonstrated. GIDL in SiGe pFinFET with Ge content of 20% is degraded by a factor of 9 compared to Si. Junction gradient and location is identified as key knobs to improve GIDL. It is demonstrated that leakage in SiGe pFET can be reduced to the same level as Si pFET by increasing RTA temperature along with L_{gate} optimization.

Acknowledgments

This work was performed by the Alliance Teams at various IBM Research and Development Facilities. The authors gratefully appreciate partner companies Executives for their support.

References

[1] S. Krishnan et al., IEDM tech dig., p.634, 2011. [2] C. Ortolland et al., IEDM tech dig., p.236, 2013. [3] R. Carter et al., IEDM tech dig., p.27, 2016. [4] M. Cho et al., IEEE EDL, p.1211, 2013. [5] P. Srinivasan et al., IRPS, 6A.3.1, 2014. [6] N. H. Thoan et al., JAP, 013710, 2011. [7] P. Hashemi, et al., Symp VLSI Tech., p.18, 2014. [8] D. Guo et al., Symp VLSI Tech., p.14, 2016. [9] G. Tsutsui et al., IEDM tech dig., p.456, 2016. [10] G. Tsutsui et al., IEDM tech dig., p.456, 2016. [11] J. Oh et al., IEDM tech dig., p.22, 2009. [12] L. Witters et al., Symp VLSI Tech., p.194, 2017. [13] V. A. Tiwari et al., IEEE ICEE, 2014. [14] K. Balakrishnan et al., DRC, p.233, 2014. [15] P. Kerber et al., IEEE EDL, p.6, 2013.

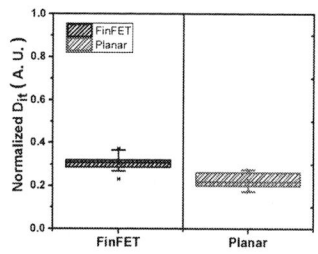

Fig. 1. SiGe D_{it} comparison between fin and planar FET on a same wafer.

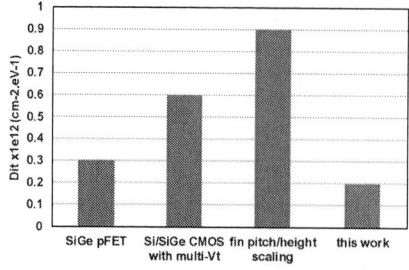

Fig. 2. SiGe pFET D_{it} transition. No change in IL and high-k processes among all the cases.

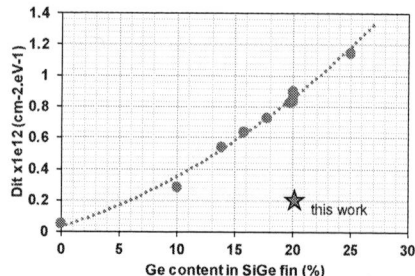

Fig. 3. pFET D_{it} – Ge content in SiGe fin.

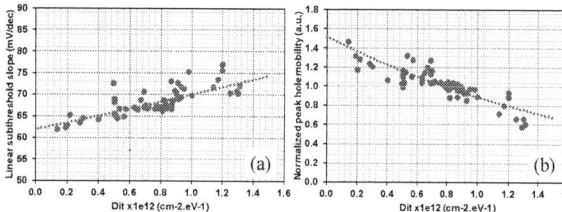

Fig. 4. Long channel SiGe pFET (a) SS and (b) peak hole mobility dependence on D_{it}.

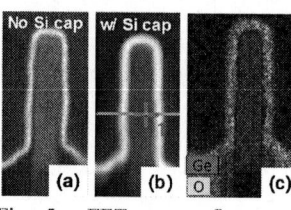

Fig. 5. pFET across fin cross section TEM images of (a) no Si cap, (b) with Si cap. (c) EDX elemental mapping of image (b).

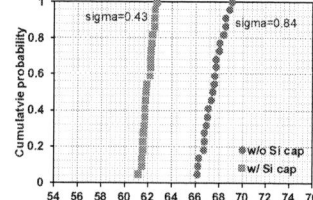

Fig. 6. CDF plot of pFET long channel linear SS.

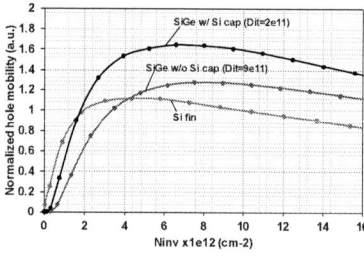

Fig. 7. Hole mobility – N_{inv} comparing SiGe and Si fin.

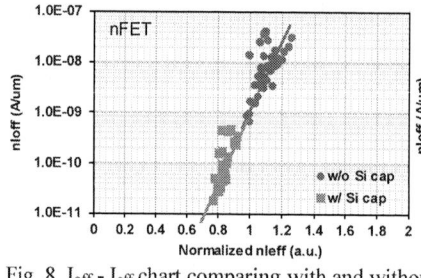

Fig. 8. I_{off} - I_{eff} chart comparing with and without Si cap passivation.

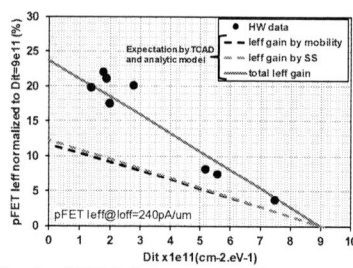

Fig. 9. pFET DC I_{eff} as a function of D_{it}.

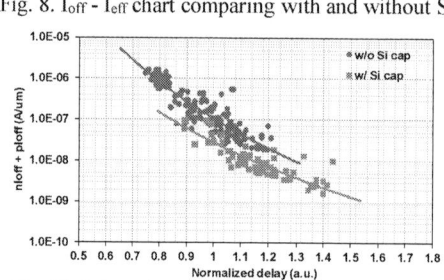

Fig. 10. RO readout comparing w/ and w/o Si cap.

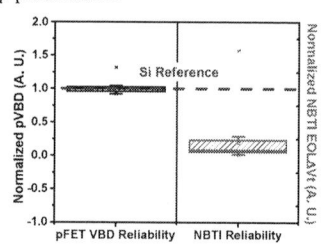

Fig. 11. Reliability readout with Si cap.

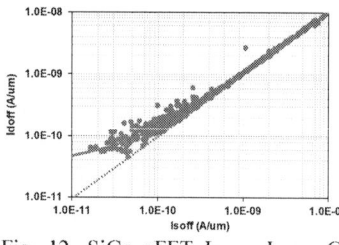

Fig. 12. SiGe pFET I_{doff} - I_{soff}. Ge content of SiGe fin is 20%, V_{ds}=0.8V.

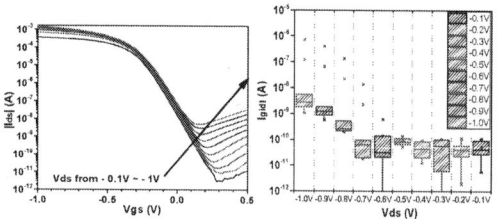

Fig. 13. SiGe pFET (a) I_{ds} - V_{gs}, (b) I_{gidl} – V_{ds}. Noise level of I_{sub} is below 1e-10A.

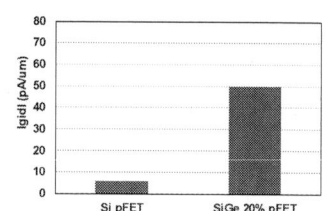

Fig. 14. pFET I_{gidl} comparison between SiGe and Si.

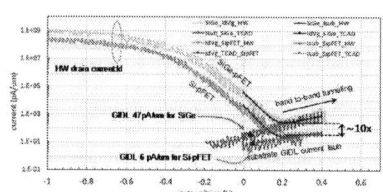

Fig. 15. GIDL calibration on TCAD.

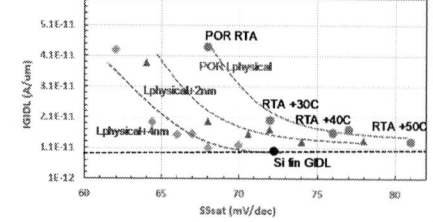

Fig. 16. GIDL reduction knobs (TCAD).

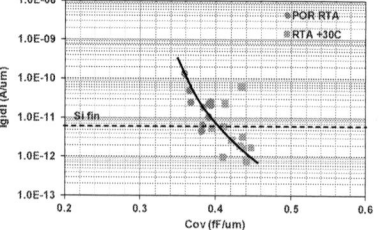

Fig. 17. I_{gidl} - C_{ov}. GIDL is reduced by higher RTA temperature.

First Direct Experimental Studies of $Hf_{0.5}Zr_{0.5}O_2$ Ferroelectric Polarization Switching Down to 100-picosecond in Sub-60mV/dec Germanium Ferroelectric Nanowire FETs

Wonil Chung[1], Mengwei Si[1], Pragya R. Shrestha[2,3], Jason P. Campbell[3], Kin P. Cheung[3] and Peide D. Ye[1]*

[1]School of Electrical and Computer Engineering, Purdue University, West Lafayette, IN 47907, USA

[2]Theiss Research, La Jolla, CA 92037, USA

[3]National Institute of Standards and Technology, Gaithersburg, MD 20899, USA

*Email: yep@purdue.edu

ABSTRACT

In this work, ultrafast pulses with pulse widths ranging from 100 ps to seconds were applied on the gate of Ge ferroelectric (FE) nanowire (NW) pFETs with FE $Hf_{0.5}Zr_{0.5}O_2$ (HZO) gate dielectric exhibiting steep subthreshold slope (SS) below 60 mV/dec bi-directionally. With applied gate bias pulses (V_G = -1 to -10 V), high-mobility Ge drain current was monitored as a test vehicle to capture the polarization switching of HZO. It was found that HZO could switch its polarization directly by a single pulse with the minimum pulse width of 3.6 ns. The polarization switching triggered by pulse train with pulse width as short as 100 ps was demonstrated for the first time.

INTRODUCTION

To further reduce device's SS for lower power operation, negative capacitance FETs (NC-FETs) and use of ferroelectric oxide have been recently studied intensively [1]–[4]. Embedded FE oxides such as HZO within the gate stack of a conventional structure has proven to be successful in SS reduction below the 60 mV/dec limit on various channel materials [5]–[10]. HZO specifically has been reported to be reliable in terms of its switching and polarization capability [8], [11]–[15]. The time response of ferroelectric polarization switching is crucial to evaluate the working speed of HZO based ferroelectric FETs (Fe-FETs) and NC-FETs. However, time response properties of the HZO in FE-FETs were rarely studied. Specifically, at ultrafast sub-10ns regime (towards GHz working frequency), time response of HZO has not been extensively studied yet. In this work, we report the *direct experimental observation* of HZO polarization switching under deep sub-10ns and extend the study using pulse train with 100 ps pulses.

EXPERIMENT

Process flow for the fabrication of Ge FE NW FET with FE HZO gate stack was elaborated in our previous report [6] except the NW release step after fin formation. GeOI wafer was implanted by BF_2^+ ions. Dimensions of fins were controlled by SF_6-based dry etching. HF solution was used to release the Ge NWs from SiO_2. 1 nm Al_2O_3 was deposited by atomic layer deposition (ALD), followed by post-oxidation (O_2, 500 °C) to form ultrathin GeO_x. Subsequent HZO (10 nm) and Al_2O_3 (1 nm) were deposited by ALD and post deposition annealed at 500 °C. Source and Drain areas were etched and deposited with Ni for optimum contacts. 3D structure and SEM images of the final devices are shown in Fig. 1 and 2 respectively.

RESULTS AND DISCUSSION

Firstly, the HZO film was analyzed with XRD as shown in Fig. 3 revealing its non-centrosymmetric, orthorhombic crystal structure which causes the FE property [10]. Ferro-electricity of HZO can be further confirmed with P-V measurement (Fig. 4) showing a clear ferroelectric loop [5]–[7].

Fig. 5 is a typical I_D-V_G curve of the fabricated Ge FE NW pFET. Steep sub-60mV/dec SS for five orders of I_D changes are observed bi-directionally. Fig. 6 (a) and (b) show I_D-V_D curves swept in forward and reverse direction, respectively. Negative differential resistance (NDR) is observed owing to negative drain-induced-barrier-lowering (DIBL) [5]. Fig 7 (a) and (b) depict the instrument set-up for generation and measurement of ultrafast pulses with logarithmically increasing widths from 100 ps to few seconds. Agilent 81110

Pulse Generator (PG) was used with current amplifier and a Lecroy oscilloscope (Fig. 7 (a)) for pulse width > 3.6 ns. For picosecond pulses, as shown in Fig. 7 (b), AVTECH PG was used triggered by the Agilent 81110 PG with a period of 2 µs. Lecroy oscilloscope operating at 80 GS/s sampling rate was used to monitor the sub-ns pulse precisely. V_G and V_D applied to the gate are shown in Fig. 8 (a) and (b). To switch the polarization to off-state, as shown in Fig. 8 (a), initialization pulse V_G = 5 V was applied (>100 ms). Different V_G levels (-1 ~ -10V) were pulsed to the gate with various pulse widths and change in I_D was monitored precisely with oscilloscope in real time through a current amplifier. V_G and V_D were 0 V and -50 mV respectively during the measurements carried out between two adjacent V_G pulses to minimize the effect on the polarization. For lower V_G range (-1 ~ -4 V), Keysight B1530A Waveform Generator/Fast Measurement Unit (WGFMU) was used and the measurement time was fixed at 100 µs. Fig. 9 shows that it takes much shorter time to switch the polarization when the V_G pulse approaches -4 V. As pulse width was decreased to sub-µs regime, transient current fluctuation was observed when the polarization switching occurred as seen in Fig. 10 (a) and (b). Therefore, the current was read after stabilization to ensure it was the current caused by changed polarization state only. At higher V_G (-5 ~ -10 V), sub-10ns switching was also observed directly by experiment and it switched fastest at V_G = - 10 V (Fig. 11). Off-state current was plotted at t = 1 ns to show clear I_D transition from off to on-state. The fastest switching by a single pulse observed in our Ge FeFETs with 10 nm HZO was 3.6 ns (Fig. 12). Considering the RC delay present in the measurement set-up and the large probing pads, the intrinsic time response of polarization is expected to be faster than 3.6 ns. To further investigate the effect of sub-ns pulses, Fig. 7 (b) set-up was configured and 100 ps pulses with period of 2 µs (Fig. 13 (a), V_G = -6 V) were generated in the form of pulse train (duty cycle = 0.005 %). Although a single 100 ps pulse (Fig. 14) did not trigger noticeable polarization switching, as the pulses accumulated, polarization switching could be detected (Fig. 15). Further studies on the effect of duty cycle with picosecond pulse widths could be valuable for deeper understanding of dynamics of polarization switching in HZO devices for various applications including FeRAM.

CONCLUSION

High mobility Ge FE NW pFETs were applied as test vehicles for the first time to study the time response of FE HZO gate stack down to 100 ps. It was found that a single deep sub-10ns pulse or even a pulse train with 100 ps pulses were enough to initiate polarization switching in HZO. This work opens the route to studying the dynamics of FE HZO through direct experiments down to 100 ps and even beyond. The work is supported by SRC and Lam Research. U.S. Government is not endorsing any of the equipment mentioned in the paper.

References: [1] S. Salahuddin et.al., *Nano Lett.*, 2008. [2] Z. Krivokapic et.al., *IEDM*, 2017. [3] H. Ota et. al., *IEDM*, 2016. [4] P. Sharma et.al., *VLSI*, 2017. [5] M. Si et. al., *Nat. Nanotechnol.*, vol. 13, p. 24–28, 2018. [6] W. Chung et. al., *IEDM*, 2017. [7] M. Si et. al., *IEDM*, 2017. [8] M. H. Lee et. al., *IEDM*, 2016. [9] C.-J. Su et. al., *VLSI*, 2017. [10] J. Muller et. al., *Nano Lett.*, 2012. [11] K.-Y. Chen et. al., *VLSI*, 2017. [12] H. J. Kim et. al., *Nanoscale*, vol. 8, p. 1383, 2016. [13] J. Muller et. al., *EDL*, vol. 33, no. 2, p. 185-187, 2012. [14] Y. Chiu et. al., *IRPS*, 2015. [15] S. Oh et. al., *EDL*, vol. 38, no. 6, p. 732-735, 2017.

978-1-5386-4219-1/18 $31.00 © 2018 IEEE

2018 Symposium on VLSI Technology Digest of Technical Papers

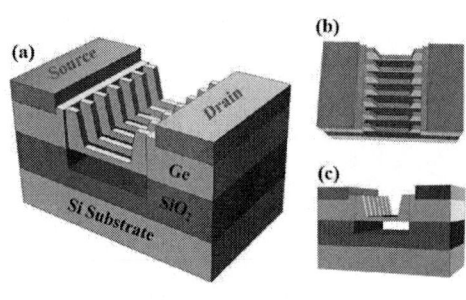

Fig. 1. 3D structure of a Ge NW FET viewed from **(a)** top-right, **(b)** top and **(c)** front. Under the nanowires, SiO_2 was etched and kept hollow.

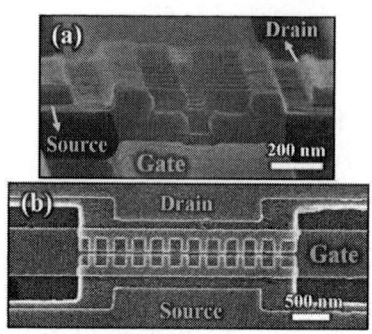

Fig. 2. SEM images of fabricated device viewed from **(a)** side and **(b)** top. 11 parallel NWs make one device.

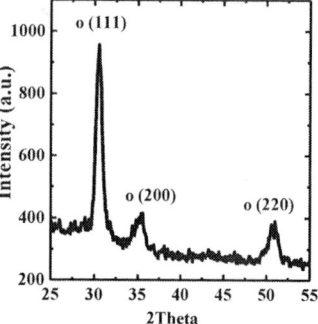

Fig. 3. XRD peaks of the $Hf_{0.5}Zr_{0.5}O_2$ film used in the device exhibits orthorhombic structure.

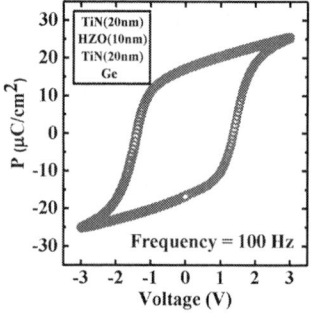

Fig. 4. Polarization-Voltage (P-V) graph of 10 nm HZO film measured at frequency of 100 Hz.

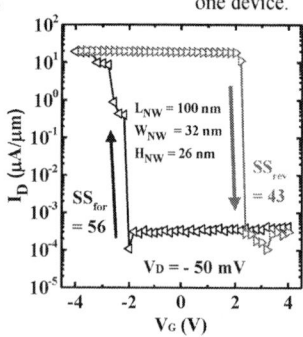

Fig. 5. I_D-V_G exhibits bi-directional SS < 60 mV/dec for 4~5 orders of I_D changes.

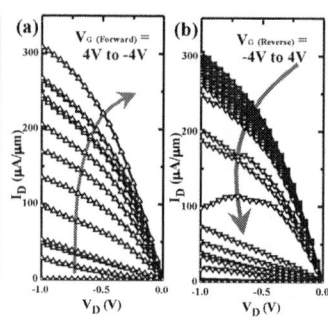

Fig. 6. I_D-V_G of **(a)** forward and **(b)** reverse sweeps. Negative Differential Resistance (NDR) can be observed in reverse sweep direction.

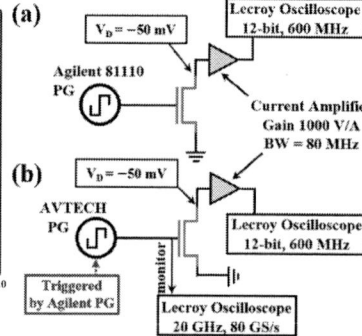

Fig. 7. Ultrafast pulse generation and measurement set-up for **(a)** pulse widths > 3.6 ns and **(b)** picosecond pulses.

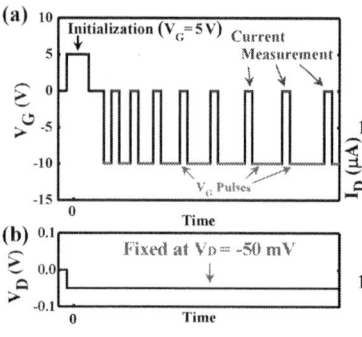

Fig. 8. Pulse time line of **(a)** V_G and **(b)** V_D. Currents were measured in between pulses with V_G = 0 V.

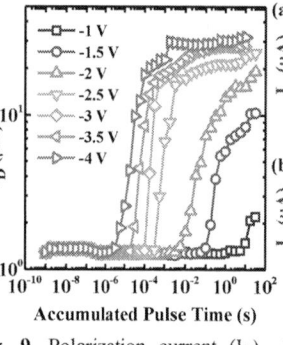

Fig. 9. Polarization current (I_D) over accumulated pulse time with different V_G pulses. Keysight B1530A was used for these pulses.

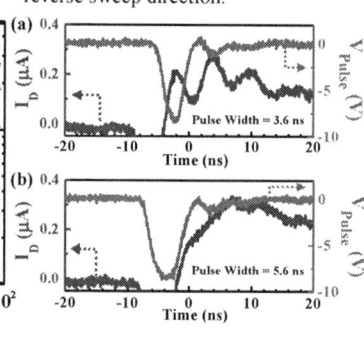

Fig. 10. Polarization switching can be monitored by I_D. Time responses with pulse widths of **(a)** 3.6 ns and **(b)** 5.6 ns are shown. I_D was taken after the fluctuation was stabilized.

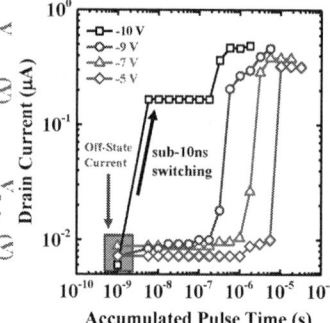

Fig. 11. V_G was pushed down to -10 V to probe the polarization switching limit which showed minimized time under 10 ns. Off-state current was plotted at 1 ns for reference.

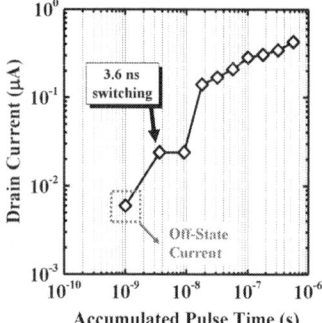

Fig. 12. The fastest switching was observed at accumulated V_G pulse time of 3.6 ns. Off-State current was plotted at 1 ns for reference.

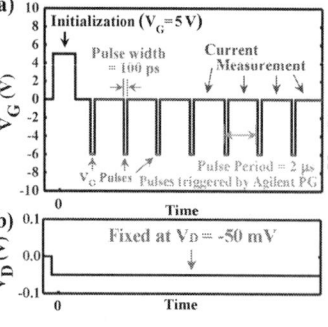

Fig. 13. (a) V_G and (b) V_D settings for sub-ns pulse measurements with Fig. 7 (b) configuration. I_D was kept constant at -50 mV and 100 ps pulses were triggered every 2 μs (duty cycle = 0.005 %).

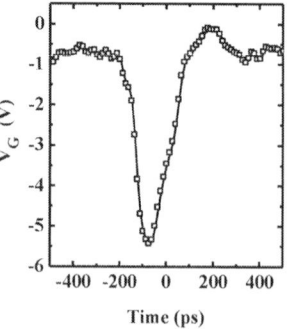

Fig. 14. 100 ps pulse generated with rise time of 60 ps was measured at the rate of 80 GS/s. Maximum voltage was -6 V. To avoid loss of voltage due to cables, the PG was placed very close to the device.

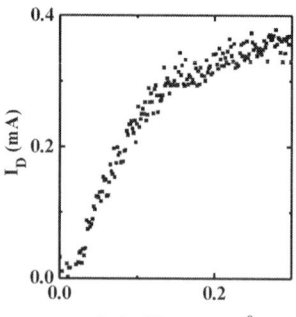

Fig. 15. Polarization switching was monitored by applying the 100 ps pulse shown in Fig. 14 in the form of a continuous pulse train.

978-1-5386-4219-1/18 $31.00 © 2018 IEEE

Gap in pagination due to formatting issues.

Pages 91-92

10μW/cm²-Class High Power Density Planar Si-Nanowire Thermoelectric Energy Harvester Compatible with CMOS-VLSI Technology

M. Tomita[1], S. Oba[1], Y. Himeda[1], R. Yamato[1], K. Shima[1], T. Kumada[1], M. Xu[1], H. Takezawa[1], K. Mesaki[1], K. Tsuda[1], S. Hashimoto[1], T. Zhan[1], H. Zhang[2] Y. Kamakura[3], Y. Suzuki[4], H. Inokawa[4], H. Ikeda[4], T. Matsukawa[5], T. Matsuki[1,5], and T. Watanabe[1]

[1]Waseda Univ., Tokyo, Japan, [2]Gunma Univ., Kiryu, Japan, [3]Osaka Univ., Osaka, Japan,
[4]Shizuoka Univ., Hamamatsu, Japan, [5]AIST, Tsukuba, Japan. E-mail: tomita_motohiro@watanabe.nano.waseda.ac.jp

Abstract

A best benchmark of Si-nanowire (NW) thermoelectric (TE) power generator has been achieved by our proposed planar device architecture compatible with CMOS process technology. The TE power density corresponds to 12 μW/cm², which is recorded at an externally applied temperature difference of only 5 K. The demonstration opens up a pathway to cost effective autonomous internet of things (IoT) application utilizing environmental and body heats.

Introduction

Miniaturized TE energy harvester is anticipated to open up the field of application of portable, wearable, and distributed sensor network systems toward the IoT society (**Fig. 1**) [1,2]. Si-NWs emerge as a promising candidate for the TE material [3] thanks to the quite low thermal conductance while keeping the high electric conductance, together with the high affinity with the CMOS technology and the low environment load unlike Pb and Te.

Conventional demonstrations of planar Si-based TE generator [4-6] employed long Si-NWs about 10-100 μm which are suspended on a cavity to cutoff the bypass of the heat current to secure the temperature difference across Si-NWs. However, the cavity structure weakens the mechanical strength of the device and increases the fabrication cost (**Fig. 2a**).

In this paper, we experimentally demonstrate an effectiveness of a novel design concept of planar and short Si-NW TE generator without cavity structure, which uses a steep temperature gradient formed in the vicinity of the main heat current (**Fig. 2b**) [7]. The power density is confirmed to be scalable by shortening Si-NW to sub-μm length and it exceeded those of the conventional planar Si-based TE generators.

Fabrication of Si-NW TE Generator

The planar Si-NW TE generators were fabricated on an SOI wafer (SOI/BOX=50nm/145nm, 745μm-base Si) by conventional Si CMOS process techniques (**Fig. 3**). The SOI layer was patterned into Si-NWs with a width of W_{NW} <100nm, followed by phosphorous doping (I/I, 5.0×10^{15} cm⁻², 10keV). Forming gas anneal was conducted at 400°C [8]. The device comprises 70 lines of Si-NWs with 500 nm pitch. The length of Si-NW (L_{NW}) ranged from 0.25 to 1 μm. A 400nm-Al/TiN/Ti electrode was formed on the Si-pad. The thickness of the base Si-substrate was varied from 745 to 50 μm by thinning with backside grinding.

Results and Discussion

Simulation of TE performance

Temperature difference of hundreds mK is maintained across sub-μm short Si-NWs, that simulated by finite element method (FEM) using COMSOL Multiphysics® (**Fig. 4**). For designing our TE generator, we numerically estimated the dependence of the TE power density on the Si-NW length and Si-substrate thickness (**Fig. 5 and 6**). In this report, the TE power density is normalized by the area of the product of twice of the NW length and the NW pitch (**Fig. 2**), supposing that the pads and NW arrays have a same footprint. The simulation shows that the TE power density is proportional to L_{NW}^{-2} in the range of L_{NW} > 0.1 μm, namely, the TE performance is improved by shrinking the device length. Furthermore, the TE power density increases dramatically by thinning the Si substrate to sub-100μm because the series thermal resistance of the substrate is drastically suppressed (**Fig. 7**).

Demonstration of TE generator

Fig. 8 shows the relation between measured TE power density vs. L_{NW}, together with the internal resistance of Si-NW. The TE power density is enhanced by shortening the Si-NW and the power density of 1.0 μW/cm² is obtained at L_{NW} = 0.25 μm. The observed power density is much larger than that expected merely from the decrease in the internal resistance of Si-NW, which is shown as P_{calc} in **Fig.8**. The enhancement was attributed to the increase in the effective Seebeck coefficient due to the enhanced phonon drag effect [9] in shorter Si-NWs (**Fig. 9**).

By thinning Si-substrate, the TE power density was further enhanced drastically (**Fig. 10**). The TE power density of the 50 μm-thick substrate device reached 12 μW/cm². The enhancement agrees with the FEM simulation. **Fig.11** shows a benchmark plot of the TE power densities of Si-based TE generators of this work and previous reports [4-6,10-12]. The TE performance demonstrated in this work surpasses other planar suspended type devices [4-6] by 10 times. The simulation shows that our TE generator has a potential to enhance the power furthermore by shortening Si-NWs, and it can exceed the highest record achieved by a vertical device operated with internal resistive heater [12].

Conclusion

We realized the high TE power density of 12μW/cm² by shortening the Si-NW length to sub-μm scale and suppressing the series thermal resistance by thinning the Si-substrate. The demonstrated power can drive various IoT devices which, for example, collect information a few times and transmit a few Mbit information once every hour (**Fig. 1**) [1,2].

Acknowledgments

This work was supported by the JST-CREST (JPMJCR15Q7), by NIMS Nanofabrication Platform, and by AIST-SCR. We are grateful to Mr. Kawaguchi in Hamamatsu Photonics K.K. for offering stealth dicing.

References

[1] K. Nair *et. al.*, Proc. Int. Conf. Green Computing and IoT (2015) 589. [2] J. Dieffenderfer *et. al.*, IEEE J. Biomed. Health Inform. **20** (2017) 1251. [3] A. I. Boukai *et. al.*, Nature **451** (2008) 168. [4] X. Yu *et. al.*, J. Micromech. Microeng. **21** (2012) 105011. [5] J. Xie *et. al.*, J. Microelectromech. Syst. **19** (2010) 317. [6] M. Strasser *et. al.*, Sens. Actuator A **97-98** (2002) 535. [7] T. Watanabe *et. al.*, EDTM Conf. Proc. Tech. Papers (2017) 86. [8] S. Hashimoto. *et. al.*, Appl. Phys. Lett. **111** (2017) 023105. [9] F. Salleh *et. al.*, Appl. Phys. Lett. **105** (2014) 102104. [10] B. Xu *et. al.*, IEEE Electron Device Lett. **35** (2014) 596. [11] K. J. Norris *et. al.*, Dr. Thesis, UCSC (2015). [12] B. M. Curtin *et. al.*, J. Electron. Mater. **41** (2012) 887.

2018 Symposium on VLSI Technology Digest of Technical Papers

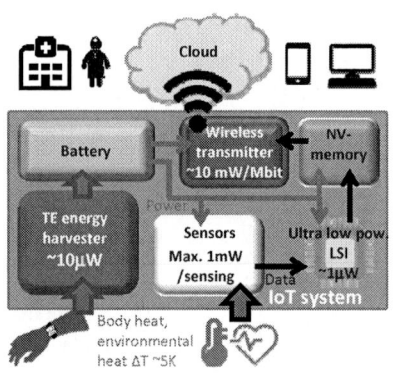

Fig. 1 Example of wearable sensor module for healthcare IoT system.

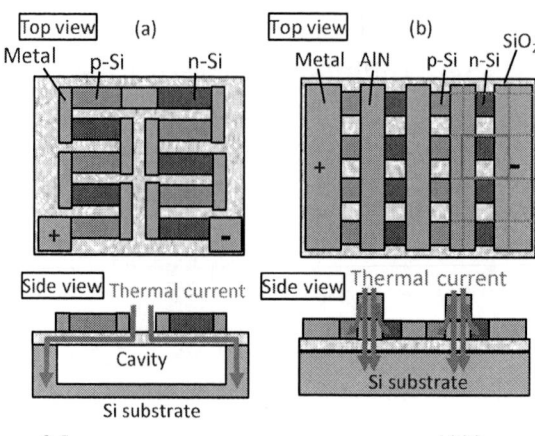

Fig. 2 Schematics of Si-NW TE generators. (a) A conventional typical device structure with long Si-NW suspended on a cavity. The thermal current flows along the Si-NW. (b) Our proposed TE device architecture with short Si-NWs and without cavity structure. The thermal current flows perpendicular to Si-substrate and it forms exuded steep temperature gradient in Si-NWs. The red rectangle indicates the unit area of the device structure to normalize TE power density.

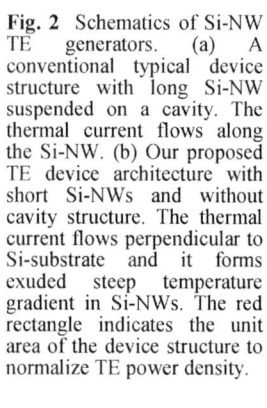

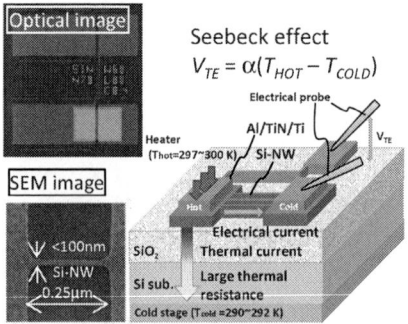

Fig. 3 Fabricated Si-NW TE generator without cavity structure. Si-NWs were patterned by ArF immersion lithography.

Seebeck effect
$$V_{TE} = \alpha(T_{HOT} - T_{COLD})$$

Fig. 4 Temperature difference across Si-NW estimated by FEM. Inset is a temperature profile in a NW taking the reference at hot side edge.

FEM simulation
$T_{hot} - T_{cold} = 5K$
Si-sub thickness = 745μm

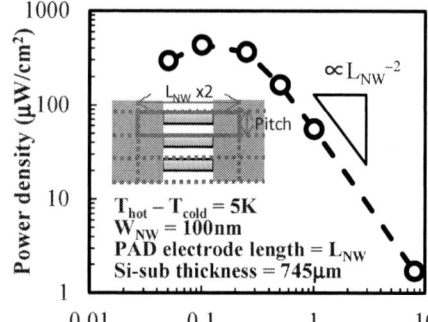

Fig. 5 Simulated power density vs. Si-NW length of proposed TE generator estimated by FEM simulation.

$\propto L_{NW}^{-2}$
$T_{hot} - T_{cold} = 5K$
$W_{NW} = 100nm$
PAD electrode length = L_{NW}
Si-sub thickness = 745μm

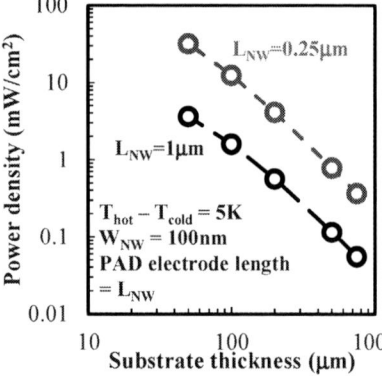

Fig. 6 Si-substrate thickness dependency of TE power density estimated by FEM simulation.

$L_{NW} = 0.25μm$
$L_{NW} = 1μm$
$T_{hot} - T_{cold} = 5K$
$W_{NW} = 100nm$
PAD electrode length = L_{NW}

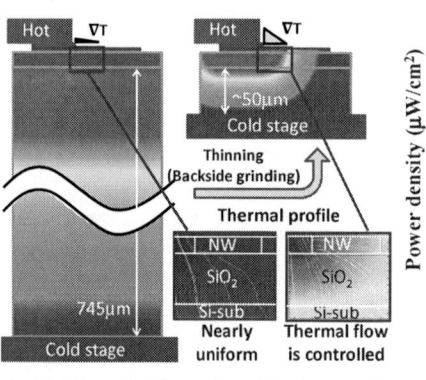

Fig. 7 Schematic illustration of the impact of thinning the Si-substrate on the temperature gradient formed in Si-NW.

Fig. 8 Experimental dependency of TE-power density and internal resistance on the Si-NW length with 100 nm width.

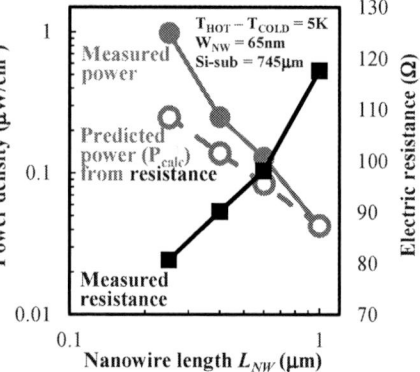

$T_{HOT} - T_{COLD} = 5K$
$W_{NW} = 65nm$
Si-sub = 745μm

Measured power
Predicted power (P_{calc}) from resistance
Measured resistance

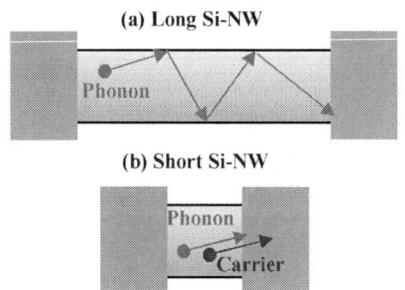

Fig. 9 Schematic illustration of behavior of phonons in Si-NWs. (a) In long Si-NW, phonon drag effect is diminished by the frequent surface scattering. (b) In short Si-NWs, the phonon drag is effective to enhance the TE power.

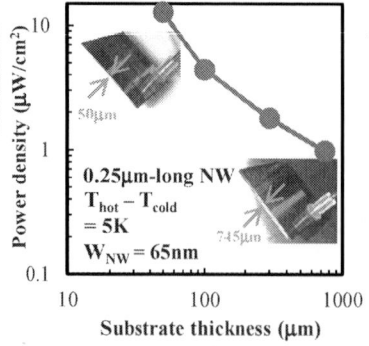

Fig. 10 Experimental dependence of TE power density on the Si-substrate thickness. Si-NWs are 0.25 μm long and 100 nm wide.

0.25μm-long NW
$T_{hot} - T_{cold} = 5K$
$W_{NW} = 65nm$

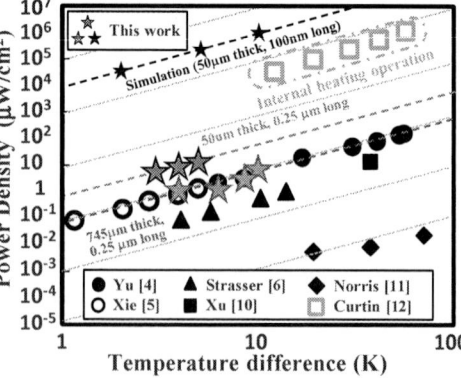

Fig. 11 TE power benchmark. The result of present work surpasses other demonstrations using externally applied temperature difference.

This work
Simulation (50nm thick, 100nm long)
Internal heating operation
50nm thick, 0.25 μm long
745μm thick, 0.25 μm long

● Yu [4]   ○ Xie [5]   ▲ Strasser [6]   ◆ Norris [11]   ■ Xu [10]   □ Curtin [12]

978-1-5386-4219-1/18 $31.00 © 2018 IEEE

A low-power and high-speed True Random Number Generator using generated RTN

James Brown, Rui Gao, Zhigang Ji*, Jiezhi Chen*[2], Jixuan Wu[2], Jianfu Zhang, Bo Zhou, Qi Shi, Jacob Crowford, and Weidong Zhang

Faculty of Engineering and Technology, Liverpool John Moores University, Liverpool, UK, email: z.ji@ljmu.ac.uk

[2] School of Information Science and Engineering, Shandong University, Jinan, P. R. China, email: chen.jiezhi@sdu.edu.cn

Abstract: A novel True Random Number Generator (TRNG), using random telegraph noise (RTN) as the entropy source, is proposed to address speed, design area, power and cost simultaneously. For the first time, the proposed design breaks the inherent speed limitation and generates true random numbers up to 3Mbps with ultra-low power. This is over 10 times faster than the state-of-the-art RTN-TRNG [6]. Moreover, the new design does not require selection of devices and thus avoids the use of large transistor array and laborious post-selection process. This reduces the circuit area and the cost. The proposed TRNG has been successfully validated on three different processes and they all passed the National Institute of Standards and Technology (NIST) tests, making it a suitable candidate for future cryptographically secured applications in the internet of things (IoT).

Introduction: True random number generators (TRNGs) harvest physical randomness as entropy sources and are heavily used in cryptography and security [1]. However, their power consumption and design complexity are often high [1-5]. The recently-proposed TRNG using Random Telegraph Noise (RTN) [6-8] has been considered as an ideal solution for future IoT applications (**Fig.1**), due to its simplicity, low power and robustness against temperature and supply voltage variations. Their practical use, however, is hindered by two major deficiencies: 1) *Device Selectivity*: RTN-TRNGs is based on one nano-scaled transistor exhibiting clear RTN signal by one pre-existing trap. However, the percentage of such devices in one wafer is very low (**Fig.2a&b**) [9]. A large transistor array is usually needed in the design, out of which one transistor will be selected manually by tuning the circuit after fabrication. This leads to larger design area and higher cost [10-11]. 2) *Low speed*: The speed of the RTN-TRNG has strong correlation with τ, the sum of the times to capture and emission (τc, τe) [8]. The opposite voltage dependence of τc and τe imposes an inherent limit for speed. *In this work, a novel RTN-TRNG is proposed to tackle the above drawbacks without extra design penalty. The design is successfully applied on three different processes (Table II) and maximum NIST-validated bit rate of 3Mbps has achieved.*

RTN generation and characterization: In addition to pre-existing traps, some traps generated by electrical stress can also induce RTN [12-13]. This provides a pathway to 'insert' RTN into any nano-scaled device. One nFET from process A1 is used as an example in the following. After hot carrier stress for 50s, a new trap can be generated, which still exists after accelerated recovery (**Fig.3a&b**). This generated trap shows clear RTN (**Fig.3c-e**). It is found in our process, A1, that although RTN only exists in 17% of the fresh devices, clear RTN can be observed in more than 80% of the devices after hot carrier stress (**Fig.4**). Therefore, the bulky transistor array and the post-selection [6] can be avoided when the TRNG entropy is taken from these generated traps. Similar to the pre-existing ones, these generated traps also show strong voltage dependence (**Fig.5a**) and are highly stable (**Fig.5b**). Their profile extracted from the RTN measurements [14] suggest that they could be away from the Si/dielectric interface, which further supports their nature of generation [15] (**Fig.6a&b**).

RTN acceleration with AC operation: All the existing RTN-TRNGs operate under DC condition and are slow. The voltage tuning is usually applied to optimize its bit rate. Due to the opposite voltage dependence,

no matter what gate voltage is applied, either τc or τe will increase, hindering further improvement (**Fig.7a**) [6-8]. To tackle this dilemma, we propose to operate RTN under AC condition: The current is sensed under VgL for half cycle after VgH is applied for the other half (**Fig.8**). This allows τc and τe being controlled by VgH and VgL independently (**Fig.7a**). The reduction of τc and τe is only limited by measurement accuracy (for the lowest-allowable VgL) and device reliability (for the highest-allowable VgH). Compared with DC operation, AC operation can easily accelerate RTN by hundreds of times (**Fig.7a**). A clear difference can be observed for the same nFET when operating under DC (**Fig.9a-c**) and AC (**Fig.9d-f**) respectively. It has been reported that τc and τe can also be reduced when the applied frequency increases [16]. Such frequency dependence can also be observed in our measurements (**Fig.9g-i**). This can further reduce τc and τe, leading to a faster TRNG. The amplitude of RTN is also found to be large under AC condition (**Fig.9a-i**), because the sensing voltage, VgL, is already in the subthreshold region where strong percolation is expected [17]. What is worth noting is that such sub-threshold sensing scheme also naturally reduces power consumption.

RTN-TRNG design and validation: Since RTN is produced at VgL under AC condition, two nFETs are used (**Fig. 10**). By applying 180°-shifted gate biases, two RTNs are generated every half cycle. After amplification and digitization, they can be combined together through a transmission gate. For a given trap, it is known that the sum of the times to capture and emission ($\tau c+\tau e$), when averaged, is a constant against time (Eqns in **Fig.11**). Therefore, by toggling only at rising edge of the RTN, the new trace can be obtained with 1s and 0s of equal probability, making it truly random without complicated post-processing [8]. After sampling at a given clock frequency, the random number stream is generated with high entropy (**Fig.12**). It passes all the NIST tests at the maximum speed of 2Mbps (**Fig.13a&b**), over 10 times faster than state-of-the-art one [6, 18].

Yield and applicability to process: The yield is estimated by evaluating 14 TRNGs. Over 90% pass at 1Mbps and almost 50% even reach 3Mbps making it readily usable in practice (**Fig.14**). The same design is also applied on another two processes, A2 and A3 (**Table II**). They also passed the NIST tests (**Table III**).

Conclusions: A novel RTN-TRNG design is demonstrated. By injecting randomness into the transistors through electrical stress and operating under AC domain, the new design provides a solution to address speed, design area, power consumption, reliability and cost simultaneously. The proposed TRNG has passed the NIST tests, making it readily applicable for cryptographically secured applications.

Acknowledgement: This work is supported by Engineering and Physical Science Research Council of UK (EP/L010607/1) and China key Research and Development Program (2016YFA0201802). The authors would like to thank imec for providing samples for this research.

References: [1] M. Bucci *et al*, p. 403, IEEE Trans. Comput 2003. [2] C. Tokunaga *et al*, JSSC 2008. [3] S. Srinivasan *et al*, p. 203, VLSI-C 2010. [4] N. Liu *et al*, p. 203, VLSI-C 2011. [5] K. Yang *et al*, p. 280, ISSCC 2014. [6] R. Brederlow *et al*, p. 79, ISSCC 2006. [7] T. Figliolia *et al*, p.17, ISCAS 2016. [8] A. Mohanty *et al*, p.2248, TVLSI 2017. [9] C. Chen *et al*, p. 190, IRPS 2011. [10] M. D. Giles *et al*, p. 501, VLSI 2015. [11] N. Tega *et al*, p.630, IRPS 2011. [12] M. Toledano-Luque *et al*, p.978, VLSI 2012. [13] R. Gao *et al*, p.778, IEDM 2016. [14] T. Nagumo *et al*, p.628, IEDM 2010. [15] C. Lu *et al*, p. 936, TED 2014. [16] R. Wang *et al*, p. 978, VLSI-T 2012. [17] L. Gerrer, *et al*, p. 226, ESSDERC 2015. [18] NIST, Pub 800-22, 2001

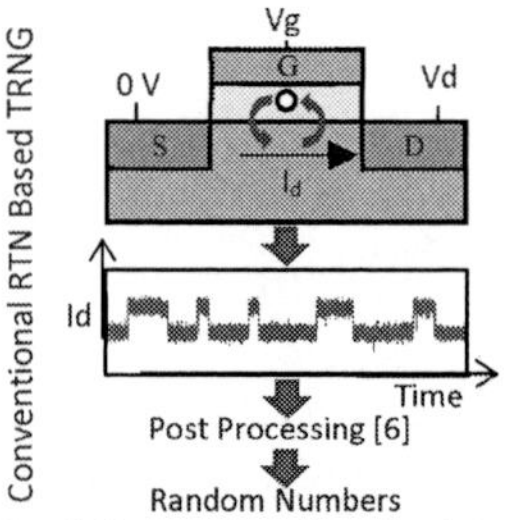

2018 Symposium on VLSI Technology Digest of Technical Papers

Fig. 1 Conventional method for true random number generation using RTN in nano-scaled transistors.

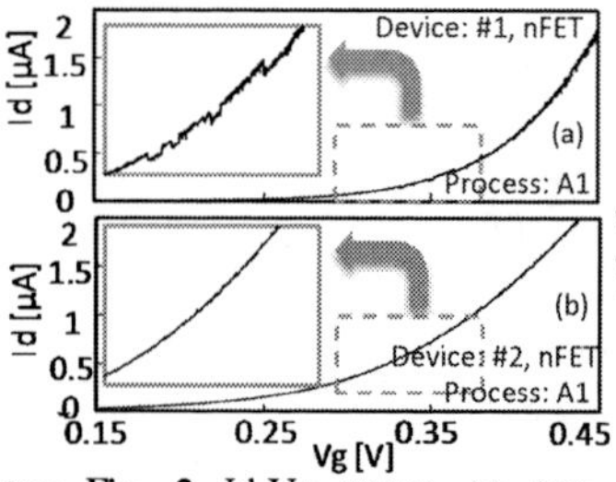

Fig. 2 Id-Vg sweep on two devices (a) with and (b) without RTN. Process A1 is used.

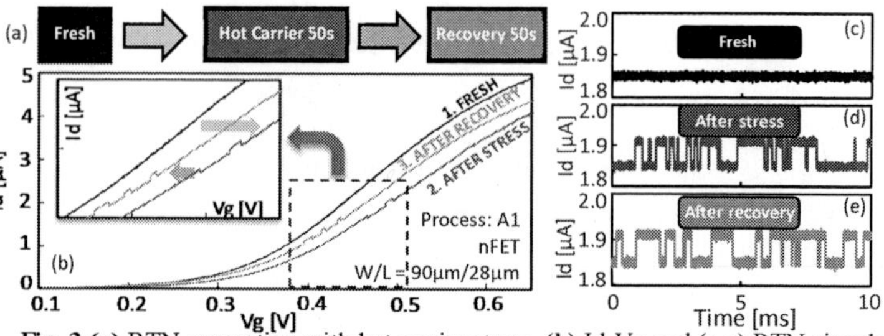

Fig. 3 (a) RTN generation with hot carrier stress. **(b)** Id-Vg and **(c-e)** RTN signals under constant Id=1.8µA after each step. Stress: Vg=Vd=1.9V,50s,125°C. Recovery: Vg=Vd=-1.5V/0V,50s,125°C. All measurements are at 27°C.

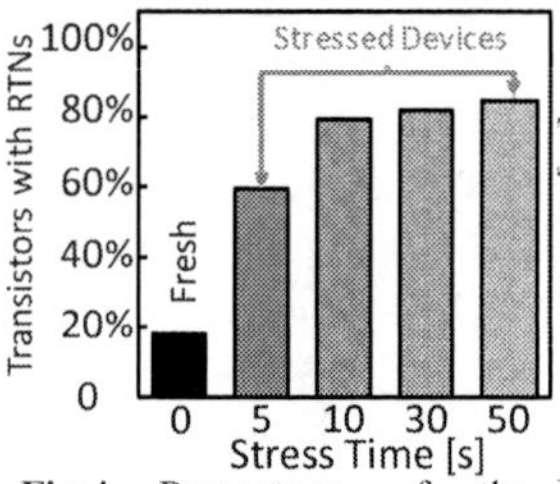

Fig.4 Percentage of the devices with observable RTN after different time of stress.

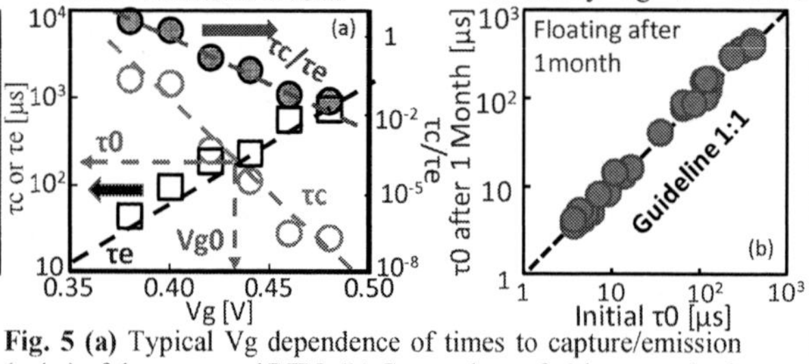

Fig. 5 (a) Typical Vg dependence of times to capture/emission (τc/τe) of the generated RTN. **(b)** Comparison of τ0 between fresh and floating over 1month after stress.

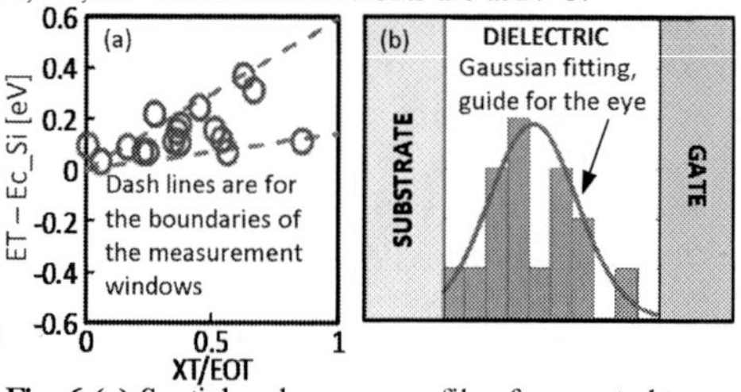

Fig. 6 (a) Spatial and energy profile of generated traps using method in ref. 14. **(b)** The histogram of trap numbers against the spatial location.

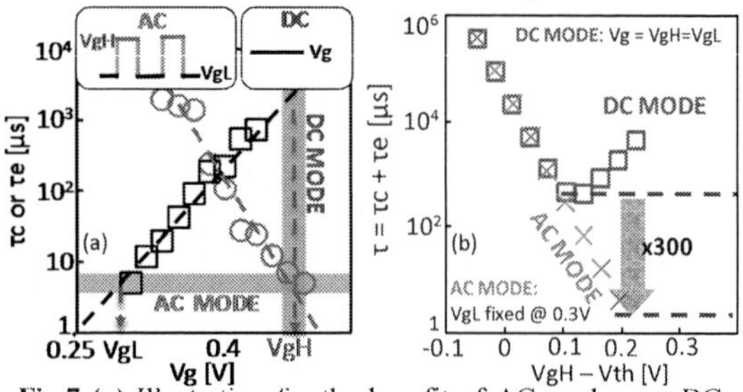

Fig.7 (a) Illustration for the benefit of AC mode over DC mode to accelerate RTN. **(b)** The comparison of the time constant, τ = τc+τe, between DC and AC operation modes.

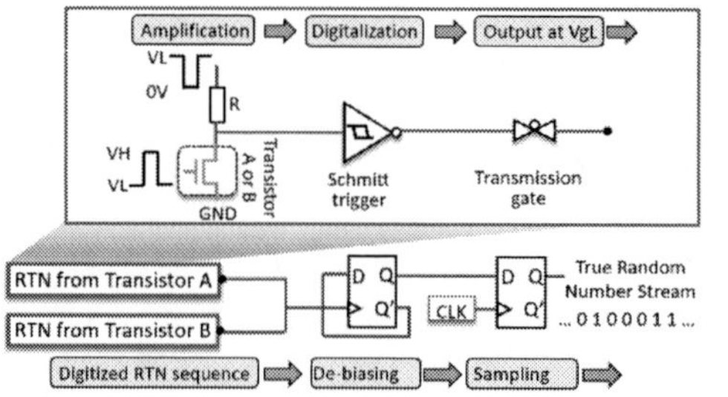

Fig.10 The circuit schematic for the proposed RTN-based TRNG. Two transistors are used to generate RTN alternatively within one voltage period.

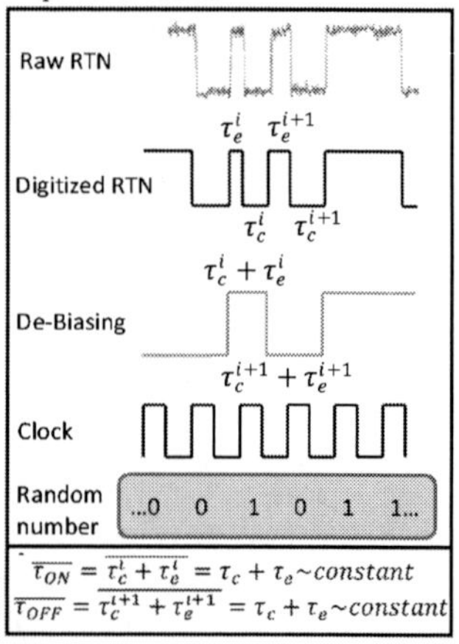

Fig.11 Procedure to convert RTN signal to random bit stream. By using the intrinsic property of RTN (Equations), true randomness is obtained without post-processing.

$$\overline{\tau_{ON}} = \overline{\tau_c^i + \tau_e^i} = \tau_c + \tau_e \sim constant$$
$$\overline{\tau_{OFF}} = \overline{\tau_c^{i+1} + \tau_e^{i+1}} = \tau_c + \tau_e \sim constant$$

Fig. 8 Illustration of the procedure for measuring RTN at AC operation mode.

Fig. 9 The measured RTNs under DC **(a-c)**, different VgH with frequency of 100kHz **(d-f)** and different frequencies, with VgH = 0.68V. All the measurements are taken at 27°C. For AC mode, VgL = 0.28V is always used.

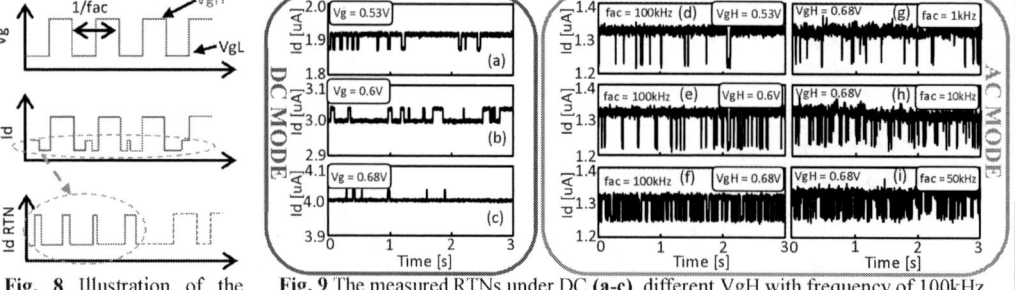

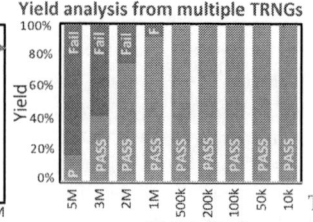

256x256 bits generated from one TRNG

Table. I Results summary from the NIST tests for process A1 where α = 0.01, bit rate = 2MHz, bit sequence length = 40k, and number of bit streams = 100.

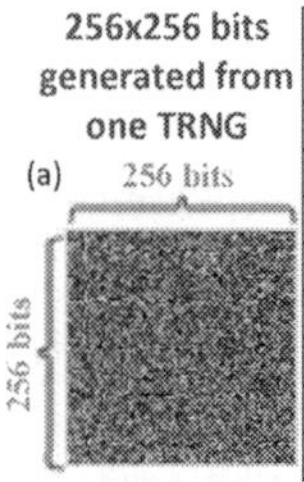

Test ID	Test Name	p Value	Result
T1	Monobit test	0.5342	PASS
T2	Frequency tests (block)	0.2133	PASS
T3	Cumulative sums Test	0.7399	PASS
T4	Run Test	0.5341	PASS
T5	Longest Run (block)	0.9915	PASS
T6	DFT Test	0.1222	PASS
T7	Non-overlapping test	0.3505	PASS
T8	Entropy Test	0.3505	PASS
T9	Serial Test	0.2133	PASS

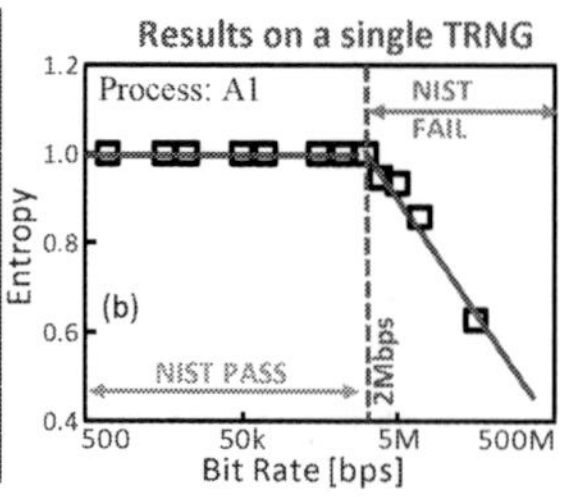

Fig. 12 (a) Bitmap of the random bits generated with 2Mbps bit rate. **(b)** Entropy of the random a bit stream generated from process A1 at different bit rates.

Fig. 13 NIST test yield at different bit rates. 14 TRNGs are randomly chosen for the analysis. Process A1.

	A1	A2	A3
Technology	22nm	28nm	45nm
W/L	90nm/28nm	90nm/70nm	
EOT	1.0nm	1.3nm	1.45nm
Dielectric	HKMG	HKMG	HKMG
Structure	FinFET	Planar	Planar

Test ID	p value	
	Process A2	Process A3
T1	0.0431 (Pass)	0.7399 (Pass)
T2	0.9114 (Pass)	0.9114 (Pass)
T3	0.7399 (Pass)	0.1223 (Pass)
T4	0.2133 (Pass)	0.9915 (Pass)
T5	0.3505 (Pass)	0.7399 (Pass)
T6	0.7399 (Pass)	0.9114 (Pass)
T7	0.9114 (Pass)	0.5341 (Pass)
T8	0.2133 (Pass)	0.7399 (Pass)
T9	0.5341 (Pass)	0.3505 (Pass)

Table. II (Top) Summary of the three different processes used in this work. **Table. III (Bottom)** Summary of the NIST tests for processes A2 and A3. Where α = 0.01. The bit stream is generated at 2Mbps.

978-1-5386-4219-1/18 $31.00 © 2018 IEEE

Ultrahigh-Sensitive and CMOS Compatible ISFET Developed in BEOL of Industrial UTBB FDSOI

Getenet Tesega Ayele[1, 2, 3, 4], Stephane Monfray[1], Serge Ecoffey[3, 4], Frederic Boeuf[1], Romain Bon[1], Jean-Pierre Cloarec[2], Dominique Drouin[3, 4], Abdelkader Souifi[2]

[1]STMicroelectronics, 850 Rue Jean Monnet, 38920 Crolles, France. [2]INL, 20 Av. Albert Einstein, 69100 Villeurbanne, France. [3]Institut interdisciplinaire d'innovation technologique (3IT), Université de Sherbrooke, QC, Canada. [4]Laboratoire Nanotechnologies Nanosystemes (LN2) CNRS UMI-3463, Université de Sherbrooke, QC J1K 2R1, Canada.
E-mail: getenet-tesega.ayele@insa-lyon.fr

Abstract

The industrialization of ion-sensitive field-effect transistors (ISFETs) has been constrained due mainly to the limited sensitivity, and inclusion of bulky reference electrode. With this paper, we report an ultrahigh-sensitive and CMOS compatible ISFET in which the need for the reference electrode is eliminated. Based on an industrial UTBB FDSOI device in BEOL, we obtained an ultrahigh sensitivity of 730 mV/pH which is 12-times higher than the Nernst limit. Integrating the sensing area and the control gate in the BEOL of UTBB FDSOI transistors with a capacitive divider circuit, and using the back biasing feature of such devices, we could eliminate the necessity of the reference electrode making our sensor highly scalable and ideal for the IoT. This is the first demonstration of an integrated pH sensor in the BEOL of FDSOI platform. The measurements on fabricated sensors have also been validated by modeling and simulation.

Introduction

Ion-sensitive field-effect transistors (ISFETs) have recently been mentioned as one of the three next-generation-sequencing (NGS) platforms for DNA sequencing [1]. They have also wide potential application in chemical, biological, medical, environmental, agricultural, food processing, and other industrial process monitoring. Despite this, after nearly 4 decades of research and development, the commercialization of ISFETs is still at infancy stage due mainly to the limited sensitivity, and inclusion of the reference electrode.

Employing UTBB FDSOI industrial transistors, we developed an ultrahigh-sensitive ISFET where the pH detection is made at the front gate while the signal is recorded at the back gate. On the other hand, the capacitive divider structure has been utilized for front gate biasing through a control gate which is fundamental for stable ISFET functioning. Hence, the reference electrode has been completely avoided making our sensor highly scalable and CMOS compatible. Fig. 1 shows schematics of our ISFET, while the fabrication process flow is presented on Fig. 2.

Our proposed approach has been validated by the consistent results obtained from our modeling, simulation, and prototype fabrication and characterization.

Modeling and Simulation of the FDSOI ISFET

In UTBB FDSOI devices, a small change in front gate voltage results in an amplified shift in threshold voltage at the back gate, due to the strong electrostatic coupling between the two gates [2]. Developing our sensor based on such devices, we obtained a sensitivity amplification which is equivalent to the ratio of gate oxide capacitance and BOX capacitance [3, 4]. We utilized industrial UTBB FDSOI transistors

manufactured by STMicroelectronics which have an amplification of 13 (indicated on Fig. 3). This gives a theoretically expected sensitivity of 780 mV/pH (for a pH-sensing film that has Nernstian response).

We simulated the FDSOI ISFET utilizing TCAD Sentaurus. From the simulation result (presented on Fig. 4), an amplification of 13 was obtained which confirms both the theoretically expected and the measured device responses.

Fabrication and Characterization

We fabricated the ISFET through BEOL processing of industrial UTBB FDSOI transistors. Hence, the sensing gate and control gate were integrated on top of the gate metal. Al_2O_3 layer for pH sensing is deposited and patterned in BEOL pads, and the control gate is processed on top of the pH-sensing layer. All electrodes were then encapsulated with a protective insulator, so that only the Al_2O_3 layer is exposed to the electrolyte for pH sensing. SEM-image of the fabricated ISFET's cross-section and the electrical connection layout are shown on Fig. 5 and Fig. 6 respectively.

Characterizing at 3 pH values, we obtained a sensitivity of 730 mV/pH (presented on Fig. 7) which is in confirmation with the modeling and simulation. This highly enhanced sensitivity and the steep subthreshold slope make the performance of our sensor suited for both fixed voltage (0.9 decade/pH) and fixed current readout circuits. In addition, these pH sensing results were taken in 1-minute after dispensing the pH solution on the sensor proving the fast response time of the ISFET.

The challenge of homopolymer repeats in ISFET based DNA sequencing platforms can also be addressed with the tuning-through-back-biasing feature of UTBB FDSOI devices. We present the measured tuning feature of such devices on Fig. 3. Downscaling the development of ISFETs deep into the 28 nm technology, our sensor is ideal for DNA sequencing platforms also that need ultrahigh density ISFET-array.

Conclusion

We fabricated and tested an ISFET which is the first demonstration of an integrated pH sensor in the BEOL of FDSOI platform allowing sensing area as small as 29 μm x 70 μm. Benchmarked with the state-of-the-art (Fig. 8), our sensor is superior in terms of both sensitivity and scaling. This pH sensor is developed with a complete elimination of the need for a reference electrode. Therefore, our sensor enables CMOS integration, ultrahigh-sensitive detection, and very large density sensor-array.

References

[1] M. A. Quail, *et al., BMC Genomics*, vol. 13, p. 341, 2012.
[2] S. Monfray, *et al., ESSDERC*, pp. 76–79, 2015.
[3] G. T. Ayele, *et al., ESSDERC*, pp. 264–267, 2017.
[4] L. Rahhal, *et al., Solid-State Electronics*, 2017.

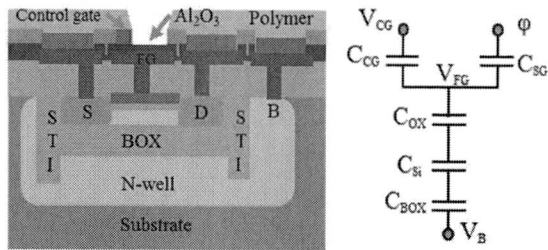

Fig. 1 Schematic diagram (left) and equivalent electrostatic model (right) of the FDSOI ISFET.

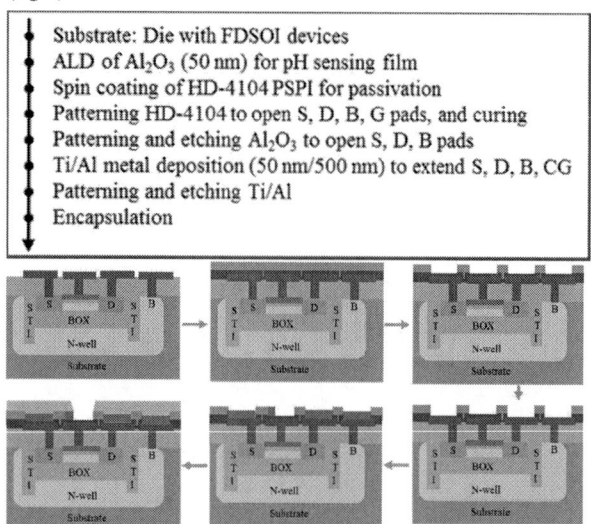

- Substrate: Die with FDSOI devices
- ALD of Al_2O_3 (50 nm) for pH sensing film
- Spin coating of HD-4104 PSPI for passivation
- Patterning HD-4104 to open S, D, B, G pads, and curing
- Patterning and etching Al_2O_3 to open S, D, B pads
- Ti/Al metal deposition (50 nm/500 nm) to extend S, D, B, CG
- Patterning and etching Ti/Al
- Encapsulation

Fig. 2 Process flow for fabrication of the ISFET in FDSOI BEOL.

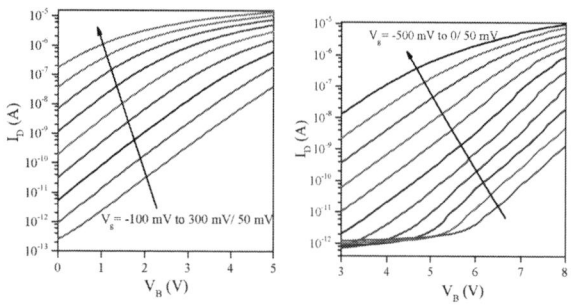

Fig. 3 Measured intrinsic amplification and V_{th}-tuning features of FDSOI devices for sensing applications. A 50 mV change in front gate voltage results in a 650 mV shift in back gate voltage (amplification = 13). The linear operating-regime can also be tuned with back gate bias depending on the measurement pH range.

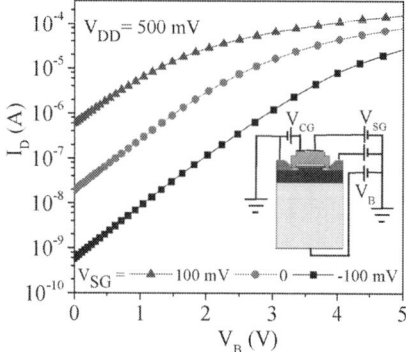

Fig. 4 Simulation result of the ISFET: back gate voltage shifts by 1.3 V for 100 mV change in sensing gate voltage (amplification = 13).

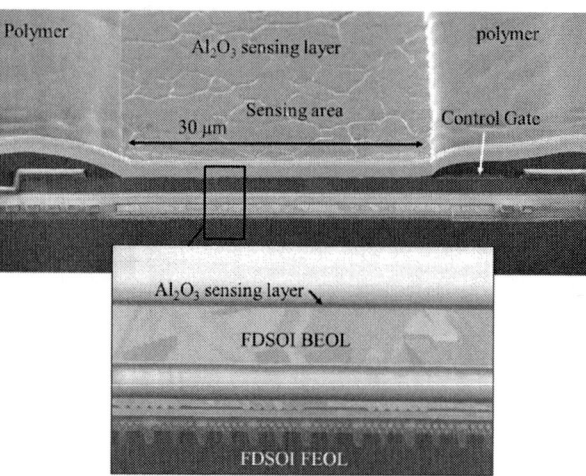

Fig. 5 SEM-image of the fabricated FDSOI ISFET's cross-section.

Fig. 6 Top down SEM-image (left), optical image (middle), and electrical connection layout (right) of the fabricated ISFET.

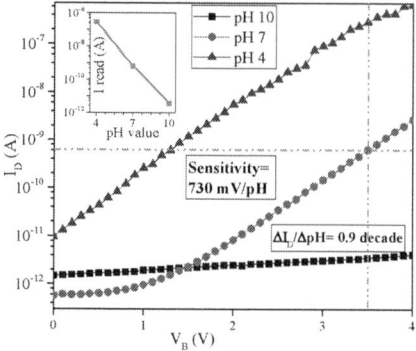

Fig. 7 Measured pH-sensitivity of the ISFET: 730 mV/pH for fixed current reading, and 0.9 decade/pH change in I_D for fixed voltage readout.

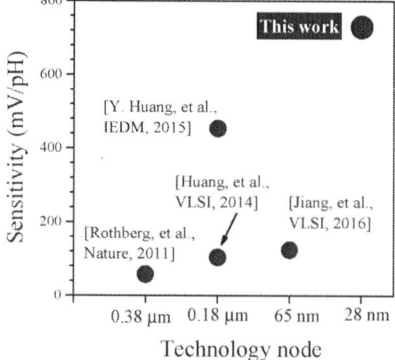

Fig. 8 Benchmark with state-of-the-art ISFETs in terms of sensitivity and technology node.

978-1-5386-4219-1/18 $31.00 © 2018 IEEE

RX-PUF: Low Power, Dense, Reliable, and Resilient Physically Unclonable Functions Based on Analog Passive RRAM Crossbar Arrays

Mohammad Reza Mahmoodi[*], Hussein Nili[*], and Dmitri. B. Strukov

Electrical and Computer Engineering Department, UC Santa Barbara, CA 93106-9560, USA

Abstract

We propose a novel architecture ("RX-PUF") for physically unclonable functions (PUF) based on analog RRAM crossbar array circuits. RX-PUF takes advantage of unique RRAM properties, such as *I-V* nonlinearity, and its device-to-device (d2d) variations and tunability. As a proof of concept, we have prototyped a 600 kb challenge response pair (CRP) PUF using 250 nm half-pitch (*F*) 20×20 crossbar arrays with passively integrated devices. The RX-PUF prototype features excellent physical characteristics, e.g. ~1600 F^2/bit density and up to 41 fJ/bit energy efficiency. Its functional performance, improved by utilizing hidden input, is also very promising. The measured bit error rate (BER) was 0.7% at RT and ≤ 5.3% at 100°C, even without using any error correction methods. The measured responses showed near-ideal uniformity (50.04%) and inter-HD (50.12%) and passed all relevant NIST randomness tests. The preliminary results showed also very high resilience of RX-PUF against machine learning (ML) attacks.

Introduction

PUFs are promising for identification and key generation tasks due to their potential for low-cost implementation, high-throughput and low-power operation, and resilience against counterfeiting and cloning. Various PUFs based on CMOS and emerging device technologies were proposed [1,2]. The main drawback of CMOS implementations is typically large area, which in turn leads to inferior energy and throughput. Some of the proposed RRAM-based PUFs rely on reprogramming the state for key generation, which is not practical given the severe switching endurance limitations of RRAM. Other RRAM approaches, that utilize devices in the digital mode, typically suffer from biases in the output. Correlations in the outputs reduce PUF complexity and could be easily detected by ML modeling [3]. Analog-grade RRAM enables very complex and resilient PUFs, though the main concern is device intrinsic noise which may significantly impact BER.

RX-PUF Implementation

The RX-PUF was prototyped using 20×20 crossbar arrays of passively integrated Pt/Al$_2$O$_3$/TiO$_{2-x}$/Pt RRAM devices [7] (**Fig. 1**). In RX-PUF (**Fig. 2**), the applied challenge uniquely determines sets of *n* rows and *m* columns, that are biased to V_{bias} and ground, accordingly, using peripheral CMOS circuitry. All remaining unselected lines are kept floating. The currents flowing into two subsets of *m*/2 columns are then sensed and compared against each other to generate a single output bit. Multiple response bits are grouped together to generate longer output key. The key idea of our approach is that the sensed currents on each column are sums of the currents via selected *n* devices and the sneak path currents that are determined by the states of all floated devices. Different selection of rows and columns results in the redistribution of sneak-path currents, which is hard to predict or model due to RRAM's *I-V* nonlinearities and their process-induced d2d variations. Furthermore, the contribution of both types of currents is

carefully balanced for each column by tuning device' conductances to specifically chosen desired (Gaussian-distributed along columns and rows) values, which reduce bias in the output and increase the noise margin. Such desired values are randomly generated by solving an optimization problem and could be pre-computed. The tuning procedure is only performed once to implement specific PUF instance. The same PUF circuit can be reconfigured many times by re-tuning its devices to a new distribution. To further improve functional performance, a two-step scheme is implemented. First, the auxiliary sense amplifier (SA), hardwired to the first and last columns, generates an "AUX" bit by comparing input currents in these lines. In the next step, a second output bit is generated by main SA, that serve all but the mentioned two columns (**Fig. 2a**). This bit is then xored with AUX bit to produce the final response bit. In the implemented prototype, *n* = 5 (out of *N* = 20 total), while *m* = 2 (out of *M* = 20-2=18 total).

Results and Discussion

Fig. 3 shows the programmed conductances for the studied PUF instance. The implemented tuning was very crude (with conductances of ~ 20% devices significantly lower than the desired values) to emulate various device non-idealities, e.g. crossbar arrays with large variations in switching thresholds (**Fig. 1b**). We then collected output currents (**Fig. 4**) and the corresponding responses (**Fig. 5**) on all 600 kb CRPs, and BER on the worst-case 10 kb responses under wide ambient temperature range (**Fig. 6**). The measured data clearly shows the benefit of AUX approach, which allowed to reduce output bias due to imperfect tuning (**Fig. 5**), and successfully pass NIST randomness test (**Fig. 7**). The inter-HD for 64 and 128 keys (**Fig. 8**) showed near-ideal values, while the detailed analysis of such keys showed almost no correlations for the AUX approach (**Fig. 9**). The resilience against machine learning attacks was assessed with 40×500×500×1 multi-layer perceptron classifier. The classifier was trained on a subset of specific size of the observed CRPs and then tested on another mutually exclusive observed set (**Fig. 10**). Even with large fraction of the training data, the classifier predicted output with close to ideal 50% accuracy for the PUF with AUX approach. The circuit consumes on average 212 fJ per response bit for V_{bias}= 0.3 V, with ~20% of the energy dissipated in RRAM array. The energy efficiency can be further improved to 41 fJ/bit by using V_{bias}= 0.1 V without significantly degrading other metrics. **Table 1** summarizes the results and shows that proposed approach compares very favorably with the state-of-the-art. We expect that a moderate increase in crossbar array size may exponentially improve resilience even against modelling with very deep classifiers, while scaling-down device feature sizes would lead to sub-fJ operation.

References

[1] S.Jeloka *et al.* VLSI'17; [2] Y.Pang *et al.* VLSI'17; [3] X.Xi *et al.* VLSI'17; [4] K.Yang *et al.* ISSCC'15; [5] A.Alvarez *et al.* ISSCC'15; [6] S.Mathew *et al.* ISSCC'14; [7] G.Adam *et al.* TED, 2017.

[*]These authors contributed equally to this work.

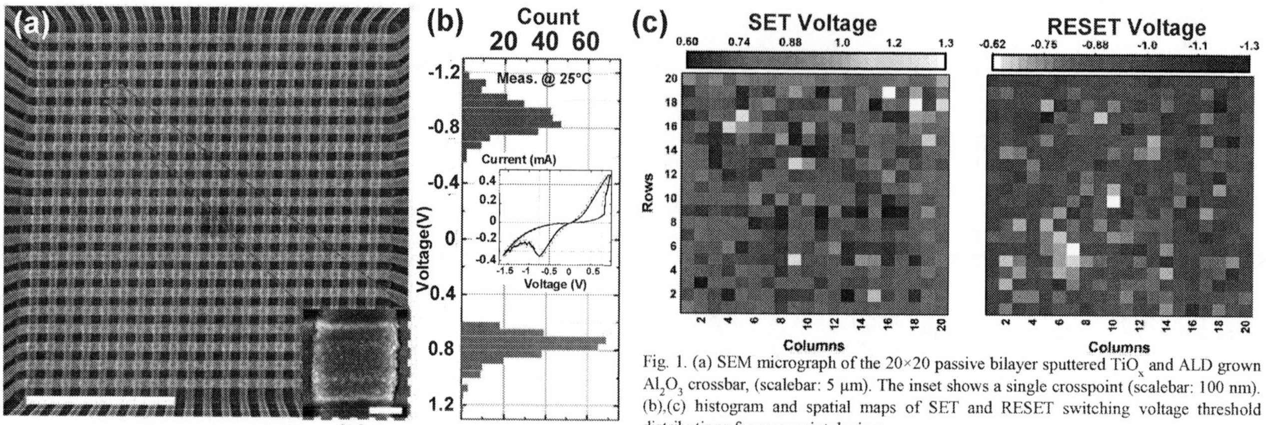

2018 Symposium on VLSI Technology Digest of Technical Papers

Fig. 1. (a) SEM micrograph of the 20×20 passive bilayer sputtered TiO_x and ALD grown Al_2O_3 crossbar, (scalebar: 5 μm). The inset shows a single crosspoint (scalebar: 100 nm). (b),(c) histogram and spatial maps of SET and RESET switching voltage threshold distributions for crosspoint devices.

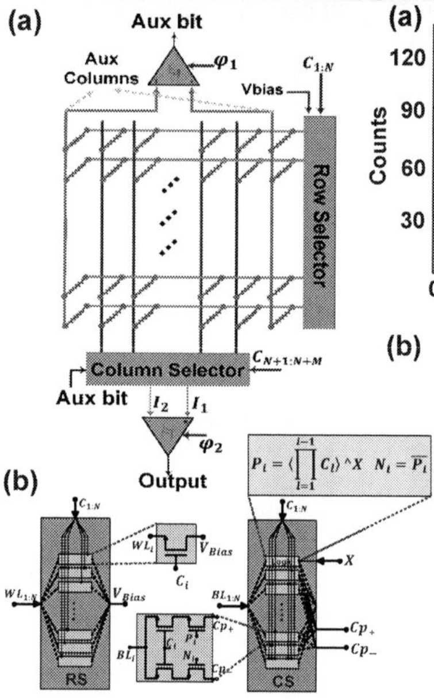

Fig. 2. (a) PUF architecture and (b) design of column/row selectors. The input is encoded by $N+M$ bits, with 1s for to the selected rows/columns.

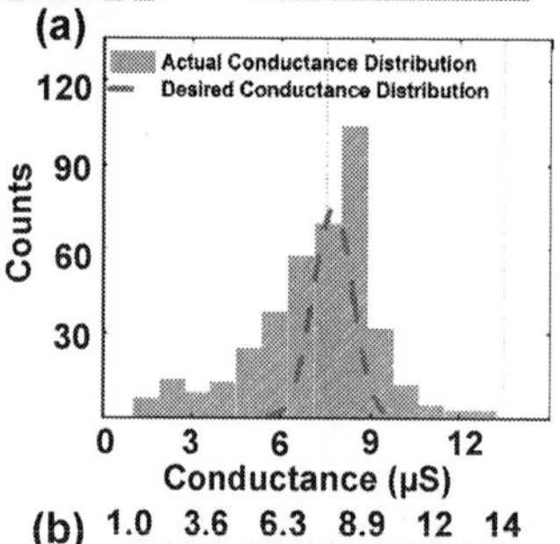

$$P_i = \langle \prod_{i=1}^{i-1} C_l \rangle^\wedge X \quad N_i = \overline{P_i}$$

Fig. 3. (a) Histogram and (b) map of cells' conductances (at 0.3 V) for the demonstrated PUF.

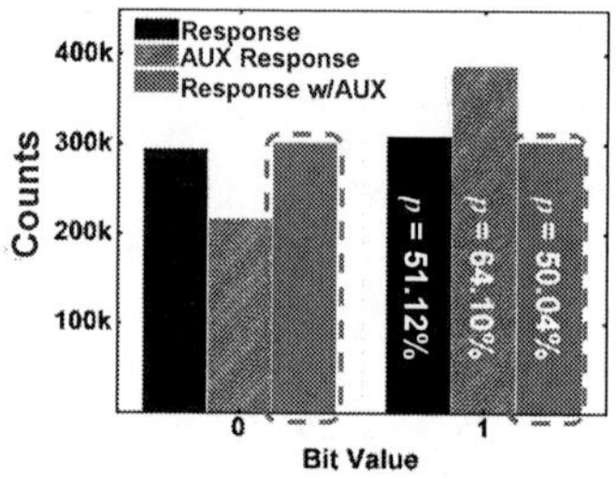

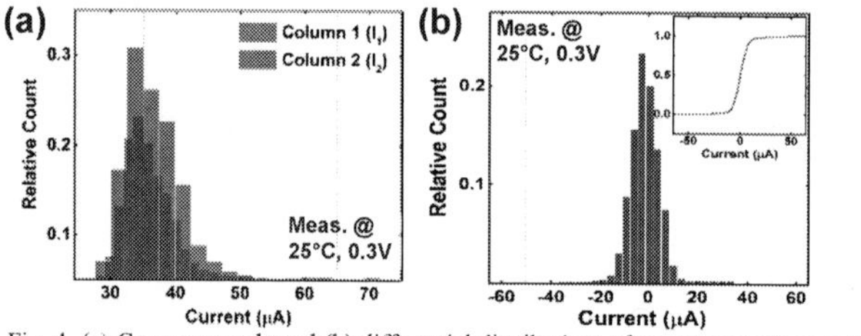

Fig. 4. (a) Common-mode and (b) differential distributions of output currents sensed by main SA over 600 kb responses. The inset shows ecdf.

Fig. 5. Distribution of PUF response (at 0.3 V @ 25°C) with and without AUX approach.

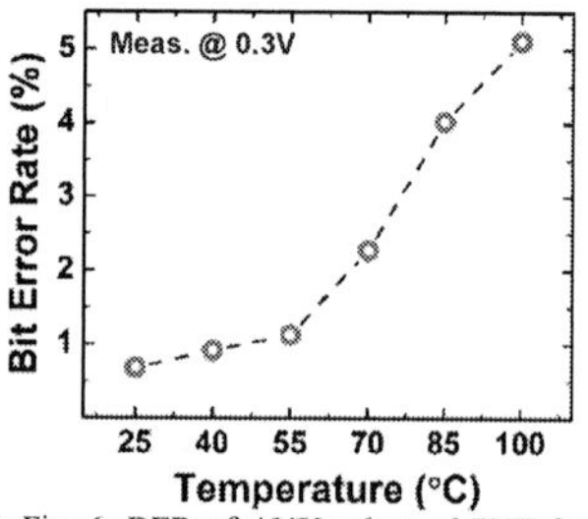

Fig. 6. BER of AUX-enhanced PUF for 10k worst-case CRPs.

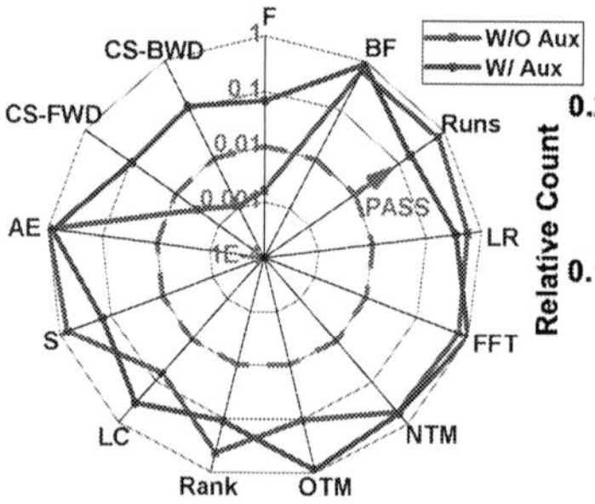

Fig. 7. Results of NIST randomness tests using measured 600kb responses for the PUF implementation with and w/o AUX.

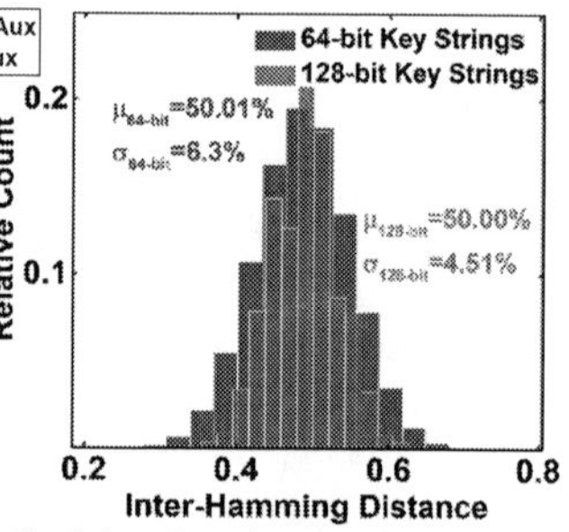

Fig. 8. Inter-hamming distance (at 0.3 V @ 25°C) for 64 and 128-bit keys formed by grouping sequentially generated 1-bit responses.

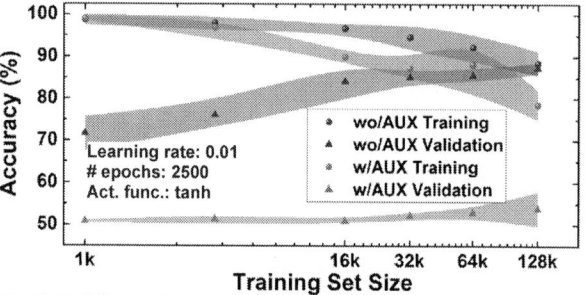

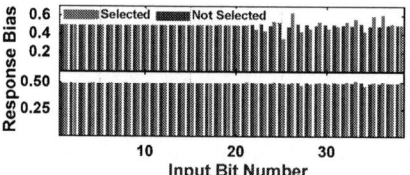

Fig. 9. Input-output correlations: Each column shows PUF output (at 0.3 V @ 25°C), averaged over all CRPs with the same fixed value of specific input bit. The bottom / top panel shows implementation with / w.o. AUX approach.

Fig. 10. Modeling attack by MLP network, tested on 5k validation set, as a function of training set size (in bits). The training was performed using a gradient descent method with momentum. Symbols show average prediction accuracy, while the line thicknesses are drawn according to the max and min values obtained over 5 runs.

Table I. Comparison with previous works (* w/o applying correction methods)

		VLSI'17 [1]	VLSI'17 [3]	ISSCC'15 [4]	ISSCC'15 [5]	ISSCC'14 [6]	This work
Technology		FDSOI (28nm)	CMOS (130nm)	CMOS (40nm)	CMOS (65nm)-SA	CMOS (22nm)	ReRAM/CMOS (200nm/55nm)
Number of CRPs		1.17×10^{11}	3.7×10^{19}	5.5×10^{28}	NA	250K	~2.37M
Worst-case BER*		~11%	9%	9%	~21%	~30%	5%
Area Efficiency (F^2/b)		970	-	-	12K	9.6K	1.4K
Energy Efficiency		97fJ/b	11pJ/b	17.7pJ/b	163fJ/b	190fJ/b	41-213fJ/b (0.1V-0.3V)
Temperature Range (°C)		0~80	-20~80	-25~125	25~85	25~50	25~100
NIST Randomness Test		NA	NA	NA	PASS	FAIL	PASS
ML Attack	Prediction Error	15%	40%	Not Tested	Not Tested	Not Tested	53%
	Training Size	$\sim 10^{-4}$%	$\sim 10^{-13}$%	-	-	-	~5%

Gap in pagination due to formatting issues.

Pages 101-102

A Methodology to Improve Linearity of Analog RRAM for Neuromorphic Computing

Wei Wu, Huaqiang Wu*, Bin Gao*, Peng Yao, Xiang Zhang, Xiaochen Peng[#], Shimeng Yu[#], and He Qian

Institute of Microelectronics, Tsinghua University, Beijing, China. [#]Arizona State University, Tempe, AZ, USA
*E-mail: wuhq@tsinghua.edu.cn, gaob1@tsinghua.edu.cn

Abstract- The conductance tuning linearity is an important parameter of analog RRAM for neuromorphic computing. This work presents a novel methodology to improve the conductance tuning linearity of the filamentary RRAM. An electro-thermal modulation layer is designed and introduced to control the distribution of electric field and temperature in the filament region. For the first time, a HfO_x based RRAM is demonstrated with linear analog SET, linear analog RESET, 50ns speed, $10\times$ analog tuning window, $100k\Omega$ on-state resistance, and high temperature retention for multilevel states. The excellent performances of the analog RRAM devices enable high accuracy online learning in a neural network.

Keywords: analog RRAM, synapse, online learning

Introduction

Different from digital memory applications, neuromorphic computing requires RRAM devices capable of analog switching behaviors, which is particularly important for online learning [1]. Various types of analog RRAM devices were reported [2-6], however, performance of the reported analog RRAM is still far from the desired. One of the key challenges of analog RRAM is how to achieve linear conductance tuning under successive identical pulses [7-8]. Large nonlinearity (defined in Fig.1) degrades learning accuracy in a neural network significantly (Fig.2). Unfortunately, the switching mechanism of the conventional filamentary RRAM implies that the nonlinearity is intrinsic. Conductance changes fast at the point that conductive filament (CF) is just connected or ruptured (Fig.3) due to enhanced field in the gap. An effective method is still lacking to realize linear analog switching while keeping other attributes such as speed and retention unaffected. In this work, we propose a novel electro-thermal coupling methodology to improve the linearity of filamentary analog RRAM. Based on the method, excellent linearity for both SET and RESET and outstanding device performances are realized.

Results and Discussion

Methodology: The origin of nonlinearity in a filamentary RRAM is investigated (Fig.3). At the beginning of RESET process, the local temperature and electric field are high due to high current and small length of CF gap. Thus, oxygen vacancies (Vo) migrate fast, resulting in fast decrease of conductance. Then Vo migration gradually slows down as the CF gap increases, resulting in slower conductance decrease. To improve the RESET linearity, the changing rate of electric field should be reduced. For SET process, the time point that CF is connected causes fast conductance change, while before and after that point, conductance changes more slowly. To improve the SET linearity, the impact of CF connection should be suppressed. Based on this physical picture, an electro-thermal coupling methodology is proposed. An electro-thermal modulation layer (ETML) is designed to cap on the switching layer (SWL). The ETML works not only as a thermal enhanced layer which can promote to form uniform Vo distribution [6], but also as an electric field modulator which can decrease the changing rate of electric field in the CF gap region.

To validate the proposed methodology, a 1Kb 1T1R RRAM array with TiN/ETML/HfOx/TiN stack is fabricated (Fig.4). A HfOx layer is deposited by ALD serving as SWL, and a TaOx layer serves as ETML. Oxygen composition of the TaOx layer is adjusted to get different resistivity and thermal conductivity.

RESET Linearity: At the beginning of RESET, voltage mostly drops on the ETML. As the conductance of SWL decreases, voltage drop transfers from ETML to SWL, but electric field in CF gap region still decreases due to the increase of gap length (Fig.5). Higher resistivity of ETML helps to reduce the decreasing rate of electric field and thus could improve the RESET linearity (Fig.5&6). From the simulation results based on the parameters of HfO_2 and assumed 10nm CF region [9], we find $\sim30m\Omega\cdot cm$ is the optimized parameter for ETML. If the resistivity is much larger than this value, CF gap region cannot get enough voltage, leading to RESET failure. From the experimental results, we also find that the RESET linearity can be improved with the optimized resistivity parameter (Fig.7), demonstrating the proposed methodology.

SET Linearity: Thermal conductivity of ETML decides the distribution of Vo [9]. High thermal conductivity results in single and strong CF, whereas low thermal conductivity makes Vo distribute uniformly, forming multiple weak CFs (Fig.8&9). Uniform Vo distribution is good for linear analog switching, since in this case each Vo almost contributes equally to the total conductance (Fig.10). The ETML has relatively low thermal conductivity [10], and thus it is easier to get linear SET for different samples with ETML (Fig.11). However, when replacing ETML as a metal layer with high thermal conductivity, large nonlinearity is observed (Fig.11a). These results indicate ETML is also effective for the SET linearity.

Device Performance: For the optimized analog RRAM (ETML~$30.4m\Omega\cdot cm$), good linearity is observed on different devices for both SET and RESET (Fig.12). The variation and fluctuation is not critical to the learning accuracy since the neural network can adapt during the iterative training. The variations can also be suppressed by combining two devices as one synaptic weight (Fig.13). Good endurance and retention of linear analog switching are demonstrated (Fig.14&15). This ETML/HfOx device also shows very low operating current and large dynamic tuning window of analog switching, which is crucial for scaling up the size of neural network. To evaluate the performance, a neural network for MNIST task is simulated. The ETML/HfOx device with optimized parameters shows the best performance (Fig.16). The overall performance metrics for different analog RRAM devices are summarized in Table I.

Conclusion

Key achievements: i) the mechanisms of nonlinearity in filamentary analog RRAM is elucidated; ii) an electro-thermal coupling method is proposed to improve linearity; iii) a fab-friendly ETML/HfOx based analog RRAM with good linearity and excellent performances is developed. This methodology provides valuable guidelines for designing RRAM based neural networks with high online learning accuracy.

Acknowledgements: This work is supported in part by the MOST of China (2016YFA0201801), the ICFC, and NSFC (61674089, 61674092, 61674087).
References: [1] H. Wu, IEDM 2017, 274. [2] S. H. Jo, Nano Letters 2010, 1297. [3] L. Gao, Nanotechnology 2015, 455204. [4] S. Park, IEDM 2013, 625. [5] J. Woo, EDL 2016, 994. [6] W. Wu, EDL 2017, 1019. [7] D. Lee, IEDM 2015, 91. [8] P.-Y. Chen, IEDM 2017, 135. [9] B. Gao, IEDM 2017, 91. [10] C. Landon, APL 2015, 107, 023108.

2018 Symposium on VLSI Technology Digest of Technical Papers

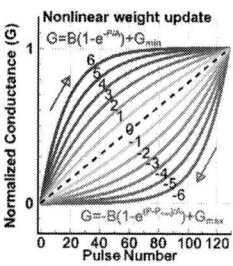

Fig.1 Analog conductance tuning with nonlinearity values.

Fig.2 Learning accuracy loss as a function of nonlinearity values.

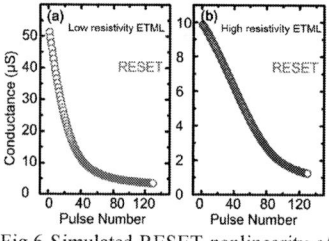

Fig.3 Mechanisms of SET and RESET nonlinearity.

Fig.4 Cross section TEM image of TiN/ETML/HfO$_x$/TiN RRAM.

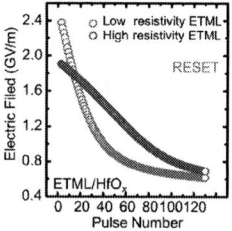

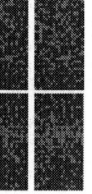

Fig.5 Simulated electrical field evolution in CF gap region with increased pulse number during RESET.

Fig.6 Simulated RESET nonlinearity of ETML/HfO$_x$ based RRAM with different ETML resistivity. Nonlinearity decreases when resistivity is higher.

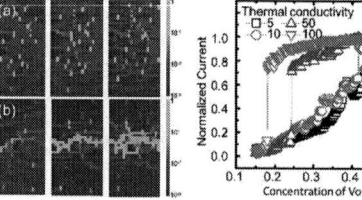

Fig.7 Measured RESET nonlinearity and variation of ETML/HfO$_x$ based RRAM. The resistivity of ETML is (a) 0.8 mΩ·cm, (b) 4.4 mΩ·cm, and (c) 30.4 mΩ·cm. The nonlinearity decreases and variation increases with the increasing of ETML resistivity.

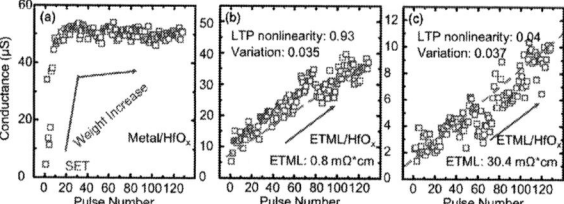

Fig.8 Simulated Vo distribution in ETML /HfO$_x$ and metal /HfO$_x$ RRAM.

Fig.9 Simulated current density distribution in ETML/HfO$_x$ and metal /HfO$_x$ RRAM.

Fig. 10 Simulated read current as a function of Vo concentration for different RRAM stacks.

Fig.11 Measured SET nonlinearity and variation of (a) Ta/HfO$_x$ based RRAM and ETML/HfO$_x$ based RRAM. The resistivity of ETML is (b) 0.8 mΩ·cm (c) 30.4 mΩ·cm. RRAM devices with different ETML resistivity all shows good linearity.

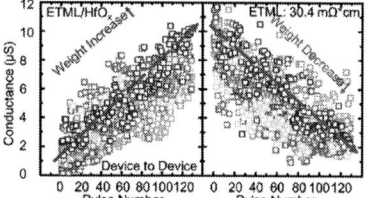

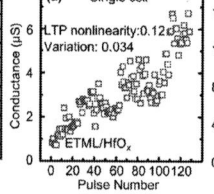

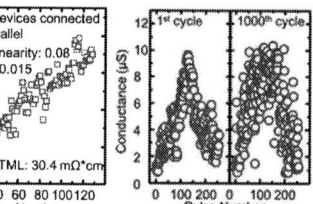

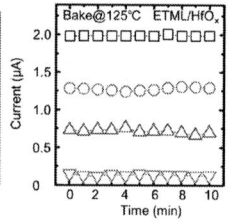

Fig.12 Analog SET and RESET behaviors of different ETML/HfO$_x$ devices. The variations are in a reasonable range.

Fig.13 Variation and nonlinearity of single cell and two cells connected in parallel. Variation is reduced by combining devices.

Fig. 14 After cycling, no degradation occurs on the analog switching.

Fig.15 No degradation occurs on different states under high temperature.

Table.1 Extracted device parameters and ideal analog RRAM

Type of analog RRAM	Ag:a-Si [2]	TaO$_x$/TiO$_2$ [3]	PCMO [4]	AlOx/HfO$_x$ [5]	ETML/HfO$_x$ (0.8mΩ·cm)	ETML/HfO$_x$ (4.4mΩ·cm)	ETML/HfO$_x$ (30.4mΩ·cm)	ETML/HfO$_x$ (2 parallel)	Ideal analog RRAM
Nonlinearity (SET)	2.40	0.66	3.68	1.94	0.96	0.93	0.04	0.08	0
Nonlinearity (RESET)	-4.88	-0.69	-6.76	-0.61	-3.26	-2.63	-0.63	-0.63	0
On-state resistance	26 MΩ	5 MΩ	23 MΩ	17 KΩ	10KΩ	30kΩ	100kΩ	50kΩ	High
ON/OFF ratio (analog region)	12.5	2	6.8	4.4	10	10	10	10	Large
SET pulse	3.2V/300μs	3V/40ms	2V/1ms	0.9V/100μs	1.7/50ns	1.5V/50ns	1.6V/50ns	1.6V/50ns	Low voltage Fast speed
RESET pulse	2.8V/300μs	3V/10ms	2V/1ms	1V/100μs	1.5V/50ns	1.5V/50ns	1.5V/50ns	1.5V/50ns	Low voltage Fast speed
Variation	3.5%	<1%	<1%	5%	3%	3.35%	3.7%	1.5%	0%

Fig.16 Comparison of the accuracy, normalized latency and normalized energy consumption of different analog RRAM based neural networks.

978-1-5386-4219-1/18 $31.00 © 2018 IEEE

Non-Volatile Ternary Content Addressable Memory (TCAM) with Two HfO₂/Al₂O₃/GeOₓ/Ge MOS Diodes

Yi Zhang[1], Bing Chen[1], Wenfeng Dong[1], Wei Liu[1], Shun Xu[1], Ran Cheng[1], Shiuh-Wuu Lee[1] and Yi Zhao[1,2]*

[1]College of Information Science & Electronic Engineering, Zhejiang University, Hangzhou, China
[2]State Key Laboratory of Silicon Materials, Zhejiang University, Hangzhou, China *E-mail: yizhao@zju.edu.cn

ABSTRACT

We propose and demonstrate the world-first ternary content ternary addressable memory (TCAM) cell using only two MOS diodes. The diodes are with simple HfO₂/Al₂O₃/GeOₓ/Ge-sub structure and could be fabricated by fully CMOS compatible process. Owing to the adoption of a very thin GeOₓ interfacial layer, the diodes show both excellent resistive switching and rectifying characteristics. Furthermore, TCAM cell and array are built with two diodes connected back-to-back. Finally, a well-functioning 8×16 HfO₂/Al₂O₃/GeOₓ/Ge-sub TCAM array for parallel multi-data search is demonstrated. This novel diode-based cell structure is very promising for future energy and area efficient TCAM applications.

I. INTRODUCTION

Content address memory (CAM), especially ternary CAM (TCAM), is of great interest thanks to its excellent parallel multi-data search capability [1]. The main applications of the CAM/TCAM include CPU/GPU's cache memory and data packet forwarding and classifications in the network like IP address lookup [2-4]. In these applications, both chip area and energy efficiencies are greatly needed. For example, in IP routers, CAM occupies a large portion of the chip area since the conventional SRAM-based CAM needs ten transistors (10T). While for TCAM, it needs sixteen transistors (16T) for one cell. Additionally, parallel data search is performed frequently in CAM, and thus the power consumption from searching and matching operations is another critical challenge[5]. In addition, With the fast emergence of Internet of Thing (IoT), where the power supply often being batteries, the static power consumption of the IoT chips, including CAM, warrants particular consideration. Therefore, a non-volatile CAM, if available, can effectively address both the chip area and energy efficiency requirements for conventional and IoT CAM applications. So far it is reported that non-volatile CAM cells have been realized based on magnetic RAM (MRAM) [6] and phase change Memory (PCM) [7]. However, both cells exhibit complicated structures, which are 11T3M for MRAM and 2T2M for PCM [6, 7]. Furthermore, the fabrication processes of MRAM and PCM are not fully compatible with the standard CMOS technologies [8, 9]. Therefore, fully CMOS compatible non-volatile CAM technology with excellent area and energy efficiency should present an attractive option. In this work, we proposed and experimentally realized a fully CMOS compatible non-volatile TCAM technology with the best area efficiency reported to date. The TCAM cell is highly compact consisting of only two HfO₂/Al₂O₃/GeOₓ/Ge-sub MOS diodes. The functionality in reading and parallel multi-data search, of an 8×16 TCAM array are also successfully demonstrated.

III. TCAM CELL/ARRAY DESIGN AND FABRICATION

Figure 1 shows the schematic of a NOR-TCAM array. A TCAM cell should have both the data storage and comparison functions. To achieve this purpose, SRAM, PCM and MRAM based cell structures have been proposed (Fig. 2(a)-(c)). These cell structures show complicated layout design and poor area efficiency. In the case of TCAM, the cell structure will be more complicated since some cells have to be fixed at a *don't care* value [1]. However, if we have a simple diode-like device possessing the *I-V* characteristics shown in Fig. 3, which we named "memdiode(MD)" in this work, the cell structure of TCAM can be significantly simplified by using two MDs (Fig. 2(d)). The basic idea could be described as follow: A single MD shows the switching behavior and diode like *I-V* characteristics at the on-state. When two MDs connected back to back, they can function like a TCAM cell (Fig.4). In the search operation, the match line (ML) will first be charged to the GND level and then the search signal is applied to the search line (SL) . If all states of cells are matched or at *don't care* state, the read voltage (V_{read}) will be applied to the off-state cell and the ML will be kept at GND. If one of cells does not match with the input value, the ML will be pulled up to V_{read} while the diode behavior will prevent the matched cells from pulling the ML back to GND. At the end, only the ML states at GND will be output.

In this study, we experimentally realized the MDs with a Ge MOS structure. The fully CMOS compatible fabrication process was used and the cross-sectional view of the final MD structure is shown in Fig. 5. First, N type heavily doped Ge on insulator substrates were used and etched into mesas as the match line (ML) of the TCAM. Next, a thin Al₂O₃ layer was deposited by ALD as the barrier layer for the follow-up ozone post oxidation (OPO) treatment [10]. Then, HfO₂ layer was deposited as a resistance switching layer. Finally, TiN top electrode was grown, patterned and connected as the sense line (SL). Ge contact holes were exposed by

RIE etching following by a deposition of the Ni contact. The device cross section and material have been confirmed by TEM and EDS, respectively (Fig. 6(a-b)). The array chip was bonded out for electrical characterizations.

IV. RESULTS AND DISCUSSION

A. Characteristics of HfO₂/Al₂O₃/GeOₓ/Ge MD

The fabricated HfO₂/Al₂O₃/GeOₓ/Ge-sub structure shows good resistive switching and rectifying characteristics (Fig. 7(a)). As discussed above, this MD device could be used for building effective TCAM cell and arrays. The GeOₓ interfacial layer is critical for obtaining RRAM and rectifying characteristics, since the Ge and Si substrates without GeOₓ layer does not possess any resistive switching and rectifying behaviors (Fig.7(b-c)). The possible mechanism for the desired characteristics of GeOₓ interfacial layer can be explained as follows:

1) Mechanism for resistive switching. After the forming process, there will be filament paths formed in the HfO₂ layer and oxygen ions may drift from HfO₂ layer into the oxygen-deficient GeOₓ layer. During the re-set process, these unstable oxygen ions in GeOₓ interfacial layer are pulled back to the HfO₂ layer and combine with oxygen vacancies in it. As a result, the HfO₂ layer recovers to its high resistance state and the diode shows good switching behavior. If there is no GeOₓ layer, since there is no locations to store oxygen ions, no resistive switching behaviors can be observed.

2) Mechanism for diode rectifying characteristic. In the low resistance state, the conductive filament could be regarded as a metal line connecting the electrode and the GeOₓ interfacial layer. It has been confirmed that the conduction band offset between GeOₓ and Ge (0.4~0.8 eV) is much smaller than the valence band offset (2.8~4.3 eV)[11]. Therefore, the forward current is much larger than the backward current in GeOₓ/n⁺ Ge structure (Fig.8) and the diode could own good rectifying characteristics.

B. TCAM Cell and Array Characterization

The MDs show good resistance distribution in both on and off states (Fig. 9). In addition, the MD TCAM cell shows good reliability behaviors under read disturb, reasonable rectifying behavior, and decent on/off ratio (Figs. 10 and 11). A 2×1 MD array was built and the V_{in} and V_{out} relationship was obtained (Fig. 12). It has been confirmed that MD could prevent V_{out} from being pulled back to GND. This fact suggests the possibility to build a large size functional TCAM array with the MDs as discussed in Part III. Fig. 13 shows the test kit for 8×16 TCAM array characterization. The measured resistances (V_{read} = +/-1V) of all MDs in the array are shown in Fig. 14 (a) and (b), indicating that the bits are correctly stored. Then, 8-bit data were applied to the array for searching and matching. By the measured V_{out}, it could be confirmed that the MLs are correctly selected. Finally, Table 1 shows the comparison of mainstream TCAM cell structures and the MD TCAM cell proposed in this work. Compared with other TCAM cells, the MD cell reported in this study demonstrates not only its full CMOS compatibility, but also its advantages in much simpler design and area efficiency. It is worth to point out that only the MD TCAM presented in this work is capable of setting any cell to 0, 1 or *don't care* value, while other TCAMs can only set a normal cell to 0 or 1 and the position of *don't care* cell was designed in the layout and could not be changed subsequently.

V. CONCLUSION

A TCAM cell with only two HfO₂/Al₂O₃/GeOₓ/Ge-sub MOS diodes has been experimentally demonstrated, for the first time. The HfO₂/Al₂O₃/GeOₓ/Ge-sub structure not only shows good rectifying characteristics, but also exhibits excellent resistive switching behaviors. The mechanism for both resistive switching and rectifying characteristics of the diodes has also been investigated. Finally, an 8×16 TCAM array with the MD cell structure has also been demonstrated for parallel data search. Since the HfO₂/Al₂O₃/GeOₓ/Ge-sub MOS structure is fully compatible with the standard CMOS technologies and the layout design is simple, the cell structure has super scalability down to 10 nm technology node, which is promising for future energy and area efficient TCAM applications.

Acknowledge This work was supported by National Key Research and Development Program (2017YFA0207600) and NSFC (No. 61704152). **Reference** [1] S. Jeloka. et. al., *VLSI*, C272, 2015. [2] R. Karam, et al., *IEEE Prof.*, 103, 1311, 2015. [3] K. Pagiamtzis, et al., *IEEE JSSC*, 41, 712, 2006. [4] B. Giraud, et al., *IEDM*, pp. 473, 2017. [5] O. Tyshcheneko, et al., *IEEE JSSC*, 43, 1972, 2008. [6] W. Xu, et al., *IEEE TVLSIS*, 18, 66, 2010. [7] L. Hsu, et al., *US Patent*, 7319608B2, 2008. [8] V. Nguyen, et al., *IEDM*, pp. 852, 2017. [9] H. Cheng, et al., *IEDM*, pp. 28, 2017.[10] R. Zhang, et al., *EDL*, 97, 831, 2016. [11] L. Lin, et al., *APL*, 97, 242902, 2010.

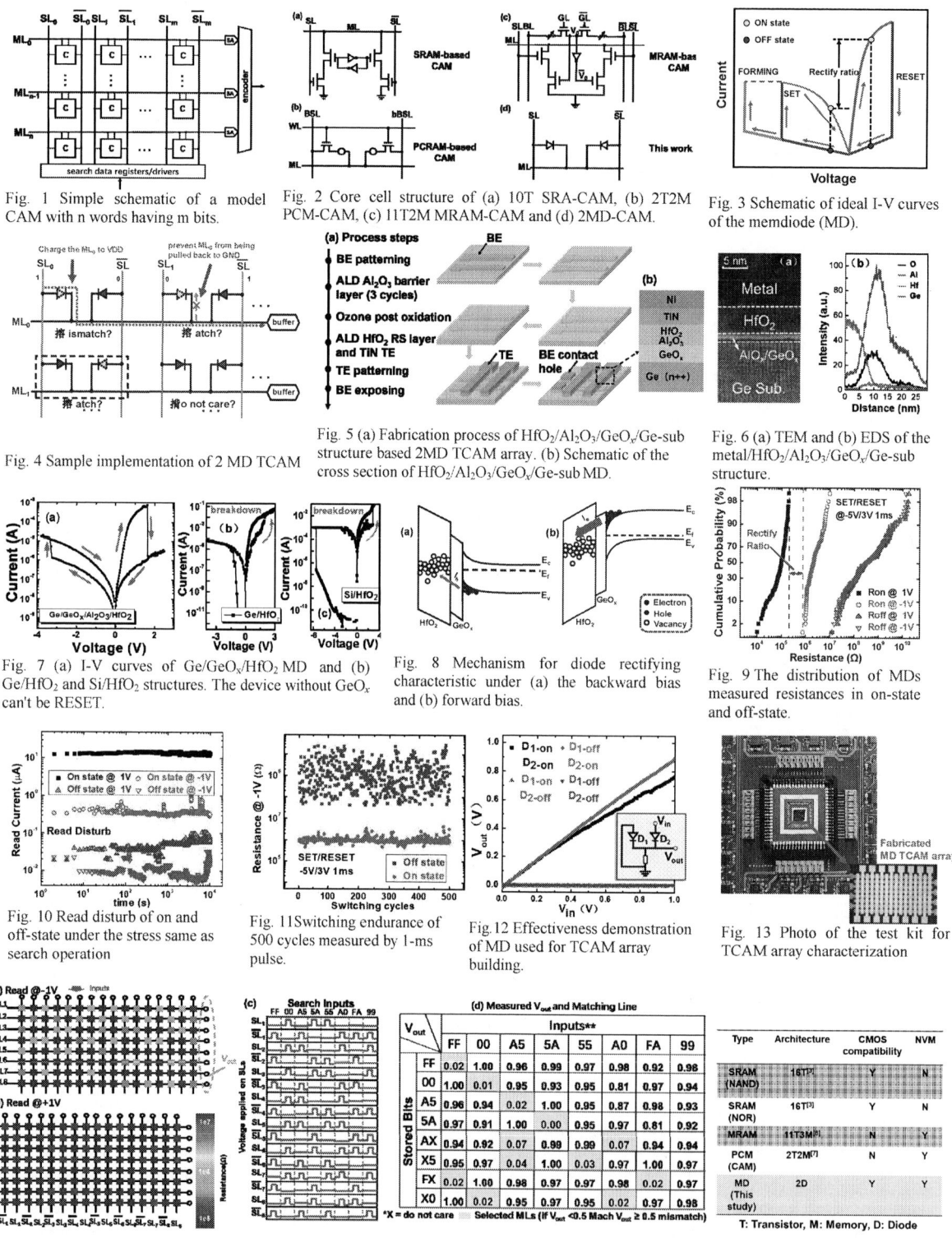

Fig. 1 Simple schematic of a model CAM with n words having m bits.

Fig. 2 Core cell structure of (a) 10T SRA-CAM, (b) 2T2M PCM-CAM, (c) 11T2M MRAM-CAM and (d) 2MD-CAM.

Fig. 3 Schematic of ideal I-V curves of the memdiode (MD).

Fig. 4 Sample implementation of 2 MD TCAM

Fig. 5 (a) Fabrication process of HfO$_2$/Al$_2$O$_3$/GeO$_x$/Ge-sub structure based 2MD TCAM array. (b) Schematic of the cross section of HfO$_2$/Al$_2$O$_3$/GeO$_x$/Ge-sub MD.

Fig. 6 (a) TEM and (b) EDS of the metal/HfO$_2$/Al$_2$O$_3$/GeO$_x$/Ge-sub structure.

Fig. 7 (a) I-V curves of Ge/GeO$_x$/HfO$_2$ MD and (b) Ge/HfO$_2$ and Si/HfO$_2$ structures. The device without GeO$_x$ can't be RESET.

Fig. 8 Mechanism for diode rectifying characteristic under (a) the backward bias and (b) forward bias.

Fig. 9 The distribution of MDs measured resistances in on-state and off-state.

Fig. 10 Read disturb of on and off-state under the stress same as search operation

Fig. 11 Switching endurance of 500 cycles measured by 1-ms pulse.

Fig. 12 Effectiveness demonstration of MD used for TCAM array building.

Fig. 13 Photo of the test kit for TCAM array characterization

Fig. 14 (a)-(b): Measured resistance mapping of the 8×16 array under +/- 1V. (c): the input search signals applied on each SLs. (d): measured V$_{out}$ and matching lines from the array when apply the search signals in (c).

Table 1 Comparison of MD TCAM cell with the conventional ones

Selector Requirements for Tera-Bit Ultra-High-Density 3D Vertical RRAM

Zizhen Jiang[1*], Shengjun Qin[1], Haitong Li[1],
Shosuke Fujii[1,2], Dongjin Lee[1,3], Simon Wong[1], and H.-S. Philip Wong[1#]

[1]Dept. of Electrical Engineering and Stanford SystemX Alliance, Stanford University, Stanford, CA 94305, U.S.A;
[2]Device Technology R&D Center, Toshiba Memory Corporation, Yokkaichi, 512-8550, Japan;
[3]Seminconductor R&D Center, 1, Samsungjeonja-ro, Hwaseong-si, Gyeonggi-do 18848, South Korea;
E-mail: *jiangzz@stanford.edu, #hspwong@stanford.edu.

Abstract: Selector requirements for tera-bit class, ultra-high-density 3D vertical resistive random access memory (VRRAM) are presented, including practical design considerations such as array efficiency (AE), pillar driver transistors (pillar drivers), and wire/metal plane resistances. We design a novel chip architecture that is different from 3D NAND: (a) separated, square and large wordplane (WP) connected by global wordplane connections (WPC) within a block to minimize influence of leakage currents, (b) compact staircase. An accurate, computationally efficient resistor network is developed to model the parasitic resistances of the architecture. Through the resistor network simulations, selector requirements for 3D VRRAM are examined. To achieve tera-bit class 3D VRRAM with density higher than the most advanced 3D NAND flash (> 4.3 Gb/mm^2), selector nonlinearity (NL) $\geq 10^2$ is required.

I. Introduction: Resistive random access memory (RRAM) offers bit-alterability, direct over-write, fast programming speed (< 10 ns), and low energy consumption ($< $ pJ) for high-density, on-chip data storage [1]. To compete with the most advanced 3D NAND flash (768 Gb and 4.3 Gb/mm^2) [2-4], suitable chip architectures for 3D VRRAM [5-8] need to be identified. Selectors are required to reduce leakage currents and increase array size, AE, and bit-density [9-11]. This paper presents an architecture for tera-bit class ultra-high-density 3D VRRAM and analyzes the selector requirements.

II. Chip Architecture: The selector requirements **must** be analyzed in the context of the chip architecture. Fig. 1 shows the floor plan of a 3D VRRAM that achieves 1 Tbit and 4.6 Gb/mm^2. There are 64 planes, each paired with a row decoder. 512 blocks are vertically aligned in each plane. Fig. 2 shows that each block has 64 layers and is further divided into 8 arrays, and each array has 4 M RRAM cells. The bitline (BL) controller and additional peripheral circuits are at the bottom of the chip. Row decoders and the BL controller guarantee random bit access. Each row decoder can arbitrarily select one wordline (WL) and one WPC. Each WPC is connected to a specific layer of WPs in a block. WPC are placed on top of the arrays, and are wide and tall with low resistivity (0.025 $\Omega\cdot\mu$m [12]). WLs and WPCs run the entire horizontal length of each plane. BLs run the entire vertical length of each plane. The BL controller and the peripheral circuits can sense one or multiple BLs. Fig. 3-4 show the 3D and top view of an array within a block. FinFETs underneath the array are used as pillar drivers. Hexagonal pillar pattern maximizes the areal density. In 3D NAND, the word line can select only four rows of cells in the same layer. In this 3D RRAM architecture, the WP can select a large array of cells. Using a larger array size or more arrays in each block can reduce the number of planes and the number of decoders, and therefore maximizes AE. Additionally, square arrays (#WL = #BL) are preferred to lower the parasitic WP resistance for accessing the worst-case selected cell. Because the WP is large and square (in contrast to 3D NAND), a compact

staircase [13] (Fig. 3) can be used to connect the WP to the corresponding WPC, which further increases AE.

III. Pillar Driver: FinFETs [14-17] are chosen for the pillar drivers due to their higher saturation current and larger pillar areal density (Fig. 5). The minimum unit cell area is determined by the gate pitch and metal pitch (since VRRAM pillar can be smaller than NAND); it can be reduced to 12 F^2 using a single-fin pillar driver using 7 nm FinFET design rules [17], potentially enabling densities of 10 Gb/mm^2. To simplify the following analysis, the more relaxed two-fin layout (Fig. 6) is applied. With the driver saturation current < 50 μA, Low Resistance State (LRS) $> V_{DD}/50$ μA is required.

IV. Modeling: A resistor network is built to simulate the worst-case scenarios (Fig. 7). The network captures all leakage paths of the arrays within a block. The WP resistances are obtained by 2D field solver (Sentaurus) simulations (Fig. 8). The selector and the RRAM (1S1R) are sandwiched between the WP and the metal pillar (Fig. 9). The 1S1R is simplified as a voltage-dependent resistor. LRS is the equivalent resistance of 1S1R at V_{DD}, while Unselected Resistance State (URS) is the equivalent resistance at $0.5 V_{DD}$ (Fig. 10) [18]. NL of 1S1R is defined as the ratio between URS and LRS. To simulate (Table I) a mega-bit array in a computationally efficient manner, we simplified the 64-layer network to 2-layer equivalent network (Fig. 11). The network is applicable to both write and read programming with $< 2\%$ error verified for sub-Mb arrays (Fig. 12).

V. Selector Requirements: Using the resistor network, we explore the number of WLs and BLs in each array and the numbers of arrays, blocks, and planes that meet the storage capacity and the bit-density requirements. The bit-density calculation takes into account the areas of the decoders, the controller, and peripheral circuits, which are estimated to be the same as those of 3D NANDs [4]. To compete with the density of 3D NAND flash (4.3 Gb/mm^2) while achieving adequate read/write margins (criteria in Table I) requires LRS = 1 MΩ and NL $\geq 10^2$ (Fig. 13), yielding 4.6 Gb/mm^2 for single-bit per cell storage. This high density is attained because of smaller pillar dimensions and more compact staircase of 3D VRRAM, as compared to 3D NAND designs. Higher NL and multi-level programming of RRAM devices can provide even higher density (6.3 Gb/mm^2). Segmented read (not reading all BLs simultaneously, trading off bandwidth/latency) and bias optimization can loosen the 1S1R requirements. The design specifications, the resistances and the resistivities used in the simulation are summarized in Table II and Table III.

VI. Conclusion: We present 1) a 3D VRRAM architecture for tera-bit class storage competitive with 3D NAND; 2) a 3D VRRAM-specific architecture that significantly loosens 1S1R requirements for high-density storage; 3) an accurate and computationally efficient resistor network. Single-bit/cell, tera-bit class high-density (> 4.3 Gb/mm^2) 3D VRRAM is achievable for LRS = 1 MΩ and selector NL $\geq 10^2$.

978-1-5386-4219-1/18 $31.00 © 2018 IEEE

2018 Symposium on VLSI Technology Digest of Technical Papers

Acknowledgement: This work is supported in part by Stanford NMTRI and the Samsung Electronics' University R&D program.
Reference: [1] Stanford Memory Trends; [2] Tanaka, ISSCC '16; [3] Kim, ISSCC '17; [4] Yamashita ISSCC, '17; [5] Baek, IEDM '11; [6] Chien, VLSI '12; [7] Chen, IEDM '12; [8] Liu, ISSCC '13; [9] Kim, VLSI '12; [10] Lee, VLSI '12; [11] Cha, IEDM '13; [12] Lee, IEDM '16; [13] Chen, IEDM '12; [14] Stanford CMOS Technology Scaling Trend; [15] Yang, EDL '08; [16] Wu, VLSI '17; [17] ASU, ASAP, 7 nm PDK; [18] Govoreanu, ICICDT '15.

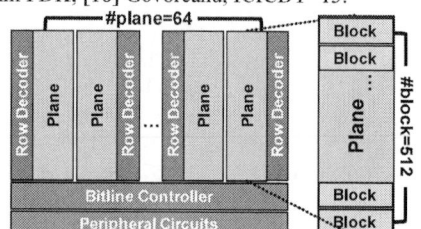

Fig. 1. Floor plan of a 3D VRRAM chip of 1 Tbit, 4.7 Gb/mm².

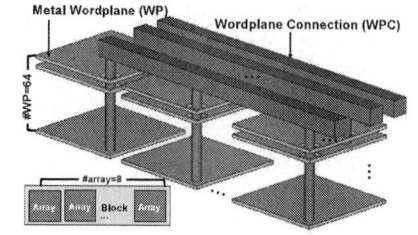

Fig. 2. Schematic of 3D arrays in a block. Low resistance WPC connects with all arrays of a layer.

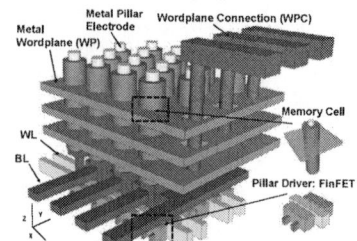

Fig. 3. View of a 3D VRRAM array.

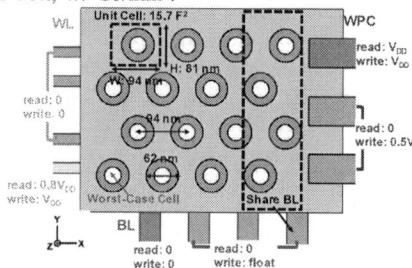

Fig. 4. Top view of 3D VRRAM array.

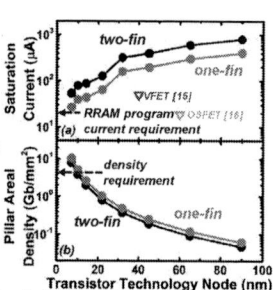

Fig. 5. Pillar driver saturation current and areal density for various transistor technology nodes. Data points for node > 22 nm indicate the equivalent transistor size scaled up from 7 nm FinFET [17]. Vertical FET (VFETs) [15] or BEOL-compatible oxide semiconductor FET (OSFETs, e.g. IGZO FET) [16] can be used if they can provide enough current for RRAM programming at the same areal density.

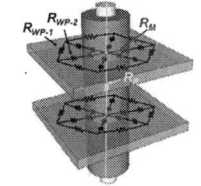

Fig. 6. Layout of pillars with two-fin transistors.

Fig. 7. Schematic of a sub-circuit module that captures the distributed resistances.

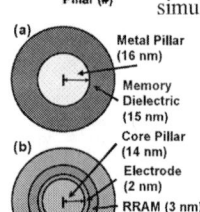

Fig. 8. WP resistances are simulated with various numbers of pillars. The resistor network captures WP resistances Sentaurus simulates.

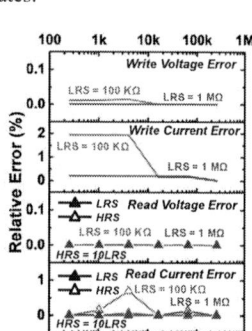

Fig. 9. Layer structures of (a) the metal pillar with dielectric and (b) the coresponding 1S1R.

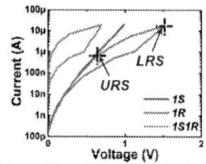

Fig. 10. Characteristics of typical selector (1S), RRAM (1R), and 1S1R devices.

Fig. 12. Relative errors of full and reduced resistor networks on the access voltage of the worst-case cell and the current through the corresponding BL respectively. The relative errors are below 2%.

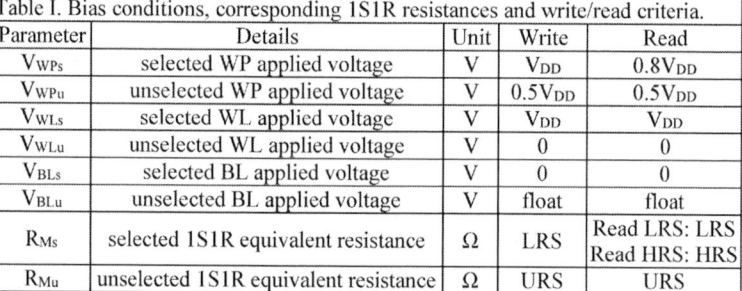

Table I. Bias conditions, corresponding 1S1R resistances and write/read criteria.

Parameter	Details	Unit	Write	Read
V_{WPs}	selected WP applied voltage	V	V_{DD}	$0.8V_{DD}$
V_{WPu}	unselected WP applied voltage	V	$0.5V_{DD}$	$0.5V_{DD}$
V_{WLs}	selected WL applied voltage	V	V_{DD}	V_{DD}
V_{WLu}	unselected WL applied voltage	V	0	0
V_{BLs}	selected BL applied voltage	V	0	0
V_{BLu}	unselected BL applied voltage	V	float	float
R_{Ms}	selected 1S1R equivalent resistance	Ω	LRS	Read LRS: LRS Read HRS: HRS
R_{Mu}	unselected 1S1R equivalent resistance	Ω	URS	URS

Write criteria: 1) $V_{WC} > 0.9V_{DD}$; 2) Pillar driver can provide enough current.
Read criteria: 1) $V_{WC} > 0.65V_{DD}$; 2) $I_{LRS} > 100$ nA; 3) $I_{LRS} > 5I_{HRS}$.
Applied voltages for the worst-case cell analysis is shown in Fig. 4. V_{WC} is the voltage drop on the worst-case cell and I_{LRS} and I_{HRS} are the currents through the pillar driver when reading the worst-case cell at LRS and HRS respectively.

Table II. Design specifications for different NL requirements of 1Tbit 3DVRRAM (LRS=1 MΩ and #WP=64).

NL	AE (%)	Density (Gb/mm²)	#plane	#block	#array	#WL	#BL
10^2	59	4.6	64	512	8	256	256
10^3	80	6.2	16	256	16	512	512
10^4	80	6.3	8	128	16	1024	1024
10^5	80	6.3	8	128	16	1024	1024

Fig. 13. LRS = 1 MΩ and NL ≥ 10^2 are required for tera-bit class 3D VRRAM with > 4.3 Gb/mm². The maximum density is limited by LRS and the areas of the row decoders, BL controller and the additional peripheral circuits.

Fig. 11. Schematic of (a) full and (b) reduced resistor networks. The unselected WPs (WP₁-WP_N-1) can be combined as one equivalent WP. The resistances of the reduced network are provided in the schematic.

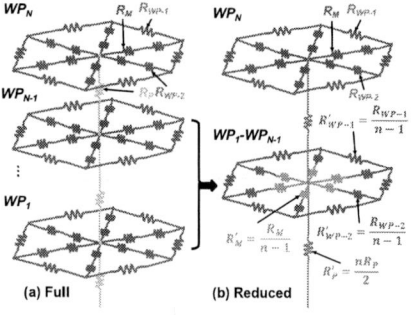

Table III. Resistance (R) and resistivity (ρ) used in the simulation. The definition can be found in Fig. 7 (WP-1, WP-2, and P) and Fig. 4 (BL and WL).

	R (Ω)	ρ (Ω·μm)
WP-1	54.4	0.2
WP-2	27.2	0.2
P	0.57	0.042
BL	1.3	0.035
WL	7.5	0.064

978-1-5386-4219-1/18 $31.00 © 2018 IEEE   108

5x Reliability Enhanced 40nm TaOx Approximate-ReRAM with Domain-Specific Computing for Real-time Image Recognition of IoT Edge Devices

Yusuke Yamaga, Yoshiaki Deguchi, Shouhei Fukuyama and Ken Takeuchi

Chuo University, 1-13-27 Kasuga, Bunkyo-ku, Tokyo, 112-8551 Japan, E-mail: yamaga@takeuchi-lab.org

Abstract

Highly reliable Approximate-ReRAM (A-ReRAM) with Pixel-to-Pixel Data Matching (P2P-DM) and Inter-Pixel error-correcting code (IP-ECC) is proposed to recognize the image accurately by deep neural network (DNN). By specializing for the image recognition applications and modulating the image data based on pixel-to-pixel features and ReRAM error characteristics, data-retention time and endurance of ReRAM increases by 5x and 3.3x, respectively.

Introduction

In future IoT edge devices, image senor, processors and ReRAM are vertically stacked by TSVs (Fig.1). ReRAM [1] works both as main memories and data storages, which requires longer data-retention time (as storages) and higher endurance (as main memories). NAND flash for DNN was proposed in [2, 3]. For the first time, this paper proposes a new type of ReRAM, A-ReRAM, that is best optimized for the image recognition applications by DNN. As a domain-specific computing, ReRAM controller is optimized by considering both application of image data and device characteristics of ReRAM. As for data, the most significant bits (MSB) of neighboring pixel data are most likely same. As for ReRAM, low resistance state (LRS) has 270x more errors than high resistance state (HRS) [4]. Two proposals are implemented in ReRAM controller (Fig. 1). First, in P2P-DM, information whether MSB of neighboring pixels are same or different are stored as HRS or LRS of ReRAM. Because the population of less reliable LRS decreases from 50% to 9%, data-retention errors decrease by 81%. Second, in IP-ECC, ECC is applied to the same order of bits among adjacent pixels and corrects errors of MSB. As a result, the data-retention time and the endurance improves by 5x and 3.3x, respectively.

Domain-Specific Image Data Characteristics

Image data captured by the image sensor are composed of sub-pixel data of red, green and blue. This paper uses CIFAR-10 dataset [5] with 60,000 image data. Each image is composed of 1024 pixels (32 width x 32 height) (Fig. 1). Each RGB sub-pixel data is 8 bit format. Average batch pixel length (BPL), which represents the number of adjacent pixels having the same data, is larger in higher-order bits because colors of adjacent pixels are similar (Fig. 2). Thus, the higher-order bits of neighboring pixels with the same color are more likely same. Fig. 3 shows the probability of having the same data between neighboring pixels (P_S). The 7th-bit shows the largest P_S and BPL.

Asymmetric Error Characteristics of ReRAM

Fig. 4 shows the 40nm TaO$_x$-based ReRAM cell [6]. By Set and Reset, HRS and LRS are switched by diffusion of oxygen vacancies (V_O) or oxygen ions (O^{2-}) between conductive filament (CF) and TaO$_x$ layer [6]. Figs. 5 and 6 show measured data-retention and endurance. Although bit-error rate (BER) of HRS stays low, BER of LRS increases as the data-retention time increases (Fig. 5(a)). While the cell current distribution of HRS hardly changes, the cell current distribution of LRS moved to HRS, which causes errors during the data-retention (Fig. 5(b)). Similarly, as the endurance increases, although BER of HRS stays low, BER of LRS increases (Fig. 6(a)). In Fig. 6(b), while the cell current distribution of HRS hardly changes, many tail bits appear in LRS at the higher endurance. Tail bits of LRS are caused by the expansion of CF and decrease in V_O density [4] (Fig. 6(c)). Thus, BERs of HRS ("0") and LRS ("1") become more asymmetric as the data-retention time and the endurance increases (Fig. 7).

Impact of Pixel Error on CNN Recognition Accuracy

Chainer example with Convolutional Neural Network (CNN) for CIFAR-10 [7] is used to recognize the image. CNN is a kind of DNNs and is composed of 13 convolution layers and 2 fully connected layers (Fig. 8). Fig. 9 shows the impact of errors in pixel data on the recognition accuracy by CNN. Average pixel error (APE) is defined as the pixel value difference between with and without random errors. When the 7th-bit has errors, APE becomes largest and the recognition accuracy degrades. Thus, among 8 bit data, the 7th-bit is the most important bit. When lower-order bits have errors, the recognition accuracy does not degrade because CNN is error tolerant. In Fig. 10, measured recognition accuracy indicates the importance of the 7th-bit. In A-ReRAM, the highest recognition accuracy is realized by minimizing errors of the 7th-bit.

Proposal 1: Pixel-to-Pixel Data Matching (P2P-DM)

To minimize errors of higher-order bits, this paper proposes P2P-DM (Fig. 11). Proposed P2P-DM encoder and decoder are composed of only 8 exclusive OR circuits. P2P-DM is implemented in ReRAM controller without circuit area/access time overhead. During the encoding, the "encoding target" next pixel data (Pixel$_n$) and the previous pixel data (Pixel$_{n-1}$) are input to the exclusive OR circuit. If Pixel$_n$ and Pixel$_{n-1}$ has the same data, "0" is stored in ReRAM. If data of Pixel$_n$ and Pixel$_{n-1}$ are different, "1" is stored in ReRAM (Fig. 11(c)). The original raw image data has 50% of "0"s and "1"s. Because P_S of higher-order bits is higher than lower-order bits (Fig. 3), higher-order bits are more likely same between adjacent pixels. In the 7th-bit, because P_S is 91%, P2P-DM increases the population of "0" from 50% to 91% (Fig. 12) and decrease the population of "1" from 50% to 9%. LRS ("1") has 270x higher BER than HRS ("0") (Fig. 7). By reducing unreliable LRS ("1") data and increasing reliable HRS ("0") data, errors of higher-order bits decrease (Fig. 13). Specifically, errors of the most important 7th-bit decreases by 81%. Note that errors of less important lower-order bits even increase, which does not degrade the recognition accuracy. Due to this error tolerance, the proposed ReRAM is called "Approximate."

Proposal 2: Inter-Pixel ECC (IP-ECC)

To correct errors of MSB, the 7th-bit, this paper proposes IP-ECC (Fig. 14). In conventional Bose-Chaudhuri-Hocquenghem (BCH) ECC, ECC is applied to the whole 8 bit pixel data. Although P2P-DM reduce errors of higher-order bits, conventional ECC cannot correct errors because errors of lower-order bits are increased by P2P-DM. To solve this problem, in proposed IP-ECC, ECC is applied to data at the same bit position, e.g. the 7th-bit, among adjacent pixels. When correcting errors of the 7th-bit, 6th to 0th-bits are not included in the codeword of ECC. Thus, errors of the 7th-bit is most effectively corrected. By combining P2P-DM and IP-ECC, proposed A-ReRAM (Case 4) extends the acceptable data-retention time (defined as no ECC failure) of the 7th-bit by 50x compared with conventional ECC (Case 1) (Fig. 15(a)). To ensure the same reliability of 7th-bit as the proposals, the conventional ECC need to decrease ECC code rate to 0.7 (Case 2). Due to the additional ECC parity, the ReRAM cell area increases by 31% in Case 2, which is unacceptably large (Fig. 15(b)). What's worse, in Case 2 the ECC decoding time becomes unacceptably large. In A-ReRAM, ECC decoding time decreases by 66% compared with Case 2 (Fig. 15(b)). Figs. 16 and 17 show the measured image recognition accuracy during the data-retention and the Set/Reset endurance cycles. Proposed A-ReRAM improves the acceptable data-retention time and the endurance by 50x (Fig. 16) and 3.3x (Fig. 17(a)), respectively. X-axis of Fig. 17(b) is the raw BER of ReRAM. A-ReRAM improves acceptable BER by 6.4x (Fig. 17(b)).

Conclusion

Table 1 summarizes this work. Proposed A-ReRAM improves the data-retention time by 50x compared with no ECC. Compared with conventional BCH ECC, the data-retention time increases by 5x. In addition, the endurance and acceptable BER increases by 3.3x and 6.4x, respectively. Because A-ReRAM achieves both longer data-retention and higher endurance, A-ReRAM can be used as both main memories and storages in IoT edge devices for the image recognition by CNN.

Acknowledgment

The authors sincerely thank Takumi Mikawa and Ryutaro Yasuhara of PSCS for their support. This work is supported by JST CREST Grant Number JPMJCR1532, Japan.

References

[1] Y. Hayakawa et al., VLSI Tech., pp. T14-T15, Jun. 2015. [2] Y. Deguchi et al., CICC, pp. 1-4, Apr. 2017. [3] K. Takeuchi, IEDM, pp. 665-668, Dec. 2017. [4] A. Hayakawa et al., IMW, pp. 24-27, May 2017. [5] A. Krizhevsky et al., Technical report, University of Toronto, Apr. 2009. [6] T. Ninomiya et al, TED, vol. 60, no. 4, 2013, pp. 1384-1389. [7] Chainer examples. Available: https://github.com/chainer/chainer/blob/master/examples/cifar/train_cifar.py

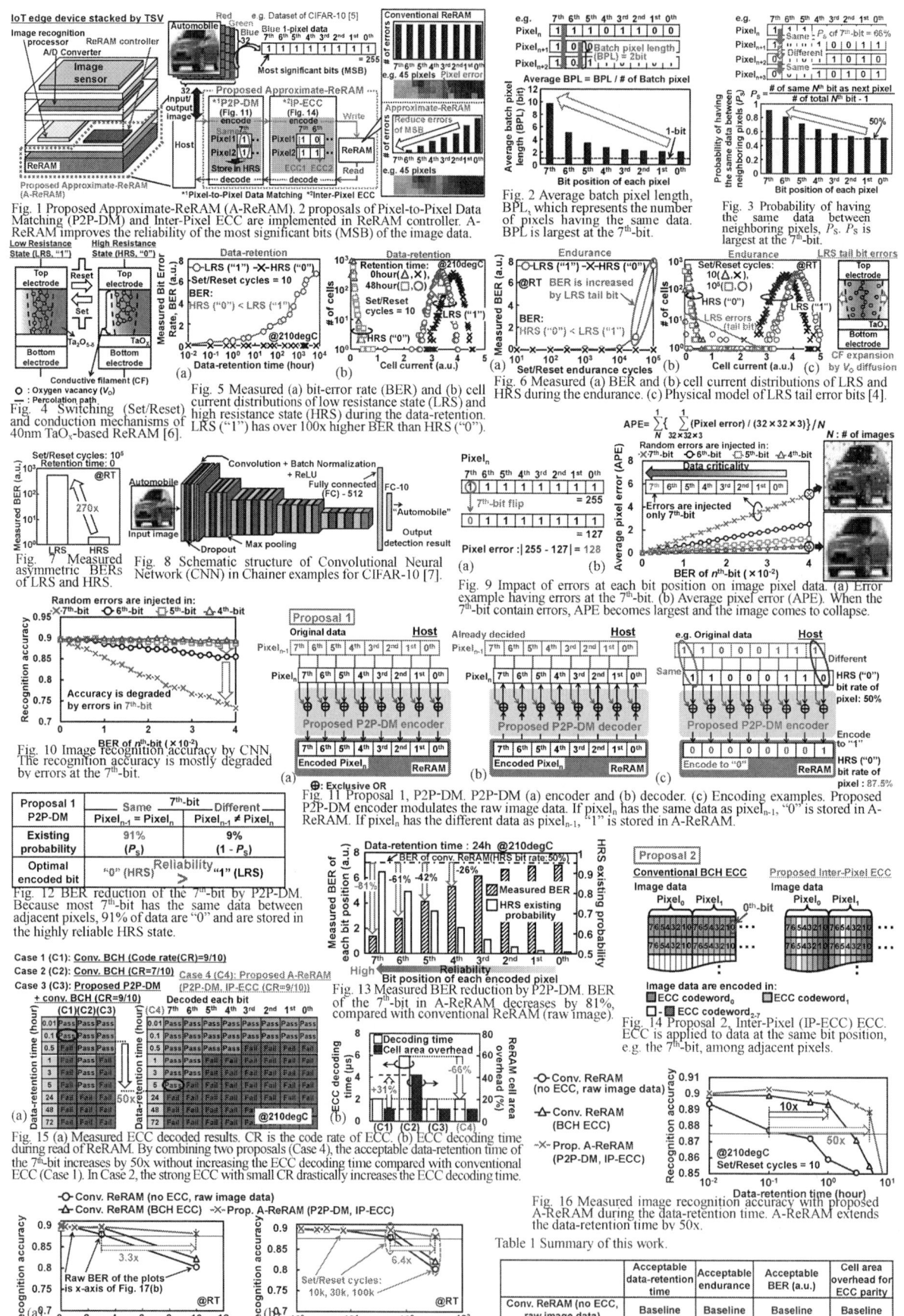

Fig. 1 Proposed Approximate-ReRAM (A-ReRAM). 2 proposals of Pixel-to-Pixel Data Matching (P2P-DM) and Inter-Pixel ECC are implemented in ReRAM controller. A-ReRAM improves the reliability of the most significant bits (MSB) of the image data.

Fig. 2 Average batch pixel length, BPL, which represents the number of pixels having the same data. BPL is largest at the 7th-bit.

Fig. 3 Probability of having the same data between neighboring pixels, P_S. P_S is largest at the 7th-bit.

Fig. 4 Switching (Set/Reset) and conduction mechanisms of 40nm TaO_x-based ReRAM [6].

Fig. 5 Measured (a) bit-error rate (BER) and (b) cell current distributions of low resistance state (LRS) and high resistance state (HRS) during the data-retention. LRS ("1") has over 100x higher BER than HRS ("0").

Fig. 6 Measured (a) BER and (b) cell current distributions of LRS and HRS during the endurance. (c) Physical model of LRS tail error bits [4].

Fig. 7 Measured asymmetric BERs of LRS and HRS.

Fig. 8 Schematic structure of Convolutional Neural Network (CNN) in Chainer examples for CIFAR-10 [7].

$$APE = \frac{1}{N} \sum_{32 \times 32 \times 3} \{ \sum (\text{Pixel error}) / (32 \times 32 \times 3) \} / N \qquad N: \text{\# of images}$$

Fig. 9 Impact of errors at each bit position on image pixel data. (a) Error example having errors at the 7th-bit. (b) Average pixel error (APE). When the 7th-bit contain errors, APE becomes largest and the image comes to collapse.

Fig. 10 Image recognition accuracy by CNN. The recognition accuracy is mostly degraded by errors at the 7th-bit.

Fig. 11 Proposal 1, P2P-DM. P2P-DM (a) encoder and (b) decoder. (c) Encoding examples. Proposed P2P-DM encoder modulates the raw image data. If pixel_n has the same data as pixel_{n-1}, "0" is stored in A-ReRAM. If pixel_n has the different data as pixel_{n-1}, "1" is stored in A-ReRAM.

Proposal 1 P2P-DM	Same 7th-bit $\text{Pixel}_{n-1} = \text{Pixel}_n$	Different $\text{Pixel}_{n-1} \neq \text{Pixel}_n$
Existing probability	91% (P_S)	9% $(1 - P_S)$
Optimal encoded bit	"0" (HRS)	"1" (LRS)

Reliability "0" (HRS) > "1" (LRS)

Fig. 12 BER reduction of the 7th-bit by P2P-DM. Because most 7th-bit has the same data between adjacent pixels, 91% of data are "0" and are stored in the highly reliable HRS state.

Fig. 13 Measured BER reduction by P2P-DM. BER of the 7th-bit in A-ReRAM decreases by 81%, compared with conventional ReRAM (raw image).

Fig. 14 Proposal 2, Inter-Pixel (IP-ECC) ECC. ECC is applied to data at the same bit position, e.g. the 7th-bit, among adjacent pixels.

Case 1 (C1): Conv. BCH (Code rate(CR)=9/10)
Case 2 (C2): Conv. BCH (CR=7/10)
Case 3 (C3): Proposed P2P-DM + conv. BCH (CR=9/10)
Case 4 (C4): Proposed A-ReRAM (P2P-DM, IP-ECC) (CR=9/10)

Fig. 15 (a) Measured ECC decoded results. CR is the code rate of ECC. (b) ECC decoding time during read of ReRAM. By combining two proposals (Case 4), the acceptable data-retention time of the 7th-bit increases by 50x without increasing the ECC decoding time compared with conventional ECC (Case 1). In Case 2, the strong ECC with small CR drastically increases the ECC decoding time.

Fig. 16 Measured image recognition accuracy with proposed A-ReRAM during the data-retention time. A-ReRAM extends the data-retention time by 50x.

Fig. 17 Measured image recognition accuracy with proposed A-ReRAM during (a) Set/Reset endurance cycles and (b) BER during the endurance stress. A-ReRAM extends the endurance and acceptable BER by 3.3x and 6.4x, respectively.

Table 1 Summary of this work.

	Acceptable data-retention time	Acceptable endurance	Acceptable BER (a.u.)	Cell area overhead for ECC parity
Conv. ReRAM (no ECC, raw image data)	Baseline	Baseline	Baseline	Baseline
Conv. ReRAM (BCH ECC)	10x	1.0x	1.0x	+11%
Proposed A-ReRAM (P2P-DM, IP-ECC)	50x	3.3x	6.4x	+11%

Gap in pagination due to formatting issues.

Pages 111-112

Comprehensive Thermal SPICE Modeling of FinFETs and BEOL with Layout Flexibility Considering Frequency Dependent Thermal Time Constant, 3D Heat Flows, Boundary/Alloy Scattering, and Interfacial Thermal Resistance with Circuit Level Reliability Evaluation

Jhih-Yang Yan[1], Chia-Che Chung[1], Sun-Rong Jan[1], H. H. Lin[2], W. K. Wan[2], M.-T. Yang[2], and C. W. Liu[1, 3, *]

[1]Graduate Institute of Electronics Engineering, National Taiwan University, Taipei, Taiwan, [2] PTD, MediaTek Inc., Hsinchu 300, Taiwan, and [3]National Nano Device Laboratories, Hsinchu, Taiwan, *E-mail: cliu@ntu.edu.tw

ABSTRACT – Thermal SPICE modeling with distributed R_{th}-C_{th} network is proposed to provide more accurate AC self-heating (SH) results than two τ_c and one τ_c models. The thermal time constant of the hotspot ($\tau_{hotspot}$) in FinFETs is frequency dependent, not a constant. The severe SH by boundary/alloy scattering and interfacial thermal resistance (ITR) is included in our SPICE. The modularized components of fins, metals, and IMDs provide device and routing flexibility, without additional FEM simulation. ITR of $Si_{80}Ge_{20}/Si_{1-x}Ge_x$ is calculated by AMM model as the lower bound for SiGe FinFETs. The intrinsic electromigration (EM) improvement of Co interconnect (5X) is countervailed (5X→2.44X) by the increasing T_{metal} due to the low thermal conductivity of Co. Different V2 placements on the power line of a ring oscillator (RO) are proposed to lower both the T_j (FinFET) and T_{metal}. The predicted EM MTTF of Co interconnect with the additional heat dissipation by V2 insertion is ~5.65X of W/Cu interconnect.

I. INTRODUCTION

SH in scaled FinFET circuits degrades both device reliability and EM [1-4]. It has been reported [7] that none of the existing SHE measurement methods [5-9] can provide precise AC response of spatial and temporal temperatures, which are essential for accurate reliability lifetime projection. Therefore, device simulation is the only way to know the exact temperature distribution and time evolution [7,10]. The computational cost makes TCAD/FEM unsuitable for circuit design. Instead of device simulation, compact thermal models (CTM) are favored with their high computational efficiency. CTMs consist of lumped R_{th}-C_{th} [11-13] were reported to simulate the transient SHEs. The two τ_c model [13] deals the AC SHE in FinFET by adding an additional lumped R_{th}-C_{th} with a small thermal time constant (~ps) to the one τ_c model of BSIM. However, the ΔT_{MAX} predicted by two τ_c model has no frequency dependence in the range of interest (Fig. 1(a)), different from our TCAD/SPICE results (Fig. 1(b)). The penalties of using lumped R_{th}-C_{th} models are the loss of accuracy and the time consumption on FEM simulations for different device geometries and circuit layouts, (e.g. different isolation spacings between n/pFETs in an inverter). The thermal resistance of BEOL ($R_{th,BEOL}$) was calculated by the gml-EMT model reportedly [14]. However, with the metal density (ϕ_m) of ~60% in products, the calculated κ_{eff}^{metal} by the gml-EMT model ($\kappa_{eff}^{metal} \sim \kappa_i \times (1 + \phi_m)/(1 + \phi_m)$ with κ_i=1.4W/m/K, and AR=10) is only 5.6W/m/K for the Cu interconnect. Using gml-EMT model overestimates the ΔT in BEOL due to the underestimated κ_{eff}^{metal}. Image charge theory [13] was reported to calculate the average ΔT in BEOL at different levels with a certain distance from the heat source. However, we find that the ΔT (Drain$_{11th stage}$ - GND$_{1st stage}$) in the M1 layers of a RO can reach ~15°C (Fig. 14). Using $T_{avg, M1}$ provided by the image charge theory overestimates the EM MTTF due to the missing temperature distribution and the layout information.

In this work, a modularized thermal SPICE model from transistor to circuit level SHE with distributed R_{th}-C_{th} network is proposed and validated by 3-D electro-thermal TCAD. Guideline for determining the grid size is offered based on the frequency dependence of $\tau_{hotspot}$. $C_{th,BEOL}$ is modeled by proposed two-step pseudo-isothermal plane model [10] and verified by simulation with high metal density. The ITRs between $Si_{0.8}Ge_{0.2}/Si_{1-x}Ge_x$ is calculated. ROs with stacked power lines are simulated with different via # and via pitch. The effective E_{a_Co} is extracted from the reported experimental data in [15] for EM lifetime prediction.

II. EQUIVALENT CIRCUIT MODEL OF A FINFET

A. Frequency dependence of $\tau_{hotspot}$ and grid size for transient/AC simulations – A FinFET is simulated with the 1.1V square wave input at 3, 6, and 9GHz by TCAD. The localized hot spot is observed in the simulated temperature distribution (inset of Fig. 2). The $\tau_{hotspot}$, which is calculated by the time when ΔT_j decreases in value to e⁻¹ in one cycle, is inversely proportional to the input frequency in the range interested (Fig. 2). The fitted slope of $\tau_{hotspot}$ versus input frequency is -0.71 ps/GHz. To have the SPICE model at an input frequency resolution of 1GHz, the thermal time constant of an element ($\tau_{element}$) should be smaller than 0.71ps.

B. Equivalent circuit of a FinFET – 3-D heat flow is considered in the equivalent circuit model of fin and metal (Fig. 3(a)). The in-plane heat flow is considered by $R_{th,vertical}$ and $R_{th,horizontal}$, and the out-of-plane heat flow is considered by the connection with neighboring regions, e.g. channel to HKMG. In a multi-fin FinFET, The STI and IMD are sandwiched by fins and metals, respectively. The heat flows from hot fins or hot metals are modeled by the equivalent model (Fig. 3(b)). The thermal conductivities of silicon fin and SiGe S/D are siginificantly reduced due to boundary scattering and alloy scattering [16-18], respectively (Table I). The grid size in the channel region is chosen 2nm x 5nm x W_{fin} which has a calculated $\tau_{element}$ ≈0.5ps with the equations in Fig 3. The frequency dependence of $\tau_{hotspot}$ is implemented by

charging different amounts of elements with different input frequencies (Fig. 4(a)). The discretized fin is modeled with device geometry and material thermal conductivity (Fig. 4(b)). The completely modularized fin (Fig. 4(c)) is then connected to the MEOL and IMD to form a FinFET.

C. Validation of the SPICE model – The lumped $R_{th0,FinFET+MEOL}$ is chosen as an indicator of the correctness of R_{th} in the discrete R_{th}-C_{th} network. The extracted $R_{th0,FinFET+MEOL}$ of multi-fin FinFETs with MEOL vs fin# in SPICE can be fitted by the same equation for SPICE and TCAD (Fig. 5) [10]. The flexibility of our modularized SPICE model on device layout has been confirmed simultaneously.

The validity of C_{th} is ensured by similar transient T_j responses in the first period with the AC input at 3 GHz, 6 GHz, and 9 GHz by TCAD (Fig. 6). The $\tau_{element}$ is confirmed to be small enough to present the frequency dependence of $\tau_{hotspot}$.

III. MODELING $C_{TH,BEOL}$ AND $\tau_{BEOL+sub}$

The $C_{th,BEOL}$ is calculated by the extended two-step pseudo isothermal plane model with identical correction factors for our non-flat isothermal plane in BEOL [10].The results are verified by TCAD simulation up to the metal density of 50% (Fig. 7). The effective thermal circuit including the FinFET, MEOL, BEOL, and substrate is shown in Fig. 8. $\tau_{FinFET+MEOL}$ and $\tau_{BEOL+Sub}$ is extracted from the envelope of transient T_j evolution by the time when ΔT_j increases to 1-e⁻¹ of the final value (Fig. 9). The $\tau_{BEOL+Sub}$ of a 1-fin-1finger FinFET for face-down configuration is ~155 μs, 7 order of magnitude larger than $\tau_{hotspot}$ and 4 order larger than $\tau_{FinFET+MEOL}$ due to the large volume.

IV. INTERFACIAL THERMAL RESISTANCE

An abrupt ΔT appears at the interface of two different materials with a heat flow due to the phonon mismatch and scattering. Interfacial thermal conductivity (ITC) values [19-21] (Fig. 10(a)) are adopted in our SPICE model using the discretized cross-sectional area, A_i (Fig. 10(b)) at the interface. With ITR, T_j increases significantly (~42°C) in a 2-fin-1-finger pFinFET (Fig. 10(c)).

A SiGe channel is one of the promising pFinFET candidates. The phonon dispersion of $Si_{1-x}Ge_x$ is calculated by the virtual crystal approximation (VCA) and adiabatic bond charge method (ABCM) (Fig. 11). with the calculated phonon dispersion, The ITC between $Si_{0.8}Ge_{0.2}/Si_{1-x}Ge_x$ is then modeled using phonon acoustic mismatch model (AMM) [22] (Fig. 12) as an upper bond in our SPICE.

V. CIRCUIT LEVEL SHE SIMULATION AND EM PREDICTION

D. Transient SHEs on the RO and BEOL – A RO is assembled with our modularized SPICE, and one of the stages is shown in Fig. 13(a). Stacked power lines (M1 and M2) and the inter-stage heat flows are considered. The V2 bundles on the M2 power line of a RO (Fig. 13(b)) and the simulated segment of the RO (Fig. 13(c)) are schematically drawn with the definitions of via # and 1/2 V2 bundle pitch (1/2 V2BP). A RO can have many stages. However, only the stages in the 1/2 V2BP are simulated with the consideration of the symmetry of heat flow across the boundaries of the segment. The highest T_j of 219.4°C is found in the last stage of the simulated segment of the RO (1/2 V2BP =11, via # =1) with $T_{chassis}$=40°C. The highest nodal temperature in the BEOL is 98.8°C at the M1/V0 interface of the drain (Fig. 14).

E. Co interconnect and EM MTTF prediction – The segment of a RO (1/2 V2BP =11, via # =1) using pure Co MEOL/V0/M1 is simulated and compared to W MEOL and Cu V0/M1 (inset of Fig. 15). The SHE is deteriorated due to the low κ_{th_Co}. $\Delta T_{j, last stage}$ =+9.2°C and $\Delta T_{M1/V0 drain, last stage}$ = +5.2°C is observed for Co case. Reducing the 1/2 V2BP and increasing the via # are effective to improve the SHE (Fig. 16(a)). $\Delta T_{M1/V0 drain, last stage}$ = -8°C is achieved by the reduction of 1/2 V2BP from 11 to 5. $T_{M1/V0 drain, last stage}$ can be further improved by -2.6°C with the increasing via # from 1 to 2. The EM MTTF is modeled by Black's empirical equation [23]: EM MTTF = $A/J^n \times \exp(E_a/kT)$. The EM MTTF of Co technology is reportedly at least 5X longer than Cu technology [15]. E_{a_Co} is calculated to be 0.9416eV with respect to E_{a_Cu}=0.9eV for EM MTTF prediction. The intrinsic EM benefits of Co interconnects is retarded to be 2.44X by the increasing temperature in BEOL as compared to W MEOL and Cu V0/M1 (Fig. 16(b)). The EM MTTF can be improved by circuit design. By the reduction of 1/2 V2BP from 11 to 5, the MTTF is extended to 4.58X. An EM MTTF improvement of 5.64X with respect to W MEOL and Cu V0/M1 is achieved by increasing the via # from 1 to 2 with SHE considered.

VI. CONCLUSION

The proposed SPICE model can predict the temperature with arbitrary layouts of device, MEOL, and BEOL. SHE-aware circuit designs can be conducted with our SPICE to efficiently achieve the robust reliability from the device to circuit level.

ACKNOWLEDGMENT

This work is partially supported by the Ministry of Science and Technology, Taiwan, R.O.C. (No. 106-2622-8-002 -001 -, 106-2622-8-002 -014 -TM, and 106-2221-E-002 -232 -MY3).

REFERENCE

[1] Liping Wang *et al.*, TED, 62.7, pp. 2106, 2015. [2] E. Bury *et al.*, IRPS, pp. XT.8.1, 2014. [3] E. Bury *et al.*, VLSI, pp. T60, 2015 [4] Hai Jiang *et al.*, EDL, 36.12, pp. 1258, 2015. [5] C. W. Chang *et al.*, IRPS, pp. 2F.6.1, 2015. [6] E. Bury *et al.*, IEDM, pp. 15.6.1, 2016. [7] M. A. Wahab *et al.*, TED, 62.11, pp. 3595, 2015. [8] F. Menges *et al.*, IEDM, pp. 15.8.1, 2016. [9] F. Stellari *et al.*, IRPS, pp. 2B.1.1, 2015. [10] J.-Y. Yan *et al.*, IEDM, pp. 35.6.1, 2016 [11] BSIM-CMG v110.0.0 manual, pp. 71, 2015 [12] H. Jiang *et al.*, VLSI, pp. T136, 2017 [13] Woojin Ahn *et al.*, IEDM, pp. 13.6.1, 2017 [14] Woojin Ahn *et al.*, TED, 64.9, pp. 3555, 2017 [15] C. Auth *et al.*, IEDM, pp. 29.1.1, 2017. [16] S. E. Liu *et al.*, IRPS, pp. 4A.4.1, 2014. [17] Steven Mittl *et al.*, IRPS, pp. 4A.4.1, 2015. [18] A. Laurent *et al.*, VLSI, pp. T48, 2016. [19] E. S. Landry and A. J. H. McGaughey, Phys. Rev. B 80, 165304 (2009). [20] National institute for material science, Japan. (http://interface.nims.go.jp). [21] J. Chen *et al.*, J. Appl. Phys. 112, 064319 (2012). [22] C. Monachon *et al.*, Annual Review of Materials Research 46 (2016): 433-463 [23] J. R. Black *et al.*, IRPS, pp. 300, 1982.

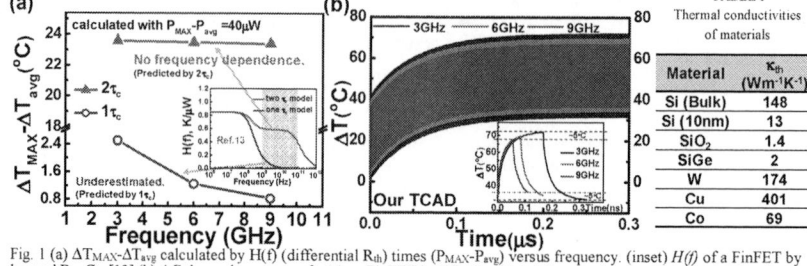

Fig. 1 (a) ΔT_{MAX}-ΔT_{avg} calculated by H(f) (differential R_{th}) times (P_{MAX}-P_{avg}) versus frequency. (inset) $H(f)$ of a FinFET by lumped R_{th}-C_{th} [13] (b) AC thermal response of a 14nm FinFET by TCAD with 1.1V square input at 3, 6, and 9 GHz. (inset) AC response in one period.

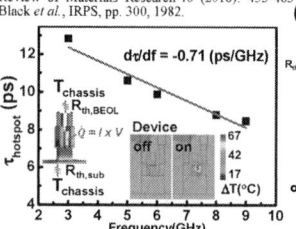

Fig. 2. $\tau_{hotspot}$ versus input frequency. (inset) Simulated FinFET with corresponding thermal resistances and the localized hotspot in AC simulation

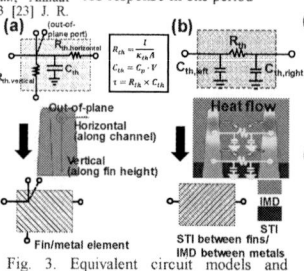

Fig. 3. Equivalent circuit models and simplified color symbols of (a) fin or metal and (b) STI between fins or IMD between metals, respectively

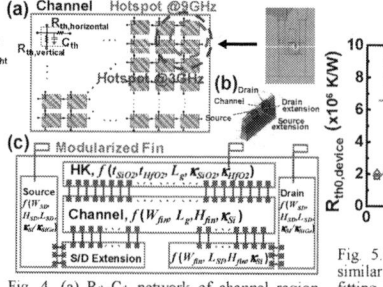

Fig. 4. (a) R_{th}-C_{th} network of channel region with schematically drawn hotspot at 3GHz and 9GHz (b) Geometry of fin including the SiGe S/D, and (c) the geometry based modularized fin

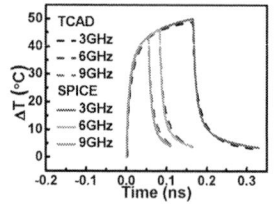

Fig. 6. Transient T_j of a 1-fin-1finger pFinFET in the first period with 0.7V square wave input at 3GHz, 6GHz, and 9GHz

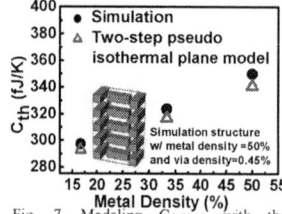

Fig. 7. Modeling $C_{th,BEOL}$ with the comparison of simulation results

Fig. 5. SPICE and TCAD yielding similar $R_{th0,FinFET+MEOL}$ with the same fitting equation for face-down configuration

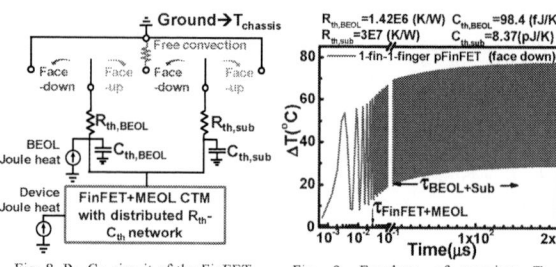

Fig. 8. R_{th}-C_{th} circuit of the FinFET, MEOL, BEOL, and substrate

Fig. 9. Envelope of transient T_j evolution at 3GHz including the BEOL and substrate

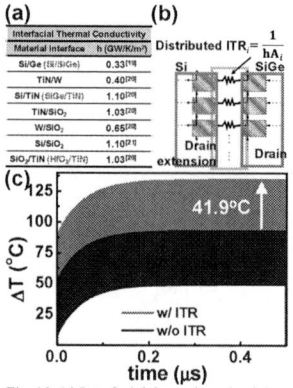

Fig. 10. (a) Interfacial thermal conductivity at material interfaces, (b) distributed ITR between two different materials in SPICE model, and (c) envelope of transient T_j evolution at 3GHz with and without ITR in the SPICE model

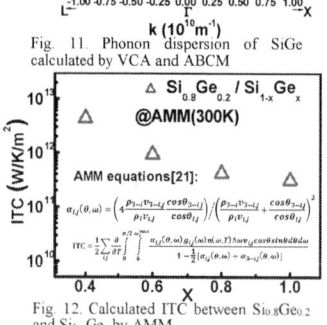

Fig. 11. Phonon dispersion of SiGe calculated by VCA and ABCM

Fig. 12. Calculated ITC between $Si_{0.8}Ge_{0.2}$ and $Si_{1-x}Ge_x$ by AMM

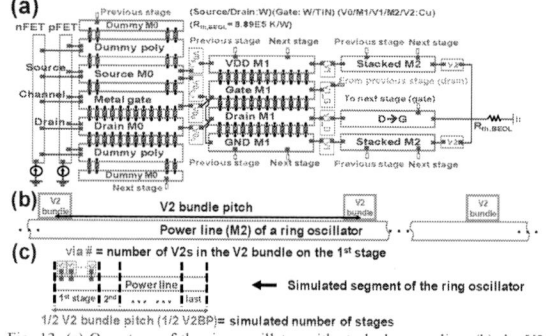

Fig. 13. (a) One stage of the ring oscillator with stacked power line, (b) the V2 bundles on the power line of a ring oscillator, and (c) simulated segment of the ring oscillator and the definitions of via # and 1/2 V2 bundle pitch (1/2 V2BP). Only the first stage has the V2s (see the left end) on the power line as shown in (c).

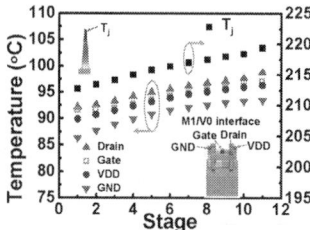

Fig. 14. T_j and the temperature at M1/V0 interfaces for a segment of the RO with 1/2 V2BP of 11 and via # of 1

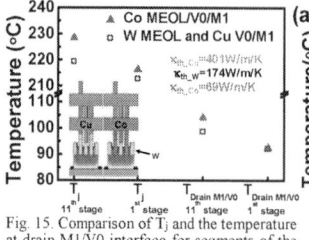

Fig. 15. Comparison of T_j and the temperature at drain M1/V0 interface for segments of the ROs with 1/2 V2BP of 11 and via # of 1 using pure Co or W MEOL and Cu M1

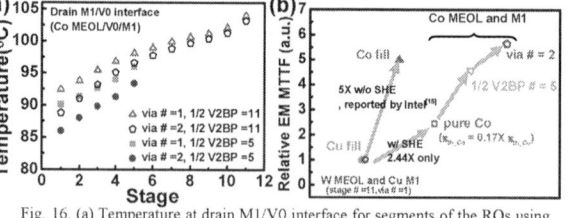

Fig. 16. (a) Temperature at drain M1/V0 interface for segments of the ROs using Co MEOL and M1 with different stage numbers and via numbers, and (b) EM MTTF prediction based on different designs as compared to Cu (3nm TaN/1.5nm Co) technology

Differentiated Performance and Reliability Enabled by Multi-Work Function Solution in RMG Silicon and SiGe MOSFETs

R. Bao, R. G. Southwick III, H. Zhou, C. H. Lee, B. P. Linder, T. Ando, D. Guo, H. Jagannathan, and V. Narayanan

IBM Research, 257 Fuller Road, Suite 3100, Albany, NY 12203
Phone: +1-518-292-7220, E-mail: rbao@us.ibm.com

Abstract

We report for the *first time* that replacement metal gate (RMG) work function metal (WFM) modulates the interface defects in Silicon and SiGe MOSFETs. Changing the effective work function (eWF) towards nFET band edge provides lower interface defects and higher mobility than eWF near the pFET band edge for both Si and SiGe substrates. Reducing the electric field across the dielectric (via eWF) improves bias temperature instability (BTI) for both n & pMOSFETs beyond expectation. Breakdown voltage increases and gate leakage decreases with increasing eWF for both n & pMOSFETs. Therefore, multi-Vt MOSFETs by RMG metal gate exhibit differentiated reliability as well as differentiated performance for both Si and SiGe channel materials.

Background and Objective

Multi-Vt gate stacks are essential for FinFET [1, 2] and gate all-round transistors [3] due to improved performance over channel doping options. One of the popular multi-Vt solutions is through metal gate tuning. The foundation for Vt tuning in RMG is by modulation of the oxygen vacancy concentration in HfO_2 via an oxygen scavenging metal. If high temperature anneals are used, oxygen scavenging metals have been shown to scavenge oxygen not only from the HfO_2 but also from the SiO_2 interfacial layer (IL) [5, 6]. So, while oxygen scavenging from the IL is clearly possible, no report exists on oxygen scavenging from the IL at metal gate deposition under RMG conditions. Understanding the impact of WFM on the IL however is very important, since it can help assess the effect of multi-Vt on device performance and reliability for cutting edge technology nodes, where new SiGe channel materials are being used to provide high mobility and reliability over Si [2, 4]. In this work, we systematically investigate the impact of RMG metal gate tuning on interfacial layer, mobility, and reliability for both Si and SiGe substrates. We will show a universal trend of interface defects modulated by metal gate and a strong correlation between performance and eWF. We will also show that reliability clearly depends on eWF.

Interface Defects and Mobility Modulated by WFM

Si nFET, Si pFET, and SiGe pFET devices were fabricated using a standard high-k RMG FinFET process. The eWF is tuned either by changing the scavenging metal thickness (WFM stack) or the bottom barrier layer thickness (multi-Vt stack), **Fig. 1**. The electron mobility (**Fig. 2a**) is clearly modulated by WFM for both WFM stack and multi-Vt stack in Si nFET. The lower the eWF, the higher the mobility. The main difference in mobility is in the low field region where coulombic scattering and phonon scattering dominate. The bottom barrier metal or scavenging metal may also modulate the mobility through stress/strain effects. Fig. 2b-c show the same trend for SiGe pFET and Si pFET. Fig. 3a shows reducing eWF clearly reduces the interface defect density, Nit, measured by charge pumping for Si nFET. Larger impact on Nit is seen when eWF is greater than 4.60eV. Fig. 3b-c show the electron mobility as a function of reciprocal of Nit and temperature. The larger slope in Fig. 3b at lower temperature

indicate coulombic scattering is playing an increasing role. The larger slope in Fig. 3 (c) at lower eWF suggests that the phonon scattering has more influence on mobility. When the eWF moves towards nFET band edge, both Dit (**Fig. 4**), measured by conductance, and Nit (**Fig. 5**) are reduced for SiGe pFET. Higher interface defects in SiGe pFET result in a larger slope (**Fig. 6**) in low field mobility and reciprocal of Nit compared to Si nFET. **Fig. 7** shows lower eWF has better subthreshold swing(SS) for both Si nFET and SiGe pFET and WFM stack and multi-Vt stack follow the same trend. More SS impact is observed when eWF is larger than 4.60eV for Si nFET and 4.70eV for SiGe pFET, well correlated to Nit trend.

Mechanism, Performance and Reliability Impact

In pWFM, the absence of a scavenging metal (**Fig. 8a**) results in an oxygen concentration in the IL and HfO_2 near the equilibrium state. When scavenging metal is added to shift the eWF towards nWFM, weakly bonded oxygen atoms in IL and HfO_2 are pulled towards/into the scavenging metal and oxygen vacancies in IL and HfO_2 are created (**Fig. 8b**) and some Si (and/or Ge) atoms are released in the interface. Subsequently, the released Si (and/or Ge) atoms can relax forming bonds with preexisting dangling bonds reducing the interface defect density. Reduced interface defects improves mobility and SS.

Fig. 9 shows Vbd and BTI for Si and SiGe. Vbd is strongly correlated to the gate leakage for both nFET and pFET. When eWF moves towards nFET band edge, Vbd is degraded. But PBTI and NBTI are improved when the eWF is moved to mid gap for both Si and SiGe beyond what is expected from a simple e-field reduction [7]. For PBTI this is related to oxygen vacancy levels in the HfO_2 making e-traps less accessible. This may also be the case for holes in NBTI [7] and/or improved IL layer as mobility, SS, and Dit/Nit measurements suggest. **Fig. 10** shows the performance corrected to the target I_{off} improves when eWF moves towards nFET band edge for both Si and SiGe. The performance could be 5 to 10% different depending on eWF (i.e. Vt) delta. The expected performance trendline with WFM multi-Vt is shown in **Fig. 11a**. Compared to the trendline when WFM only shifts Vt (**Fig. 11a**), low Vt nFET and high Vt pFET have the improved performance from high Vt nFET and low Vt pFET, respectively. **Fig. 11b** shows proof of concept data for Si pFET matching the expectation.

Conclusion

We observe that RMG metal gate modulates the interface defects and thus mobility as well as reliability for both Si and SiGe MOSFETs. WFM having lower eWF gives lower interface defect density which impact the device performance. Multi-Vt defined by RMG metal gate offers different device performance and reliability for both Si and SiGe MOSFETs.

Acknowledgement

This work was performed by the Research and Development Alliance Teams at various IBM Research and Development Facilities.

References

[1] C.H. Lin *et al.*, *IEDM*, 2014, 3.8.1. [2] D. Guo *et al.*, *VLSI*, 14 (2016). [3] J. Zhang *et al.*, *IEDM*, 22 (2017). [4] R. Xie *et al.*, *IEDM*, 47 (2016). [5] T. Ando, *Mater.* 5, 478 (2012), [6] T. Ando *et al.*, *Appl. Phys. Lett.*, 96, 132904:1 (2010). [7] B. P. Linder *et al.*, *IRPS* (2016).

2018 Symposium on VLSI Technology Digest of Technical Papers

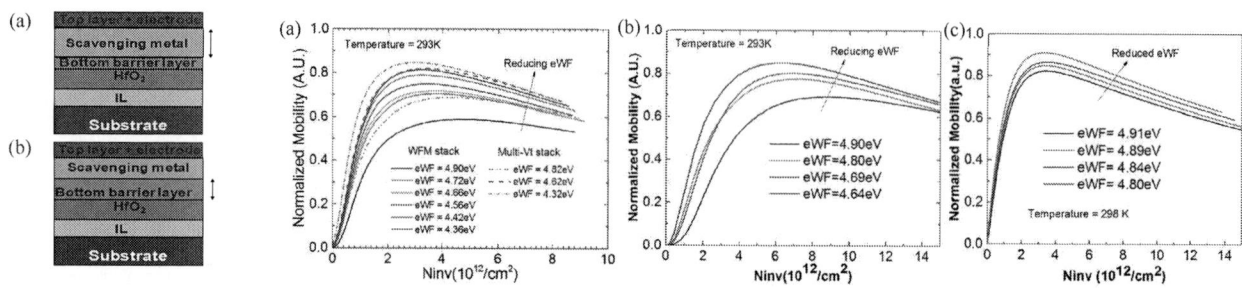

Fig. 1 eWF modulated by (a) scavenging metal thickness and (b) bottom barrier layer thickness

Fig. 2 The effect of eWF on the mobility of Si nFET (a), SiGe pFET (b) and Si pFET (c). Reduced eWF improves the mobility for all devices.

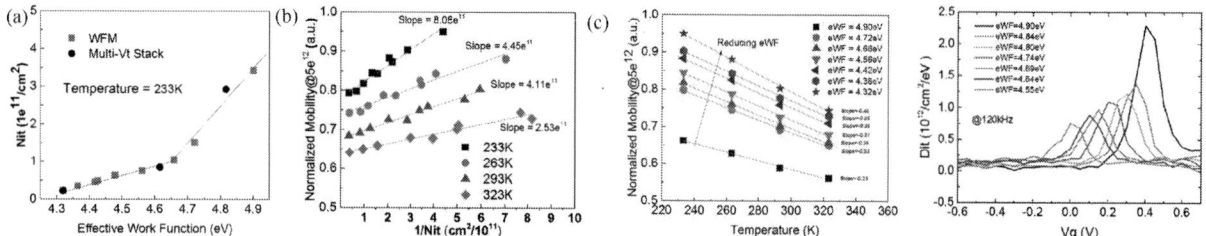

Fig. 3 The relationship of Nit and eWF (a), low field mobility and reciprocal of measured Nit (b), and low filed mobility and temperature (c) for Si nFET.

Fig. 4 Dit measured by conductance method for SiGe

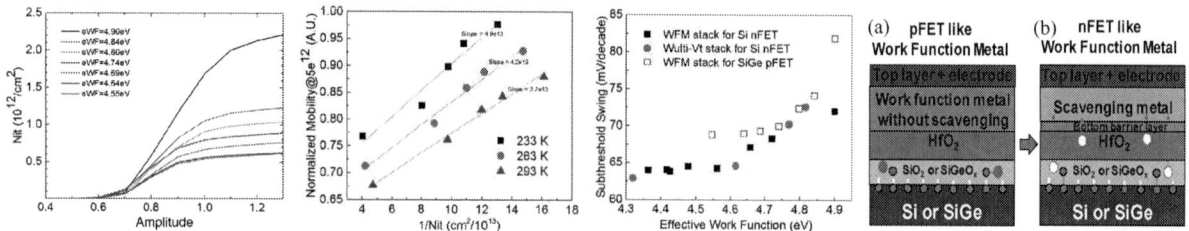

Fig. 5 Nit measured by charge pumping method for SiGe pFET

Fig. 6 The relationship of low field mobility and reciprocal of measured Nit for SiGe pFET

Fig. 7 Linear subthreshold swing for both Si nFET and SiGe pFET

Fig. 8 The mechanism for interface defects modulated by work function metal gate due to scavenging metal

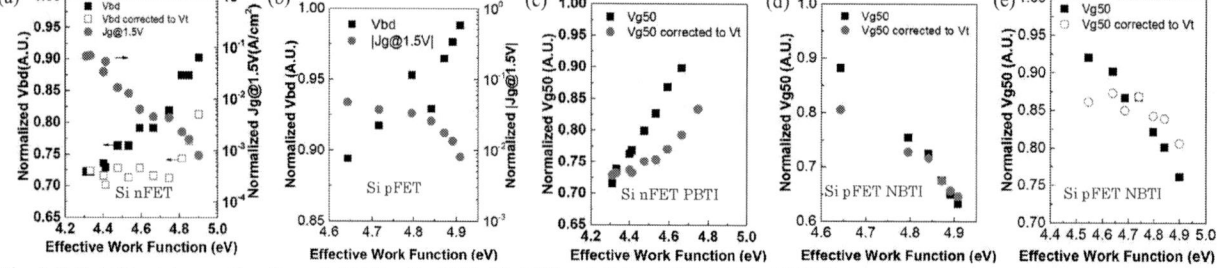

Fig. 9 Reliability data as a function of eWF for Si nFET, Si pFET and SiGe pFET. (a) Si nFET Vbd and gate leakage, (b) Si pFET Vbd and gate leakage, (c) Si nFET PBTI, (d) Si pFET NBTI, and (e) SiGe pFET NBTI. Vg50 is the measured voltage across the gate to move the Vt 50mV.

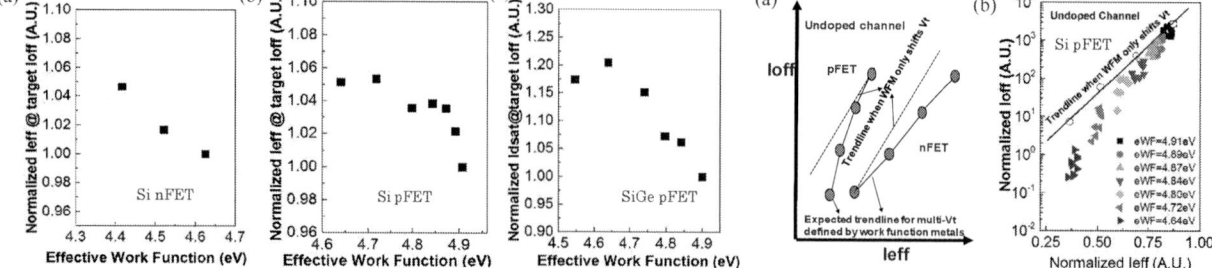

Fig. 10 Device performance as a function of eWF (a) Si nFET, (b) Si pFET, and (c) SiGe pFET. WFM having lower eWF improves the device performance, especially for SiGe pFET

Fig. 11 Expected performance trendline with multi-Vt (a) and proof of concept experimental data (b) for undoped channel

978-1-5386-4219-1/18 $31.00 © 2018 IEEE   116

Process Optimization of Perpendicular Magnetic Tunnel Junction Arrays for Last-Level Cache beyond 7 nm Node

Lin Xue [a], Chi Ching, Alex Kontos, Jaesoo Ahn, Xiaodong Wang, Renu Whig, Hsin-wei Tseng, James Howarth, Sajjad Hassan, Hao Chen, Mangesh Bangar, Shurong Liang, Rongjun Wang, Mahendra Pakala [b]

Applied Materials, Inc., Sunnyvale, California 94085, USA

[a] Lin_Xue@amat.com, [b] Mahendra_Pakala@amat.com

Abstract

This paper demonstrates systematic process optimization of perpendicular magnetic tunnel junction (pMTJ) by hardware, unit-process, and material stack design. TMR of 200% at RA 5 Ohm·μm^2, H_{SAF} ~ 8 kOe, and 10-time tunability of Hc were achieved at the film level. After patterning, 10^{-6} write error rate was reached at 0.4 pJ, V_{BD} was as high as 1600 mV at 20 ns pulse width, and excellent device stability against 400°C BEOL baking was demonstrated. The device performance along with the process capability to make MTJ array at 88 nm pitch provides opportunities for LLC applications.

Introduction

STT-MRAM has become one of the leading new memory technologies as it demonstrated excellent scalability and low power consumption for next-generation embedded memory to extend from eFlash [1]. Moving forward, the industry is developing STT-MRAM for last-level cache (LLC) applications [2] as SRAM scaling is expected to slow beyond 7 nm node (Fig. 1). STT-MRAM can also reduce LLC power consumption for artificial intelligence (AI) applications. LLC requires high performance of STT-MRAM cell, namely magnetic tunnel junction (MTJ), with low RA, high TMR, low write energy at low error rate, and good stability against BEOL baking. We will discuss hardware, process, and material stack development to meet MTJ performance requirements for LLC.

Process

The MTJ array module begins with film stack deposition in a Endura Clover™ PVD System of Applied Materials. The pMTJ stack of about 20 nm thick is composed of 7 functional layers. Each layer is a multilayer of laminated materials at the Angstrom level. The pMTJ array is then patterned with 193 nm immersion lithography and direct ribbon beam etching technology of Applied Materials with MTJ CD of 30~40 nm and pitch down to 88 nm (~4Gb of MTJs per die). Top-down SEM and cross-section TEM images are shown in Fig. 2. MTJ etch process was optimized to get higher TMR of 138% compared to 76% by baseline etch with film TMR = 160% at RA = 5 Ohm·μm^2 annealed at 400°C for 0.5 hour.

MTJ Film

Multiple hardware and unit-process optimizations were implemented for MTJ film PVD. A multi-cathode chamber was developed to enable sputtering from an oblique angle with wafer rotation, which allows controlling thickness and uniformity of ultra-thin films. A RF sputtering chamber was specially designed for the MgO tunnel barrier that allows good barrier integrity, stoichiometry, and particle control. High vacuum of 10^{-9} Torr was implemented for all chambers as well as the platform, which keeps materials and their interfaces free of moisture. Heating, cooling, and oxidation capabilities were also developed and integrated with the PVD system.

To deposit the MTJ film at low RA, a co-optimization of MgO barrier thickness and additional buffers underneath was implemented to reduce effective barrier thickness while keeping good TMR. As in Fig. 3, MTJ films in RA range of 1~30 were deposited, and their TMR values were measured. Although lower polarization at lower RA limits TMR, TMR of 200% was achieved at RA of 5 Ohm·μm^2, relevant to LLC applications. TMR ~ 100% was measured at RA ~ 1 Ohm·μm^2. The results showed excellent RA scalability of MTJ films.

The magnetic properties of the MTJ stack were improved by testing different materials for capping and SAF coupling. Fig. 4 shows films of four different materials of capping with Hc from 8 to 80 Oe. No obvious change of RA or TMR was measured. High SAF coupling (H_{SAF}) of the MTJ film allows low write error rate. By optimizing SAF coupling material as well as magnetic material thickness within the reference layer, H_{SAF} increased from 4 to 8 kOe (Fig. 5).

Device and Array Performance

Optimized films were patterned into devices, and device performance was characterized, which is critical to confirm and further guide MTJ film/process optimization. A MTJ film with TMR = 170% at RA = 7 Ohm·μm^2, Capping 1, and Ir 1st peak SAF coupling was patterned into MTJs with electrical CD (eCD) ~ 34 nm. 1 kbit of MTJs were electrically tested within each die. Average TMR = 150% (Fig. 6). Write error rate (WER) measurement was done at the single device level. With the high H_{SAF} achieved by Ir 1st peak, WER reached 10^{-6} with pulse width (PW) of 20 ns, while previous devices with Ru 1st peak could only reach 10^{-3} WER before reference layer was destabilized and WER increased (Fig. 7). WER was compared with different PWs as a function of write voltage (Vw) and energy (Ew). Although Vw is smaller for longer pulses at a certain WER, Ew is smaller at 5 and 10 ns than at 50 ns due to reduction of PW. A typical Ew ~ 0.4 pJ was measured at WER of 10^{-6} for 5 and 10 ns pulses.

At the array level, breakdown voltage V_{BD} was measured at both P state (V_{BD} = 1580 mV) and AP state (V_{BD} = 1620 mV) (Fig. 8). V_{BD} – Vw@10^{-6} WER > 1 V for both states with 20 ns PW leaving enough write window. 400°C baking test was done at patterned device level to study device parameter stability against baking during BEOL process (Fig. 9). TMR is stable up to 3-hour baking, although Rp increase, Hc decrease, and Vc increase were observed. All parameter changes were within 10% from after patterning to 3-hour baking. Read disturbance rate (Fig. 10) and PW dependence of Vc (Fig. 11) were measured at room temperature to extract data retention energy barrier Eb separately. Consistent results were obtained. Eb with 34 nm eCD = 44 kT, and Eb with 62 nm eCD = 91kT. 34 nm devices have good retention of LLC, while 62 nm devices can be used as embedded memory on the same chip. Eb can be further increased by using capping with higher Hc.

Conclusions

Through process optimization, excellent MTJ film properties and device performance were demonstrated with high TMR at low RA, low write energy at low WER, and stability against BEOL backing required for LLC applications.

References

[1] C. Park, et al., IEDM 2015, pp 664-667
[2] G. Jan, et al., VLSI 2016, pp 18-19

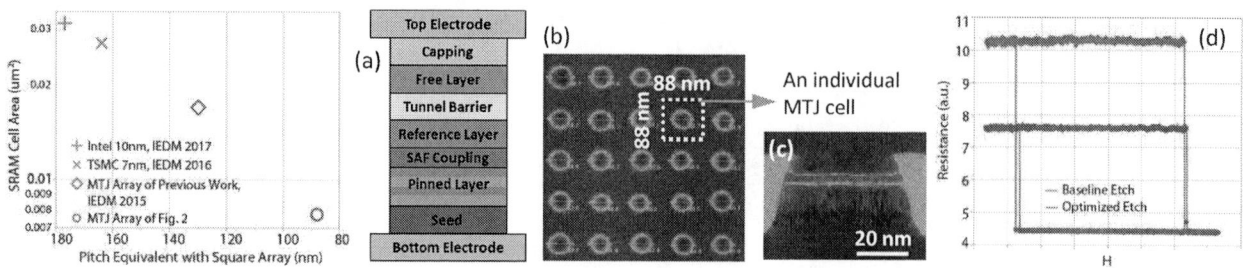

Fig. 1. STT-MRAM has excellent cell size scalability beyond advanced nodes for LLC

Fig. 2. (a) pMTJ stack structure. (b) Top-down SEM of a MTJ array with 88 nm pitch after hard mask patterning. (c) Cross-section TEM of a patterned pMTJ. (d) MTJ etch optimization to improve TMR after patterning

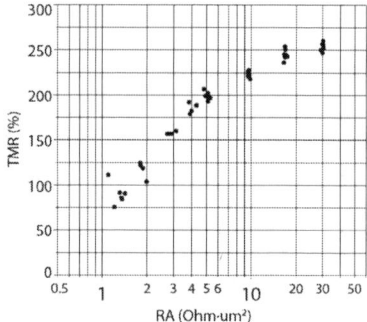

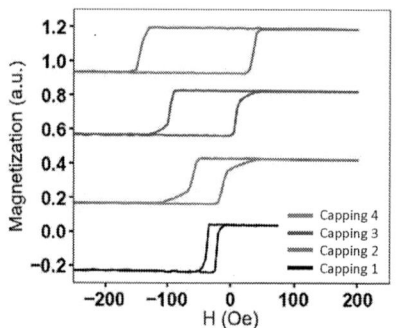

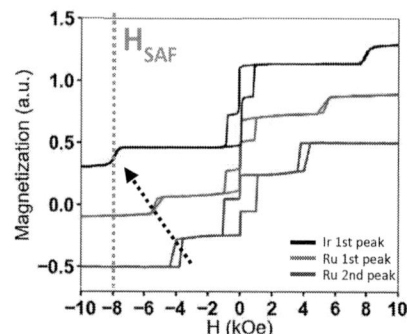

Fig. 3. RA dependence of TMR covering low RA range down to RA of 1 Ohm·μm^2

Fig . 4. 10-time Hc tunability via capping tuning

Fig. 5. H_{SAF} improvement via pinned-layer optimization

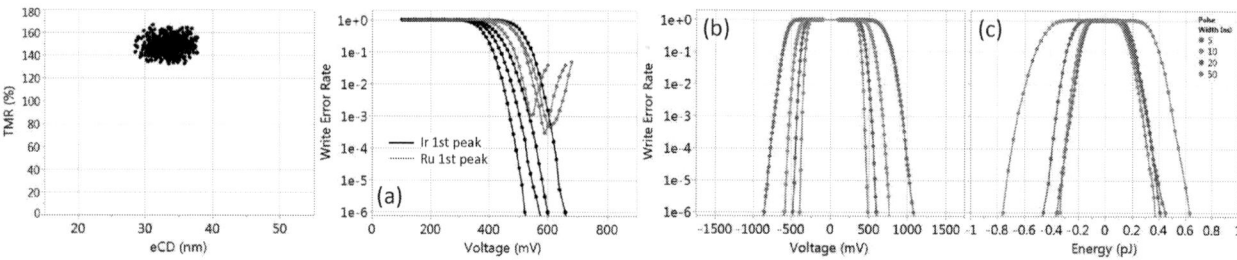

Fig. 6. TMR vs. eCD of devices from a die with eCD ~ 34 nm

Fig. 7. (a) Write error rate below 10^{-6} reached with H_{SAF} improvement. (b)-(c) Write error rate as a function of write voltage and energy with different pulse widths. Positive voltage/energy is defined as AP to P switching

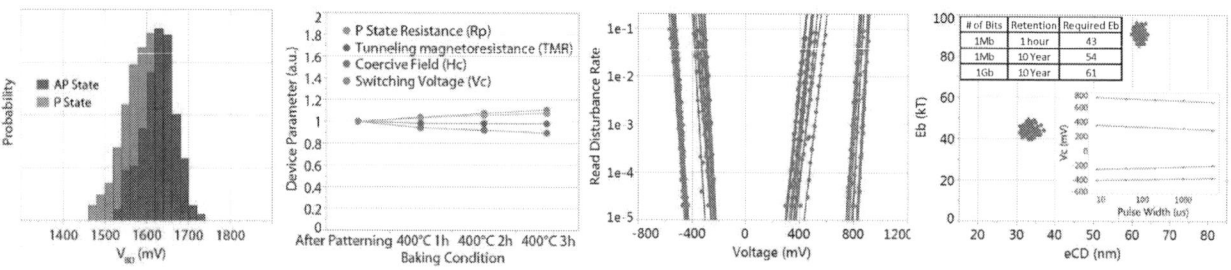

Fig. 8. Distribution of breakdown voltage with 20 ns pulse width for both P and AP states

Fig. 9. Changes of MTJ parameters as a function of total baking time at patterned device level

Fig. 10. Read disturbance measurement of MTJs with eCD about 34 nm (red) and 62 nm (green) using 20 ns pulse

Fig. 11. Data retention Eb of MTJs with eCD about 34 nm (red) and 62 nm (green) extracted by measuring Vc vs. pulse width (inset)

978-1-5386-4219-1/18 $31.00 © 2018 IEEE

Dependence of Reliability of Ferroelectric HfZrO$_x$ on Epitaxial SiGe Film with Various Ge Content

Kuen-Yi Chen, Yen-Hua Huang, Ruei-Wen Kao, Yan-Xiao Lin, and Yung-Hsien Wu*

Department of Engineering and System Science, National Tsing Hua University, 300, Hsinchu, Taiwan
*E-mail: yunhwu@mx.nthu.edu.tw

Abstract

TiN/ferroelectric-HfZrO$_x$ (HZO)/epi-SiGe (MFS) structure was employed as the platform to investigate the dependence of Ge content on reliability performance and the mechanism behind it. As compared to Si counterpart, HZO on Si$_{0.56}$Ge$_{0.44}$ exhibits not only enhanced remnant polarization (P$_r$) by 58 % but much improved reliability in terms of negligible P$_r$ degradation up to 10^9 cycles under ±4 V/100k Hz bipolar AC stress, desirable retention at pristine and cycled state up to 10^4 sec, and smaller imprint effect against time at 85 °C. The Ge content-dependent reliability performance is mainly due to the thinner sub-oxide interfacial layer (IL) with better quality since it is too thin to trap charges while less vulnerable to defect generation due to stronger bonding (fewer Vo). IL with higher κ value is also helpful to suppress E-field across it, beneficial to enhance reliability. The results suggest that as the technology advances into SiGe platform, it is more viable for MFS-based memory as the reliability issues for Si will be greatly mitigated.

Introduction

The development of HfO$_2$-based ferroelectric (FE) has ushered in a new era for green memory and logic devices which encompass two basic structures, metal-ferroelectric-metal (MFM) typically for memory applications such as FeRAM [1] or ferroelectric tunneling junction (FTJ) [2] and metal-ferroelectric-semiconductor (MFS) mainly for steep-slope negative capacitance logic devices [3] and memory-based FeFETs [4]. Besides performance, it is imperative to study the reliability performance before devices can be practically used. The reliability issues have been well explored and possible solutions were proposed for MFM structures [5]. For MFS structures, although the semiconductor material has been evolved from Si to Ge substrate [6], similar studies on Ge are in the very early stage [7, 8] and the impact of introducing Ge into Si substrate on reliability characteristics has never been reported. In this work, by using epitaxial SiGe on Si substrate with various Ge content as the platform, with MFS structure, the dependence of reliability characteristics of ferroelectric HZO on Ge content ranging from 16 % to 44 % was investigated. The importance of this work lies in two aspects. (a) Bridging the understanding gap between FE-HZO on Si and Ge substrate and (b) Pioneering the in-depth FE-HZO investigation on epitaxial SiGe film, which is the most promising successor to Si for sub-7 nm VLSI technology before Ge process matures. The results indicate that as compared to Si counterpart, FE-HZO on Si$_{0.56}$Ge$_{0.44}$ corresponds to enhanced P$_r$ and improved reliability in terms of more robust endurance up to 10^9 cycles, better retention, and less imprint effect at 85 °C. The Ge content-dependent reliability implies that the reliability issues of MFS devices on Si are alleviated as it migrates to SiGe platform.

Device Fabrication

15-nm epitaxial SiGe film with Ge content of 16 % and 44 % grown on Si substrate was adopted as the platform. After doping the SiGe by P (2×10^{19} cm^{-3}), HZO dielectric of 10 nm was formed followed by 30-nm TiN electrode deposition, both by ALD, to form the MFS structure. Then 30-sec RTA at 500 °C was used to crystallize HZO for inducing ferroelectricity. The same substrate doping and MFS structure was also prepared on Si substrate as reference. Besides electrical measurement to confirm the ferroelectricity, extensive reliability characteristics including endurance, retention and imprint effects were studied for different samples. Physical analyses were also carried out to investigate the mechanism of different behaviors.

Results and Discussion

(A). Conformation of ferroelectricity of HZO on epitaxial SiGe

XRD patterns (**Fig. 1**) reveal the HZO crystallinity for different substrates. Although HZO on all kinds of samples exhibit orthorhombic phase (o-phase), HZO on substrate with higher Ge content shows better crystallinity due to smaller FWHM of the o-phase diffraction peak. All samples were found to have ferroelectricity by measuring polarization (P) and capacitance (C) vs. voltage (V) and doing piezo-response force microscopy (PFM) analysis. **Fig. 2** shows the results for HZO on Si$_{0.56}$Ge$_{0.44}$ after wake-up. Asymmetric coercive voltage (V$_C$)

is due to different top/bottom electrode. **Fig. 3** displays the enhanced P$_r$ by 58 % for Si$_{0.56}$Ge$_{0.44}$ as compared to Si.

(B). Ge content-dependent reliability performance

Fig. 4 shows the dependence of cycling on P-V curves for devices on Si$_{0.56}$Ge$_{0.44}$ by ±4 V/100k Hz bipolar stress. **Fig. 5** displays the endurance performance under the same stress but with different frequencies for all kinds of samples. Abrupt leakage increase indicates generation of a large amount of defects that completely pin domain walls. For Si, although reliability improves as stress frequency increases from 1k to 100k Hz, significant wake-up and fatigue effects are still observed. However, devices on Si$_{0.56}$Ge$_{0.44}$ are almost free from the issues up to 10^9 cycles (100k Hz). **Fig. 6** shows the Arrhenius plot of wake-up for activation energy (Ea) extraction [9]. Ea for Si is 0.58 eV, which is close to the oxygen vacancy (Vo) migration energy near Si/HfO$_2$ interfaces [10]. Si$_{0.56}$Ge$_{0.44}$ with higher Ea of 1.17 eV implies fewer Vo at interface. **Fig. 7** shows the retention of P$_{SW, OS}$ (Switched Polarization, Opposite State) for all samples under pristine state at 25/85 °C. Severer retention degradation is found at 85 °C for Si due to imprint effect, which is inferred by the significant OS read current peak shift (**inset**). The degradation is greatly mitigated for Si$_{0.56}$Ge$_{0.44}$ with stable P$_{SW, OS}$ up to 10^4 sec at 25 °C. However, P$_{SW, SS}$ (Same State) is independent of substrate/temperature (**Fig. 8**). Retention for cycled devices are shown in **Fig. 9** and Si$_{0.56}$Ge$_{0.44}$ still shows stable P$_{SW, OS}$ against time even with 10^6 cycles. Imprint effect is characterized by V$_C$ shift at 25/85 °C vs. retention time and the results for all kinds of samples are shown in **Fig. 10**. Vc shift with retention time is observed for Si and worse degradation is found at 85 °C, consistent with P$_{SW, OS}$. The imprint effect is alleviated by using Si$_{0.56}$Ge$_{0.44}$. As the Ge content in the substrate increases, HZO reliability improves accordingly. It indicates that HZO/substrate interface plays the essential role.

(C). Origin of superior reliability for higher Ge content

The fatigue effect during cycling is due to newly generated defects and injected charges that pin the domain walls. Retention loss is resulted from leakage followed by charge trapping while imprint effect is caused by internal E-field due to trapped charge. The reliability improvement by increasing Ge content can be explained by the thinner sub-oxide interfacial layer (IL) with better quality between HZO/substrate that helps suppress trapping/generation of charges. The inference is evidenced by the XPS Ge 2p and Si 2p spectra (**Fig. 11**). For Ge 2p spectra, the peak at 1220.5 eV corresponds to Ge-O bond in the form of GeO$_x$ (x=1.5~2). However, the GeO$_x$:Ge peak intensity ratio is rather weak and independent of Ge content, indicating less GeO$_x$. It is caused by interdiffusion of ZrO$_2$ in HZO and GeO$_x$, consistent with the result that ZrO$_2$/Ge tends to form nearly zero GeO$_x$ IL after 500 °C annealing [11]. Interdiffusion means Ge incorporation into HZO which is helpful to form o-phase [11] as seen in XRD and explains a larger P$_r$ due to Zr-Ge bond at interface [6]. For Si 2p spectra, compared to Si, SiO$_x$ component on Si$_{0.56}$Ge$_{0.44}$ is also found but with much weaker intensity and stronger oxide quality (BE closer to higher oxidation state Si^{4+} implies more stoichiometric bonding with fewer Vo, consistent with the larger Ea of Si$_{0.56}$Ge$_{0.44}$). EDS line scan also proves Zr-Ge bond for Si$_{0.56}$Ge$_{0.44}$ due to incorporation of Ge at HZO/IL interface (**Fig. 12**). The thinner IL is confirmed by the inset TEM. IL in MFS is critical for reliability [12]. Better reliability for Si$_{0.56}$Ge$_{0.44}$ is due to its IL that is too thin to trap charges and of stronger bonding to suppress defects generation. Smaller E-field across the IL (higher κ with GeO$_x$) also explains the better reliability.

Conclusion

Reliability of FE-HZO was evaluated in MFS structure on epitaxial SiGe film with various Ge content. For devices on Si$_{0.56}$Ge$_{0.44}$, negligible P$_r$ degradation up to 10^9 cycles under bipolar AC stress, desirable retention at pristine and cycled state, and small imprint effect at 85 °C are obtained, much improved as compared to Si counterpart due to IL with thinner thickness, robust quality (fewer Vo) and higher dielectric constant, making less charge trapping and generation during test. The Ge content-dependent reliability suggests that the reliability issues will be alleviated as VLSI technology enters new era of SiGe platform.

2018 Symposium on VLSI Technology Digest of Technical Papers

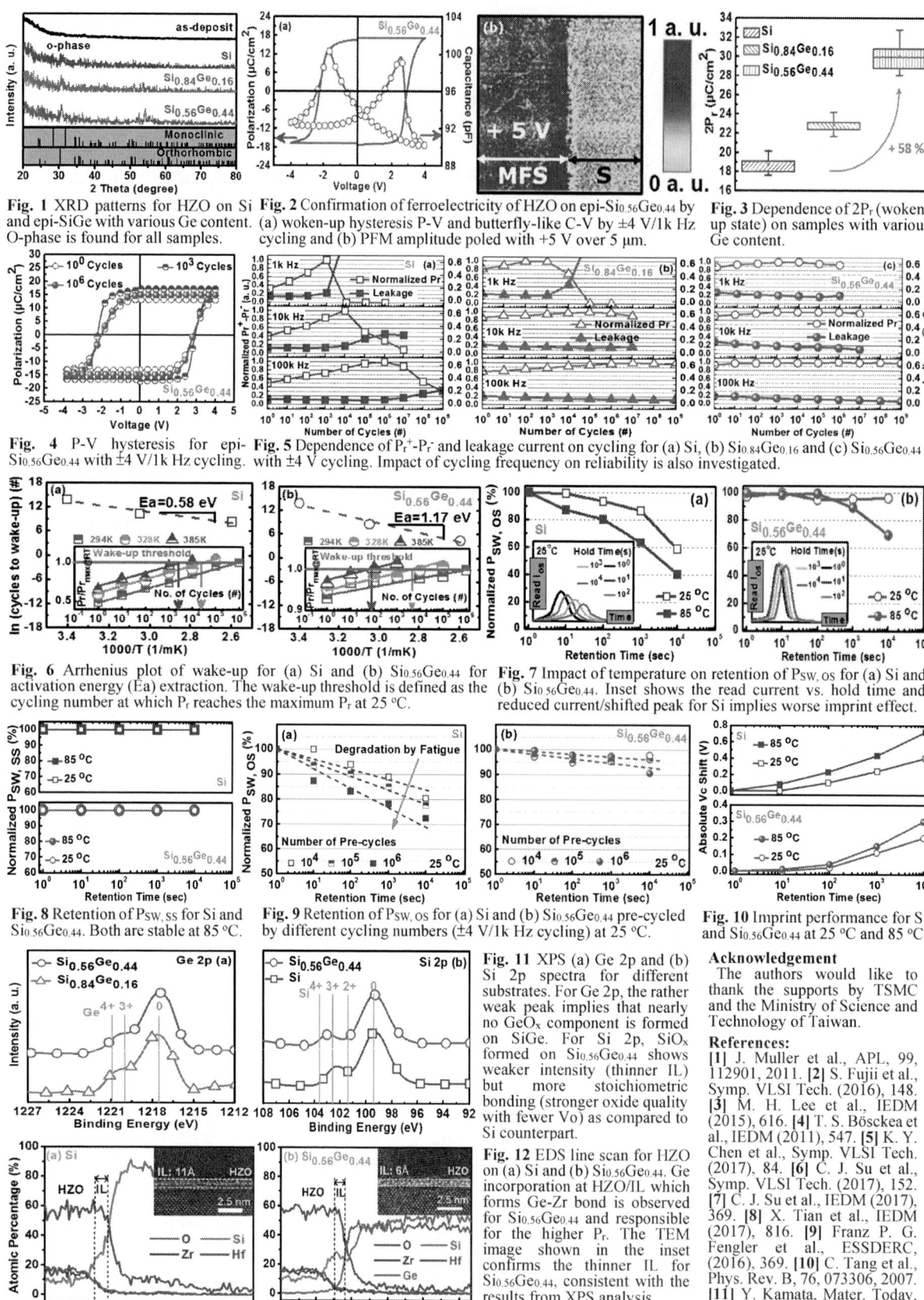

Fig. 1 XRD patterns for HZO on Si and epi-SiGe with various Ge content. O-phase is found for all samples.

Fig. 2 Confirmation of ferroelectricity of HZO on epi-Si$_{0.56}$Ge$_{0.44}$ by (a) woken-up hysteresis P-V and butterfly-like C-V by ±4 V/1k Hz cycling and (b) PFM amplitude poled with +5 V over 5 μm.

Fig. 3 Dependence of 2P$_r$ (woken-up state) on samples with various Ge content.

Fig. 4 P-V hysteresis for epi-Si$_{0.56}$Ge$_{0.44}$ with ±4 V/1k Hz cycling.

Fig. 5 Dependence of P$_r^+$-P$_r^-$ and leakage current on cycling for (a) Si, (b) Si$_{0.84}$Ge$_{0.16}$ and (c) Si$_{0.56}$Ge$_{0.44}$ with ±4 V cycling. Impact of cycling frequency on reliability is also investigated.

Fig. 6 Arrhenius plot of wake-up for (a) Si and (b) Si$_{0.56}$Ge$_{0.44}$ for activation energy (Ea) extraction. The wake-up threshold is defined as the cycling number at which P$_r$ reaches the maximum P$_r$ at 25 °C.

Fig. 7 Impact of temperature on retention of P$_{sw, os}$ for (a) Si and (b) Si$_{0.56}$Ge$_{0.44}$. Inset shows the read current vs. hold time and reduced current/shifted peak for Si implies worse imprint effect.

Fig. 8 Retention of P$_{sw, ss}$ for Si and Si$_{0.56}$Ge$_{0.44}$. Both are stable at 85 °C.

Fig. 9 Retention of P$_{sw, os}$ for (a) Si and (b) Si$_{0.56}$Ge$_{0.44}$ pre-cycled by different cycling numbers (±4 V/1k Hz cycling) at 25 °C.

Fig. 10 Imprint performance for Si and Si$_{0.56}$Ge$_{0.44}$ at 25 °C and 85 °C.

Fig. 11 XPS (a) Ge 2p and (b) Si 2p spectra for different substrates. For Ge 2p, the rather weak peak implies that nearly no GeO$_x$ component is formed on SiGe. For Si 2p, SiO$_x$ formed on Si$_{0.56}$Ge$_{0.44}$ shows weaker intensity (thinner IL) but more stoichiometric bonding (stronger oxide quality with fewer Vo) as compared to Si counterpart.

Fig. 12 EDS line scan for HZO on (a) Si and (b) Si$_{0.56}$Ge$_{0.44}$. Ge incorporation at HZO/IL which forms Ge-Zr bond is observed for Si$_{0.56}$Ge$_{0.44}$ and responsible for the higher P$_r$. The TEM image shown in the inset confirms the thinner IL for Si$_{0.56}$Ge$_{0.44}$, consistent with the results from XPS analysis.

Acknowledgement

The authors would like to thank the supports by TSMC and the Ministry of Science and Technology of Taiwan.

References:

[1] J. Muller et al., APL, 99, 112901, 2011. [2] S. Fujii et al., Symp. VLSI Tech. (2016), 148. [3] M. H. Lee et al., IEDM (2015), 616. [4] T. S. Bösckea et al., IEDM (2011), 547. [5] K. Y. Chen et al., Symp. VLSI Tech. (2017), 84. [6] C. J. Su et al., Symp. VLSI Tech. (2017), 152. [7] C. J. Su et al., IEDM (2017), 369. [8] X. Tian et al., IEDM (2017), 816. [9] Franz P. G. Fengler et al., ESSDERC, (2016), 369. [10] C. Tang et al., Phys. Rev. B, 76, 073306, 2007. [11] Y. Kamata, Mater. Today, 11, 30, 2008. [12] E. Yurchuk et al., IRPS (2014), 2E.5.1.

978-1-5386-4219-1/18 $31.00 © 2018 IEEE

Modeling of FinFET Self-Heating Effects in multiple FinFET Technology Generations with implication for Transistor and Product Reliability

H. C. Sagong, K. Choi, J. Kim, T. Jeong, M. Choe, H. Shim, W. Kim, J. Park, S. Shin, and S. Pae

Foundry Business, Samsung Electronics, Gi-Heung, Korea, 446-771, email: hc.sagong@samsung.com

Abstract

We report the characterization and modeling of FinFET self-heating (FSH) and its reliability impact across multiple FinFET process technology generations. With technology node scaling, taller and narrower Fin shape allows higher performance. However, increased FSH and potential reliability issues must be well understood and mitigated. This paper presents FSH effects across multiple technology nodes and characterization, and modeling efforts used in design will be presented. The results on transistor and product level demonstrate excellent reliability performance beyond 10yrs

Introduction

FinFET's 3D geometry allows vertical scaling while other 2D footprint areas are further squeezed (with dimension scaling) [1]. Fig.1 shows higher aspect ratios of FinFET scaling across multiple technology nodes. The scaling propels the Fin to be more vertical and taller. Some of the areas to think about in terms of reliability and design are (i) fundamental mechanics of having such high aspect ratio structures, and (ii) heat effects due to increasing thermal resistance, in addition to conventional transistor reliability. In this regard, extensive reliability characterization on 14nm and 10nm FinFETs have been were reported [2-3]. This paper focuses more on the overview of FSH and its reliability characteristics in device and circuit level, and product HTOL.

Results and Discussions

Fig.2 depicts concept of heat dissipation in planar transistor and 3D FinFETs. In planar device, heat generated during device operation is dissipated through available bulk (and contacts). Meanwhile on 3D FinFETs, the heat generated at the top of the Fin gets dissipated through the Fin bottom and then into the substrate bulk (as well as contacts). Intuitively, the taller Fin height and narrower Fin bottom width become more difficult to remove the generated heat. Fig.3 shows the FSH across technologies (including DOEs) at fixed power. FSH was characterized using the pulsed IV setup discussed in [3]. Fig.4 shows FSH vs. power across different technology nodes (note that within the technology, Fin shape DOE was performed). It is apparent that as technology is scaled, somewhat "universal" behavior was observed using the below FSH index, (in Fig.7)

$$FSH\ index = \frac{b \times H}{(a \times BCD) \times (c \times CCD)}$$

where, a, b, and c are the corresponding weighting factors according to the Fin process technology, BCD is Fin bottom CD, CCD is Contact CD, and H is Fin height, all in 'nm's.

In addition to Fin scaling, embedded eSiGe is used in advanced logic technology nodes. Higher Ge% content helps improve PFET mobility, thereby its performance. However, it is known that Ge has poorer thermal conductivity than Si, therefore a speculation of PFET being worse in terms of FSH over the NFET has been investigated. In [4], it was found that there's no thermal conductivity change among the Ge content (x=) of 0.3~0.8%, which covers the typical ranges of the Ge% content used in the embedded $eSi_{1-x}Ge_x$ source/drain. Fig.5

shows FSH ratio between N vs. PFETs across many technology nodes (and DOEs). It is shown that there's no apparent FSH increase on PFET w.r.t. the NFET, despite the increase in %Ge contents on advanced nodes. This is also predicted from the TCAD simulations shown in Fig.6a. Extensive additional FSH simulation was carried out with changes in Fin height (H, in Fig.6b), BCD, and Contact sizes. Table I summarizes the FSH simulation results. Fig.7 shows universal line of FSH index, which can be used to predict the future technology nodes and trends for FSH and characterized design kit models can be updated accordingly.

It is discussed that the FSH only matters during device operation and can manifest itself as HCI increase (more severe in PFETs than in NFETs were discussed in detail [5]). Fig.8 shows that both NFET and PFET BTI don't really depend on the Fin shape nor the FSH. Fig.9 shows that I/O NFET and PFET HCI are degraded without FSH de-embedding. But it shows similar aging levels after de-embedding the FSH (from the Fin DOEs). Note that the junctions for various DOEs have been optimized for HCI and performance. The result is shown to illustrate the FSH effect on stress aging. FSH effects have not been very apparent on the Ring Oscillators with their very high frequency operation and voltages around $1/2 \cdot V_{dd}$ during switching transitions even for long durations. For analog circuits with higher duty, FSH modeling can be done with reduced temperature due to longer gate lengths (resulting in lower power, shown in Fig.10). Fig.11 shows FSH induced temperature increases at the metal interconnect layers above the FinFET transistor. Metal 4-point Kelvin structures at various metal layers were placed on top of Fin devices to characterize the FSH. Through simulation, worst-case interconnect structures for both EM and FSH could be identified and additional contacts or dummies (in terms of metal routing) could be added where area doesn't constrain. By characterizing the duty and AC, FSH on high speed logic could be mitigated or can be built into the design considering the aging portion.

Circuit thermal simulation (using Heatwave and HSPICE) indicates that FSH can be reduced and saturated over time under the high frequency (2GHz) operation and 1V shown in example. The simulated thermal image shows the heat profile with heat peak on the active Fin regions and at the Fin top region (Fig.12). Fig.13 shows IP blocks with nominal voltage (lower than 1V), frequency, and duty scenario have about 5~6C (degrees) for the 10nm FinFET nodes. The temperature sensing of an AP product while operating firmware to check for any localized hot spots was carried out to check for FSH, showed low risk. More than several thousands of units have gone through HTOL stress and portions of units up to 1000hrs built on 10nm process exhibiting excellent reliability behavior. The intrinsic TDDB lifetime and voltage acceleration characterization (to be discussed) could support the technology with much severe conditions. The HTOL was ran with two different settings: One with conventional 1MHz HTOL setup while the other setup was with HTOL running at multi-GHz speed, referred to as ASH in [6]. The high-speed ASH showed

978-1-5386-4219-1/18 $31.00 © 2018 IEEE

similar ΔVmin-shifts vs. the conventional stress (in Fig.14) and was still dominated by the NBTI, and not HCI.

Conclusions

Extensive FSH measurements, simulations and reliability characterizations have been conducted across FinFET technology nodes. The higher aspect ratio increases FSH as technology scales. With proper design methodology, such as modeling FSH, accounting for realistic use, and further optimized process, FSH effects can be mitigated in design with scaling FinFETs into 7nm and beyond. Despite taller and narrower Fins, the thermal mechanical integrity was not compromised. Robust HTOL results on SRAMs, IP blocks, and products built in 10nm reliability were demonstrated.

References

[1] H. Cho, *VLSI*, 2016, pp. 1-2 [2] M. Jin, *IEDM*, 2016, pp. 15.1.1-15.1.4. [3] S. Pae, *IEDM*, 2015, pp. 20.6.1 - 20.6.4. [4] H. Stohr, *Z. Anorg. Allgem. Chem*, 241, 1954, p. 305. [5] M. Jin, *IRPS*, 2016, pp. 2A-2-1-2A-2-5. [6] J. Park, *IRPS*, 2013, pp. 3E.2.1-3E.2.4.

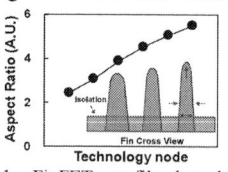

Fig.1. FinFET profile has been scaled to taller and narrower direction for better performance and short channel effects. (Process technology nodes: 14nm to 7nm as shown in Fig.3.)

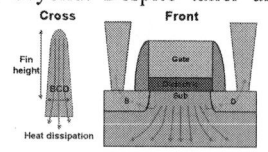

Fig.2. FinFET self-heat, during transistor operation gets dissipated via several nano-meters of Fin width to bulk substrate and also through the contacts.

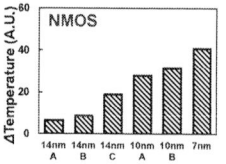

Fig.3. NMOS FinFET self-heating temperature increased with technology scaling (@same power). ΔT_{SH} is characterized with pulsed IV on RF test structures. PMOS exhibits similar trend (not shown).

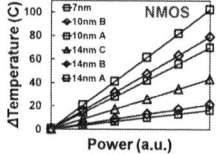

Fig.4. NMOS FinFET self-heat (ΔT_{SH}) vs. power (@$V_{dd}=V_{gs}=V_{ds}$). Note that this should represent the worst-case DC measurements at full I_{dsat}. (A, B, and C refers to process splits within the given technology nodes.)

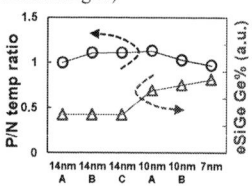

Fig.5. FinFET self-heat temperature (ΔT_{SH}) ratio of PMOS to NMOS FinFET does not show any particular trend across technology nodes studied here. NMOS and PMOS have same Fin dimensions. Also highlighted are the %Ge increase used in eSiGe process as more strain is needed to boost PMOS performance.

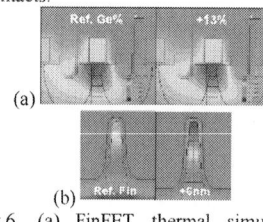

Fig.6. (a) FinFET thermal simulation showing that the amount of Ge% change in eSiGe has negligible effect on the temperature dissipation. This can be explained by the fact that the thermal conductivity is constant, used in advanced technologies. (b) TCAD simulation shows that self-heat temperature increases with Fin height (H) increase, showing +0.46%/+1nm sensitivity.

Table I. Summary of FinFET self-heat temperature simulation by changing process parameters, such as Fin Height (FH), Bottom Fin CD (BCD), Contact CD (CCD), and %Ge used in the source-drain epi (+ sign indicates ΔT_{SH} increase, - sign is for ΔT_{SH} decrease.)

Factors	Sensitivity
FH	+0.46% / +1nm
BCD	-1.63% / +1nm
CCD	-1% / +1nm
Ge%	No change

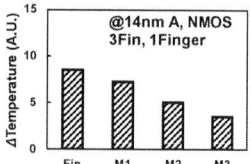

Fig.7. FinFET self-heat temperature has shown a "universal" linear relationship with FSH index. The ΔT_{SH} temperature with a new Fin profile (open symbol for example) can be estimated and modeled.

$$FSH\ index = \frac{b \times H}{(a \times BCD) \times (c \times CCD)},$$

Where a=1.6, b=0.5, and c=1.0 respectively for this study.

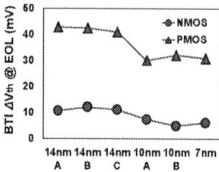

Fig.8. Logic (thin-gate oxide) NMOS and PMOS BTI degradation across the multiple FinFET process technology generations. FinFET self-heating does not influence the BTI (no transistor conduction, just the channel inversion, on-state). BTI VAFs is well correlated with the slope of IV curves where stress is performed (to be discussed).

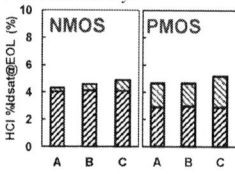

Fig.9. 14nm I/O NMOS/PMOS FinFET HCI degradation before and after self-heat de-embedding. Stress was performed at $V_{gs}\sim1/2V_{ds}$, 125C. Note that the degradation is almost similar after de-embedding indicating same intrinsic reliability behavior. The higher degradation (C>B>A) corresponds to the portion of extra degradation due to FinFET self-heating during stress. NMOS HCI E_a is much lower.

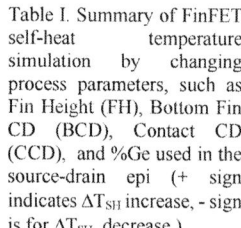

Fig.10. Due to reduced power, longer channel devices have reduced FinFET ΔT_{SH}. This helps mitigate ΔT_{SH} increase in analog circuits that might exhibit higher duty of operation.

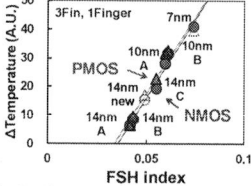

Fig.11. 14nm ΔT_{SH} is also dissipated (note the temperature is further reduced at upper metals) through metal layers in the BEOL interconnects. Test structures with metal probes on top of Fin devices can be used to probe the resistance change that can be converted to temperature change.

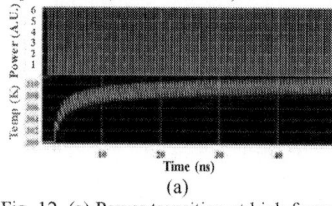

Fig. 12. (a) Power transition at high frequency operation (2GHz and 1V switching) is too fast to induce sufficient FinFET self-heating, resulting in the reduced and saturated heat. (b) In the steady state, thermal image of 3fin transistor shows that self-heat is generated at the fin top and dissipated through the bulk.

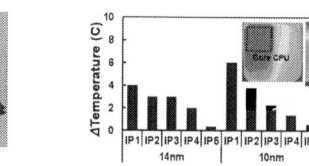

Fig.13. Aging simulation of IP blocks evaluated in self-heating shows at most 5-6°C increase considering their duty during operation. (Inset) 10nm core CPU under accelerated heavy-use GHz operation shows max +20°C increase which is expected (and is not directly due to self-heating).

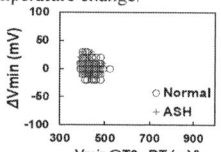

Fig. 14. Die-to-die level ΔVmin-shift of up to 1000hrs stress on 10nm product. Die to die shift was within 30-50mV. The high frequency (ASH) stress showed no difference vs. 1MHz (Normal).

978-1-5386-4219-1/18 $31.00 © 2018 IEEE

Gap in pagination due to formatting issues.

Pages 123-124

All-Electrical Control of a Hybrid Electron Spin/Valley Quantum Bit in SOI CMOS Technology

L. Hutin[1*], L. Bourdet[2], B. Bertrand[1], A. Corna[2], H. Bohuslavskyi[1,2], A. Amisse[1,2], A. Crippa[2], R. Maurand[2], S. Barraud[1], M. Urdampilleta[3], C. Bäuerle[3], T. Meunier[3], M. Sanquer[2], X. Jehl[2], S. De Franceschi[2], Y.-M. Niquet[2*], M. Vinet[1]

[1] CEA, LETI, Minatec Campus, F-38054 Grenoble, France

[2] CEA, INAC, F-38054 Grenoble, France [3] CNRS, Institut Néel, F-38042 Grenoble, France

*e-mail: louis.hutin@cea.fr; yann-michel.niquet@cea.fr

Abstract

We successfully demonstrated experimentally the electrical-field mediated control of the spin of electrons confined in an SOI Quantum Dot (QD) device fabricated with a standard CMOS process flow. Furthermore, we show that the Back-Gate control in SOI devices enables switching a quantum bit (qubit) between an electrically-addressable, yet charge noise-sensitive configuration, and a protected configuration.

Introduction

Following the emergence of Silicon spin qubits as serious contenders in the race for quantum computation [1], we have recently demonstrated two-axis control of the first hole spin qubit in Si transistor-like structures using a CMOS technology platform [2,3]. Tunnel barriers are defined by protecting the SOI film from self-aligned ion implantation between dense Gates (64nm pitch) by larger-than-usual SiN spacers (typically 30nm), thus leading to a linear arrangement of Gates along an intrinsic NanoWire [4] (**Fig. 1**). At very low temperatures (~1K and below), each Gate defines a QD with a discrete energy spectrum, which can be used to confine a small number of charges controlled by the Coulomb blockade effect (**Fig. 2**). Making a qubit out of a QD entails the ability to initialize and manipulate a two-level quantum state of a single charge, such as spin-down $|\downarrow\rangle$ and spin-up $|\uparrow\rangle$.

Inducing Electron Spin Resonance (ESR) with an RF magnetic-field is the most straightforward approach to spin control (**Fig. 3**), although the excitation is hardly applied locally, as opposed to an all-electrical control scheme. The latter however requires a way to couple the spin of a charge to its orbital motion. Unfortunately in Si electrons, unlike holes, have generally weak intrinsic Spin-Orbit Coupling (SOC). Electrical Dipole Spin Resonance (EDSR) can in principle be achieved by placing the charge in a magnetic-field gradient produced by a micromagnet [5], though more compact and scalable alternatives are desirable.

Device and definition of quantum states

Our test device for EDSR demonstration (**Fig. 4**) consists of a Two-Gate nFET-like structure with Gates partially wrapping around the [110]-oriented SOI NanoWire (W=30nm; T_{Si}=12nm). We consider two QDs, QD1 and QD2 confined in the "corners" defined by G_1 and G_2. If both are in the same spin state (e.g. parallel spins, which is the ground state in a finite magnetic field B), Pauli's exclusion principle prevents charge movement from QD1 to QD2, and hence I_{DS} current from flowing. However, a spin rotation obtained by applying a resonant RF E-Field to G_1 would lift the Pauli Spin Blockade and enable a non-zero current. Spin degeneracy is lifted by means of an externally-applied static magnetic field, the splitting energy being $E_Z=g.\mu_B.B$ where g is the Landé g-factor (g≈2 for electrons in Si) and μ_B the Bohr magneton. The principle of resonant spin transitions, and the corresponding expected ESR signal are shown on **Fig. 5**. Yet, the additional valley degree of freedom needs to be considered. The conduction band of bulk Si features six degenerate Δ valleys. Structural and electrical confinement in our device, however, leaves two low-lying valleys v_1 and v_2, projected in Γ and separated by an energy Δ_V (**Fig. 6**). From these two valleys, four distinct states can be resolved upon applying a static magnetic field: $|v_1, \downarrow\rangle, |v_1, \uparrow\rangle, |v_2, \downarrow\rangle$ and $|v_2, \uparrow\rangle$.

Corner Dots and spin-valley mixing

Of particular interest are the two states $|v_1, \uparrow\rangle$ and $|v_2, \downarrow\rangle$, which may be mixed under the condition that an inter-valley spin-orbit (SO) coupling C_{v1v2} in the Hamiltonian is non-zero. As illustrated in **Fig. 7**, this criterion is fulfilled if the mirror symmetry of the electron wavefunction with respect to the (XZ) plane is broken. The partially overlapping Gate leading to the "Corner Dot" confinement is therefore the key to spin-valley-orbit mixing in this case.

As B is increased and the spin splitting $E_Z=g.\mu_B.B$ approaches the valley splitting Δ_V, the $|v_1, \uparrow\rangle$ and $|v_2, \downarrow\rangle$ energies may cross either cross (no coupling) or anticross ($C_{v1v2} \neq 0$). In the former case (**Fig. 8a**), only spin-preserving inter-valley transitions can be expected in response to pure E-field excitations. In the latter case, due to states mixing near the anticrossing, B-dependent spin/valley transition diagonals may add-up to the EDSR signal (**Fig. 8b**). A color plot of I_{DS} measured in a cryostat at T=15mK vs. E-field frequency and B clearly shows spin resonance lines (**Fig. 8c**). This is to our knowledge the first experimental measurement of micromagnet-free resonant E-field manipulation of electron spins in Si QDs [6].

Programming a valley state, encoding a spin state

Since the splitting between v_1 and v_2 is related to charge confinement close to an interface, it is possible to tune Δ_V by modulating the vertical electric field. This was shown in [7] using coplanar side Gates on bulk Si, but SOI offers the possibility of using the Back-Gate potential V_b. We calculated the $\Delta_V(V_b)$ energy dependence using a Tight Binding model for the valley and the SO coupling at the atomistic level [8]. The results are shown in **Fig. 9** together with corresponding plots of the electron wavefunction. The tunability of Δ_V can be leveraged as schematized on **Fig. 10**: adiabatically changing V_b allows following the lower branch past the anticrossing and transitioning continuously from $|v_1, \uparrow\rangle$ to $|v_2, \downarrow\rangle$. If one defines the qubit basis states $|0\rangle$ as $|v_1, \downarrow\rangle$ and $|1\rangle$ as this hybridized lower branch, V_b enables to switch between a pure spin regime and a pure valley regime. The advantage of a valley qubit is the all-electrical addressability of inter-valley transitions, the downside being sensitivity to charge noise and hence shorter decoherence times. Conversely, when in spin regime, the qubit is scarcely addressable electrically but benefits from a longer lifetime.

This approach leads to circumventing a trade-off between qubit manipulation speed and coherence time, thus improving the number of operations/error. Advantageously, the qubit rotation speed is maximal when the charge is pulled away from the interfaces, which is more difficult to achieve by using only coplanar Front Gates [9]. **Fig. 11** shows the simulated chronograms of the electrical RF Gate1 excitation signal (ν = 23.66 GHz), the resulting Rabi oscillations of the qubit (f_{Rabi} = 80 MHz) in valley mode, and the eventual spin rotation as V_b adiabatically ramps past the anticrossing back to spin mode. We accounted for local surface roughness variability to estimate the impact of Δ_V fluctuations on the optimal operating V_b range (**Fig. 12**), which can be individually calibrated for each qubit with separate Back-Gates.

Conclusions

We induced spin transitions in MOS Gate-confined electrons in a Si NW using only E-field excitations and without resorting to co-integrated micromagnets. The underlying mechanism is based on the interplay between Spin-Orbit Coupling (SOC) and the multi-valley structure of the Si conduction band, and is enhanced by the "Corner Dot" device geometry. By offering the ability to break and restore the confinement symmetry at will, the SOI Back-Gate permits fast programming in valley mode, and information storage in spin mode. This functionality could alleviate the trade-off between fast manipulation and long coherence time, thereby improving the outlook for compact, scalable and fault-tolerant quantum logic circuits.

References

[1] M. Veldhorst et al., Nature 526 (2015) [2] L. Hutin et al., VLSI 2016 [3] R. Maurand et al., Nature Comm., 7 (2016) [4] S. De Franceschi et al., IEDM 2016 [5] E. Kawakami et al., Nature Nanotechnol., 9 (2014) [6] A. Corna et al., NPJ Quantum Inf. (2018) [7] C.H. Yang et al., Nature Comm., 4 (2013) [8] Y.-M. Niquet et al., Phys. Rev. B, 79 (2009) [9] W. Huang et al., Phys. Rev. B, 95 (2017)

Acknowledgement

The authors acknowledge financial support from the EU under Project MOS-QUITO (No. 688539).

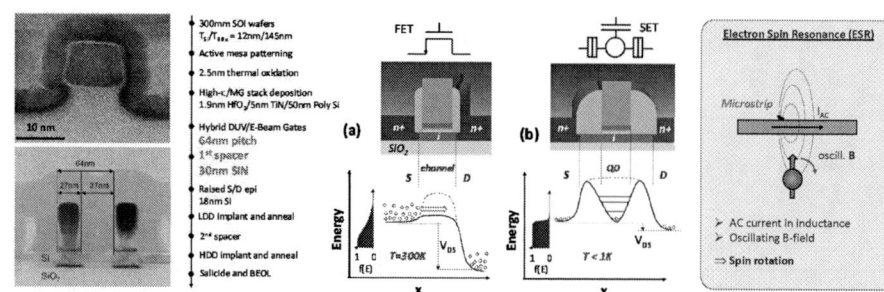

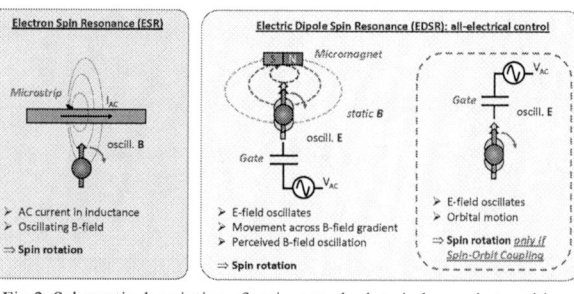

Fig.1: Top left: STEM view along the Gate wrapping around the Si channel. Bottom left: STEM view of two Gates in series (64nm pitch) showing the width of the 1st spacer. Right: simplified process flow.

Fig.2: Energy profile along the channel of (a) an SOI FET at 300K in which carriers flow continuously above a lowered barrier (b) an SET operating in the Coulomb Blockade regime at low T due to large tunnel barriers beneath the spacers and energy quantization in the Gate-defined Quantum Dot (QD).

Fig.3: Schematic description of various methods to induce spin transitions for a localized charge. Magnetic field manipulation is physically straightforward but requires flowing an AC current through a microstrip in the vicinity of the targeted spin. Creating a B-field gradient through a micromagnet enables indirect control via electric field. A more scalable, all-electrical micromagnet-free approach is possible in the case of strong spin-orbit coupling, which in Si usually applies to holes but not electrons.

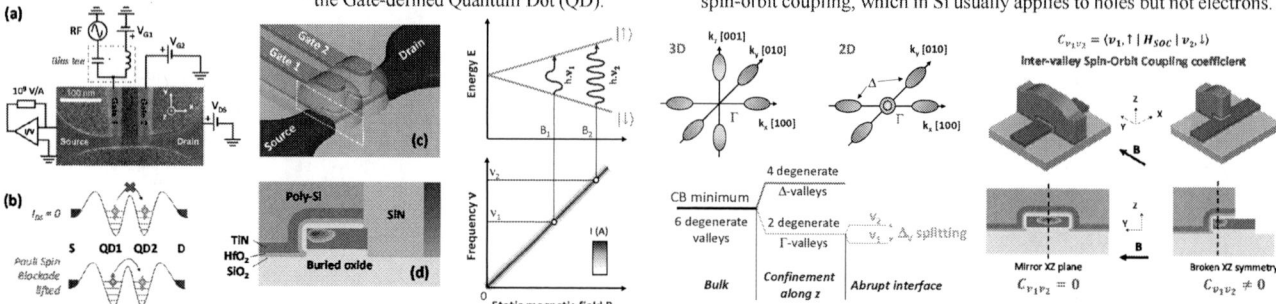

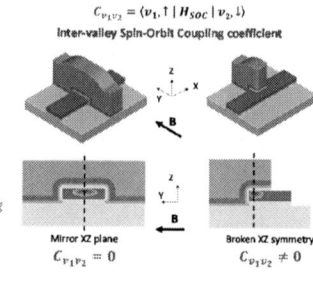

Fig.4: (a) Top-view SEM of the two-Gate device after Gate patterning, and setup description. (b) Spin-filtering mechanism across the Double QD based on the Pauli Spin Blockade rectifying the Drain current. (c) Schematic view of the partially wrapping Gates. (d) Cross-section along a Gate and representation of the asymmetrical electron wavefunction along the mesa edge, or "Corner Dot".

Fig.5: Principle of Zeeman splitting of the degenerate spin states, and of resonant transitions. Bottom shows a typical E(D)SR signal: current is prevented by the Pauli Spin Blockade except on the transition line.

Fig.6: Valley splitting in a 2D-confined configuration in Silicon. The originally sixfold degenerate Δ valleys split into four Δ in the plane of confinement, and two Γ. The abrupt interface further splits the two Γ valleys into v_1 and v_2 by an energy noted Δ_V in the following.

Fig.7: Impact of device geometry on inter-valley Spin-Orbit Coupling. The coupling term C_{v1v2} is non-zero if the symmetry of the electron wavefunction is broken the (XZ) plane. This condition is fulfilled in the case of Corner Dots.

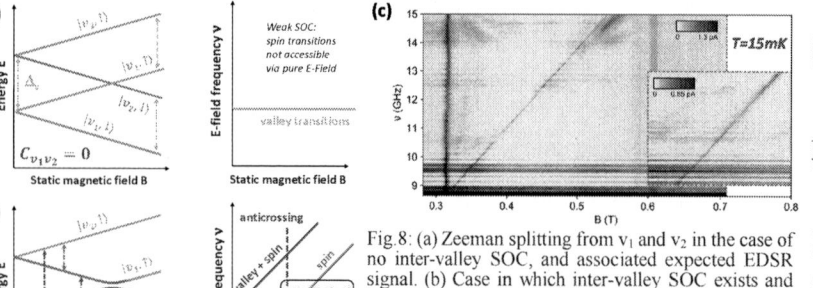

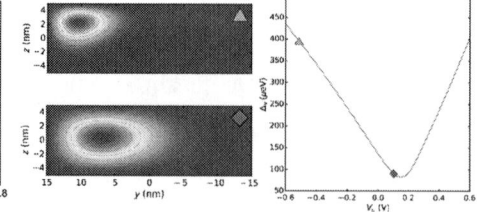

Fig.8: (a) Zeeman splitting from v_1 and v_2 in the case of no inter-valley SOC, and associated expected EDSR signal. (b) Case in which inter-valley SOC exists, and states anti-cross, and expected EDSR. The dotted frame symbolizes the measured region. (c) Experimentally measured EDSR signal, showing spin and spin/valley transitions.

Fig.9: Simulated influence of the SOI Back-Gate voltage V_b on the Δ_V valley splitting for an ideal device (no surface roughness). Δ_V is maximal when the charge is confined against an interface. Positive V_b tends to pull the wavefunction towards the center of the NanoWire, away from the interfaces. A further V_b increase results in increasing Δ_V again, due to charge confinement against the interface with the buried oxide.

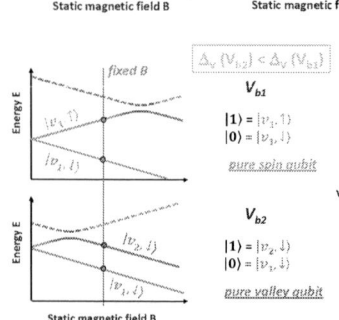

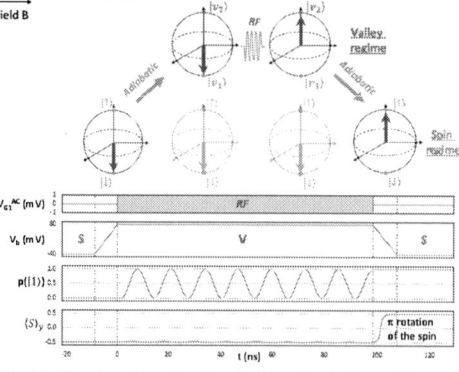

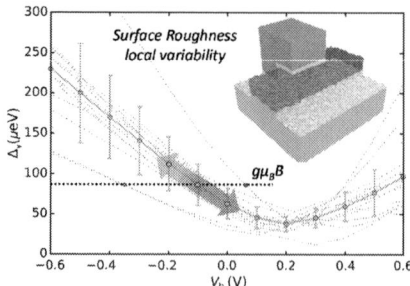

Fig.10: Energy diagram showing two V_b configurations. At a given B, changing V_b adiabatically enables to switch between a spin qubit and a valley qubit regime, by operating respectively left and right of the anti-crossing.

Fig.11: Simulated purely electrical manipulation of the spin of a confined electron. A V_b ramp brings the qubit in the valley regime, in which it can oscillate (f_{Rabi} = 80MHz) in response to an RF E-field excitation (here ν=23.66 GHz). As the V_b ramp is reversed, the $|1\rangle$ eigenstate transitions from $|v_2,\downarrow\rangle$ to $|v_1,\uparrow\rangle$, thus leading to a π rotation of the spin.

Fig.12: Impact of local surface roughness variability on the $\Delta_V(V_b)$ dependence (RMS 0.4nm). The spreading tends to be less severe near the Δ_V minimum, so the magnetic field can be chosen to operate close to this point. As $\Delta_V=g\mu_B.B$ defines the anticrossing, traveling up the curve leads to the spin regime, and down to the valley regime. Separate back-Gates for each qubit would enable to adjust individually the V_b range to toggle between the spin and valley regimes.

978-1-5386-4219-1/18 $31.00 © 2018 IEEE

High-Density and Fault-Tolerant Cu Atom Switch Technology Toward 28nm-node Nonvolatile Programmable Logic

R. Nebashi, N. Banno, M. Miyamura, Y. Tsuji, A. Morioka, X. Bai, K. Okamoto,
N. Iguchi, H. Numata, H. Hada, T. Sugibayashi, T. Sakamoto, and M. Tada

NEC Corporation, Tsukuba, Ibaraki, Japan.
E-mail: r-nebashi@ak.jp.nec.com

Abstract

Key device/circuit technologies for realizing a 28nm-node atom switch programmable logic (AS-PL) have been developed. An advanced polymer solid-electrolyte (PSE) reduces set voltage down to 1.6 V while ensuring ON-state and OFF-state reliabilities under current and voltage stress at 125°C. A fine-grain redundancy in a cross-bar array also contributes to reduce supply voltage by 6%. A routing-based wear leveling improves programming cycles by nine times. The developed technologies allow us to design the 28nm-node AS-PL with a 32% higher performance and 11% lower power.

Introduction

Programmable logics (PLs), e.g., FPGAs in energy-constrained IoT devices require a high energy efficiency to enlarge device lifetime. Nonvolatile switch technologies such as resistive RAM [1] and atom switch (AS) [2] are strong candidates to improve energy efficiency. The AS-PL has achieved 2x logic density (Fig.1), 3.8x operation speed, and 3x power efficiency. However, the logic capacity of the nonvolatile-switch based PL is still small and limits their addressable market. Extending the logic capacity toward 1M LUTs in a 28nm-node poses several challenges (Fig.2). A first challenge is to reduce the set voltage (Vset). Continued reduction is needed to meet less-tolerant 28nm-node transistors. Variability and defects also become increasingly important to ensure sufficient chip yield with increasing the number of integrated switches up to sub-Gbit. The other challenge is an improvement of cycle endurance of the AS-PL chip. This paper presents key technologies to overcome these challenges, which are demonstrated on a 40nm-node CMOS.

Atom Switch Programmable Logic (AS-PL)

The AS with a superior ON/OFF resistance ratio is integrated between the M4 and M5 layers on CMOS (Figs.3,4). The post AS process uses a standard BEOL process. Totally nine Cu layers can provide sufficient routing/power resources in the 28nm-node AS-PL. The routing switch that we used is a complementary atom switch (CAS) where two ASs are serially connected to enhance OFF state reliability and a select transistor is used for programming [3]. The AS turns to ON or OFF by forming or annihilating a conductive Cu bridge between PSE and Ru-alloy electrode. An advanced PSE is introduced and the Vset is sufficiently reduced (Fig.5). Both of OFF-state (Fig.6) and ON-state (Fig.7) lifetimes are estimated at 125°C for 10 years, in which the allowable maximum voltage of 1.8 V and current of 100 μA are extracted. These characteristics satisfy requirements for the 28nm-node AS-PL design.

A 34k-LUT AS-PL chip is fabricated with 51 Mbit of ASs (Fig.8). The AS-PL has logic cells, each of which comprises multiple logic elements (LEs) with their own LUT, and multiplexers (MUXs) for signal routings among the LEs (Fig.9).

Fine-grain Redundancy

The chip features a fine-grain redundancy circuit comprising a spare MUX and selector every 8 MUXs. When a defective AS makes the MUX disable, a cross-point connecting spare MUX is programmed and transfers the signal. In addition, AS-based memories are also implemented at each logic cell to activate the shift function of the selectors. The developed redundancy circuit is implemented with small overhead of ~13% in the logic cell area.

To demonstrate an effect of the fine-grain redundancy, the chip yield is evaluated first, where the redundancy is disabled. The yield increases with supply voltage for writing (Vddw) (Fig.10). The measurement results are reasonably consistent with the estimations from the Vset distribution. Second, the yield enhancement by the redundancy is clarified at the failure rate of 10^{-6} on average. All failure bits are repaired when the redundancy scheme is applied (Fig.11). The measured results are well fit on the estimations from a Poisson distribution. The redundancy can reduce the minimum Vddw by 6% in a 28nm-node PL chip with 1-Gbit ASs (Fig.12).

Routing-based Wear Leveling (WL)

We introduced a wear cost, which was calculated from a cost function based on the count of set/reset cycles for each AS (N_{AS}). In a routing step, nets are routed by minimizing a total cost so that the set/reset count of ASs are leveled evenly across the entire chip (Fig.13). For comparison, 100 different configuration data with WL or without WL are generated by the developed tool. Then, the distribution of the set/reset count of ASs (N_{AS}) is derived after 100 programming cycles of PL (N_{PL}). The WL method reduces maximum N_{AS} from nine to one. Assuming that an allowable limit of N_{AS} is around 10^3 [3], we get N_{PL} of 60k with no penalty in delay and power. Otherwise, N_{PL} of 100k is obtained with the expense of a 39% delay and 32% power overhead, respectively.

28nm-node AS-PL

A chip performance of delay and power in the 28nm-node AS-PL is simulated by SPICE including the aforementioned circuit technologies (Fig.14). The routing path of the chain pattern is configured to transfer data between neighborhood logic cells. The cell-to-cell delay and power consumption at the 28nm-node are expected to be improved by 32% and 11% as compared to the 40nm-node one [2].

Conclusion

Table I summarizes the results obtained. The developed technologies enable the ultra-low power design of AS-PL in 28nm-node.

Acknowledgement

A part of this work was supported by NEDO.

References

[1] J. Cong et al., NANOARCH, pp. 1-8 (2011). [2] X. Bai et al., VLSI Tech., pp. 28-29 (2017). [3] M. Tada et al., IEDM, pp. 689-692 (2011). [4] Y. Tsuji et al, VLSI Circ., p. 16 (2016). [5] M. Miyamura et al., FPGA, p. 236 (2015). [6] M. Miyamura et al., ISSCC, p. 228 (2011).

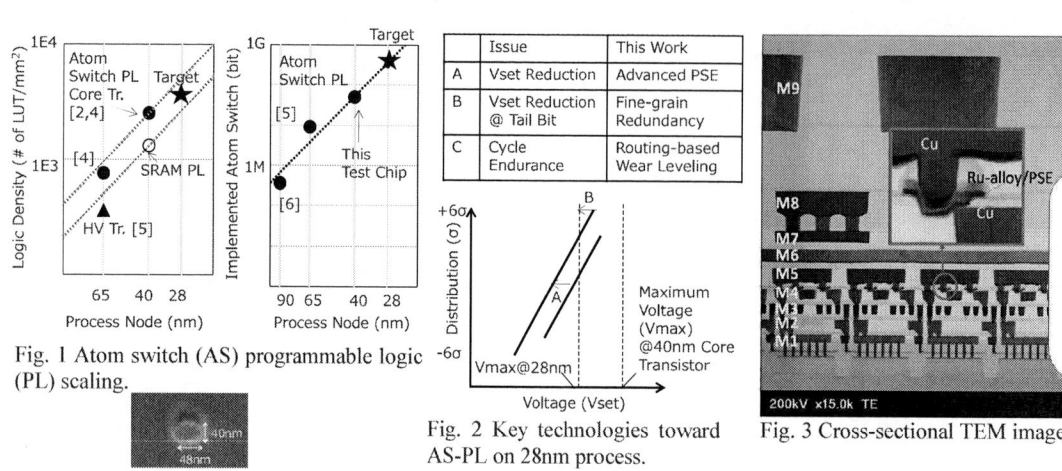

Fig. 1 Atom switch (AS) programmable logic (PL) scaling.

Fig. 2 Key technologies toward AS-PL on 28nm process.

Fig. 3 Cross-sectional TEM images of atom switch.

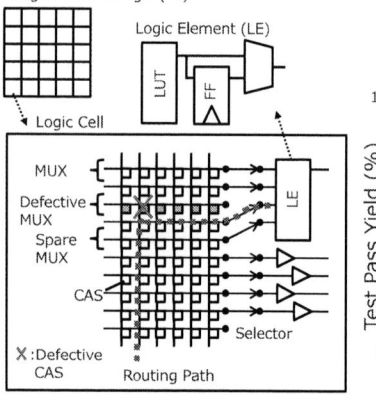

Fig. 4 Top-view SEM image of AS.

Fig. 5 Switching characteristics of atom switch with advanced PSE.

Fig. 6 Off-state lifetime versus DC stress voltage for CAS with advanced PSE.

Fig. 7 On-state lifetime versus DC current for CAS with advanced PSE, and On-state lifetime probability distribution.

Fig. 8 Die photo, shmoo of application (image compression) and table of test chip. No fail bit and no degradation in delay were observed after 150°C 1h baking.

Process	40 nm CMOS with 9 Metals
IP	1.5 Mbit BRAM, PLL
# of LUTs	34k
# of ASs	51 Mbit

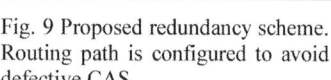

Fig. 9 Proposed redundancy scheme. Routing path is configured to avoid defective CAS.

Fig. 10 Test pass yield dependence on supply voltage. Mapping pattern with 0.8Mb was programmed.

Fig. 11 Yield improvement by redundancy. Mapping pattern with 0.1Mb was programmed.

Fig. 12 Estimated minimum supply voltage for 28nm-node AS-PL.

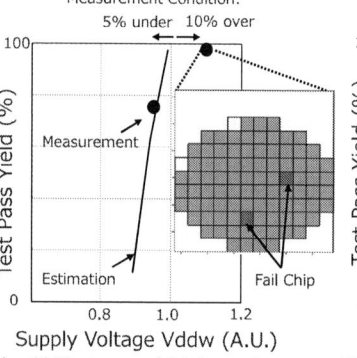

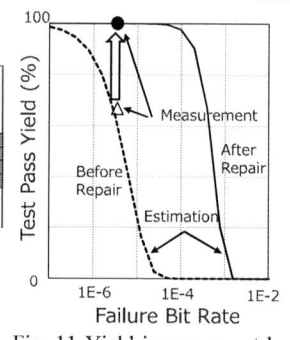

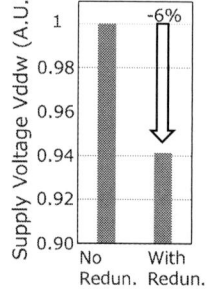

Fig. 13 (a) Concept and (b) tool flow of routing-based wear leveling (WL). (c) Histogram of set/reset count of AS. (d) Static timing analysis (STA) simulation results.

Fig. 14 SPICE simulation for data transfer on 28nm-node AS-PL.

Table I. Summary.

	VLSI`17[2]	This Work
Target Technology-node	40nm-CMOS	28nm-CMOS (Demonstrated in 40nm)
Device	1P7M	1P9M
# of Atom-switch in test chip	4Mb	51Mb
Off-state Reliability (Voltage stress)	-	Vstress_max >1.8 V @ 6σ, 125℃, 10 years ※estimated
On-state Reliability (Current stress)	-	I_max_ave > 100 μA @ 6σ, 125℃, 10 years ※estimated
Supply Voltage (Vddw)	-	6% Reduction @1Gbit (Fine grain redundancy)
Cycle Endurance	1k cycles per device	60k cycles per chip (Routing base wear leveling)
Delay(ns) ※simulated	0.66	0.45 (-32%)
Power(μA/MHz/cell) ※simulated	0.26	0.23 (-11%)

A Threshold Switch Augmented Hybrid-FeFET (H-FeFET) with Enhanced Read Distinguishability and Reduced Programming Voltage for Non-Volatile Memory Applications

M. Jerry[1*], A. Aziz[2*], K. Ni[1], S. Datta[1], S. K. Gupta[2], N. Shukla[3]

[1]University of Notre Dame, Notre Dame, IN 46656, USA; [2]Purdue University, West Lafayette, IN 16802, USA; [3]University of Virginia, Charlottesville, VA-22902, USA; Phone: +1-434-924-6587; Email: ns6pf@virginia.edu

Abstract– In this work, we demonstrate a novel Hybrid-FeFET (H-FeFET) that leverages the threshold switching characteristics of Ag/HfO_2 to overcome the fundamental trade-off between memory window (MW)/read current ratio ($I_{read,1}/I_{read,0}$), and program voltage (V_{prog})/maximum electric-field in standard FeFETs for non-volatile memory application. The H-FeFET incorporates the threshold switch (TS) in the source of the FeFET, and is designed to exhibit a ferroelectric state-dependent volatile HRS to LRS transition ($I_{ON}/I_{OFF} > 10^7$) – during read, the TS turns ON only if the FeFET is in the low-V_T SET state, and remains OFF if the FeFET is in the high-V_T RESET state, thus, selectively suppressing the RESET read current. Leveraging this principle, the H-FeFET: (a) Demonstrates 77% higher MW and 1000× larger $I_{read,1}/I_{read,0}$ compared to the FeFET, at iso-V_{prog} (DC); (b) Enables 25% reduction in V_{prog} at iso-$I_{read,1}/I_{read,0}$ during pulse operation-facilitated by the 8× improvement in $I_{read,1}/I_{read,0}$; (c) Exhibits 2.5× reduction in programming power at iso-$I_{read,1}/I_{read,0}$ in the H-FeFET-based AND array architecture, as shown by simulations. Thus, the H-FeFET overcomes the FeFET design challenges while retaining its existing advantages, making it a promising candidate for non-volatile memory applications.

Introduction. The recent discovery of a ferroelectric phase in doped Hafnia (HfO_2) [1] has generated immense interest in evaluating 1T-FeRAM [2][3] as a potential candidate for high-speed and dense non-volatile memory owing to its scalability and CMOS compatibility. However, practical demonstration of a 1T-FeRAM **(Fig. 1(a))** is impeded by the trade-off between the memory window (MW) (and read-current ratio $I_{read,1}/I_{read,0}$), and the maximum E-field (across the interlayer- IL) and V_{prog}. As shown in **Fig. 1(b)(c)**, a larger MW entails a larger V_{prog}, and thus, higher E-field across the IL- likely to result in its breakdown [4]. To overcome this challenge, we propose and demonstrate a *Hybrid-FeFET (H-FeFET) that leverages the volatile threshold resistance switching exhibited by $Ag/HfO_2/Pt$ to augment the FeFET read current ratio*. Consequently, this also enables V_{prog} reduction.

H-FeFET Operating Principle. The H-FeFET operation involves ferroelectric state-dependent transition [5] of the Ag/HfO_2 TS. During the read operation, the Ag/HfO_2 TS is designed such that it turns ON (LRS) only when the FeFET is in the low-V_T SET state; no transition is induced when the FeFET is the high-V_T RESET state. Such *state-dependent* TS behavior can be engineered because in the low-V_T SET state, the FeFET channel resistance is much smaller than the HRS of the TS. Consequently, almost all the V_{DS} drops across the TS, facilitating the HRS to LRS transition [6]. In contrast, when the FeFET is in the high-V_T RESET state, the higher channel resistance ensures that the voltage drop across the TS is insufficient to induce the transition. Thus, the RESET state read current is suppressed with negligible reduction in the SET read current **(Fig. 2)** resulting in an enhanced $I_{read,1}/I_{read,0}$ ratio.

H-FeFET Characteristics. Fig. 3(a) shows the experimentally measured DC I-V characteristics of the Ag/HfO_2 TS exhibiting volatile switching with ON/OFF ratio $> 10^7$. The I_{DS}-V_{GS} characteristics of the corresponding H-FeFET are shown in **Fig. 3(b)** along with the stand-alone FeFET for comparison. It is can be observed that the H-FeFET exhibits a memory window (MW) of 1.6V (@1µA), and $I_{read,1}/I_{read,0} > 10^5$ which corresponds to a 77% and 1000× improvement **(Fig. 3(c))**, respectively, compared to the stand-alone FeFET. **Fig. 4** shows the measured (DC) I_{read} for the SET and RESET states over 1000 samples, which demonstrate excellent read current stability. We also evaluate the H-FeFET performance during pulse-programing operation. **Fig. 5(a)** shows the programming scheme as well as the evolution of the TS during the program/read operation which illustrates the state-dependent TS transition. **Fig. 6(b)** compares the I_{DS}-V_{GS} characteristics during read for the H-FeFET& FeFET, and shows the suppression of the RSESET state in H-FeFET. It is to be noted that the since RESET state read current is suppressed below 1µA, the effective MW at 1µA would be greater than V_{prog}. **Fig. 7** shows the evolution of the $I_{read,1}/I_{read,0}$ for the H-FeFET and FeFET as a function of V_{prog}. It is evident that the H-FeFET not only offers a larger $I_{read,1}/I_{read,0}$ at iso-V_{prog} but also enables reduction of the V_{prog} at iso- $I_{read,1}/I_{read,0}$.

1T-H-FeRAM Array Analysis. We evaluate the implementation of the 1T-H-FeFET based AND arrays using the adapted 45nm technology node model **(Fig. 8(a))**. The corresponding program/read scheme for the H-FeFET is shown in **Fig. 8(b)**. The integration of the TS entails that unlike in FeFET-based arrays, a finite V_{DS} is required on the bit-line while programming the cell. The TS not only enhances the $I_{read,1}/I_{read,0}$ ratio of the target cell, but also minimizes the current through the unassessed cells. **Fig. 9** shows the V_{prog} & program power for the respective devices in the AND array at iso-$I_{read,1}/I_{read,0}$ ratio; the H-FeFET exhibits a 3.1×/ 1.25×, 2.5× reduction in SET/ RESET program voltage, and program energy (SET + RESET), respectively.

Conclusion. In summary, we have demonstrated for the first time, a novel Ag/HfO_2-TS augmented H-FeFET that exhibits (a) 1000× improvement in $I_{read,1}/I_{read,0}$; (b) 25% V_{prog} reduction at iso-$I_{read,1}/I_{read,0}$. - in comparison to the stand-alone FeFET; (c) Further, using simulations, we show the ability to achieve 2.5× reduction in program energy at scaled nodes. Thus, our work provides a new route to enabling scaled ferroelectric-based NVM technology.

References: [1] J. Muller *et al.*, IEDM, 2013 [2] Mulaosmanovic et al. IEDM, 2015 [3] S. Dünkel *et al.*, IEDM 2017; [4] K. Ni submitted to TED (2017); [5] N. Shukla *et al.*, Nat. Comm. 7812 (2015) [5] A. Aziz *et al.*, ISLPED 2015; [6] A. Aziz *et al.*, TED 64.3 1350-1357. (2017). [6] J. Chow, et al., Symp. VLSI Circuits. Dig. Tech. Pap., 2004. [7] M. Ullmann et al., Integrated Ferroelectrics, 34:1-4, 155-164 (2001).

2018 Symposium on VLSI Technology Digest of Technical Papers

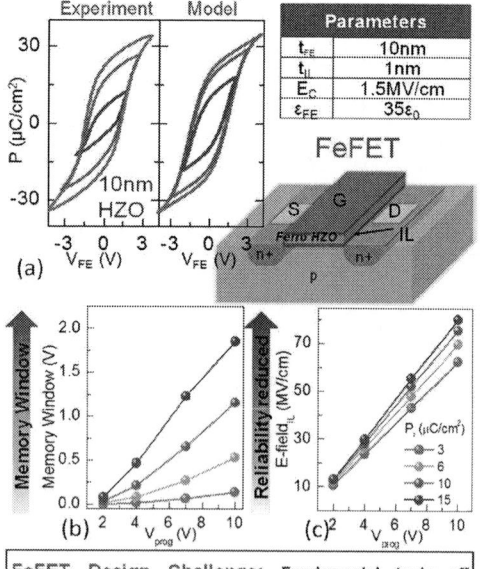

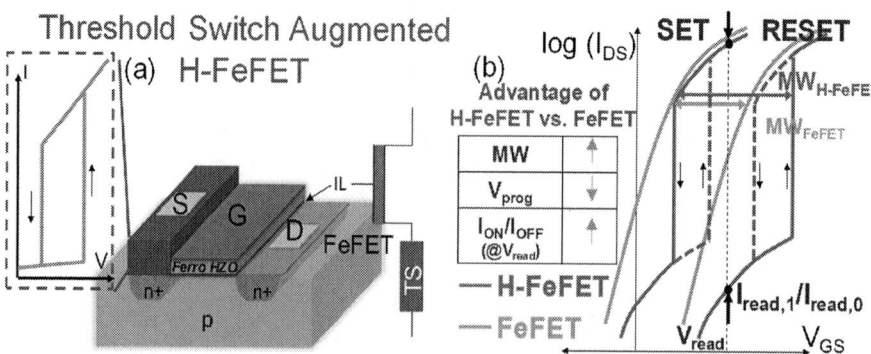

Fig. 1 (a) Schematic of FeFET along with the experimental and simulated P-V_{FE} characteristics (b) Simulated Memory Window (MW), and (c) E-field (across the interlayer IL) as a function of program voltage (V_{prog}) to illustrate the FeFET design challenge. A larger MW can be enabled by increasing V_{prog} but this also significantly increases the E-field across the IL (model adapted from [6]).

FeFET Design Challenge: Fundamental trade off between MW (*performance*) vs. V_{prog} & max E-field across interlayer (*lower reliability & endurance*).
Objective of Work: Demonstrate Hybrid-FeFET (H-FeFET) with an integrated threshold switch (TS) that demonstrates amplified MW, and overcomes the FeFET trade-off.

Fig. 2. (a) Proposed H-FeFET where a TS is integrated with a standard FeFET. (b) H-FeFET operating principle: the TS shows HRS to LRS transition only if the FeFET is in the low-V_T SET state; the TS does not transition if the FeFET is in the high-V_T RESET state, thus minimizing the RESET read current. This amplifies $I_{read,1}/I_{read,0}$ and MW.

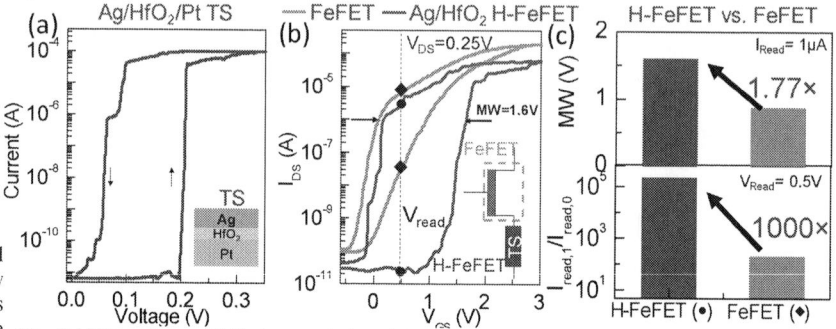

Fig. 3 (a) Experimental I-V characteristics of the Ag/HfO₂/Pt based threshold switch used in the H-FeFET. (b) Experimentally measured switching characteristics of H-FeFET (green) and stand-alone FeFET (orange). (c) 77% improvement in MW, & 1000× improvement in the $I_{read,1}/I_{read,0}$ is observed.

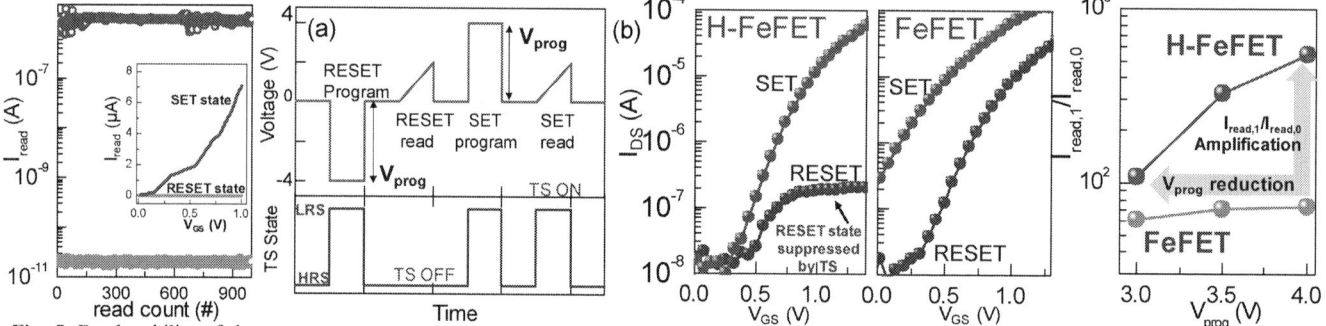

Fig. 5. Read stability of the SET & RESET states over 1000 samples. Inset: DC I_{DS}-V_{GS} read characteristics of the SET & RESET state.

Fig. 6. Pulse program operation of H-FeFET. (a) Schematic of pulse program/read scheme and the evolution of the TS state. (b) I_{DS}-V_{GS} read characteristics of the H-FeFET and FeFET for the SET & RESET; the TS suppresses the RESET state current in H-FeFET which increases the $I_{read,1}/I_{read,0}$ ratio (V_{DS}=0.8V). Since the RESET read current is suppressed in H-FeFET to below 1µA, the MW at 1µA is larger than V_{prog}; at I_{read}=200nA, 3.2× higher MW is observed.

Fig. 7. Maximum $I_{read,1}/I_{read,0}$ vs. V_{prog} for H-FeFET & FeFET. The H-FeFET enables higher $I_{read,1}/I_{read,0}$ at *iso*-V_{prog}; and lower V_{prog} at iso- $I_{read,1}/I_{read,0}$.

H-FeFET Architecture

ACC: Accessed cell;
HAC/HAR: Half Accessed Column/Row;
UA: Unassessed Cell

(b)		Accessed Cell	Un-accessed Cells
Write 1 (SET)	WL:	0→0.8V	WL: 0V
	BL:	0.3V	BL: 0V
	SL:	0V	SL: 0V
Write 0 (RESET)	WL:	0→0.65V→-2V	WL: 0V
	BL:	0.3V	BL: 0→-2V
	SL:	0V	SL: 0→-2V
READ	WL:	0→0.3V	WL: 0V
	BL:	0.3V	BL: 0V
	SL:	0V	SL: 0V

Fig. 8 (a) Schematic of the 1T-FeFET-based AND array evaluated in this work using adapted 45nm technology model; a similar architecture has also been evaluated for 1T-FeFET; (b) Program/Read scheme for the 1T-H-FeFET array. The scheme enables the TS to amplify $I_{read,1}/I_{read,0}$ ratio of the target cell, but also reduce (sneak) current through the other cells.

Fig. 9. Comparison of program voltage and program energy (SET+RESET) for the FeFET and H-FeFET in AND array configuration at *iso*-$I_{read,1}/I_{read,0}$

978-1-5386-4219-1/18 $31.00 © 2018 IEEE 130

A Circuit Compatible Accurate Compact Model for Ferroelectric-FETs

Kai Ni[*], Matthew Jerry[*], Jeffrey A. Smith, and Suman Datta

University of Notre Dame, Notre Dame, IN 46556, USA; [*]equally contributing authors;

Email: mjerry@nd.edu & kni@nd.edu

Abstract: In this work we develop a compact model of ferroelectric field-effect-transistors (FeFET) for memory applications, enabling their exploration at the circuit and architecture level. In contrast to Landau-Khalatnikov (L-K) based approaches, the presented model is founded on the combination of a nucleation dominated multi-domain Presiach theory of ferroelectric switching with a conventional transistor model. The model successfully reproduces the evolution of the FeFET memory window as a function of the program and erase conditions (amplitude, pulse width, and history). To calibrate the model, we fabricated 10nm thick $Hf_{0.4}Zr_{0.6}O_2$ (HZO) MFM capacitors and FeFETs and characterized the polarization switching dynamics. Our results highlight the importance of accounting for the switching history, minor loop trajectory, and coupled time-voltage response of the ferroelectric to quantitatively reproduce the measured FeFET characteristics.

Introduction: Due to CMOS compatibility and remarkable scalability, ferroelectric HfO_2 has created a resurgent interest in FeFET for nonvolatile memory (NVM) [1]. The successful integration of FeFET into advanced technology nodes further highlights the promise for embedded NVM applications [2]. To further evaluate FeFET performance and technology applications, it is necessary to have a compact model describing the physical device behavior. Although several compact models for negative-capacitance FeFETs (NC-FET) have been proposed [3], [4] they are based on phenomenological L-K equation where the ferroelectric behavior is described by a power-law relationship between the polarization (Q_{FE}) and voltage (V_{FE}). This relationship describes an unstable transition region between the two stable polarization states, which shows negative capacitance behavior enabling sub-60 mV/decade subthreshold swing device. While such models are well suited for simulation and optimization of steep slope NC-FETs, they are not appropriate for FeFET memory applications for several reasons. The L-K equation describes a single-domain ferroelectric material and assumes a single coercive field for the entire ferroelectric thin film. Such an assumption is unrealistic even for ultra-scaled devices [5]. Additionally, when the ferroelectric is combined in series with the MOSFET capacitance within the L-K framework, the trajectory of the $Q_{FE}(V_{FE})$ relationship remains controversial. In this work, we develop a FeFET compact model based on the multi-domain Preisach theory that accurately replicates the experimental characteristics of fabricated FeFETs.

Modeling Framework: The modeling framework is shown in Fig. 1(b), where the FeFET is composed of two sub-components. The first is the conventional MOSFET, which is modeled using BSIM 4 compact model [6]. This captures the charge-voltage relationship of the MOSFET, $Q_{MOS}(V_{MOS})$. The other component is the ferroelectric, described by the dynamic Preisach model, which gives the ferroelectric charge-voltage relationship, $Q_{FE}(V_{FE})$. As the ferroelectric is integrated within the MOSFET gate stack, the following two equations governing charge conservation and voltage division must be solved simultaneously: $Q_{FE}(V_{FE}) = Q_{MOS}(V_{MOS})$ and $V_G = V_{FE} + V_{MOS}$. The accuracy of this approach lies within the correct determination of Q_{FE} which is dependent on the

ferroelectric history, minor loop trajectory, and switching dynamics (Fig. 1(c)). To capture these effects, the Preisach model combines a static model, describing (1) the saturation loop (Fig. 3), (2) the history dependence and minor loops (Fig. 5), and (3) a delay unit, describing quantitatively the ferroelectric switching dynamics (Fig. 4).

MFM Capacitor Results: Fig. 3 shows the static Preisach model, in which the aggregate effect of multiple independent domains (with their own coercive field) is highlighted by the continuous response of the P-V loop (Fig. 3(c)). The ferroelectric history and minor loop paths are modeled by tracking the evolution of the turning points [7]. To describe those inner loops, a simple linear scaling from the saturation loop is applied (Fig. 5(a)) where $m_{\uparrow/\downarrow}$ and $b_{\uparrow/\downarrow}$ are the linear scaling constants and $\uparrow/\downarrow$ indicate the forward/reverse branch of Q_{FE}-V_{FE} loop. Fig. 5 shows the modeled and measured Q_{FE}-V_{FE} loops for three different applied voltage waveforms. The model accurately recreates the experimental data by tracking the turning points (A, B, C, etc.). The time and voltage dependent switching dynamics are modeled by an RC time delay (Fig. 4(a)), since the static Preisach model assumes an instantaneous response of the ferroelectric which is unphysical, as it takes a finite amount of time to switch a domain. The delay unit accounts for this by computing a delayed voltage (V_{eff}) experienced by the ferroelectric to which the ferroelectric responds (Fig. 4(b)) [8]. Fig. 4(c) shows that, the simple RC model reproduces the measured P_r as a function of applied pulse magnitude and pulse duration.

FeFET Results: The calibrated ferroelectric model for MFM capacitors is directly applied to the FeFET framework as described in the Modeling Framework section. The resulting compact model allows the FeFET characteristics to be solved for any arbitrary bias conditions and pulse sequence. This is exemplified in Fig. 6(a-d) where the MW as a function applied pulse width and program/erase voltages is well captured by the model. Further, the model captures the partial polarization switching which enables a gradual tuning of V_T (Fig. 7(a-c)) and hence analog response of the channel conductance with success erase pulses (Fig 7(d-e)).

Conclusion: We demonstrate a circuit compatible FeFET model capable of reproducing both the binary and analog memory response of experimentally measured FeFETs. The model is based on multi-domain Preisach theory (in contrast to previous L-K based approaches) and enables accurate exploration of FeFET devices at the circuit and architecture level.

References: [1] M. H. Park et al., Adv. Mater., vol. 27, no. 11, Mar. 2015. [2] S. Dünkel et al. , IEDM, 2017. [3] A. Aziz, et al., EDL, vol. 37, no. 6, 2016. [4] H. P. Chen, et al., TED, vol. 58, no. 8, 2011. [5] H. Mulaosmanovic et al., IEDM, 2015. [6] Berkeley Short Channel IGFET Models. [7] Bo Jiang, et al., VLSI Technology, 1997 [8] J. Chow, et al.,VLSI Circuits. Dig. Tech. Pap., 2004.

Acknowledgment: This project was supported by the National Science Foundation under grant 1640081, the NERC, a wholly-owned subsidiary of the SRC, through Extremely Energy Efficient Collective Electronics (EXCEL), an SRC-NRI Nanoelectronics Research Initiative under Research Task IDs 2698.001 and Global Research Corporation (GRC), Task ID 2657.

978-1-5386-4219-1/18 $31.00 © 2018 IEEE

FeFET Compact Model

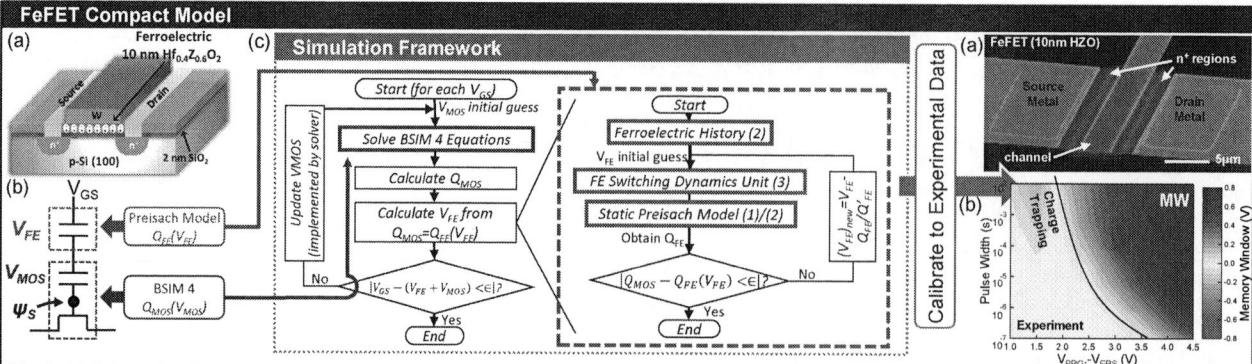

Fig. 1: (a) Schematic of FeFET. (b-c) A FeFET compact model is developed by jointly solving two sub-components for $Q_{MOS}(V_{MOS})=Q_{FE}(V_{FE})$. The baseline MOSFET (modeled with BSIM 4 [6]) is used to compute $Q_{MOS}(V_{MOS})$ and a dynamic history aware Preisach model computes $Q_{FE}(V_{FE})$. The Preisach model captures core aspects of the ferroelectric: (1) saturation loop, (2) ferroelectric switching dynamics, and (3) history and minor loop trajectory.

Fig. 2: (a) 10nm $Hf_{0.4}Zr_{0.6}O_2$ FeFETs are fabricated and used for model calibration. (b) Measured memory window matrix.

(1) FE Saturation Loop

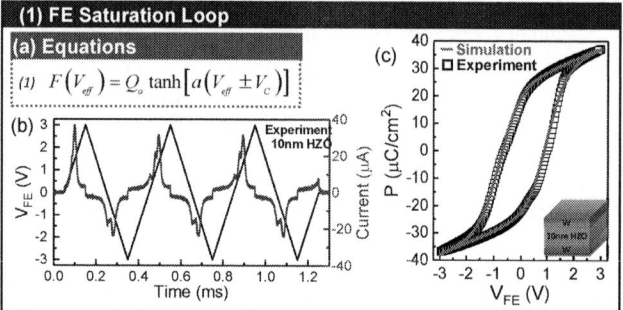

Fig. 3: (a) Multi-domain effects within the saturation loop are captured using the standard tanh function, which provides an excellent fit to the (b-c) experimental MFM data (measured on 10nm $H_{0.4}Z_{0.6}O_2$ capacitor).

(3) FE Switching Dynamics

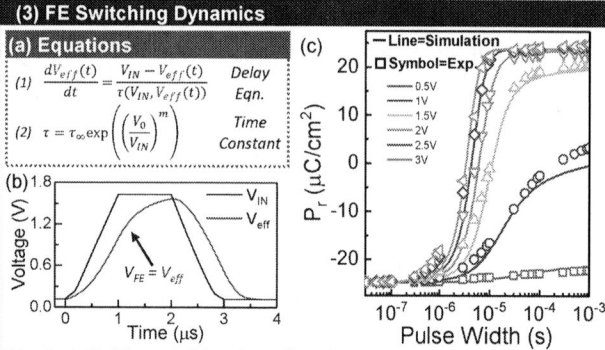

Fig. 4: (a-b) The switching dynamics of MFM capacitors as a function of applied pulse-width and pulse-amplitude are captured using a RC type delay, where the computed V_{eff} is the voltage that appears across the ferroelectric. (c) An excellent match is observed between simulation and experiment

(2) FE History and Minor Loop Trajectory

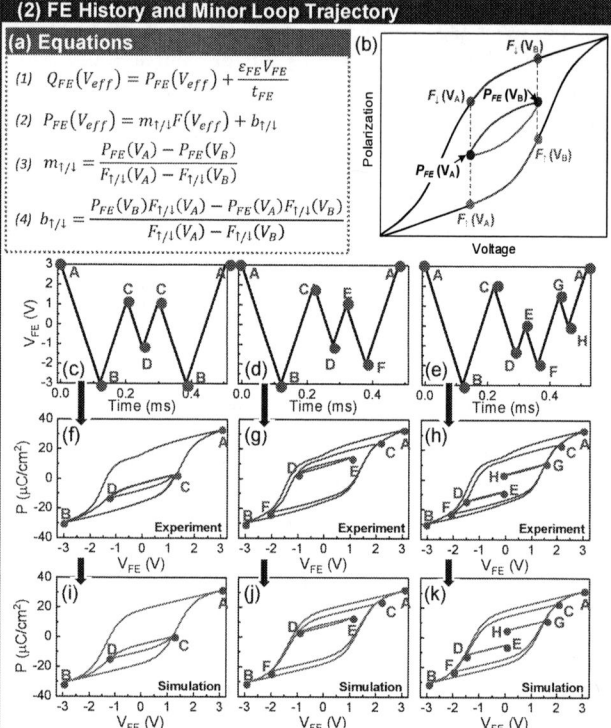

Fig. 5: (a-b) Minor loops are simulated by a Linear scaling from the saturated polarization-voltage hysteresis loop. ↑/↓ indicates forward/reverse branch respectively. (c-k) The model successfully captures the turning point and trajectory of the experimentally measured minor loops.

FeFET Response: Binary and Analog

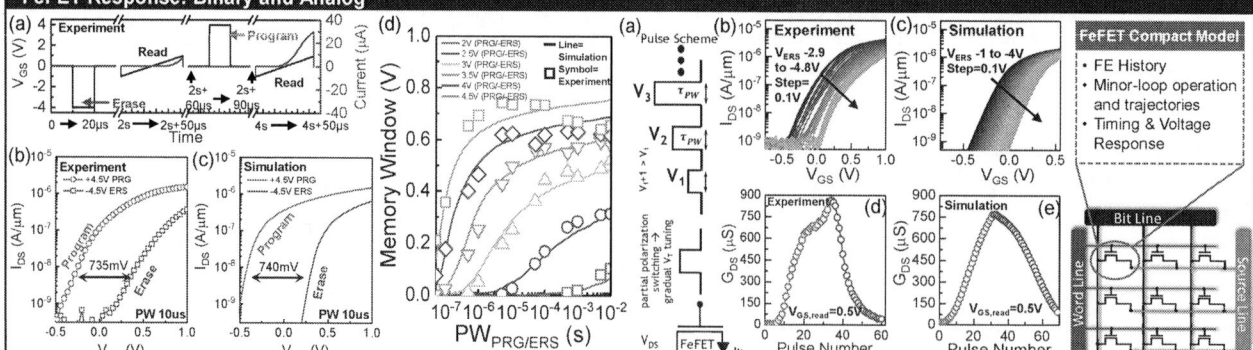

Fig. 6: (a-b) Measured memory window (MW) of fabricated FeFETs (SEM shown in Fig. 2(a)). Upon calibration, the FeFET compact model is able to accurately capture the measured trade off between pulse-width and pulse-amplitude as highlighted in (c-d).

Fig. 7: (a) Analog memory behavior for variable pulse amplitude. (b-c) Measured I_{DS}-V_{GS} and conductance (G_{DS}) (V_{DS}=50mV) characteristics are accurately captured by the FeFET compact model.

FeFET Compact Model
- FE History
- Minor-loop operation and trajectories
- Timing & Voltage Response

Fig. 8: The compact model enables FeFET array and circuit macro simulations.

978-1-5386-4219-1/18 $31.00 © 2018 IEEE

Record 47 mV/dec top-down vertical nanowire InGaAs/GaAsSb tunnel FETs

Alireza Alian, Salim El Kazzi, Anne Verhulst, Alexey Milenin, Nicolò Pinna, Tsvetan Ivanov, Dennis Lin, Dan Mocuta, Nadine Collaert

imec vzw, Kapeldreef 75, 3001 Leuven, Belgium

Abstract

Pocketed vertical nanowire InGaAs/GaAsSb tunnel FETs (TFET) with sub-threshold swing (SS) reaching 47 mV/dec are demonstrated. The achieved sub-threshold performance is the steepest reported so far for a top-down TFET in the III-V material system. Smooth vertical wires with diameters as narrow as 30 nm are achieved using a CH_4 based dry etch process. Drive current at 0.35 V supply voltage approaches 0.7 µA/µm for a fixed I_{off} of 1 nA/µm.

Introduction

TFETs are interesting devices for low power applications [1] however, their potential is not fully realized yet. TFETs suffer from low drive current and SS mostly not steeper than the thermal limit. Heterojunction III-V tunnel FETs, specially staggered gap InGaAs/GaAsSb devices, are one of the promising structures to realize the full strength of the TFET [2]. Except for a recent report [3] on grown InAs/GaAsSb vertical nanowires, no significantly steeper than 60 mV/dec SS performance is reported for such a device at room temperature [4,5,6,7,8]. This work reports substantially steeper than 60mV/dec SS III-V nanowires which deliver the best reported SS for a TFET fabricated through a top-down approach in the III-V material system. Furthermore, this is the first demonstration of sub-60 mV/dec DC performance for an InGaAs/GaAsSb TFET. A dry etch process for smooth etching of the heterostack is developed as the key enabler to realize the nanowire TFETs.

Processing details

The III-V layers as shown in figure 1 are grown [9] by MBE. Wires are dry etched at a temperature of 100°C in a CH_4 based plasma [10] following the ebeam lithography definition using a HSQ resist. A hard mask is used to improve the dry etch selectivity. A smooth and vertical profile was obtained in the heterostack as shown in figure 2a. The gate oxide of 1nm Al_2O_3 + 2nm HfO_2 (expected EOT is 0.8 nm) is deposited by ALD following a diluted HCl and $(NH_4)_2S$ treatment. The $(NH_4)_2S$ etches GaAsSb resulting in an undercut at the source side which shrinks the actual tunnel junction perimeter by 30nm (figure 2b). Top and bottom spacers are patterned using resist planarization and etch back. Devices are annealed in forming gas at 350°C.

Results

Figure 3 shows the TEM image of a device with 50nm diameter at the top of the wire. The actual tunnel junction diameter is 30 nm located at the narrowest part of the wire as confirmed by the EDS elemental maps in figure 3. The device parameters in subsequent figures are normalized by the actual tunnel junction diameter measured from TEM. Figure 4 shows the SIMS profile of Be and Si in a calibration sample grown prior to the actual TFET stack. The Be profile is steep at 1.5 nm/dec, however, some Be diffusion (~2 nm) in the InGaAs layers occurs and can reduce the effective pocket doping. Vice versa, Si may diffuse in the GaAsSb layer due to the amphoteric nature of Si in this matrix, the effective source doping may drop if Si becomes an n-dopant upon diffusion. Figure 5 shows (a) the transfer characteristics of selected sub-60 mV/dec performing devices at 0.2 V V_d as well as (b) the transfer and (d) output characteristics and (c) the extracted SS for a 30 nm tunnel junction diameter single nanowire device. There is a significant variability on the I_{on}

and V_{th} of the measured devices likely due to doping/diameter variations. For the example device, SS reaches a minimum of 47 mV/dec and stays sub-60 mV/dec for drain voltages up to 0.5 V. NDR presence in the output characteristics confirms that the operation of the device is dominated by tunneling. Diameter dependence of the device parameters (SS, I_{on}, g_m and PVCR) are plotted in figure 6 before and after FGA. The FGA significantly improves the device performance as it reduces InGaAs Dit [13]. Sub-60 mV/dec performance is only observed for 30 and 50 nm junction diameter devices. The diameter scaling seems to be one of the key metrics to enable sub-60 operation of the devices so narrower wires may lead to even steeper SS. However, as can be seen in figure 5a, the on-current drops for some of the wires as V_g is increased. This current drop is plotted in figure 7 as a function of nanowire diameter and shows that the drop becomes significant at smaller diameters. The current drop is attributed to the source depletion effect and it may suggest that either the effective doping concentration reduces as the diameter shrinks or the narrow source starts to be fully depleted. As a result, higher source doping and a smaller gate-source overlap will be required to enable further scaling of the wire diameter.

Figure 8 shows the temperature dependence of the transfer characteristics. The behavior is almost temperature independent except for a large 0.26 V positive V_{th} shift at -40°C compared to the room temperature behavior. This shift is consistently observed without a clear explanation so far.

Semi-classical simulations of the BTBT current of nanowire TFETs with varying junction diameters and a similar shape to the figure 1d are shown in figure 9a and 9b. A source/pocket doping gradient dropping towards the nanowire surface was incorporated to be able to reproduce the experimentally observed drive currents. As the diameter shrinks, the source depletion arising from the gate/source overlap, becomes significant and the current drops consistent with experimental observations of figure 7. On the other hand, the SS is improving for the smaller diameters which suggests an optimum junction diameter below 50 nm for the highest drive current, consistent with the experimental observations of figure 6b. Quantum mechanical simulations predict a significant performance boost for a 20nm wire diameter and increased source/pocket doping concentrations to $5x10^{19}$ cm^{-3} to avoid source depletion effects. The benchmark of the sub-60 performance of the devices is shown in figure 10. Average sub-60 SS is the effective SS over the whole sub-60 mV/dec performing window. The devices of this work show the steepest average sub-60 mV/dec SS among the III-V-only TFETs.

References

[1] D. H. Morris et al, VLSI 2016 p. 222. [2] A.S. Verhulst et al, IEDM 2014, p.717 [3] E. Memisevic et al, IEDM 2016, p.500 [4] A. Alian et al, Appl. Phys. Lett. 109, 243502 (2016). [5] D. H. Ahn et al, VLSI 2016, p. 224. [6] B. Ganjipour et al, ACS Nano 2012, p. 3109. [7] T. Gotow et al, J. Applied Physics 122, 174503, 2017 [8] G. Dewey et al, IEDM 2011, 33.6.1–33.6.4. [9] S. El Kazzi et al, J. Crystal Growth 424 (2015) 62–67. [10] A.Milenin, et al, MNE2017 [11] K. Tomioka et al, Appl. Phys. Lett. 104, 073507 (2014). [12] K. Tomioka et al, VLSI 2012, p. 47. [13] A. Alian et al, ECS JSS, 1 (6) p.310 (2012)

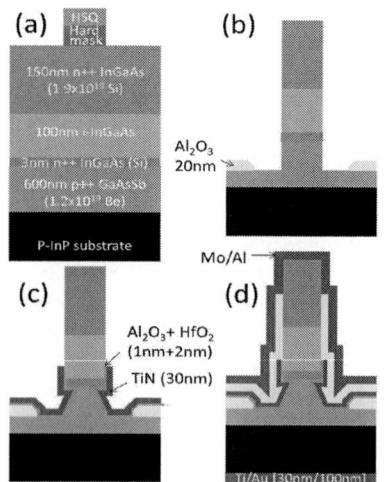

Figure 1. Device structure and process flow. (a) MBE growth of the layers and ebeam definition of the wires. (b) dry etch of the wires, strip of the hardmask, formation of the bottom spacer (c) gate stack deposition and opening of the wire top. (d) top spacer formation and top/back contact deposition.

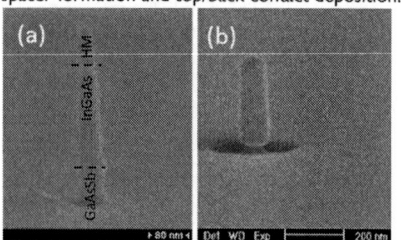

Figure 2. (a) Wires as narrow as 30nm diameter after the dry etch. (HM: hardmask). (b) wires after gate stack deposition. $(NH_4)_2S$ treatment etches GaAsSb and creates an undercut at the bottom of the wire.

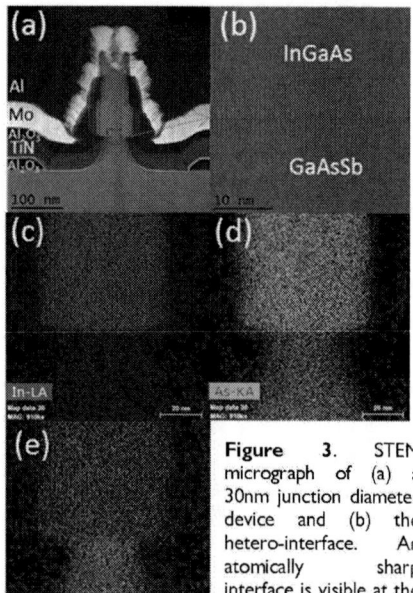

Figure 3. STEM micrograph of (a) a 30nm junction diameter device and (b) the hetero-interface. An atomically sharp interface is visible at the hetero junction. EDS elemental analysis maps of (c) In , (d) As and (e) Sb confirm the heterojunction position being the narrowest part of the wire.

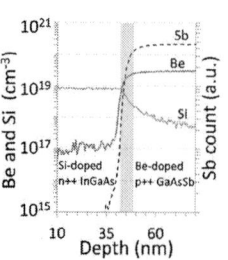

Figure 4. SIMS profile of a calibration sample grown prior to the actual TFET stack [9]. Gray area is the approximate position of the hetero-junction. Hall measurements show 100% Si activation but Be is only 33% activated ($1e19cm^{-3}$ active concentration). The target doping concentrations of the actual TFET sample are $1.9e19cm^{-3}$ Si in InGaAs for both the pocket and the drain layer and $1.2e19 cm^{-3}$ Be in the GaAsSb source layer.

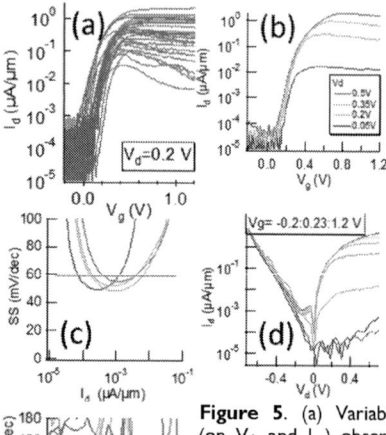

Figure 5. (a) Variability (on V_{th} and I_{on}) observed among the sub-60 performing devices. (b) the transfer and (d) the output characteristics and (c) the corresponding SS behaviour as well as (e) the drain conductance slope of a 30 nm junction diameter single nanowire device. The minimum conductance slope is 84 mV/dec.

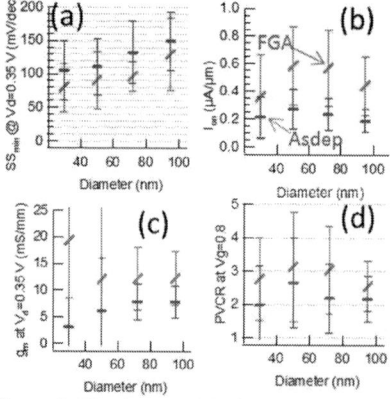

Figure 6. SS_{min}, I_{on}, g_m and PVCR vs. tunnel junction diameter before and after anneal. Values are measurements over around 200 devices. The anneal significantly improves the device performance. SS_{min} is calculated over a 40 mV V_g window. I_{on} is extracted at a fixed I_{off} of 1 nA/μm and a supply voltage of 0.35 V.

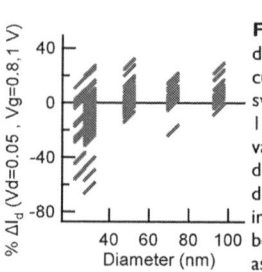

Figure 7. Relative drop in the drain current as V_g is swept from 0.8 V to 1 V. A negative value means the drain current decreases as V_g increases. The drop becomes significant as the diameter is reduced. Source depletion due to the gate source overlap is suspected to cause this behavior.

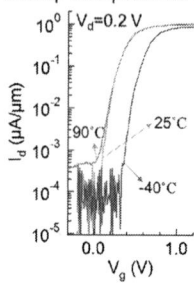

Figure 8. Temperature dependence of the transfer characteristics. A significant 0.26V positive V_{th} shift was recorded at -40°C but no shift at 90°C compared to the room temperature behavior. Apart from the V_{th} shift, the sub-threshold behavior is not temperature dependent. Off state leakage floor increases at 90°C.

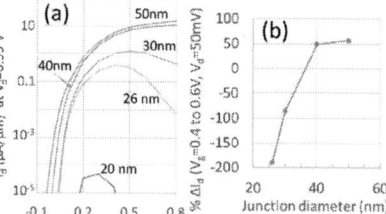

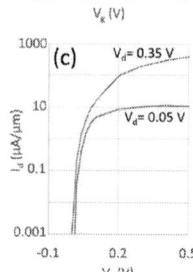

Figure 9. Simulations of BTBT current. (a) Transfer characteristics for different junction diameters show that the source depletion becomes significant as the diameter shrinks causing the current to drop at higher V_g. This drop is quantified in (b). (c) Prediction (via quantum mechanical simulations) for a 20nm wire diameter and increased source/pocket doping concentrations to $5x10^{19}$ cm^{-3}. A significant performance boost is expected.

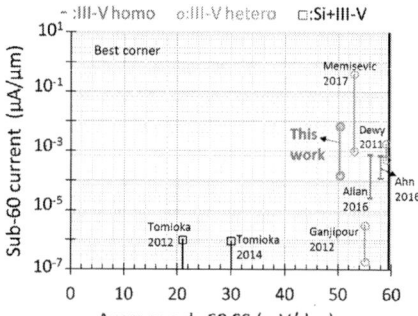

Figure 10. Benchmark of the sub-60 performance (V_d<0.5 V) of this work against the literature. [3, 4, 5, 6, 8, 11, 12]. The devices of this work show the steepest average sub-60 SS of a III-V-only TFET. The devices have the steepest SS among the III-V nanowire TFETs fabricated through a top-down approach.

978-1-5386-4219-1/18 $31.00 © 2018 IEEE   134

Gap in pagination due to formatting issues.

Pages 135-136

Improving Performance, Power, and Area by Optimizing Gear Ratio of Gate-Metal Pitches in Sub-10nm Node CMOS Designs

Yongchan Ban, Xuelian Zhu, Jan Petykiewicz, Jia Zeng

GLOBALFOUNDRIES, 2600 Great America Way, Santa Clara, California 95054, USA
yongchan.ban@globalfoundries.com

Abstract

This paper presents improvements in performance, power, and area (PPA) obtained by optimizing the gear ratio (GR) between the Gate and vertical metal layer pitches in standard cells in sub-10nm node CMOS SoC designs. Changing the GR from 1:1 to 3:2 leads to better pin accessibility, routability, and higher cell density. This in turn enables a gate pitch relaxation and associated improvements in cell delay. Implementation of 3:2 GR ultra-dense cells in an SoC CPU block results in up to 17% higher performance, 4% smaller logic size, and 8% lower dynamic power at typical PVT conditions.

Introduction

The metal 1 (M1) layer is typically used as the first routing layer and gives input/output pins for connecting standard cells with back-end interconnect metals. With the advent of the unidirectional patterning, the bidirectional M1 is decomposed into the horizontal M0 layer and the vertical M1 layer [1-2] as shown in Fig.1. Cell dimensions are driven by the minimal contacted poly pitch (CPP, ~cell width) and horizontal metal pitch (~cell height). Given expected sub-10nm node pitches, a design-technology co-optimization (DTCO) effort is necessary to improve pin accessibility and routability for better PPA.

The ratio of pitches between Gate and M1 is commonly referred to as the gear ratio. It has traditionally been 1:1 (the same pitch for Gate and M1) in unidirectional designs due to aggressive shrinking of those layers. While the minimal CPP is dictated by process impacts on parametric variations, the M1 pitch (M1P) does not suffer from similar limitations and can be readily decreased to match other critical metal pitches. Reducing the M1 pitch to achieve a 3:2 M1P:CPP GR, as in Fig.2, can improve routability [3]. This paper demonstrates the PPA benefits and examines the manufacturability tradeoffs of an optimized GR for our advanced node CMOS technology.

Optimization of Gate-to-Metal Gear Ratio

Two high-symmetry flavors of 3:2 GR configuration are possible, as shown in Fig.3; "M1 on-top-of PC" where some M1 polygons can be placed directly over gate polygons (PC), and "M1 in-between PC" where some M1 polygons are located directly between two PCs. In addition, each symmetry option requires two versions of each cell with different M1 offset (Fig.4) in order to avoid an M1 coloring problem in sequential locations. Thus, a 3:2 GR library must have 2 times as many cells as an 1:1 library.

The two different M1 offset cell types can have varying cell delays. Fig. 5 reports an average delay gap of 0.04% across all the cells in an "M1 in-between PC" library. In sequential cells, up to 0.50% variation in the rising arc is observed. The maximum delay difference in combinational cells in all timing arcs is about 0.25%. Yet, all delay differences are within the cell characterization criteria.

We investigate the best symmetry option for 3:2 GR cells with considerations of manufacturing difficulty and design benefits. Examination of a complex cell (AOI22x1) reveals that the "M1 in-between PC" option has relatively smaller pin accessibility in between M0 and M1, but shows the best manufacturability; meanwhile, the "M1on-top-of PC" option has more pin accessibility in routing, at the cost of manufacturing challenges, e.g. V0 (M0 to M1) enclosure rules.

Enhancing PPA in SoC with the Best Gear Ratio

We applied the 3:2 GR cells to an SoC CPU block, OR1200, and compared PPA with ultra-dense design kits. 45 cells with 1:1 GR (1x M1P and CPP) and 90 (45x2 M1 shift) cells with 3:2 GR (0.71x M1P and 1.07x CPP) were used in Fig.7, and all PPA were measured at the TT/0.75v/25c corner. We choose the "M1 in-between PC" option for 3:2 GR cells since it is more compatible with current patterning capabilities. 13 metal stacks with 3 1x metal layers and a dual Vdd/Vss power rail architecture were applied. Due to CPP relaxation, the 3:2 GR library is about 7% bigger in cell size, but 1.6% faster in delay.

We first measured the maximum achievable frequency in OR1200. The 1:1 GR library achieves up to 2.4GHz, while the 3:2 GR library reaches up to 2.8GHz, shown in Fig.8. The key components of the performance benefits with 3:2 GR are 1) slight better cell delay (around 1.6%), 2) much better routability and utilization; 3:2 GR has a less logic depth at the critical path as shown in Fig.9. 3) smaller wire loads; the 3:2 GR shows smaller interconnect net delay portion for critical path groups in Fig.9. The 3:2 GR has a similar or lower total wire length, despite its bigger cell size in Fig.10. M1 metal usage in the 3:2 GR is around 7% higher at the same frequency due to its superior pin accessibility and routability. About 16% smaller via count is reported with the 3:2 GR in Fig.11. The numbers of V1 and V2 are very similar, but those of V3 and above are reduced with the 3:2 GR. This implies that the 3:2 GR has better routability, fewer detours and lower wire loads.

The utilization (cell density) in the 3:2 GR library is about 8% higher in Fig.12. Although 3:2 GR cells are about 7% larger, the overall logic area for the chip becomes about 4% smaller (block level scaling) at 2GHz target (Fig.13).

The total dissipated power is similar in the entire frequency range without consideration of chip area, as displayed in Fig.14. At the same performance (2GHz) and the same area, the 3:2 GR design shows about 8% less in total dynamic power, as shown in Fig.15. Even though the pin capacitances of 3:2 cells are slightly (1%) larger than those of 1:1 cells, the effective dynamic power dissipation capacitance in 3:2 cells is smaller due to faster switching as well as the shorter wire load.

Conclusion

We present the benefit to PPA from a 3:2 M1-to-PC gear ratio in sub-10nm node CMOS designs. Experimental results with an SoC block and ultra-dense library show that the 3:2 GR cell library promises sizeable improvements to power and performance. For future nodes, it is important to develop technology architecture definitions that have a good M1-to-PC gear ratio and allow M1 to PC offset.

References

[1] L. Liebmann et al., in Proc. VLSI Tech. Digest, pp.112, 2016.
[2] L. Clark et al., in *Microelectronics Journal*, 53, pp.105-115, 2016.
[3] L. Lu, in Proc. ACM ISPD, pp. 63, 2017.

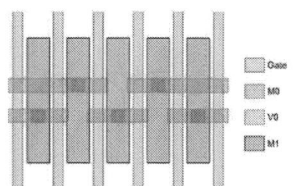

Fig. 1. The bidirectional M1 layer is decomposed into the unidirectional M0 and M1 layers in sub-10nm node technology.

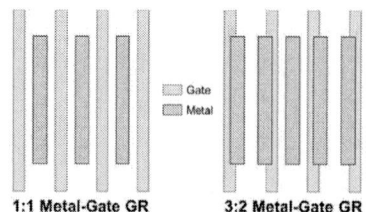

Fig. 2. Metal to Poly (Gate) gear ratio; M1 pitch in 3:2 GR is 1.5x smaller than that of PC.

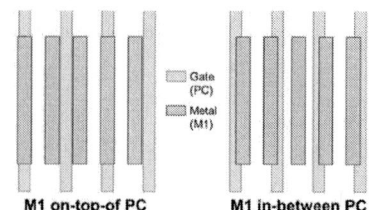

Fig. 3. 3:2 GR configuration has two high symmetry options for metal shift.

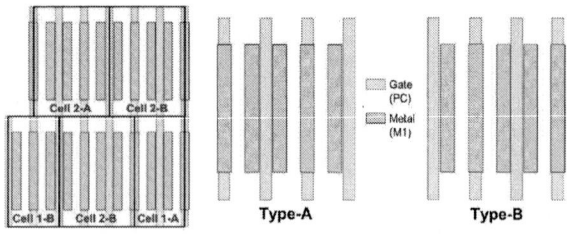

Fig. 4. For 3:2 GR, each cell needs two versions of M1-offset for cell abutting in placement.

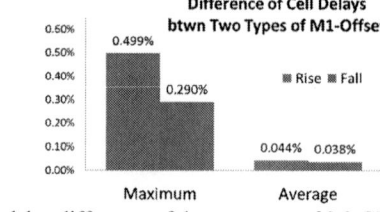

Fig. 5. The delay difference of the two types of 3:2 GR cells shows around 0.04% on average and up to 0.50% for the overall cells.

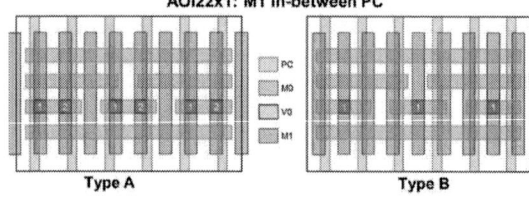

Fig. 6. The best 3:2 GR option can be found by considering DTCO. "M1 in-between PC" shows the best manufacturability yet relatively smaller number of pin accessibility in between M0 and M1, while "M1on-top-of PC" has the better pin accessibility with challenging Via enclosures.

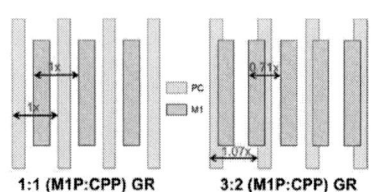

Fig. 7. 3:2 GR library shows around 7.1% larger cell size, yet about 1.6% faster in delay due to CPP expansion.

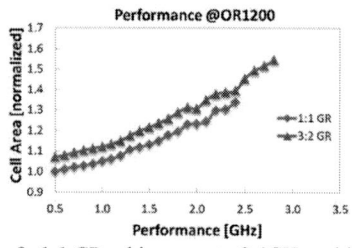

Fig. 8. 1:1 GR achieves up to 2.4GHz, while 3:2 GR reaches up to 2.8GHz.

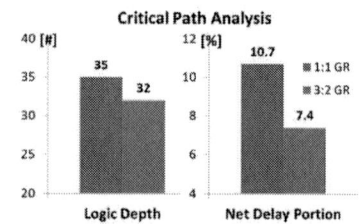

Fig. 9. 3:2 GR has less logic depth and smaller net delay portion in the critical path.

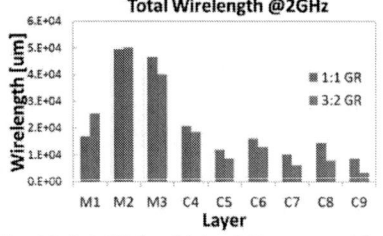

Fig. 10. 3:2 GR has higher M1 usage and less total wire length despite its bigger cell size.

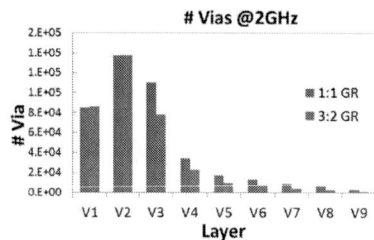

Fig. 11. About 16% smaller Via count is reported with a 3:2 GR.

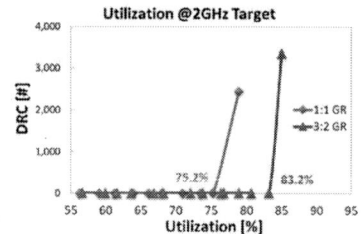

Fig. 12. 3:2 GR library shows about 8% better utilization (cell density) at the same frequency.

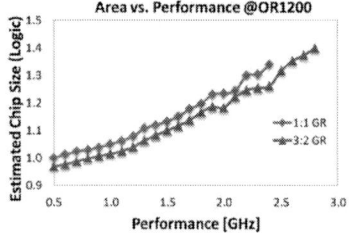

Fig. 13. The overall logic area in chip is about 4% smaller in 3:2 GR at a 2GHz target.

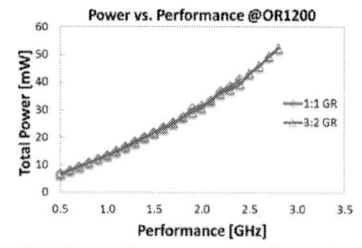

Fig. 14. The total powers are similar (slightly higher in 1:1 cells) in the entire freq. range.

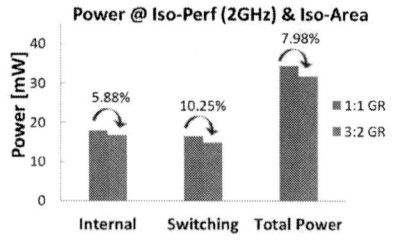

Fig. 15. At the same performance and area, 3:2 GR shows about 8% less total power.

978-1-5386-4219-1/18 $31.00 © 2018 IEEE

Achieving High-Scalability Negative Capacitance FETs with Uniform Sub-35 mV/dec Switch Using Dopant-Free Hafnium Oxide and Gate Strain

Chia-Chi Fan[1], Chun-Hu Cheng[2,*], Chun-Yuan Tu[1], Chien Liu[3], Wan-Hsin Chen[3], Tun-Jen Chang[1], and Chun-Yen Chang[1,#]

[1]Dept. of Electronics Engineering, National Chiao Tung Univ., [2]Dept. of Mechatronic Engineering, National Taiwan Normal Univ., [3]Dept. of Electrophysics, National Chiao Tung Univ., Taiwan ; E-mail: *chcheng@ntnu.edu.tw, #cyc3562@gmail.com

Abstract

For the first time, we successfully demonstrated that the 4-nm-thick dopant-free HfO_2 NCFETs using gate strain can implement an energy-efficient switch of a low gate overdrive voltage and a nearly hysteresis-free sub-40 mV/dec swing. The gate strain favorably rearranges oxygen vacancies and boosts orthorhombic phase transition. Furthermore, the dopant-free HfO_2 NCFET can be further improved by in-situ nitridation process. The 4-nm-thick nitrided HfO_2 NCFETs achieve a steep symmetric sub-35 mV/dec switch, a sustained sub-40 mV/dec SS distribution, and excellent stress immunity during NC switch. The high-scalability and dopant-free NCFET shows the great potential for the application of future highly-scaled 3D CMOS technology.

Introduction

To reach the power requirement (sub-1 µW) of Internet of Thing (IoT) devices, we have proposed negative capacitance transistors (NCFETs) based on doped hafnium oxides ($HfAlO_x$, $HfZrO_x$) and gate strain to implement steep sub-60 mV/dec switch with ultralow off leakage [1],[2]. However, the ferroelectrtivity and NC effect of doped hafnium oxides may be restrained in the application of sub-5nm node. This is due to the poor tunability of dopant (<5% Al or 50% Zr) at a highly scaled capacitance equivalent thickness (CET). An inappropriate doping in hafnium oxides will lead to unstable para/ferroelectric transition. In this work, we proposed dopant-free HfO_2 NCFETs with gate strain and nitridation engineering. The optimal characteristics achieve a symmetric sub-35mV/dec subthreshold swing (SS), a sustained sub-40 mV/dec SS distribution, and excellent stress immunity.

Device Fabrication

A 1-nm chemical oxide was grown and then 4-nm, 7-nm, and 10-nm-thick $HfAlO_x$ (4% Al) and dopant-free HfO_2 were deposited by ALD on Si substrate. To reduce oxygen vacancy (O_{vac}) in orthorhombic HfO_2 (o-HfO_2), a remote NH_3 plasma was employed. After depositing TaN gate with an appropriate gate strain [1],[2], self-aligned As^+ was implanted and activated. These dopant-free NCFET devices were subjected to constant voltage stress (CVS) to investigate NC reliability. The NC switch behavior was simulated by FPC.

Results and Discussion

Fig. 1 shows the I_{DS}-V_{GS} characteristics of 4-nm-thick $HfAlO_x$ NCFET. The $HfAlO_x$ NCFET shows a steep sub-60mV/dec SS (SS_{for}=32mV/dec, SS_{rev}=35mV/dec), a hysteresis of 80 mV, and a sub-60 mV/dec SS range over 2 decade of I_{DS}. Compared to our previous work [1], the off-state leakage and hysteresis significantly increases in this scaled 4-nm-thick $HfAlO_x$ NCFET. We also find that devices yield featuring sub-60 mV/dec SS is obviously degraded, which can be ascribed to poor dopant (4%) tenability in this scaled $HfAlO_x$. To avoid the dopant issue at a scaled CET, the development of dopant-free HfO_2 is necessary. Recently, the dopant-free HfO_2 MIM capacitors have been reported by modulating O_{vac} during deposition [3]. Here, we further investigate dopant-free HfO_2 NCFETs using gate strain engineering. As shown in Fig. 2 and Fig. 3, we clearly observe

that sub-40mV/dec SS can be measured with physical thickness of HfO_2 scaling from 10 nm to 4 nm. This results demonstrate that the local gate strain enhances with thickness scaling, which effectively rearranges O_{vac} and boosts the orthorhombic phase transition [1]. The energy landscape of Fig.4 explains that the 4-nm-thick HfO_2 owns lower Gibbs free energy than 7-nm-thick one, which offers stronger NC effect. The strong NC effect is beneficial to lower SS, enabling transistor switch at a low gate overdrive voltage. Thus, the gate strain favorably promotes NC effect at a scaled thickness. However, the O_{vac} correlated with phase transition also bring the leakage issue at a highly scaled thickness, as shown in Fig. 5. To lower the impact of O_{vac}, an in-situ nitridation was employed to guarantee the HfO_2 quality. Fig. 6 shows the P-E loop of o-HfO_2 MIM capacitor with nitridation, which confirms the existence of ferroelectricity of nitrided-HfO_2. Figs. 7 shows the TEM, EDS and FFT of scaled 4-nm nitrided HfO_2 NCFETs. The EDS profile proves the nitrogen signal in o-HfO_2. Fig. 8 and Fig. 9 shows the I_{DS}-V_{GS} characteristics of 4-nm nitrided HfO_2 NCFETs. This nitrided NCFET exhibits a nearly hysteresis-free switch (Hyst.=35 mV, SS_{for}=34mV/dec, SS_{rev}=32mV/dec) and a low off-state leakage of 1pA/µm improved by approximately one order of magnitude under nitridation process. Notably, no apparent phase transformation and impact on NC switch after nitridation, as evidenced by GI-XRD of Fig. 10 and I_D-V_G curve of Fig. 8. Fig. 11 shows V_{DS} dependence of SS. Compared to SS with V_{DS} dependence of control HfO_2 NCFET, the nitrided HfO_2 NCFETs shows a sustained sub-40 mV/dec switch from V_{DS}=0.1 to 0.5V for both forward and reverse directions. Fig. 12 shows the SS_{min} statistical distributions of dopant-free HfO_2 NCFETs with and without nitridation. The results indicate that the nitrided HfO_2 NCFETs has much better SS distribution (SS_{avg}<40mV/dec) than that of control HfO_2. It may be inferred that the uniformity of NC switch can be stabilized by in-situ HfO_2 nitridation, even at a scaled 4 nm. Fig. 13 and Fig. 14 present the evolution of the relative SILC ($\Delta I_G/I_{Go}$) and stress time dependence of SS under 2V and 3V CVS. The nitrided HfO_2 NCFET with gat strain not only exhibits a low SILC degradation rate but also has superior stress immunity against CVS. This is because the nitrogen atom are favorably coupled with O_{vac}, which can suppress the electron charge traps at O_{vac} [4]. The formation energies for charged traps can be expressed as $\Omega^Q = E^Q + Q\mu_e$, where E^Q is total energy, Q is charged traps and μ_e is electron chemical potential [4]. Table 1 summarizes the important features of state-of-the-art NCFETs. The dopant-free HfO_2 NCFETs with gate strain and nitridataion engineering show high-scalability, energy-efficient uniform NC switch and excellent stress immunity.

Conclusions

A uniform SS_{avg} of <35mV/dec swing with hysteresis free and excellent stress immunity for NC switch are achieved in high-scalability HfO_2 NCFET with gate strain and nitridation.

References

[1] C. C. Fan et al, IEDM, 2017, p. 561. [2] Y. C. Chiu et al, VLSI, 2016, p. 150. [3] A. Pal et al, Appl. Phys. Lett., vol. 110, p. 022903, 2017. [4] A. Pal et al, IEEE EDL, vol. 28, p. 363, 2007. [5] M. H. Lee et al, IEDM, 2015, p. 616.

2018 Symposium on VLSI Technology Digest of Technical Papers

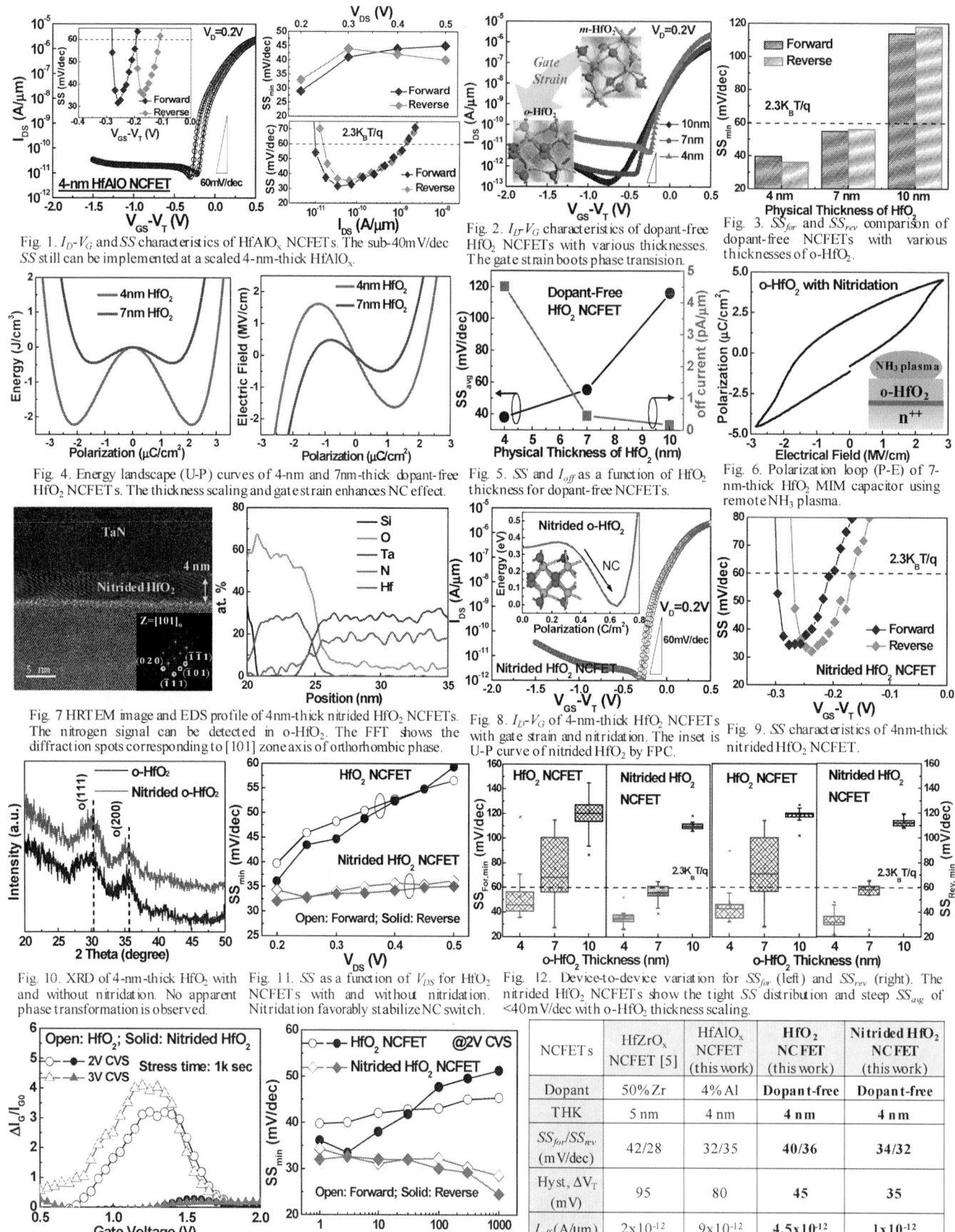

Fig. 1. I_D-V_G and SS characteristics of HfAlO$_x$ NCFETs. The sub-40mV/dec SS still can be implemented at a scaled 4-nm-thick HfAlO$_x$.

Fig. 2. I_D-V_G characteristics of dopant-free HfO$_2$ NCFETs with various thicknesses. The gate strain boots phase transision.

Fig. 3. SS_{for} and SS_{rev} comparison of dopant-free NCFETs with various thicknesses of o-HfO$_2$.

Fig. 4. Energy landscape (U-P) curves of 4-nm and 7nm-thick dopant-free HfO$_2$ NCFETs. The thickness scaling and gate strain enhances NC effect.

Fig. 5. SS and I_{off} as a function of HfO$_2$ thickness for dopant-free NCFETs.

Fig. 6. Polarization loop (P-E) of 7-nm-thick HfO$_2$ MIM capacitor using remote NH$_3$ plasma.

Fig. 7 HRTEM image and EDS profile of 4nm-thick nitrided HfO$_2$ NCFETs. The nitrogen signal can be detected in o-HfO$_2$. The FFT shows the diffraction spots corresponding to [101] zone axis of orthorhombic phase.

Fig. 8. I_D-V_G of 4-nm-thick HfO$_2$ NCFETs with gate strain and nitridation. The inset is U-P curve of nitrided HfO$_2$ by FPC.

Fig. 9. SS characteristics of 4nm-thick nitrided HfO$_2$ NCFET.

Fig. 10. XRD of 4-nm-thick HfO$_2$ with and without nitridation. No apparent phase transformation is observed.

Fig. 11. SS as a function of V_{DS} for HfO$_2$ NCFETs with and without nitridation. Nitridation favorably stabilize NC switch.

Fig. 12. Device-to-device variation for SS_{for} (left) and SS_{rev} (right). The nitrided HfO$_2$ NCFETs show the tight SS distribution and steep SS_{avg} of <40mV/dec with o-HfO$_2$ thickness scaling.

Fig. 13. SILC ($\Delta I_G/I_{G0}$) spectrum during CVS. The SILC for 2V and 3V are 3.2 and 4.1, respectively.

Fig. 14. Stress time dependence of SS. The SS_{min} of nitrided HfO$_2$ NCFET exhibits excellent stress immunity.

NCFETs	HfZrO$_x$ NCFET [5]	HfAlO$_x$ NCFET (this work)	HfO$_2$ NCFET (this work)	Nitrided HfO$_2$ NCFET (this work)
Dopant	50% Zr	4% Al	Dopant-free	Dopant-free
THK	5 nm	4 nm	4 nm	4 nm
SS_{for}/SS_{rev} (mV/dec)	42/28	32/35	40/36	34/32
Hyst, ΔV_T (mV)	95	80	45	35
I_{off} (A/μm)	2x10^{-12}	9x10^{-12}	4.5x10^{-12}	1x10^{-12}
THK, CET Scalability	Medium	Low	High	High

Table 1 Characteristics comparison for state-of-the-art NCFETs.

978-1-5386-4219-1/18 $31.00 © 2018 IEEE

The Complementary FET (CFET) for CMOS scaling beyond N3

Ryckaert J., Schuddinck P., Weckx P., Bouche G.[†], Vincent B.[††], Smith J.[†††], Sherazi Y., Mallik A., Mertens H., Demuynck S.,
Huynh Bao T., Veloso A., Horiguchi N., Mocuta A., Mocuta D., Boemmels J.

imec, Kapeldreef 75, B-3001 Leuven, Belgium; Email: ryckj@imec.be, [†]GLOBALFOUDRIES, local assignee at imec, Leuven, Belgium,
[††] Coventor, 1000 CentreGreen Way, Suite 200, Cary, NC, 27513, [†††] Tokyo Electron, Albany, New York, USA

Abstract

The complementary FET (CFET) device consisting of a stacked n-type vertical sheet on a p-type fin is evaluated in a design-technology co-optimization (DTCO) framework. Through a double level access it offers a structural scaling of both standard cells (SDC) and SRAM by 50%. The proposed process flow requires accurate control of the elevation dimension for manufacturability. Based on TCAD analysis, the CFET can eventually outperform the finFET device and meet the N3 targets in power and performance. To achieve that, the dominating parasitic resistance of the deep vias needs to be reduced by the introduction of advanced MOL contacts featuring thin barriers.

Introduction

The limitations of feature size scaling has lead technologists to explore other avenues for scaling [1]. Among these, SDC track height has become a scaling knob for foundries since N14. Fin depopulation as well as limited intra-cell routing required FEOL integration innovations such as fin height increase, self-aligned contacts and contact over active gate (COAG). The prospect of fin depopulation down to a single fin per device finds a natural extension to devices based on a stack of the 2 channel types called CFET [2-3]. It originates from the complementary nature of CMOS logic where both nFET and pFET are controlled by the same gate (Fig 1). Two pairs of stacked S/D electrodes are used to provide access to device pins forming a 5-terminal structure. This paper describes the area scaling benefit of CFET for SDC and SRAM, then discusses an integration flow with a sensitivity analysis and a cost assessment. Finally, a TCAD evaluation compares the CFET to finFET in terms of power and performance.

CFET standard cell

SDC area in advanced finFET nodes is mostly driven by transistor access terminals. The area gain with CFET doesn't lie in the reduction of the active footprint by stacking p and n regions. CFET allows instead a simplification of the transistor terminals access for two reasons: firstly, the complementary nature of CMOS requiring both pull-up and pull-down networks to be controlled by the same input gates and, secondly, the advantageous interconnect network around these devices. In advanced technologies with unidirectional interconnects, two orthogonal layers are needed to provide intra-cell connectivity. This network usually suffers from fixed n and p regions confined to the North and the South of the cell. The stacked active of the CFET is positioned at the center of the cell with the port escape of each device being provided at will to North or South irrespective of the device type (Fig 2.). Thereby the vertical routing resource can be absorbed in the MOL and the cell can be finished with a single horizontal interconnect layer. To fully benefit from the CFET architecture, a buried power rail is added. This allows to reduce the SDC down to 3 routing tracks where today's most advanced finFET libraries are limited to 6 tracks. Fig 3 shows two complex cell layouts. Not only the area is reduced by a factor 2, the number of internal routing layers is also reduced to one horizontal layer for most of the cells in the library (with the exception of the flip-flop). Thereby, the finishing metal being horizontal, the cell opens to a first vertical Place-and-Route layer substantially improving overall port accessibility.

CFET SRAM

The output pass gate of the 6T SRAM makes it a non-CMOS structure. This has forced a modification of the flow to enable a compact SRAM cell. Fig 4 shows the strategy for SRAM layout scaling. The 2 SRAM inverters are implemented with a CFET by folding the nFET on the pFET. However, the pass-gate nFET cannot shift to the same fin track due to the presence of the cross-coupled connection. Recognizing that the pass-gate consists of a single device in the CFET stack, an additional set of steps is used to recess the top device leaving the bottom device capped and isolated from any connection on top. This way the cross-coupling can be enabled with the bottom device being accessed by the WL gate contact at the border of the cell. The pass-gate becomes a p-FET device and requires modification of its periphery. This approach allows for the reduction of the cell height to 4 fin tracks reducing the area by 50% compared to the 8 fin tall layout.

CFET integration and process sensitivity

The proposed flow starts with a standard bulk wafer. Although several stacked channel configurations are possible, we assume a vertical nFET sheet stacked over a pFET fin. This choice exploits the finFET flow and benefits from strain engineering in the bottom pFET. The flow (Fig 5) builds from the conventional gate-all-around (GAA) sequence [4] where 2 Si channels are separated by a SiGe sacrificial layer. Notice the addition of a module for a buried power rail after the fin formation. Once fins and gates are formed with the gate SiCO spacer, the S/D is recessed. A first challenging step is the formation of an inner spacer in a large vertical SiGe region. Then the core steps of the CFET follow. The approach used here consists in building both devices sequentially. A sealing step based on a sacrificial spacer covers the top device while the p-type S/D region is epitaxially grown on the bottom device followed by the formation of a metallic bottom electrode in a fill-CMP-etch back sequence (Fig. 6) that may introduce process variations. An isolating dielectric fill allows the nFET to be fabricated in a similar set of steps where the epitaxial growth of the n-type S/D region is seeded from the meanwhile exposed NW. The 2 electrodes need to be accessed from the buried metal layer. Therefore shallow and deep vias are patterned during the bottom and top electrode formation respectively. The flow assumes dual damascene fill. A COAG provides the gate contact while the S/D access to both terminals is created by shallow (top access) and deep (bottom access) contacts. The flow relies on accurate control of the elevation during the S/D contact formation to avoid short or defect between the stacked devices. A process control analysis (Fig 7) using Coventor SEMulator3D® [5] identified the MOL metal recess as well as SoC fill and recess accuracy as critical for yield. Improving process control through newer deposition techniques such as bottom-up fill and selective deposition may become critical for manufacturing. A cost analysis was performed based on the CFET flow for N5 patterning assumptions. The die cost is compared in Fig 8 to a finFET technology dimensionally scaled to N3. As observed, the relative reduction in die cost favors a structural CFET scaling over a dimensional finFET scaling path.

CFET inverter (INV) performance analysis

The CFET consisting of 2 stacked Si channels should exhibit similar intrinsic performance to the equivalent unfolded finFET design. However, the extra parasitic RC introduced by the compact structure need to be mitigated. A parasitic analysis based on COMSOL® [6] is used to model the device extrinsic components. The deep vias from the power rails and from the top interconnects are the highest contributors to resistance. It is therefore essential to reduce these extra access resistances by enabling thin barriers in the MOL. Ru is expected to lower the barrier thickness while maintaining acceptable bulk resistivity [7]. Thereby the CFET INV performance becomes comparable to the densest finFET INV with similar footprint. Fig 9 shows the power-performance of CFET compared to finFET for various MOL assumptions. This experiment suggests that the performance drop is dominated by parasitic R, where parasitic C is similar in both architectures. Moreover, CFET minimizes the gate drain parallel capacitance since in the CFET the gate runs orthogonal to the contact while in finFET the active contact runs parallel to the gate p-n routing. In this way, the gate extension beyond the fin can be minimized reducing the Miller capacitance in CFET. Fig. 10 shows that with the extension reduced to 10nm the CFET even outperforms the finFET and meets N3 target performance.

Conclusion

A DTCO analysis of the CFET demonstrates its scaling potential. 50% area scaling is obtained thanks to the reduction of tracks in SDC and a novel cross-couple scheme for the SRAM. The process flow requires a control of the elevation of the layers deposited in the S/D contact. A parasitic extraction from TCAD identified deep vias as major source of performance loss highlighting the need for advanced MOL schemes such as the utilization of Ru. In this way, CFET provides the necessary paradigm change for the next device architecture by overcoming the cost and performance degradation of traditional scaling giving a new horizon to Moore's Law.

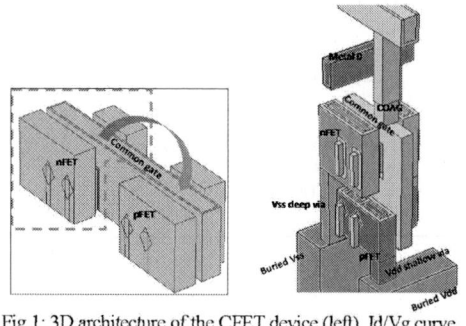

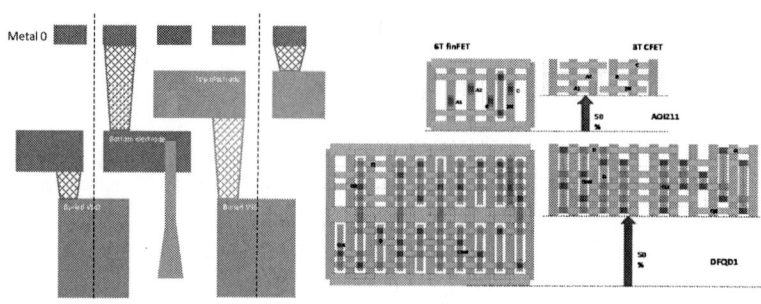

Fig 1: 3D architecture of the CFET device (left). Id/Vg curve of the 30nm tall vertical sheet (nFET) and finFET (pFET) Isolation distance is set to 20nm. Patterning assumptions are 42nm CPP, 24nm fin pitch and 24nm metal pitch.

Fig 2: Cross-section schematic of a CFET device in a 3T standard cell. Shallow and deep vias are needed to access the Power rail and the electrodes

Fig 3: Layout comparison between a 2 fin 6 track finFET cell and 1 fin 3 track CFET cells. Top shows an AOI211 while bottom shows a flip-flop. Note that power rail is buried in the CFET.

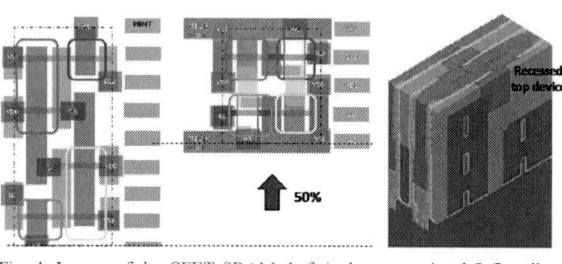

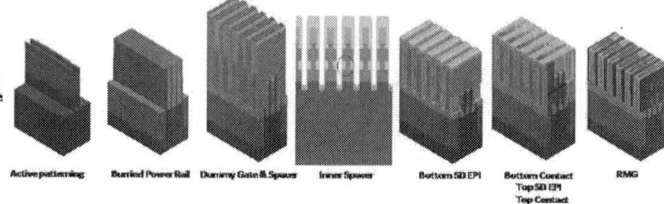

Fig. 4: Layout of the CFET SRAM. Left is the conventional 8 fin tall SRAM while middle shows the CFET SRAM. The right figure shows a X-section in the gate trench highlighting the recessed top device of the pass gate allowing the cross-couple to run above it.

Fig 5: Essential steps in the CFET process flow. Flow follows the standard GAA flow until the contact S/D. A bottom epi and contact is performed by sealing the top device. After dielectric isolation the top device (epi and contact) can be formed. The FEOL flow ends with the Replacement gate module.

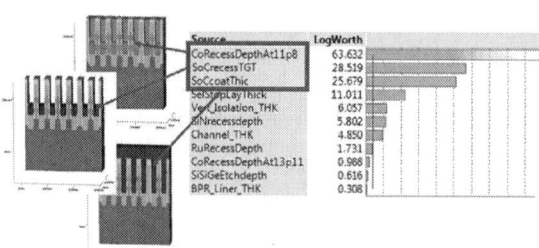

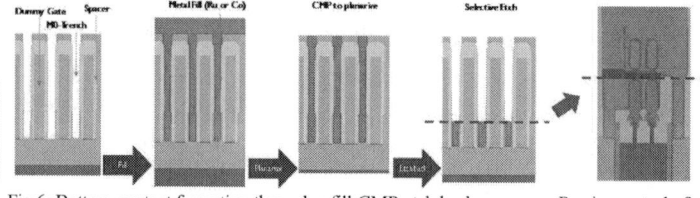

Fig 6: Bottom contact formation through a fill-CMP-etch back sequence. Precise control of the contact elevation is critical for the CFET process.

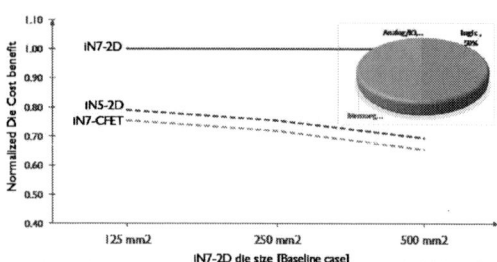

Fig 7: Process sensitivity analysis using Coventor SEMulator3D® identifies the metal recess (here Co) the Spin-on Carbon (SoC) thickness and SoC recess as the critical steps in the flow

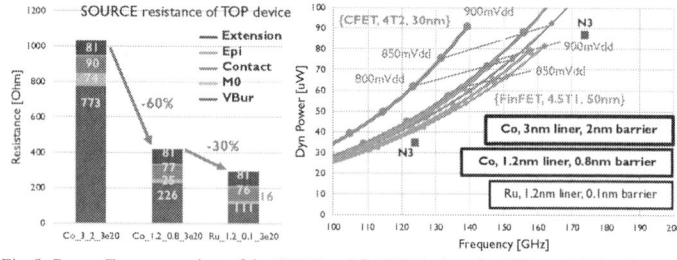

Fig. 9: Power-Frequency plots of the CFET and finFET devices for different MOL schemes. For a fair footprint comparison at N3 we assume a 4.5T single fin standard cell architecture for finFET and a 4T dual fin for CFET. The red squares define the power/performance scaling target for N3. A single fin per device is assumed for the finFET while a pair of stacked vertical sheets are assumed for the CFET. The test bench is based on a RO with FO3 and BEOL load. On the left is the source resistance breakdown.

Fig. 8: Relative cost analysis comparing a dimensional finFET scaling to a structural CFET scaling from N5 to N3. Assumptions are 50% Logic – 40% SRAM – 10% Analog/IO for chip distribution. A 6T library is assumed for finFET while a 4T is assumed for the CFET.

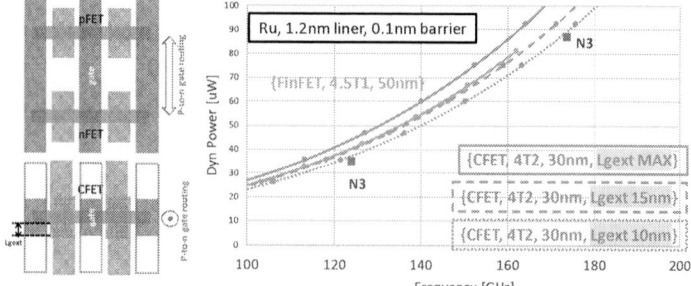

Fig. 10: Power-Frequency plots comparison under gate extension beyond fin optimization. Up to 7% performance improvement is observed in CFET.

References

[1] Liebmann L., IEEE Symposium on VLSI Technology, 2016, pp 112
[2] Kuhn K., Patent US 20120138886 A1
[3] Wu X., IEEE Transactions on Electron Devices 2005, pp 1998-2003
[4] Mertens H., IEEE Symposium on VLSI Technology, 2016, pp 158
[5] Coventor SEMulator3D®, http://www.coventor.com/
[6] COMSOL multiphysics, www.comsol.com
[7] Varela Pedreira O., IEEE International Reliability Physics Symposium, 2017

Power-performance Trade-offs for Lateral NanoSheets on Ultra-Scaled Standard Cells

M. Garcia Bardon[1], Y. Sherazi[1], D. Jang[1], D. Yakimets[1], P. Schuddinck[1], R. Baert[1], H. Mertens[1], L. Mattii[2], B. Parvais[1,3], A. Mocuta[1], D. Verkest[1]

[1] Imec, Kapeldreef 75, 3001 Leuven, Belgium; [2] Cadence Design System, San Jose, USA; [3] VUB, Brussels, Belgium

marie.garciabardon@imec.be

Abstract

In this paper, the performance of standard cells scaled down to 4.5 metal tracks based on Lateral NanoSheets is investigated for 3nm technology node targets using relevant logic benchmarks and power-aware metrics. The cell layout and parasitics in 4.5T cells set strong constraints on the NanoSheets geometry. The optimized NanoSheets could still outperform FinFETs by 9 to 20% frequency depending on circuit context, reaching 3nm node targets. An extra 21% performance improvement is expected with device level boosters enablement.

Introduction

As pitch scaling slows down, decreasing the height of standard cells by reducing the number of metal tracks has become a knob to maintain node to node area targets [1]. Ultra-scaled 5.5 tracks (5.5T) and 4.5T cells could give 15% and 30% extra area benefit over the foreseen 6.5T cell of the 5nm node. The shrink in active area leads to fin depopulation down to 2 and 1 fin only, challenging device strength [2-5]. In this context, Lateral NanoSheets (LNS) appear as promising option to replace FinFETs due to higher effective width per footprint and better gate control [6,7]. Wide sheets with 10 nm vertical sheet spacing have been demonstrated [8]. However, a large W_{eff} raises the question of dynamic power scaling due to capacitance increase. LNS also have the advantage of variable EUV-defined width, enabling leakage (IDDQ) and performance tuning at design time. The trade-offs for frequency, active and leakage power need to be precisely evaluated on ultra-scaled cells in circuit context.

This paper analyzes 4.5T cells with LNS of varying sheet widths (W, 11 to 21 nm), number of stacked sheets (NNS, 1 to 5), and vertical pitch (PNS, 18 down to 12 nm, resulting in sheet spacing of 13 to 7 nm), searching the optimal geometry while assessing the impact of different fan-outs (FO1 to FO10) and interconnect lengths, and benchmarking them against FinFETs cells, namely 5.5T2 (2 fins) and 4.5T1 (1 fin) (Fig.1). The sensitivity to specific device boosters (contact resistivity, spacer dielectric constant, channel stress) is then evaluated. The analysis uses hardware influenced TCAD-based models for the devices (Fig.2) and 3-D TCAD for the intra-cell parasitics extracted from layouts of patterning compliant standard cells.

4.5T cells layouts and scaling boosters

Cells have been drawn with aggressive 3nm node ground rules (Table1), assuming single Cobalt local interconnect (M0A), fully-self-aligned gate contact, and supervias from Metal1 to Gate and M0A (Fig.3). In 4.5T cells, only three metal tracks are left for internal routing. The shortage in metal tracks results in heavier use of M0A as intra-cell routing, in particular on complex cells. The pin access congestion at PnR is solved by unfolding complex cells into double height with minor or zero area penalty. The M0A tip-to-tip of 10.5 nm between adjacent cells is the main spacing bottleneck, requiring tight cut schemes. Single sheet stacks of maximum 21 nm width can be placed within these 4.5T cells. LNS assume wrapped around contact except for the widest sheet that have space for top contact only, with M0A cut on sheet edge.

Power Performance evaluation

LNS benefit from better SS and DIBL than FFs (Fig.4) and from a wider contact area, hence lower source series resistance. For three stacked sheets, this contributes to larger Ion current than in 4.5T1FF, for all sheet widths considered (Fig.5). This advantage does not propagate uniformly to AC level due to capacitances. The AC metric and targets used are shown on Fig.6. The reference 5.5T2 FF is 9% slower than 3nm node targets (defined as +40%

speed at constant power and -60% power at constant speed from 7nm node). The single FF 4.5T1 reaches the same performance curve as 5.5T2 but requires a higher operating voltage.

Sheet width, Number of stacked sheets, Vertical pitch- The frequency gain of 4.5T LNS over 5.5T2FF is shown in Fig.7, extracted at constant Vdd, then at constant power. At constant Vdd, the gain increases with NNS and W, then saturates. Fig.6 shows that despite this apparent performance improvement, increasing NNS leads back to the 5.5T2 power-frequency curve. Reporting speed gain at Vdd could hence lead to erroneous device selection. At constant power, the highest speed gain is for NNS 2 or 3, then degrades strongly. For higher NNS, W larger than 11 nm degrades speed even further. Despite this limit, many LNS geometries have better performance than 5.5T2FF. But if 3nm node targets need to be reached, the design space reduces: the only options are 3 stacked sheets of width 21 nm with vertical pitch 14 nm or smaller, or W 16 nm with PNS 12 nm (Fig.8, Fig.9). Two sheets of W 21 nm at PNS 18 nm reach targets but at the cost of Vdd increase similarly to the single FinFET option .

Capacitance breakdown and role of M0A- The previous limits can be understood by the relative increase in effective capacitance and current from AC simulations (Fig.10). In FO3, Ceff is dominated by the parasitic capacitance between M0A and Gate (Fig.11), amplified by Miller effect and increasing with NNS and PNS. Sheet width increases the gate and overlap capacitances mainly. The impact on total Ceff of increasing NNS, W, or PNS leading to a performance turning point depends on the capacitances balance, hence on Fan-Out, wire length, and intra-cell routing complexity. For example, for FO1, in INV cells the major Ceff component is the BEOL, but in XNOR M0A to Gate dominates since M0A has to be heavily used for routing (Fig.11b).

Fan-out and interconnect RC- The analysis is extended varying fan-outs and BEOL wire lengths (Fig.12). LNS have maximum gain in frequency over FF in two design spaces: for the largest FO at any wire length, and for FO1 with wires shorter than 1 um, where there is maximum benefit of LNS, with up to 20% higher speed than FF. Long wires increase the capacitive load but also induce important BEOL resistance penalty on Ieff (Fig.13). The analysis of SoC design after Place and Route for 4.5T cells shows extensive use of low FO and short wire lengths (Fig.14), where LNS show more benefit.

Sensitivity to device boosters- The potential for further step up is evaluated in Fig.15 by introducing several device boosters to the most aggressive LNS options and a more relaxed LNS options and on the 4.5T1 FF. If tight PNS can be enabled, LNS21 could reach 21% speed increase over 3nm targets with all step up elements considered.

IDDQ-performance optimization - Finally, the benefit of tuning W at design time is evaluated on Fig.16. Half a decade IDDQ window with 30% speed change is expected if W can be tuned from 11 nm to 21 nm. Two initial process defined Vt flavors would cover two decades of leakage power by changing W.

Conclusions

Lateral NanoSheets on 4.5T cells outperform 5.5T2 and 4.5T1 FF but need to be co-optimized with the intra-cell and inter-cell interconnect to account for power. 4.5T cells make heavy use of local interconnect for routing. Because of this, 4.5T LNS have a reduced geometry design space and are most efficient with low fan-outs and short wire lengths. Yet, with three stacked sheets of width 21 nm at tight vertical pitches of 14 nm or less, LNS can outperform FinFETs and reach performances beyond 3nm node targets while offering Vt adjustment on chip.

978-1-5386-4219-1/18 $31.00 © 2018 IEEE

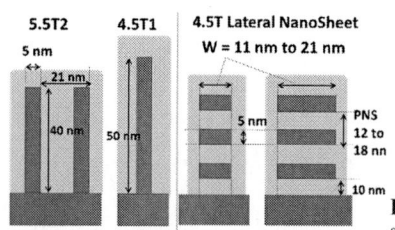

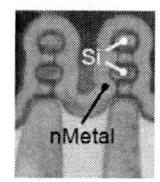

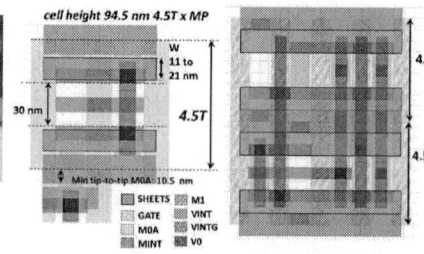

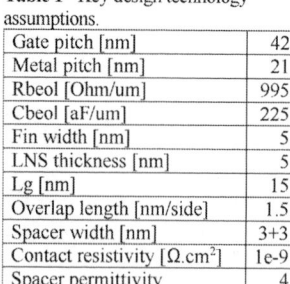

Fig.1- Lateral NanoSheets geometries studied and FinFETs references. Baseline is nSi-pSi channel with regrown S/D epi stressors for all.

Fig.2- TEM cross-section of fabricated devices of width 12nm [9]

Fig.3- 4.5T INVD1 and XNOR2D1 cells with LNS. M0A extends to middle tracks for P/N connection [10].

Table 1- Key design technology assumptions.

Gate pitch [nm]	42
Metal pitch [nm]	21
Rbeol [Ohm/um]	995
Cbeol [aF/um]	225
Fin width [nm]	5
LNS thickness [nm]	5
Lg [nm]	15
Overlap length [nm/side]	1.5
Spacer width [nm]	3+3
Contact resistivity [$\Omega.cm^2$]	1e-9
Spacer permittivity	4

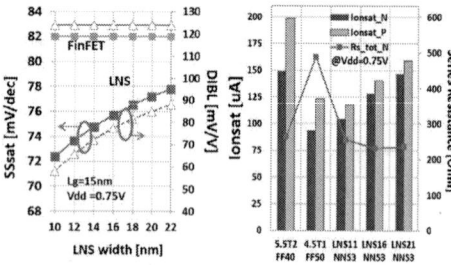

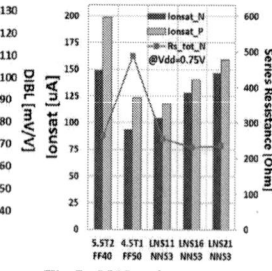

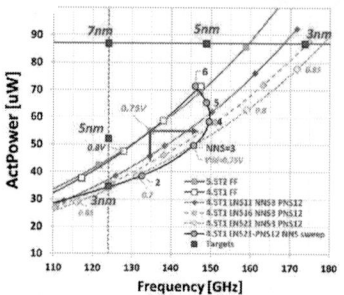

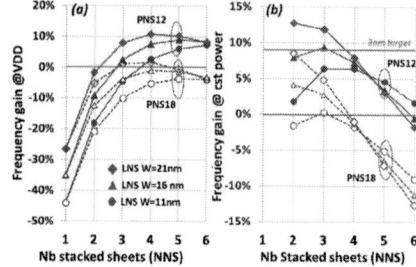

Fig.4- Larger sheet width benefits Weff but degrades the gate control (SS, DIBL). FF (fixed fin width 5 nm) indicated for comparison.

Fig.5- ION and source resistance obtained from TCAD. All devices have same High Performance Ioff=4 nA obtained by Vt shift.

Fig.6- Benchmark on RO INV FO3 with BEOL wire load 1.4 um. 5.5T2 used as reference to extract frequency gain at constant power and power at speed (arrows). For all devices, Ioff=4 nA at Vdd=0.65V.

Fig.7- Frequency relative to 5.5T2 FF for varying NNS, W, PNS *(a)* extracted at same Vdd=0.75V and *(b)* extracted at constant power. On *(b)*, a gain lower than zero is slower than 5.5T2 FF, and higher than 9% reaches target.

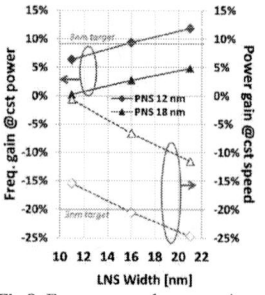

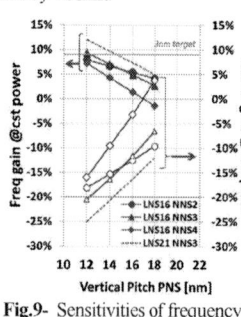

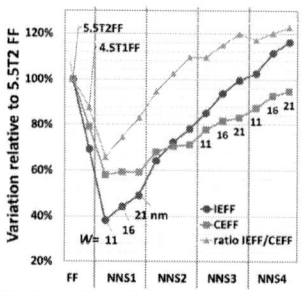

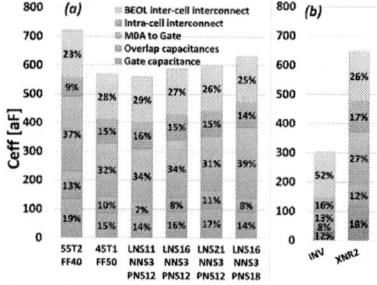

Fig.8- Frequency and power gains variation with LNS width for NNS=3, at constant IDDQ.

Fig.9- Sensitivities of frequency and power gains to PNS.

Fig.10- Ieff and Ceff from RO relative to 5.5T2 for increasing Weff (NNS, W). Ieff is affected by DIBL and Rbeol.

Fig.11- Ceff breakdown from RO simulations (a) for INV FO3, NNS=3, and (b) for INV FO1 versus XNOR2 FO1 at NNS=2.

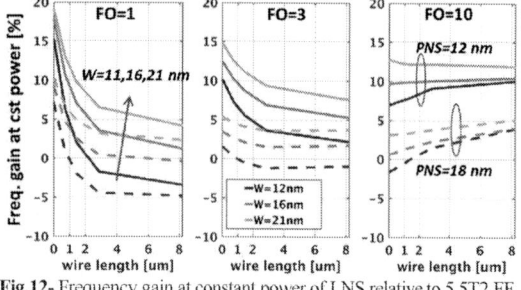

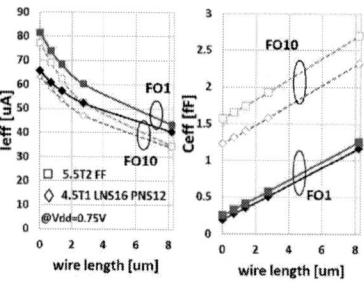

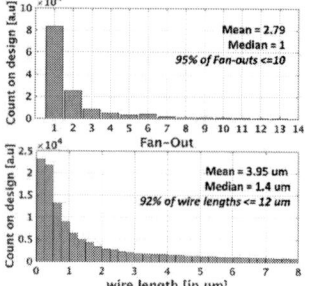

Fig.12- Frequency gain at constant power of LNS relative to 5.5T2 FF for different Fan-Outs and BEOL wire lengths. LNS width 11, 16, 21 nm for vertical pitch 12 and 18 nm, at NNS=3.

Fig.13- Ieff degrades strongly with BEOL interconnect resistance favoring short wires on design. 4.5T LNS benefit of smaller Ceff at large FO.

Fig.14- Distribution of Fan-outs and interconnect lengths after Place and Route on Mobile Core using 4.5T library

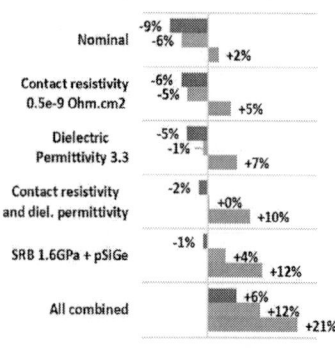

4.5T1 FF 50 nm
4.5T LNS16 NNS3 PNS18
4.5T LNS21 NNS3 PNS12

Fig.15: Performance step up with device boosters: expected frequency relative to 3nm node target frequency on nominal benchmark (FO3 INV with 1.4um wire) for improved contact resistivity, spacer permittivity constant, use of Strain Relaxed Buffer with pSiGe channel, and a combination of all, for 4.5T1 FF and LNS.

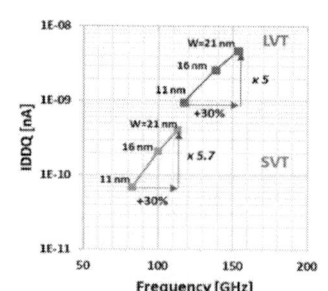

Fig.16: Increase in IDDQ and frequency from process defined Vts due to sheet width tuning at design time. Ioff is set for W=21 nm to 4 nA for LVT and 0.4 nA for SVT.

References

[1] L. Liebmann et al, VLSI, p. 11-1, 2016

[2] S. C. Song et al, VLSI, p. T198, 2015

[3] M. G. Bardon et al, IEDM, p. 28-2, 2016

[4] D. Yakimets et al, IEDM, p. 20-4, 2017

[5] P. Feng et al, TED, vol. 38(12), 2017

[6] Y. M. Lee et al, IEDM, p. 29-3, 2017

[7] D. Jang et al, TED, vol.64(6), 2017

[8] N. Loubet et al, VLSI, p. T230, 2017

[9] H. Mertens et al, IEDM, p. 37-4 , 2017

[10] Y. Sherazi et al, SPIE, 2018

Gap in pagination due to formatting issues.

Pages 145-146

Enabling CMOS Scaling Towards 3nm and Beyond

A. Mocuta, P. Weckx, S. Demuynck, D. Radisic, Y. Oniki, J. Ryckaert

imec, Kapeldreef 75, B-3001 Leuven, Belgium;

Email: Anda.Mocuta@imec.be

Abstract

We look at several scaling boosters necessary to accomplish CMOS area scaling towards the 2nm node. We consider aspects of standard cell area scaling, transistor architecture, SRAM, and BEOL. We also demonstrate integrated flows and hardware feasibility for such scaling boosters.

Introduction

Traditional CMOS scaling involves scaling the Contacted Poly Pitch (CPP) and Metal Pitch (MP) to produce about 50% area gain per CMOS generation. As seen in Fig. 1. due to limits in patterning and in device performance [5] the scaling of critical pitches is slowing down, placing an increased burden on other elements such as reducing Standard Cell (SDC) track height, and scaling boosters which are special process/layout constructs to allow further gains in area at SDC and block level. In this abstract we highlight a few scaling boosters needed to continue the path towards 3nm and beyond.

Standard cells and device architecture

While the industry may introduce a variety of nodes and different node nomenclature, we aim to highlight requirements needed to approach 40-50% area scaling node to node, while also considering difficulties in scaling critical pitches and finding alternative solutions. Table 1 summarizes critical dimensions and area scaling across anticipated scaled nodes. In order to satisfy area scaling requirements from previous node, a standard cell height of 6 tracks will be needed at 7nm. The track height and power rail CD at 7nm/5nm are driven by a requirement to keep 4 internal tracks for signal routing and by spacing requirements at the edge of the cell as seen in Fig 2. Note that at 5nm we can still achieve area scaling close to target by scaling the MP and CPP but will require the introduction of a Self-Aligned Gate Contact (SAGC) [6] at 5nm. We have developed a process flow for SAGC and have successfully demonstrated its feasibility at 42nm CPP, as seen in Fig. 3.

At 3nm node-to-node area scaling form CPPxMP is reduced to ~34% and it becomes necessary to consider track height reduction. Fig 4 shows a few options and their scalability potential. In Fig. 4(a) a conservative option is shown (5.5T) which has the advantage of still retaining a 2fin/device but has very tight requirements for active contact tip-tip. In Fig 4(b) and 4(c) we see more aggressive 5T options that require 1 fin per device, and require a buried power rail (BPR) or a complex MOL process to make the mid track available for connections to active. We favor a buried power rail (Fig. 4(b)), from a process complexity perspective (vs. Fig. 4(c)), also because it still allows 4 internal tracks and it is an enabler to further nodes. In Fig. 5 we show a cross-section of buried power rail, where Ru lines with aspect ratio up to 7 are demonstrated, and with resistance values compatible with power rail requirements. As a single fin device becomes the only option, based on drive strength and variability requirements a nanosheet device architecture could become desirable at 3nm [8, 9, 10]. At 2nm (table 1. Fig. 6), scaling becomes the usual trade-off between MP and trach height. At 2nm BPR allows consideration of

further SDC track height reduction. With that, a main constraint coming into play include a very tight N-P space within the cell, which has to happen within about a single fin pitch. For this reason, at 2nm the complementary FET (CFET) device architecture becomes relevant. This involves a stacked NFET/PFET architecture (finFET or GAA) as shown in Fig. 7. Initially at 2-fin CFET could be used, with potential to scale to a 1fin CFET.

SRAM

Historical and projected SRAM cells scaling is shown in Fig. 8 ([ref. 11-20]) SRAM scaling is limited by a different set of ground rules as illustrated in Fig. 9 [11]. At 7nm the limiter is the gate to S/D contacts distance (Fig. 9(a)), and scaling to 5nm requires SAGC introduction, similar to logic SDC. At 5nm the gate cut/gate extension past fin is the limiting factor, and is related to RMG process requirements. At 3nm a metal gate cut (MGC) is required to enable further cell height reduction [11]. Fig. 10(a) shows a schematic cross section for cut before RMG and for MGC. A process flow has been established – Fig. 10(b). Beyond 3nm SRAM scaling can continue on the same trend as in Fig. 9 by using a CFET device architecture. An SRAM cell layout using CFET is shown in the inset of Fig 8.

BEOL

Scaling the BEOL down to the 3nm node will require innovation in patterning. Recognizing that line tip-to-tip need to be patterned with a separate block lithography, two patterning options are considered for 3nm: SAQP with EUV block and EUV LELE with 193i block. The difference in the block imaging is driven by minimization of EUV masks for cost requirements. Nevertheless, defining a block in a sea of lines separated by 10nm requires self-alignment techniques of the block mask to the adjacent lines. This can be done using the self-aligned block (SAB) technique as shown in [21]. Here the SAQP gap lines are filled with an SoC material on which etch selectivity can be obtained with respect to mandrels in amorphous C and spacers in SiO2. With the block pattern being decomposed in 2 masks, one for each line population (mandrel and gap), the necessary self-alignment can be obtained as shown in Fig. 11. BEOL scaling in conjunction with track height reduction brings an added complexity in the pin access of the standard cells. A PnR study on an ARM M0 core shows that the 6T to 5T gain at cell level is hampered due to the minimum run length (minRL) of the first metal layer, M0. This issue can be mitigated using alternative metals such as Co which allows reducing the minRL and recover the full benefit of track height reduction (Fig. 12) [22].

CONCLUSION

We have outlined a few key requirements of scaling to N3 and beyond, together with experimental demonstration for some of them. We show that SAGC, BPR, MGC, SAB, minRL are needed to enable further scaling. GAA nanosheets and CFET architectures are enablers from the device side.

Acknowledgments: We would like to acknowledge our collaboration with TEL on SAGC and SAB.

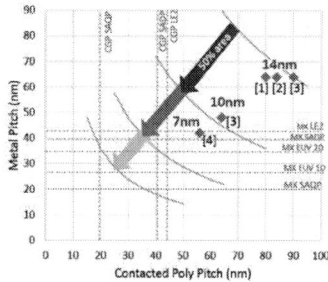

Fig. 1 CMOS Area scaling by scaling transistor gate and metal pitches.

	7nm	5nm	3nm	2nm		
Poly pitch (CPP)	56	48	42	42	42	42
Metal Pitch (MP)	42	28	21	18	16	14
CPPxMP scaling		42.86%	34.38%	14.29%	11.11%	12.50%
Track Height	6	6	5	4	4.5	5
cell height	252	168	105	72	72	70
SDC Area scaling		42.86%	45.31%	31.43%	31.43%	33.33%

Table 1 Critical dimensions for anticipated scaled CMOS nodes. Without standard cell track height scaling very limited gains are expected.

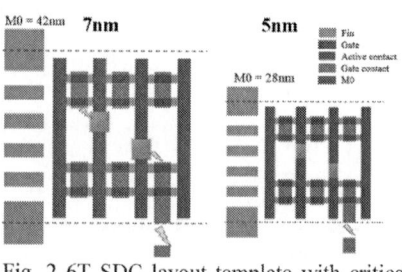

Fig. 2 6T SDC layout template with critical spacings highlighted: for 7nm active contact tip-to-tip and gate-to-active contacts; at 5nm active contact tip-tip remains critical (SAGC required).

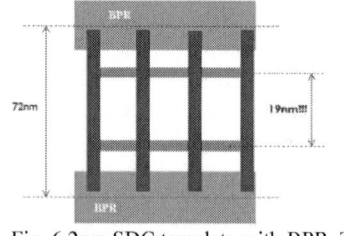

Fig. 3 SAGC Process flow and first experimental demonstration at 42nm CPP.

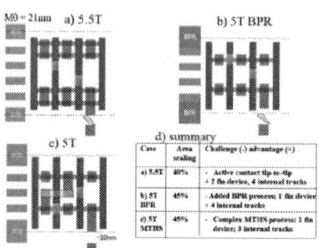

Fig. 4 SDC template options at 3nm a) 5.5T with a challenge in active tip to tip; b)5T with a BPR; c) 5T with a challenge in MOL connecting to middle track; d) summary table of pros and cons

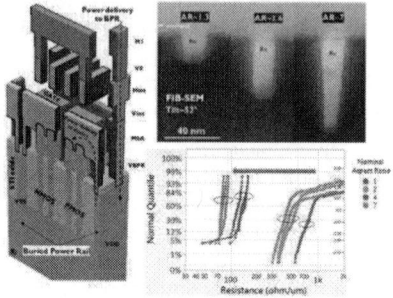

Fig. 5 BPR demonstration. Process emulation, cross-section of high aspect ratio Ru lines, line resistance distributions

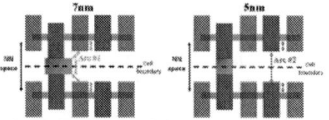

Fig. 6 2nm SDC template with BPR. The N-P separation becomes an issue

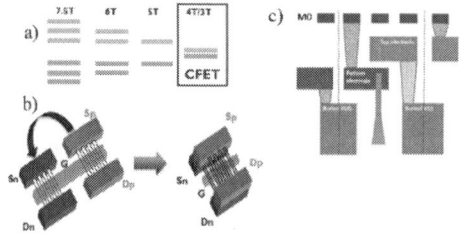

Fig. 7 CFET concept a) SDC height evolution and need for CFET at 4T/3T; b)"folding" the NFET over the PFET to accomplish the CFET; c) CFET S/D connection

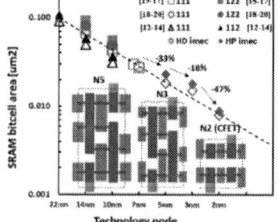

Fig. 8 SRAM area scaling across nodes including anticipated nodes. CFET enables SRAM scaling beyond N2.

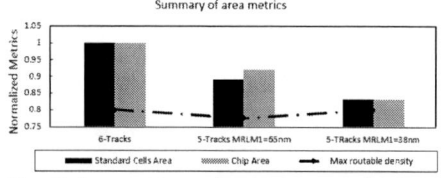

Fig. 9 SRAM limiting ground rules at 7nm/5nm. For 7nm arc #1 is important describing critical distances between contacts. For 5nm arc#2 is important. Arc#2 resolution at 3nm requires MGC.

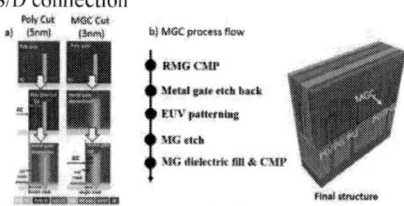

Fig. 10 Metal gate cut (MGC) a) compared to gate cut at poly and b) process flow

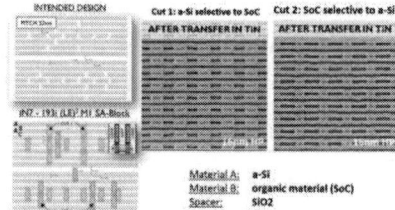

Fig. 11 SAB experimental demonstration (alternate lines cut)

Fig. 12 Area scaling and maximum routable density. Reduced minRL recovers the benefits of SDC scaling at block level.

References: [1]S.-Y.Wu *et al*. IEDM 2013; [2] C.-H. Lin *et al*. IEDM 2014;[3] H.-J. Cho *et al*. IEDM 2016; [4] S. Narasimha *et al*. IEDM 2017; [5] M.Garcia-Bardon *et al*. IEDM 2016; [6] C. Auth *et al*. IEDM 2017; [7] Coventor process emulation software; [8] Y.M.Lee *et al*. IEDM 2017; [9] N. Loubet *et al*. VLSI 2017; [10] D. Yakimets *et al*. IEDM 2017; [11] P.Weckx *et al*. IEDM 2017; [12] E. Karl *et al*., ISSCC 2012; [13] E. Karl *et al*., ISSCC 2015; [14] https://newsroom.intel.com/ [15] Y. H. Chen *et al*., ISSCC 2014; [16] S-Y Wu, *et al*., VLSI 2016; [17] M. Clinton *et al*., ISSCC 2017; [18] T. Song *et al*., ISSCC 2014; [19]T. Song *et al*., ISSCC 2016; [20] T. Song *et al*., ISSCC 2017; [21] F. Lazzarino *et al*. SPIE 2017; [22] L. Mattii *et al*. SPIE 2018;

Smart scaling technology for advanced FinFET node

Jongwook Kye, Hoonki Kim, Jinyoung Lim, Seungyoung Lee, Jonghoon Jung, and Taejoong Song

Foundry Business, Samsung Electronics, Hwaseong-si, Korea

E-mail: jongwook.kye@samsung.com

Abstract

Because of the complexity of technology the level of engagement between technology and design has been increased more than ever before. Design technology co-optimization (DTCO) is used to describe the process of making with competitive power, performance, area, and yield (PPAY) in various applications. This paper describes smart scaling technologies for advanced FinFET node to make technology more competitive.

Overview

Turn-around-time (TAT) of technology development has been required to be decreased, while complexity of technology has been increased on the contrary, shown in Fig. 1 [1]. To provide competitive PPAY efficiently, aggressive DTCO strategies have been applied to FinFET technology. Many applications such as internet-of-things (IoT), and machine-to-machine communicated applications demand low-power and smaller-area [2]. In addition, machine learning, block chain, and data mining have challenges to reduce energy consumption due to the requirements of huge amount of computational power. Therefore, various low-power driven DTCO have been highlighted recently. Mobile application has been a driving force over the past couple of decades for technology definition with relatively short life cycles. However, it's a complete different story for other industrial applications where devices need to be supported for longer period of time. To support all those segment of markets more competitive and holistic DTCO need to be considered from the beginning of the process development to chip design.

A. Area-driven DTCO: SDB and Construct

Fig. 2 shows DTCO activity for smaller logic area in 14nm node which is the first FinFET generation. Single diffusion break (SDB) has been supported for 14nm node, and continued to support multiple FinFET generations for standard cells to get better area scaling. Also in 10nm, a structure for process and scaling improved cell (SIC) is newly devised through DTCO, which is important structure for logic shrink, shown in Fig.3. It simplifies local wire interconnection by reducing metal congestion inside standard cells.

B. Low leakage-driven DTCO: CPP and Wimpy

Reducing leakage is crucial for low power applications. One of effective ways to reduce leakage is to use a wimpy device [3]. Fig. 4 shows a contacted poly pitch (CPP) optimization for leakage power vs. area. To meet high demand on low leakage applications, slightly larger CPP is chosen. It is also effective for lower the dynamic power by optimizing gate capacitance.

C. Low power and density-driven DTCO: low track library

A primary goal of low track library is the area reduction, and power reduction is additional benefit. Shown in Fig. 5, low track library has less gate capacitance due to the less number of fins and shorter cell height, which results in dynamic power reduction. We should not forget the P&R efficiency to preserve the area benefit that we've earned through the short track library enablement.

D. Variation-aware DTCO: LLE

As technology scales down, local layout effects (LLE) becomes more dominant factor that needs to be considered for circuit design [4]. Mostly, LLE is not preferred in terms of device variability and design V_{MIN}. On the other hand, it can be utilized to boost performance with circuit and layout technique [5]. As shown in Fig. 6, due to shallow trench isolation (STI) LLE, PFET drive-ability is better as it moves away from STI and NFET drive-ability is worse as it moves from STI. At a certain optimum distance, PFET's performance improvement is greater than NFET's degradation. Using these LLE feature, "Fast INV X2" can be faster than conventional "INV X2" without additional power consumption, and performance would be closer to "INV X4".

E. Low V_{MIN}-driven DTCO: Low V_{MIN} SRAM & F/F

Low V_{MIN} is challenging in 10nm and beyond due to small voltage headroom shown in Fig. 7 [6]. Critical circuits to limit low-voltage operation in SOC are SRAM and flip-flop, since these need large voltage headroom to maintain stability. Fig. 8 shows V_{MIN} enhancement techniques in SRAM and flip-flop. SRAM assist techniques help to increase stability and improve V_{MIN}, while sacrificing PPA. Thus, it is important to select the PPA optimized circuit. A flip-flop using transmission gate (TG) is area-efficient but vulnerable to write-fail due to bi-directional data path during low-voltage operation. To prevent flip-flop write-fail, low-voltage flip-flop circuit using uni-directional signal path without TG is better design for low V_{MIN} operation and reduces power by 23%.

F. Next DTCO for Yield: Chip-level SRAM Redundancy

While we have been focused on layout & circuit-based DTCO for low power for 10nm/8nm, yield-aware chip-level DTCO has been relatively overlooked [7]. A redundancy would compensate yield-loss of SRAM, but it requires additional circuitry shown in Fig. 9, which occur latency and area overhead depending on various redundancy types shown in Fig. 10. Chip-level PPA vs Yield-aware DTCO for different stages of process maturity helps to increase cost efficiency.

Summary

Various competitive DTCO for smart scaling technology is presented. It provides an optimized technology library for different requirements, and achieves development time and cost saving.

References

[1] L. Liebmann, *et al.*, *VLSI Tech.*, pp. 1-2, 2016.
[2] T. Dry, *et al.*, *VLSI Tech.*, pp. T164-T165, 2017.
[3] J. Ryckaert, *et al.*, *CICC*, pp. 1-8, 2014.
[4] S. Yang, *et al.*, *VLSI Tech.*, pp. T70-T71, 2017.
[5] F. Sato, *et al.*, *VLSI Tech.*, pp. T116-T117, 2013
[6] T. Song, *et al.*, *JSSC*, vol. 52, no. 1, pp.240-249, 2017.
[7] X. Wang, *et al.*, *Trans. Electron Devices*, pp. 3139-3146, Oct. 2015.

2018 Symposium on VLSI Technology Digest of Technical Papers

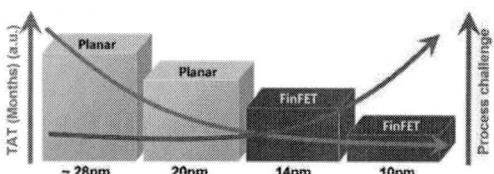

Fig. 1 Development timeline and the process-complexity

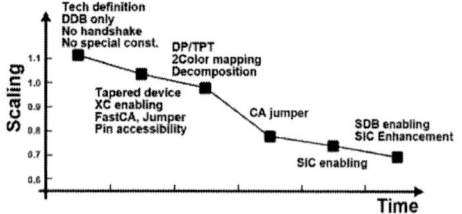

Fig. 2 Standard cell DTCO activity for smaller logic area

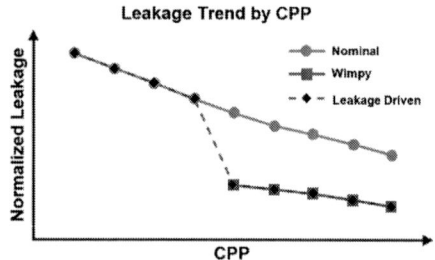

Fig. 4 CPP optimization to enable Wimpy for lower leakage

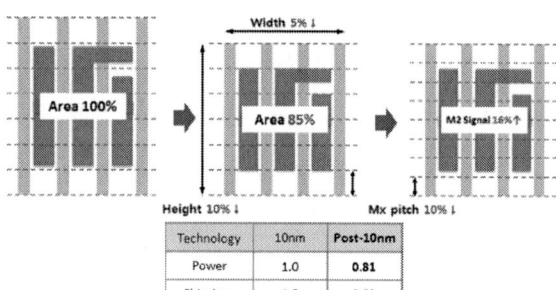

Fig. 5 Low track cell to enable low dynamic power

Technology	10nm	Post-10nm
Power	1.0	0.81
Chip Area	1.0	0.83

< Critical V_MIN limitation block in SoC >		
	SRAM Bitcell	Flip-Flop
Area portion in SOC	15 %	20 %
Stability	Write-ability & Stability	Latch stability
How to Improve V_MIN	- Large bitcell - SRAM assist	Low-voltage F/F

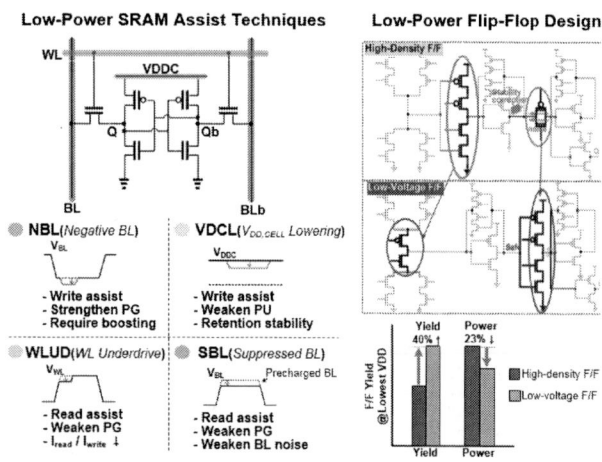

Fig. 8 SRAM assist and low-voltage F/F for low V_{MIN}

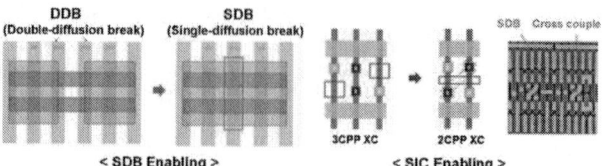

Fig. 3 SDB and SIC for area reduction

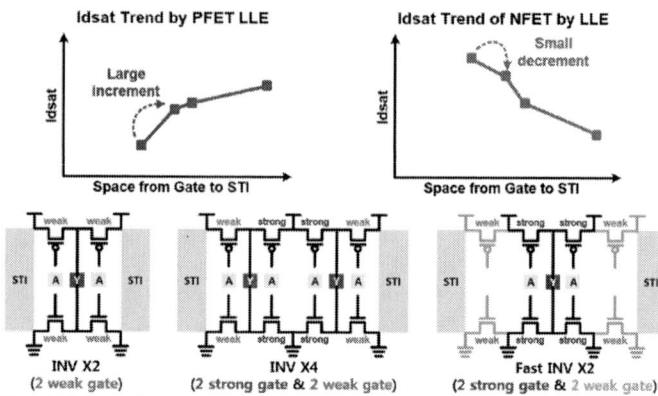

Fig. 6 high-performance and low-power INV cell by using LLE

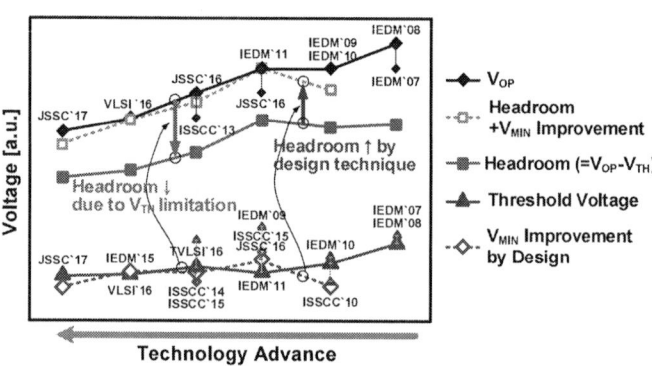

Fig. 7 Trend of the operating voltage (V_{OP}), threshold voltage (V_{TH}), and the voltage headroom (V_{OP}-V_{TH}) over technologies

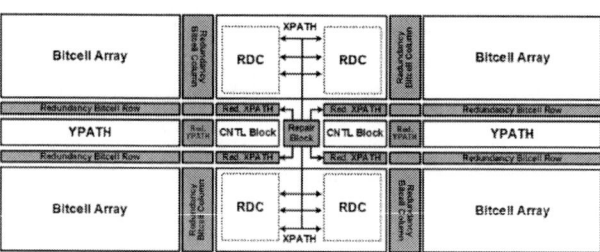

Fig. 9 SRAM architecture with redundancy circuitry

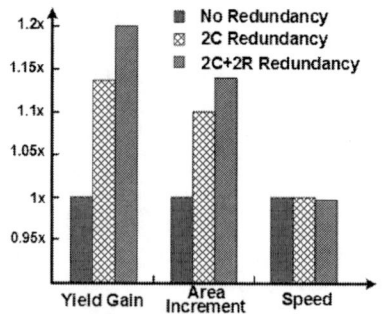

Fig. 10 SRAM redundancy for chip-level yield DTCO

978-1-5386-4219-1/18 $31.00 © 2018 IEEE

Sub-550mV SRAM Design in 22nm FinFET Low Power (22FFL) Technology with Self-Induced Collapse Write Assist

Daeyeon Kim, Jami Wiedemer, Pramod Kolar, Ayush Shrivastava, Jinal Shah, Satyanand Nalam, Gwanghyeon Baek, Xiaofei Wang, Zheng Guo, Eric Karl

Advanced Design, Logic Technology Development, Intel Corporation, Hillsboro, OR, USA, E-mail: daeyeon.kim@intel.com

Abstract

Exceptionally low minimum operating voltage (V_{MIN}) SRAM arrays have been demonstrated on 22nm FinFET low power technology (22FFL) [1]. By optimizing an undoped SRAM transistor and applying industry standard write assist techniques, 16Mb array of $0.087\mu m^2$ high-density bitcell (HDC) and 32Mb array of $0.107\mu m^2$ high-current bitcell (HCC) achieve the 95th percentile V_{MIN} of 505mV and 450mV respectively across a temperature range of -10°C to 95°C. A self-induced collapse (SIC) write assist integrated into the 6-T HDC SRAM bitcell array enables 110mV V_{MIN} reduction relative to an unassisted array at the 95th percentile with negligible power overhead.

Introduction

The ever-increasing necessity of battery-operated mobile and Internet-of-Things devices continues to emphasize the importance of low voltage operation. 22FFL [1] is introduced to provide a unique combination of low wafer cost and ease of design, high performance FinFET transistors, ultra-low leakage and ultra-low V_{MIN} for memory and logic circuits. Low wafer cost and ease of design is achieved through extensive use of single patterning in the metal stack and simplified design rules.

In this paper, the ultra-low V_{MIN} of HDC and HCC arrays on 22FFL are demonstrated using conventional write assist techniques. Furthermore, an SIC circuit for 6-T SRAM is introduced to provide V_{MIN} improvement with minimal power overhead and significantly reduced design complexity.

SRAM Bitcells and V_{MIN} with Conventional Write Assist

Fig. 1 shows the planar TEM images of HDC and HCC. The fin ratio of 1:1:1 (PU:PG:PD) is used in HDC to minimize bitcell footprint while HCC uses a larger PG and PD at 1:2:2 to lower V_{MIN} and to increase performance. Pulsed transient voltage collapse (TVC) [2] (Fig. 2) and negative bit-line (NBL) [3] (Fig. 3) write assist techniques are utilized in HDC and HCC respectively to improve write margin at low voltages. Fig. 4 shows active (read and write) and retention V_{MIN} distributions of 16Mb of HDC and 32Mb HCC arrays. HDC array shows the worst-case V_{MIN} distribution of 505mV at the 95th percentile and 430mV at median. HCC array demonstrates the V_{MIN} of 450mV and 415mV at the 95th percentile and at median respectively.

One of the key transistor features which enables exceptionally low V_{MIN} is V_T targeting based on gate length and work function, which enables improved random V_T variation (σV_T) [1]. The measured σV_T of 13mV/14mV for single fin NMOS/PMOS transistors is a 2.5X improvement over the previous 22nm [4] (Fig. 5).

Self-Induced Collapse (SIC) Write Assist

Even though TVC and NBL have been widely used to lower V_{MIN}, they lead to power [2] and area overheads. Area overhead is significant for small instances (Fig. 6) due to fixed-size active write assist. SIC, which has been used for hierarchical 8-T SRAM [5-7], is introduced for 6-T SRAM to mitigate area and power overheads. Fig. 7 shows the circuit diagram and waveforms of SIC. During write operation, the stacked PMOS transistors between VCC and VCS passively induce a certain amount of voltage drop at VCS, increasing write margin. Compared to TVC and NBL assists, it does not require active circuitry or careful timing between control signals. Improved σV_T in 22FFL makes SIC reasonably effective in V_{MIN} reduction even for HDC.

The array implementation is described in Fig. 8. To maximize the array efficiency, SIC cell for the stacked PMOS transistors as well as VCS isolation cell is modified from the actual SRAM bitcell – no additional transition space required. One VCS node is connected to only 30 bitcells to guarantee enough voltage drop at VCS and to minimize power overhead. The modular integration scheme enables 6-T SRAM to take advantage of 2.5-5.5% area overhead of SIC, compared to 3.5-13% area overhead of TVC. Also, it eliminates the necessity of device sizing and signal timing optimization in write assist circuitry for different-size instances.

V_{MIN} and Write Power Reduction with SIC

The active (read and write) V_{MIN} distributions of 3.75Mb HDC array are shown in Fig. 9. With SIC, 110mV V_{MIN} reduction is achieved compared to no write assist (NOWA) at the 95th percentile. The elimination of control signals and active circuitry result in negligible power overhead over NOWA, as seen in power measurements at 675mV and 1GHz (Fig. 10). In contrast to TVC, the local VCS in SIC is only connected to 30 bitcells instead of the full column, and only columns containing a low write margin bitcell will experience noticeable self-induced collapse, so the dynamic power consumption is minimized. For even lower V_{MIN} operation, wide-pulse TVC enables 115mV V_{MIN} reduction at the expense of additional 65% write power overhead. The trade-off between the 95th percentile write power and V_{MIN} is summarized in Fig. 11. SIC provides an attractive tradeoff by lowering V_{MIN} by 110mV with negligible power overhead, limited design complexity and more uniform behavior across different-size instances.

Fig. 12 contains the shmoo of NOWA, SIC, and wide-pulse TVC. A 2064x68 2-cycle latency HDC SRAM array with TVC achieves 2.1GHz at 720mV/-10C. At 720mV, SIC provides 12% better performance than NOWA. The die micrograph is shown in Fig. 13.

Conclusion

We demonstrated the 95th percentile V_{MIN} at -10°C/95°C of 505mV and 450mV of 16Mb HDC and 32Mb HCC arrays respectively. Using SIC, 110mV of V_{MIN} reduction with negligible power overhead relative to an unassisted array is demonstrated on an HDC array.

References

[1] B. Sell, et al., IEDM, pp. 685-688, Dec. 2017
[2] E. Karl, et al., JSSC, pp. 222-229, Jan. 2016
[3] E. Karl, et al., IEDM, pp. 561-564, Dec. 2012
[4] S. Natarajan, et al., IEDM, pp. 71-73, Dec. 2014
[5] M. Yuffe, et al., JSSC, pp. 194-205, Jan. 2012
[6] J. Kulkarni, et al., ISSCC, pp. 234-235, Feb. 2012
[7] K. Koo, et al., VLSIC, pp. 266-267, June 2015

2018 Symposium on VLSI Technology Digest of Technical Papers

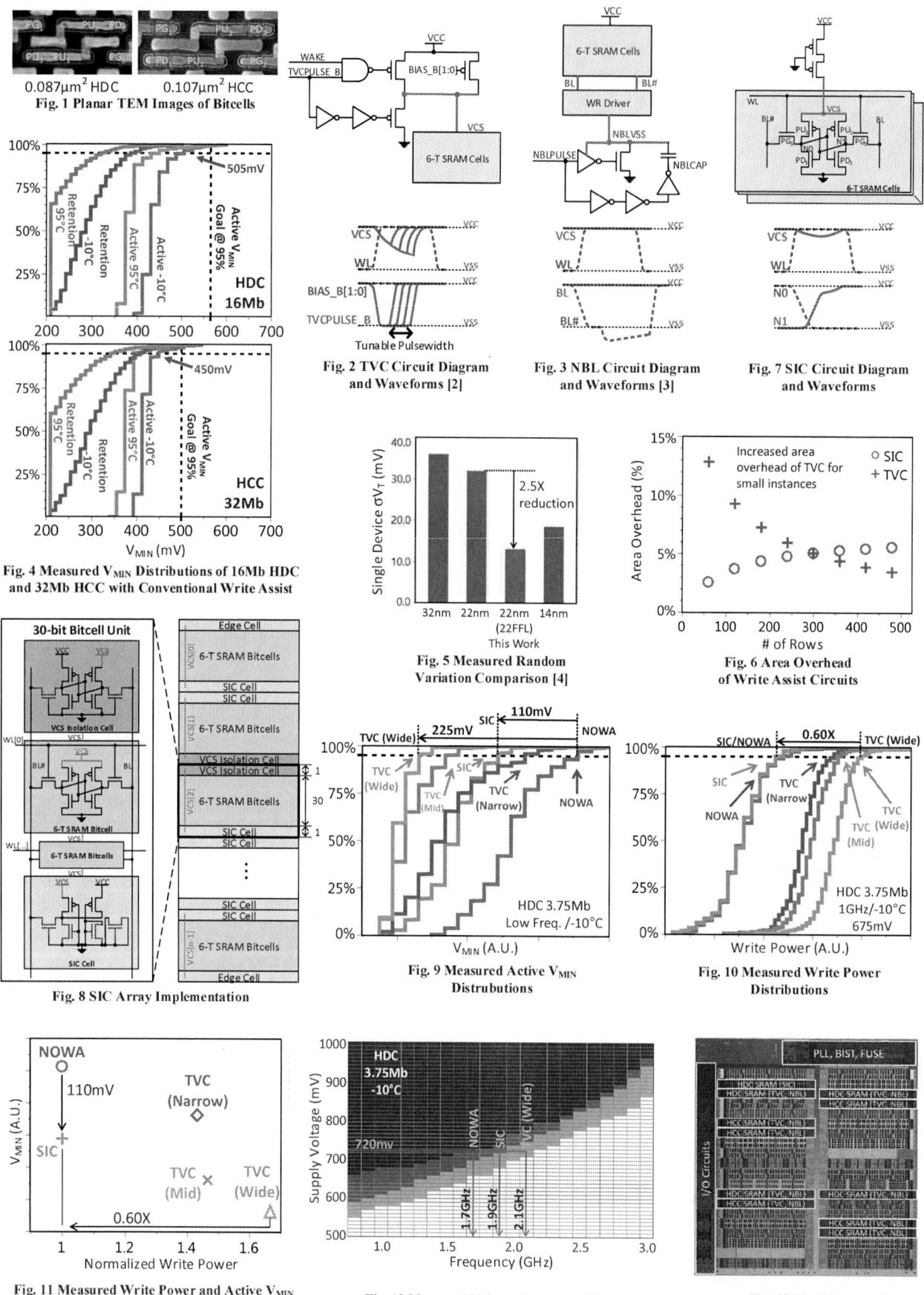

0.087μm² HDC   0.107μm² HCC
Fig. 1 Planar TEM Images of Bitcells

Fig. 2 TVC Circuit Diagram and Waveforms [2]

Fig. 3 NBL Circuit Diagram and Waveforms [3]

Fig. 7 SIC Circuit Diagram and Waveforms

Fig. 4 Measured V_MIN Distributions of 16Mb HDC and 32Mb HCC with Conventional Write Assist

Fig. 5 Measured Random Variation Comparison [4]

Fig. 6 Area Overhead of Write Assist Circuits

Fig. 8 SIC Array Implementation

Fig. 9 Measured Active V_MIN Distrubutions

Fig. 10 Measured Write Power Distributions

Fig. 11 Measured Write Power and Active V_MIN Trade-off at the 95th Percentile

Fig. 12 Measured Voltage-Frequency Shmoo

Fig. 13 Die Micrograph

978-1-5386-4219-1/18 $31.00 © 2018 IEEE   152

Design Technology Co-Optimization in advanced FDSOI CMOS around the Minimum Energy Point: body biasing and within-cell V_T-mixing

F. Andrieu[1], L. Pirro[2], R. Berthelon[1,3], J. Morgan[2], G. Cibrario[1], M. Wiatr[2], J. Hoentschel[2] and M. Vinet[1]

[1] CEA-LETI; [2] GLOBALFOUNDRIES, fab 1; [3] STMicroelectronics

Email: francois.andrieu@cea.fr ; Phone: +33438780139

Abstract- We propose an original Technology/Design Co-optimization of standard cells mixing devices of different threshold voltages (V_T-flavors) within a cell. It is successfully applied with nMOS Low-V_T (LVT) and pMOS Super-Low-V_T (SLVT) in Ultra-Low-Voltage (ULV) Fully Depleted Silicon-On-Insulator (FDSOI) LETI standard cells using diffusion breaks. It enables adjusting the V_T of pMOS subject to SiGe-channel-induced Local Layout Effect (LLE); leading experimentally to a 23% frequency gain on 22nm FDSOI technology for a 2-finger inverter Ring Oscillator (IVSX2 RO) vs. reference LVT at the same static leakage and V_{DD}=0.4V supply voltage; which corresponds to the Minimum Energy Point (MEP). This solution is combined with Forward Body Biasing (FBB), which brings +253% frequency at V_{DD}=0.4V and FBB=1.6V and improves the energy efficiency with a -13% minimum Energy Delay Product (EDP) along with a 50mV V_{DD} reduction at the minimum EDP.

Introduction- FDSOI CMOS offers an excellent performance/power/cost tradeoff for mobile, IoT and wearable applications [1]. Moreover, its excellent electrostatic control and variability, as well as its FBB capability, make it a good candidate for ULV operations [2-3]. At low V_{DD}, the transistor V_T must be perfectly controlled. LLE can impact the device centering, as it is the case for SiGe-induced LLE in advanced FDSOI technologies. Process or (continuous-RX) design solutions exist to suppress LLE and get the maximum performance from the SiGe booster [4-5]. Here, we propose an original design that mitigates SiGe-induced LLE to increase the performance while preserving low-leakages. It consists in mixing within a standard cell different V_T flavors. It is first assessed using SPICE, showing -37% delay vs. LVT on a 2-finger inverter (IVSX2) at V_{DD}=0.4V and finally validated on 22nm FDSOI, with an experimental 23% frequency gain.

Concept of within-cell V_T-Mixing (SPICE)- SiGe-channel induces a LLE, changing both the hole mobility and the pMOS threshold voltage V_{Tp} for short gate-to-STI distances (so-called SA/SB) (Fig.1-2 and [5]). It impacts all the cells that do not use continuous-RX [5] but a diffusion break (Fig.2). More precisely, using LETI SPICE model, we evaluate that $|V_{Tp}|$ increases by about 120mV between long SA and SA=94nm [4]. Such a V_T shift corresponds rather well to the difference between LVT and SLVT (Fig.1). Our original concept is thus to integrate SLVT pMOS with LVT nMOS in so-called MIX cells (Fig.2). It aims at drastically increasing pMOSFETs drive current without impacting the cell leakage, which is still limited by the LVT nMOS OFF-state current. It leads to a 20% delay gain at a given static leakage (I_{stat}) and V_{DD}=0.65V on a 1-finger inverter at CPP=104nm vs. ref LVT (Fig.3). The gain is higher for lower CPP (i.e. small SA) and shorter cells (consistently w/ Fig.1), as well as for lower V_{DD}, where V_{Tn}/V_{Tp} balance is key (Figs.3-4). Up to 60% delay improvement is obtained by IVSX1 at V_{DD}=0.4V, CPP=104nm (Fig.5). Finally, as the SiGe-induced LLE mainly impacts the pMOS at the edges of the active region, we have also studied the opportunity to swap only these edge transistors from LVT to SLVT (in a so-called "MIX 1finger"

cell, see Fig.2), exploiting further the small granularity of LVT/SLVT mixing capability in 22FD-SOI. There, after standard cell abutment, the minimum length of SLVT/LVT islands is 2 CPPs. MIX 1finger is a scriptable solution, which shows better performance than MIX for large cells (Fig.4).

Experimental 22nm RO results- 22FD-SOI integrates SiGe channel (Fig.6, [1]) and offers LVT/SLVT flavors mixable (on the same flipped-well) and compatible with FBB. First, we characterized some ULV features on LETI IVSX2 ROs (FO=3, SA=302nm). We report for the first time that the minimum EDP of this technology, corresponding to the maximum energy efficiency, is obtained at V_{DD}=0.7V for LVT (V_B=0V) (Fig.7). Reducing V_{DD} below 0.7V, the energy per cycle can be reduced even more down to MEP, which is reached below 0.4V for LVT (Fig.8). However, at such an ultra-low V_{DD}, speed is usually weak. FDSOI can solve this issue with FBB, which enables to boost the frequency by 253% at V_{DD}=0.4V and FBB=1.6V (Fig.9) and even to improve the energy efficiency with a -13% EDP along with a 50mV V_{DD} reduction at the minimum EDP (Fig.7). Fig.7 also evidences for the first time that a minimum EDP exists for various V_B at a given V_{DD}=0.65V (reached at V_B=0.8V). Then, we have assessed 2 solutions to adjust n/p ratio: asymmetrical ($|V_{Bp}|{\neq}|V_{Bn}|$) body bias and MIX designs. For SA=302nm, the sensitivity of the frequency/leakage to the body bias is similar when V_B is applied on either pMOS or nMOS (V_{Bp}/V_{Bn}) (see the symmetry of Fig.9-10a). This is due to both the similar body bias efficiency of n&pMOS and the good nMOS/pMOS V_T-centering at this geometry. This is not true for SA=94nm because of the aforementioned LLE on pMOS (Fig.10b). That is why, for SA=94nm, $|V_{Bp}|{>}|V_{Bn}|$ is rather the optimum biasing configuration because it reinforces the pMOS drive current (Fig.11). Body biasing is thus efficient to adjust the n/p ratio, provided V_{Bp}&V_{Bn} can be adjusted by the circuit. It is a block-level solution (and thus perfectly suitable for temperature compensation, e.g. [3]). On the other hand, within-cell V_T-mixing (MIX) is effective to manage LLE and is a cell-level solution. MIX cell performance exceeds SLVT one and leads to a +23% performance vs. ref LVT for IVSX2 at the same I_{stat}, same effective capacitance and V_{DD}=0.4V (Figs.12-13). However, for MIX IVSX2 cells, the $|V_{Bp}|{=}|V_{Bn}|$ operation space is not the optimum (Fig.14). This result tends to show that further optimizations are possible. Moreover, a higher gain could be expected for IVSX1 (as simulated in Fig.5b, but not taped-out yet). Anyway, MIX solution, combined w/ body biasing, successfully reduces the min EDP at V_{DD}=0.65V SA=94nm and thus improves the energy efficiency.

Conclusion- We report for the first time some 22FD-SOI ULV features measured on LETI inverter ROs: MEP below 0.4V and min EDP at $V_{DD}{\approx}0.65$V for LVT. We evidence also for the first time that an optimum V_B exists leading to -13% min EDP (vs. V_B=0). On these ULV cells using diffusion breaks, an original DTCO has been assessed based on within-cell V_T-mixing, leading to +23% frequency for IVSX2 MIX ROs vs. ref LVT at the same static leakage and V_{DD}=0.4V. Up to 60% delay improvement is expected for IVSX1 MIX cells by SPICE.

978-1-5386-4219-1/18 $31.00 © 2018 IEEE

Acknowledgements·

This work was funded by the ECSEL WAYTOGOFAST project

References :

[1] R. Carter et al., p.2760, IEDM'16. [2] S. Kamohara et al., p. 978-9, VLSI'14. [3] S. Clerc et al, p.150-1, ISSCC'15. [4] R. Berthelon et al, p.468-71, IEDM'16. [5] R. Berthelon et al, VLSI'16

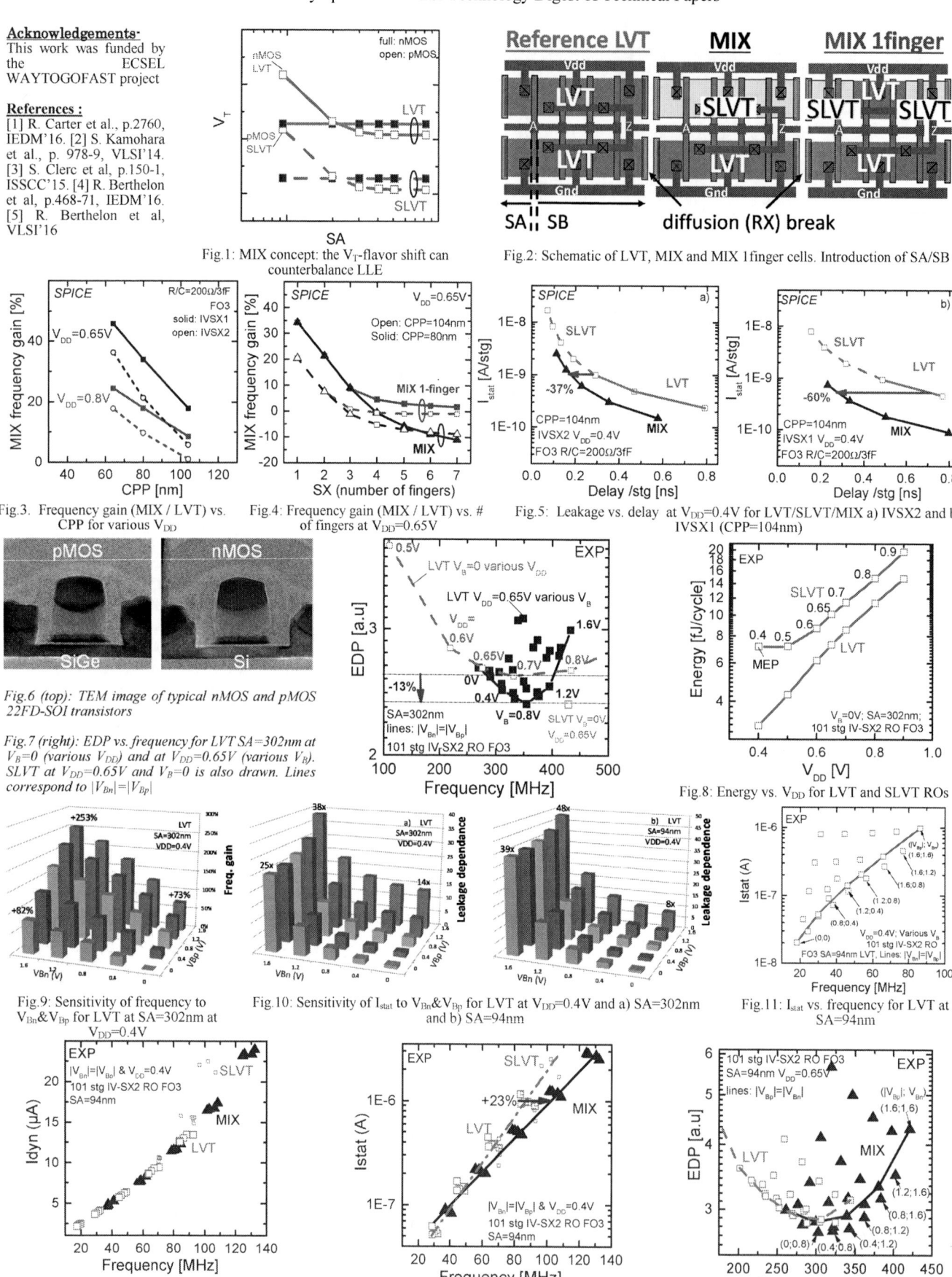

Fig.1: MIX concept: the V_T-flavor shift can counterbalance LLE

Fig.2: Schematic of LVT, MIX and MIX 1finger cells. Introduction of SA/SB

Fig.3. Frequency gain (MIX / LVT) vs. CPP for various V_{DD}

Fig.4: Frequency gain (MIX / LVT) vs. # of fingers at V_{DD}=0.65V

Fig.5: Leakage vs. delay at V_{DD}=0.4V for LVT/SLVT/MIX a) IVSX2 and b) IVSX1 (CPP=104nm)

Fig.6 (top): TEM image of typical nMOS and pMOS 22FD-SOI transistors

Fig.7 (right): EDP vs. frequency for LVT SA=302nm at V_B=0 (various V_{DD}) and at V_{DD}=0.65V (various V_B). SLVT at V_{DD}=0.65V and V_B=0 is also drawn. Lines correspond to $|V_{Bn}|=|V_{Bp}|$

Fig.8: Energy vs. V_{DD} for LVT and SLVT ROs

Fig.9: Sensitivity of frequency to V_{Bn}&V_{Bp} for LVT at SA=302nm at V_{DD}=0.4V

Fig.10: Sensitivity of I_{stat} to V_{Bn}&V_{Bp} for LVT at V_{DD}=0.4V and a) SA=302nm and b) SA=94nm

Fig.11: I_{stat} vs. frequency for LVT at SA=94nm

Fig.12: I_{dyn} vs. frequency LVT, SLVT and MIX ROs

Fig.13: I_{stat} vs. frequency for LVT, SLVT and MIX

Fig.14: EDP vs. frequency for LVT & MIX

Gap in pagination due to formatting issues.

Pages 155-156

Self-organized gate stack of Ge nanosphere/SiO₂/Si₁₋ₓGeₓ enables Ge-based monolithically-integrated electronics and photonics on Si platform

P. H. Liao,[1] M. H. Kuo,[1] C. W. Tien,[2] Y. L. Chang,[2] P. Y. Hong,[2] T. George,[1] H. C. Lin,[2] and P. W. Li[1,2]

[1]Department of Electrical Engineering, National Central University, [2]Department of Electronics Engineering & Institute of Electronics, National Chiao Tung University, Taiwan

Abstract-We report the first-of-its-kind, self-organized gate stack of Ge nanosphere (NP) gate/SiO₂/Si₁₋ₓGeₓ channel fabricated in a single oxidation step. Process-controlled tunability of the Ge NP size (5–90nm), SiO₂ thickness (2–4nm), and Ge content ($x = 0.65$–0.85) and strain engineering ($\varepsilon_{comp} = 1$–3%) of the Si₁₋ₓGeₓ are achieved. We demonstrated Ge junctionless (JL) n-FETs and photoMOSFETs (PTs) as amplifier and photodetector, respectively, for Ge receivers. L_G of 75nm JL n-FETs feature $I_{ON}/I_{OFF} > 5\times10^8$, $I_{ON} > 500\mu A/\mu m$ at $V_{DS} = 1V$, $T = 80K$. Ge-PTs exhibit superior photoresponsivity $>1,000A/W$ and current gain linearity ranging from nW–mW for 850nm illumination. Size-tunable photoluminescence (PL) of 300–1600nm (NUV-NIR) are observed on 5–100nm Ge NPs. Our gate stack of Ge NP/SiO₂/Si₁₋ₓGeₓ enables a practically achievable building block for monolithically-integrated Ge electronic and photonic ICs (EPICs) on Si.

Keywords: Ge, junctionless, phototransistor, monolithic integration.

1. Introduction

Seamless integration of Si-based EPICs in CMOS technology promises cost-effective data transfer, IoT, and quantum computing applications. While it is a formidable task for Si itself to implement EPICs, recently Ge emerges as the savior thanks to its pseudo-direct bandgap, high carrier mobility, and the CMOS compatibility. Discrete Ge MOSFETs [1–4] and photodetectors (PDs) [5–8] were demonstrated, however, monolithically-integrated Ge receivers on Si for optical/electrical signal conversion and amplification, is challenging due to the size incompatibility for μm-scale PDs v.s. nm-scale MOSFETs. In this paper, we proposed self-organized gate stack of Ge/SiO₂/Si₁₋ₓGeₓ being simultaneously fabricated in a single oxidation step of SiGe pillars over buffer Si₃N₄ on Si, enabled by a unique combination of Ge interstitial-enhanced oxidation of Si₃N₄ and Si interstitial-mediated SiO₂ "destruction-construction" mechanism [9].

2. Ge-sphere/SiO₂/Si₁₋ₓGeₓ-channel and Device Fabrication

We started with a sequential deposition of 20nm-thick Si₃N₄ and 70nm-thick poly-Si₀.₈₅Ge₀.₁₅ over SOI (100) substrates. Si₀.₈₅Ge₀.₁₅ nanopillars were lithographically patterned and then subjected to oxidation at 900°C in an H₂O ambient for forming gate stacks of Ge-NP gate/SiO₂/Si₁₋ₓGeₓ-recess channel. For JL n-FETs (Fig. 1), Ge NP and S/D were implanted by phosphorous at $5\times10^{15}cm^{-2}$, 20keV simultaneously. n⁺-poly-Si was deposited in contact with the Ge NP followed by 900°C anneal for 30min. For Ge-NP PTs (Fig. 2), S/D was implanted by phosphorous and activated by RTA at 900°C, followed by transparent ITO gate formation.

3. Results and Discussion

A. Process-controlled Tunability of Gate-stacking Structure:

Fig. 3 displays CTEM/EDX micrographs of 5–90nm Ge-NP/SiO₂/SiGe stacks formed over Si substrates. Thermal oxidation converts the Si content of Si₀.₈₅Ge₀.₁₅ pillars to SiO₂ and squeezes residual Ge radially inwards to the core of the oxidized pillars, forming a cluster of Ge crystallites. Under optimum process conditions, this Ge cluster can be made to migrate through newly-formed SiO₂, penetrate buffer Si₃N₄, and submerge the Si substrate below. Concurrent with the migration, coarsening of the Ge crystallites proceeds to complete coalescence, resulting in a single Ge NP formed for each pillar (Fig. 4). A recess Si₁₋ₓGeₓ shell, conformally matching the Ge NP, is simultaneously formed on the surface of Si substrate due to Ge migrating from the NP and dissolving within Si. This SiGe shell is separated from the Ge NP by a 2–4nm-thick SiO₂

being formed at 900°C. Process-controlled tunability of the Ge NP size, SiO₂ thickness, and SiGe shell is achieved by a dynamic balance between concentrations of Ge, Si, and O interstitials. Our self-organized heterostructure is analogous to the gate stack of poly-Si/SiO₂/Si, and the channel dimensions (L_G/W_G) are essentially determined by the area of the Ge NP in contact with Si.

B. Crystal Orientation, Strain Engineering of SiGe channel:

Our gate stack possesses feasibility of producing high Ge-content Si₁₋ₓGeₓ recess channels with high compressive stress on Si with a high degree of crystallinity. HRTEM/NBD patterns (Fig. 5) show that the SiGe shell inherits the original crystal orientation of Si (100) substrates. Increasing the Ge NP size leads to a systematic blue shift in phonon lines of Ge-Ge and Si-Ge, confirming size- tunable strain engineering for the SiGe channels (Fig. 6). Estimated Ge content and strain for the Si₁₋ₓGeₓ shells are $x = 0.65$–0.87 and -0.25–-3.2%, respectively. EDX analysis confirms the NP size-dependence of Ge content and strain for the Si₁₋ₓGeₓ shell.

C. Size-dependent Optical Bandgap of Ge NPs:

Size-tunable bandgap energy in terms of $E_{bandgap} \propto 1/D_{NP}^{1/2}$ for Ge NPs embedded within SiO₂ and Si₃N₄, respectively, are evidenced by a systematic blue shift in PL wavelength from 300–1600nm when decreasing the Ge NP size from 90–5nm (Fig. 7) due to quantum confinement and strain effects. Activation energy of detrapping carriers from Ge NPs appears to increase with reducing the NP size due to strong charge interactions.

D. Performance of Ge JL-nFETs:

Fig. 8 shows I_D-V_G and I_D-V_D curves of Ge NP/SiO₂/SiGeOI JL n-FETs. $I_{ON}/I_{OFF} >10^5$ and 5×10^8 are measured at 300K and 80K, respectively, when biased at $V_{DS} = 0.5V$. Very low I_{OFF} ($<$ pA/μm) in combination with high I_{ON} ($> 0.5mA/\mu m$ at $V_{GS} = V_{DS} = 1V$) verifies superior gate oxide integrity of our gate stacks, which are our salient features in comparison to state-of-the art Ge n-FETs (Fig. 9). [1–4]

E. Performance of Ge PTs for Visible-NIR Photodetection:

Fig. 10 shows that our Ge-NP/SiO₂/SiGeOI n-PT exhibits photoresponsivity of $>1,000A/W$ at 0.01μW or current gain linearity in the power range of nW– mW for 850nm illumination, depending on gate polarity. 3-dB frequency of 2GHz is achieved. Detectivity of 6×10^{12} cm·Hz$^{1/2}$/W is much higher than that of Ge/GeSn PDs and even comparable to InGaAs PDs (Fig. 11). [10, 11]

In summary, the primary advantages of our gate stack lie in (1) its inherent structural simplicity and the process elegance of being simultaneously produced within a single oxidation step, (2) the process-controlled capability to produce highly-stressed SiGe channels with a high degree of crystallinity, (3) size-tunable bandgap energy for Ge NPs, and (4) strain engineering for the SiGe channels. Armed with these salient features, high drive current in combination with low I_{OFF} are achievable for Ge JL n-FETs as well as high photoresponsivity and detectivity are measured on Ge NP/SiGe PTs, respectively, enabling feasibility of high-performance, functionally-diversified CMOS EPICs.

Acknowledgment: This work was supported by MOST 105-2221-E-009-134-MY3 and 106-2633-E-009-001, Taiwan, R. O. C.

References: [1] W. Chang, *IEEE EDL*, 37, p. 253 (2016). [2] H. Wu, *IEEE TED*, 62, p. 1419 (2015). [3] C. Chung, *IEDM*, p. 383 (2012). [4] C. Lee, *IEDM*, p. 416 (2010). [5] R. Going, Opt. Exp. 23, 11975 (2015). [6] J. Wang, *IEEE PTL*, 23, p.765 (2011). [7] R. Going, *IEEE JSTQE*, 20, 8201607 (2014). [8] C. Chien, **Nanoscale**, 6, 5303 (2014). [9] T. George, *JAPD*, 50, 105101 (2017). [10] S. Siontas, *APL*, 109, 053508 (2016). [11] T. Pham, *Elect. Lett.*, 51, 854 (2015)

2018 Symposium on VLSI Technology Digest of Technical Papers

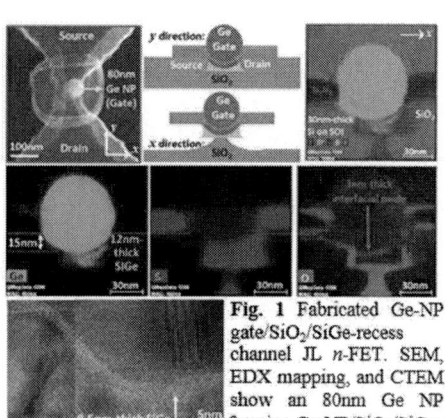

Fig. 1 Fabricated Ge-NP gate/SiO$_2$/SiGe-recess channel JL *n*-FET. SEM, EDX mapping, and CTEM show an 80nm Ge NP forming Ge-NP/SiO$_2$/SiGe channel simultaneously in a single oxidation step.

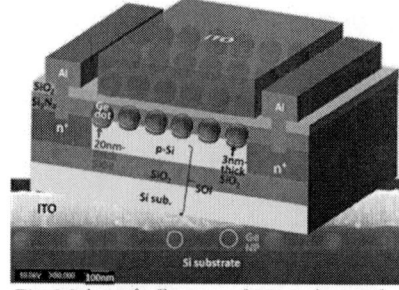

Fig. 2 Schematic diagram and SEM micrograph of photoMOSFETs comprising a gate stack of 35nm-thick SiO$_2$/90nm Ge-NP/3nm-thick SiO$_2$/ 20nm-thick SiGe-channel on (100) SOI substrate.

Fig. 3 HRTEM/EDX mappings show process-controlled tunability of 5–93nm Ge NP/SiO$_2$/ SiGe shell gate stacks over Si. A conformal, 2.6nm-thick SiO$_2$ well separates the Ge NP and recess SiGe channel.

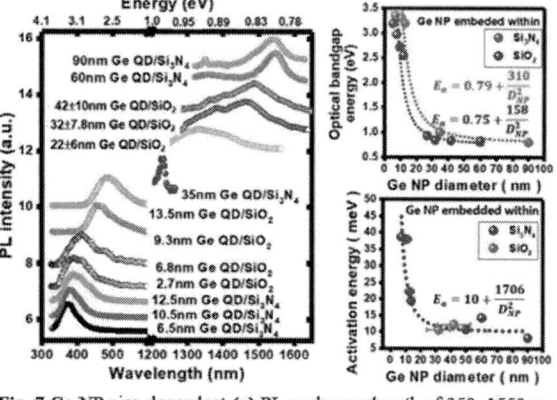

Fig. 4 Oxidation time dependent morphological and migration evolution behaviors of **(a)** coarsening, **(b)** migration and ultimately, **(c)** coalescence for Ge nanocrystallite clusters generated from thermal oxidation of the poly-Si$_{0.85}$Ge$_{0.15}$ pillars. From the original, as-formed location within the oxide the cluster migrates towards the Si$_3$N$_4$ buffer layer underneath for oxidation times of **(a)** 10, **(b)** 20, and **(c)** 30min at 900°C.

Fig. 5 Highly-stressed Si$_{1-x}$Ge$_x$ shell with a high degree of crystallinity are evidenced by sharp spots from plan-view NBD of Si$_{1-x}$Ge$_x$ shells over SOI (100) substrate.

Fig. 6 Size-dependent Ge composition and compressive strain for the Si$_{1-x}$Ge$_x$ recess channels formed on various Si substrates. **(a)** Raman spectra, **(b)** EDX line scan and **(c)** Derived Ge composition and compressive strain for the Si$_{1-x}$Ge$_x$ recess shells as high as $x = 0.85$ and $\varepsilon = -3\%$ formed on (100) SOI.

Fig. 7 Ge NP size-dependent **(a)** PL peak wavelength of 350–1550nm, **(b)** optical bandgap energy of 0.8–3.4eV, and **(c)** activation energy of 6–40meV are measured on 3–90nm Ge NPs embedded within matrices of SiO$_2$ or Si$_3$N$_4$. Notably the optical bandgap energy and activation energy are inversely proportional to the Ge NP size due to quantum confinement effect.

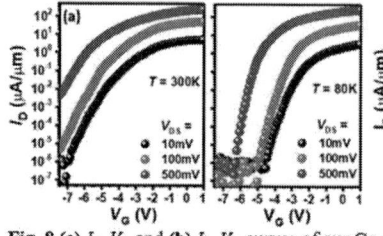

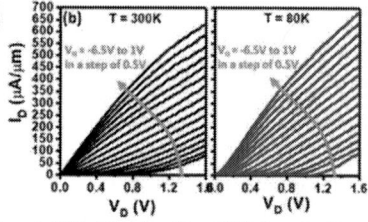

Fig. 8 (a) I_D-V_G and **(b)** I_D-V_D curves of our Ge-gate/SiO$_2$/Si$_{0.2}$Ge$_{0.8}$ JL *n*-FET show I_{ON}/I_{OFF} >10^5 and >5×10^8 measured at 80K and 300K, respectively, when biased at V_{DS} = 0.5V.

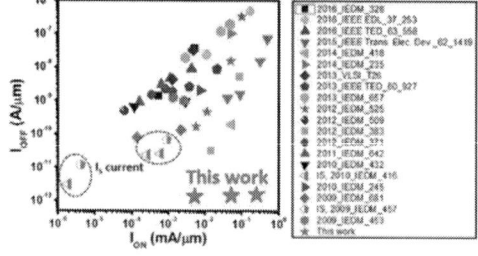

Fig. 9 In comparison with previously reported Ge *n*-FETs, our Ge-NP gate/SiO$_2$/ SiGe-recess channel JL *n*-FET features very low I_{OFF} of 1pA/μm at given I_{ON} of 0.3-0.5 mA/μm verifying superior gate oxide integrity of our gate stacks.

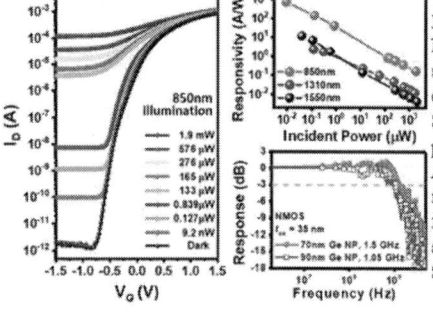

Fig. 10 (a) I_D-V_G curves of Ge-NP/SiO$_2$/SiGe MOSFET on SOI show high photosensitive current under variable-power 850nm illumination. **(b)** Large photoresponsivity of 10– 1,000 A/W for 850–1550nm illumination, and **(c)** 3dB frequency of 2GHz was achieved for our Ge-NP PTs on SOI substrate.

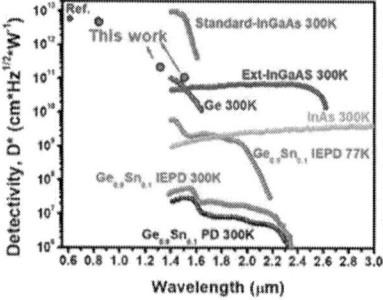

Fig. 11 Our Ge-NP/SiO$_2$/ SiGe photoMOSFETs on SOI substrate exhibit very high detectivity of 6×10^{12} cm·Hz$^{0.5}$/W, which is comparable to state-of-the art InGaAs PDs and much higher than the reported values of Ge and GeSn PDs.

978-1-5386-4219-1/18 $31.00 © 2018 IEEE

A Near- & Short-Wave IR Tunable InGaAs Nanomembrane PhotoFET on Flexible Substrate for Lightweight and Wide-Angle Imaging Applications

Yida Li, Alireza Alian[†], Li Huang, Kah Wee Ang, Dennis Lin[†], Dan Mocuta[†], Nadine Collaert[†], and Aaron V-Y Thean

National University of Singapore, Dept. of Electrical & Computer Engineering, 4 Engineering Drive 3, Singapore 117583

[†]IMEC, 75 Kapeldreef, 3001 Heverlee, Belgium

Email contact: li.yida@nus.edu.sg

Abstract: We demonstrate an InGaAs nanomembrane field-effect phototransistor with wide-band spectral response tunability, from the visible to near-infrared light. The ultra-thin InGaAs channel (15nm) device, enabled by epitaxial lift-off of InGaAs-on-InP MOSHEMT, is integrated with a fully exposed channel for photosensitivity enhancement. The photocurrent is tunable >5 orders for a gate bias range of 6 V. On-state photo-responsivities of 380 A/W to 15 A/W for 660 nm to 1877 nm light is measured, >2x more sensitive than existing silicon and III-V photodetectors [1-3]. The device shows no performance degradation when flexed down to 10-cm radius, showing suitability for conformal surface sensor applications.

Introduction

Hyperspectral sensing and imaging through haze, fog, rain and other atmospheric conditions are of special interests for a widening range of surveillance applications [4]. Low-power light-weight Near- and Short-Wave Infra-Red (IR) imagers with high sensitivities are desired for future high-mobility robotic applications like autonomous drones and smart vehicles. In addition, thin membrane-like devices which enable large-area-array implementation of surface imagers and sensors may enable ultra-light machines with wide-field imaging capability (Fig 1).

IR sensors/imagers of different semiconductors have been investigated. For example, conventional semiconductors such as Si, Ge, III-V [1-3, 5-6], and emerging two-dimensional (2D) materials such as graphene, transition metal dichalcogenides (TMD) and monochalcogenides [7-8]. However, there are tradeoffs between bandgap, form factor, and photon absorption cross-section [9]. Transistor structures with top metallization is not ideal as a photo-imager due to absorption of incident light by the contact metal.

In this work, we propose a thin-body InGaAs (In = 53%) phototransistor with a direct bandgap of 0.7 eV, which responds strongly to short-wave IR radiation 1.8 μm and below wavelength. The transistor gain allows us to boost the response for weaker radiation when the intrinsic response wanes, while the InGaAs channel can be epitaxially engineered to maximize absorption cross section. The channel is fully exposed through epitaxial lifting of the InGaAs MOSHEMT and inverting the membrane device (Fig. 3). The top gate-source-drain metal contacts were also transferred intact which becomes the bottom electrical access for the device (Fig. 2), avoiding incident light obstruction. The photocurrent can be tuned >5 orders, and responds to incident light of wavelengths from 660 nm to 1877 nm through V_g biasing. With an incident power of 65 nW, photo-responsivity of 380 A/W at 660 nm to 17 A/W at 1877 nm in the on-state is measured. Flexing the sample (10 cm radius) does not produce detrimental effect on its performance.

Device Fabrication and Characterization

A. Fabrication

InGaAs MOSHEMT is first fabricated on a rigid InP substrate, with mesa isolation [10]. Prior to transfer, an addition step of routing electrodes contacts to field area is done to allow for access to contact after the transfer process. Fig 3 illustrates the transfer and bonding processes from rigid to flexible (SU-8 <50 μm), and then onto polydimethylsiloxane (PDMS) layer ~500 μm thick.

B. Characterization and Results

Raman and Photoluminescence (PL) spectra of the inverted InGaAs MOSHEMT on flexible (Fig. 4 & 5 respectively) confirms the preservation of the InGaAs channel. PL confirms the bandgap of the exposed channel to be ~0.7 eV (~1.8 μm). The phototransistor is placed on a customized probe-station where light wavelengths 660/1553/1877 nm is guided through a fiber placed ~8 mm above. The incident power on the phototransistor is ~65 nW and the device is tested in both unflexed and flexed (~10 cm bending radius) state. Fig 6 shows the I_d-V_g curves of the phototransistor without illumination (dark). The only degradation after the phototransistor transfer appears to be the increased contact resistance as the subthreshold swing (~200 mV/dec) and on/off ratio (~10^6) remain unchanged (Fig 7). Fig 8 shows the photocurrent ($I_{d-illuminated}$ – I_{d-dark}) response of the phototransistor as a function of V_g under a fixed V_{ds}. The photocurrent is tunable across >5 orders with V_g biasing alone. Fig 9 to 11 is a plot of the photocurrent contour map as a function of V_{ds} and V_g with illumination wavelengths indicated.

The photo-responsivity of the phototransistor across positive V_g (Fig 12) relates to its tunability in conjunction with the photocurrent. An illumination pulse train response (Fig 13) shows the time-domain response of the photocurrent (subtracted from base "dark" current) phototransistor, in both the light "on" and "off" state. Though noisier and smaller in magnitude (corresponds to lower responsivity – Fig. 12), wavelengths up till 1.8 μm is still readily detected. The response of illumination > 1.8 μm is not tested and deserved further investigation. As a final demonstration, the performance of the phototransistor while flexed (bending radii down to~10 cm) shows no degradation (Fig 14) as well as its photo-responsivity compared to un-flexed state. The benchmark (Fig. 15) shows that the channel-exposed InGaAs phototransistor is competitive to 2D materials while being superior to conventional Si/Ge/III-V based photodetectors in the region of < 1.8 μm.

Conclusion

We have built an InGaAs phototransistor with exposed channel on flexible substrate. The device shows high optical response tunability over 5 orders for 6 V bias range, for near and short-wave IR (660 nm to 1877 nm). Such a device is competitive to 2D TMDs and superior to conventional Si/Ge/III-V based photodetectors in the region of < 1.8 μm. With its flexibility, it can be suitable for lightweight surface mounted sensors applications.

Acknowledgements

The work is supported in part by Singapore's National Research Foundation, Hybrid Integrated Flexible Electronic Systems (HiFES) Program (hifes.nus.edu.sg), and E6Nanofab at the National University of Singapore (NUS).

978-1-5386-4219-1/18 $31.00 © 2018 IEEE

2018 Symposium on VLSI Technology Digest of Technical Papers

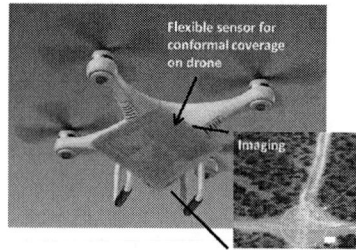

Fig. 1 An illustration of drone conformally covered with large area sensors for wide-field imaging [4]

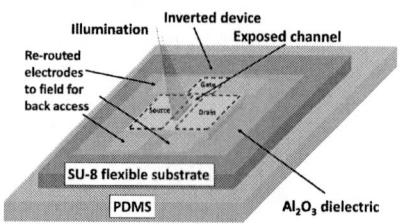

Fig. 2 Device schematic of inverted phototransistor in this work. The exposed channel serve to capture all incident light by maximizing absorption cross-section

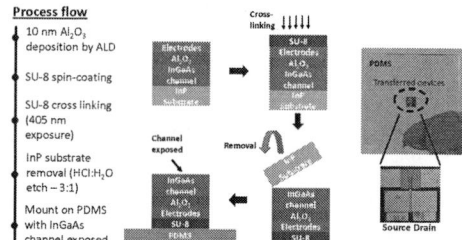

Fig. 3 Process flow of transferring MOSHEMT on rigid to flexible with schematic. Photo images and microscope images of transferred devices are shown on the right

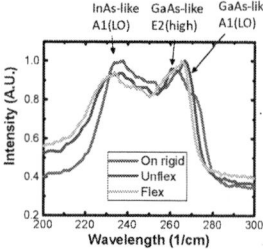

Fig. 4 Raman spectra of InGaAs channel before and after transfer, as well as flex/un-flex mode with a bending radius of ~10 cm

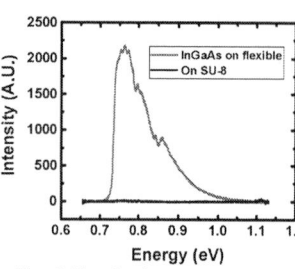

Fig. 5 Photoluminescence spectra of transferred InGaAs channel on SU-8 and only SU-8 substrate as comparison.

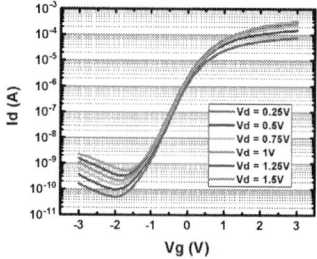

Fig. 6 Transfer curve of tested device after transfer (dark) with V_d bias from 0.25 V to 1.5 V

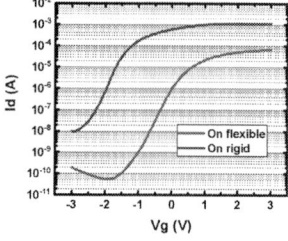

Fig. 7 Transfer curve comparison of devices on rigid/flexible. SS remains unchanged at 200 mV/dec but with reduction in I_d probably due to increase in $R_{contact}$

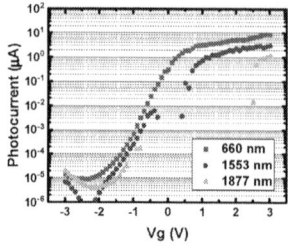

Fig. 8 Photocurrent extracted versus V_g and wavelength as indicated with 65 nW incident power

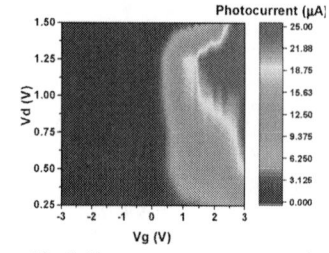

Fig. 9 Photocurrent contour map of phototransistor as a function of different bias under 65 nW incident power (660 nm)

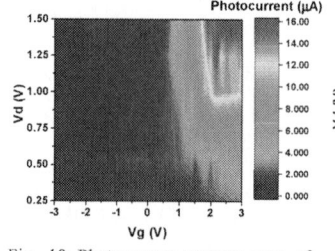

Fig. 10 Photocurrent contour map of phototransistor as a function of different bias under 65 nW incident power (1553 nm)

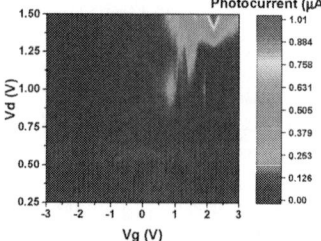

Fig. 11 Photocurrent of phototransistor as a function of different bias under 65 nW incident power (1877 nm)

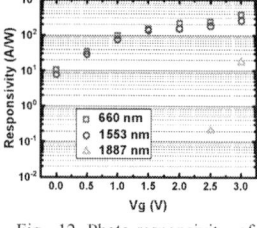

Fig. 12 Photo-responsivity of the device across positive V_g under 65 nW incident power (wavelengths indicated in legend)

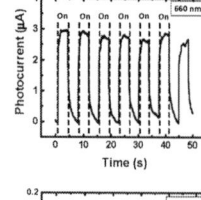
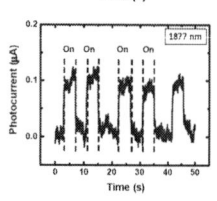

Fig. 13 Pulse train response of the phototransistor under optical illumination with wavelengths of 660/1553/1877 nm as indicated. The photocurrent is subtracted against the background I_d in the "off" state. The time when the light is "on" is marked out in the plots.

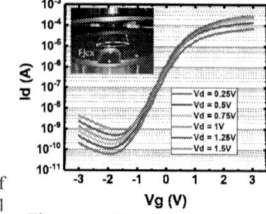

Fig. 14 Transfer curve of phototransistor in a flexed mode (radii 10 cm). No observable difference as compared to Fig. 6. Inset shows photo of testing setup.

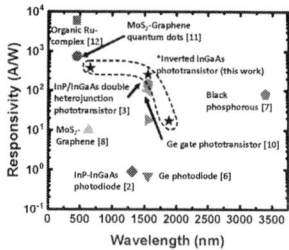

Fig. 15 Benchmark graphs comparing responsivity of various types of photodetectors versus wavelengths

References

[1] L. Li et al, Proc of 16th Solid-State Sens., Act. and Microsys. Conf., 2011
[2] Y. Yoneda et al, Proc of OFC/NFOEC Conf., 2008
[3] J. Kim et al, Solid State Electronics, 51(7), 2007
[4] P. W. T. Yuen et al, The Imaging Sci. J., 58 (5), 2010
[5] M. Casalino et al, Sensors, v. 10, 2010
[6] N. N. Feng et al, Optics express, 18 (1), 2010

[7] M. Huang, Adv Mat., 28 (18), 2016, pp. 3481 3485
[8] H. Xu et al, Small, 10 (11), June 2014
[9] B. E. A. Saleh, et al, Fundamentals of Photonics (Wiley, New York, 1991)
[10] Alian et al, IEDM 2013, pp. 437-440
[11] C. Chen et al, Sci Rep, 5 (11830), 2015
[12] X. Liu, Appl. Mat. & Int., 8, 2016

978-1-5386-4219-1/18 $31.00 © 2018 IEEE

Integration of 2D Black Phosphorus Phototransistor and Silicon Photonics Waveguide System Towards Mid-Infrared On-Chip Sensing Applications

Li Huang[1,2,*], Bowei Dong[1,*], Xin Guo[3], Yuhua Chang[1], Nan Chen[1], Xin Huang[1,2], Hong Wang[3], Chengkuo Lee[1,#], Kah-Wee Ang[1,2,#].

[1]Department of Electrical and Computer Engineering, National University of Singapore, 4 Engineering Drive 3, Singapore 117583
[2]Centre for Advanced 2D Materials, National University of Singapore, 6 Science Drive 2, Singapore 117546
[3]School of Electrical and Electronic Engineering, Nanyang Technological University, 50 Nanyang Avenue, Singapore 639798
Phone: +65 6516-2575, Fax: +65 6779-1103, #Email: eleakw@nus.edu.sg, elelc@nus.edu.sg; *Co-first authors.

Abstract

We demonstrate the first black phosphorus phototransistor integrated with Si photonics waveguide system towards mid-infrared (MIR) sensing. At a wavelength of 3.78 μm, the black phosphorus phototransistor achieves a high responsivity of 0.7 A/W under a small drain bias of −1 V at room-temperature. Additionally, the device offers gate and drain bias tunability to suppress dark current while simultaneously optimize photo-response performance. Our results reveal the potential of black phosphorus for MIR detection to enable the realization of integrated on-chip systems for MIR sensing applications.

I. Introduction

Mid-infrared (MIR) nanophotonics have attracted great research interests over the past decade due to its promising potential for label-free and damage-free sensing [1]. Despite the development of single photonic components for MIR, the integration of waveguide with lasers and photodetectors remains an issue that hinders the realization of integrated on-chip MIR systems [2]. In particular, photodetector-waveguide integration is constrained by the feasible detection materials.

Various materials such as HgCdTe alloys [3], GeSn [4], III-V [5] and II-VI [6] compounds have been proposed for MIR photodetection. Nonetheless, they are restricted by either growth challenge, low operation temperature, or short cut-off wavelength. Black phosphorus (BP), by virtue of its narrow band gap and layered lattice structure with in-plane anisotropy, stands out among the other candidates due to its superior properties including broadband MIR absorption, potential CMOS compatibility and polarization sensitivity [7],[8].

For the first time, we demonstrate a BP phototransistor integrated with Si photonics waveguide for MIR photodetection at a wavelength of 3.78 μm. Tunability from the transistor's gate bias and drain bias are exploited to optimize the photo-response performance. The power dependence of responsivity is also investigated.

II. Device Structure and Fabrication

The schematic of the device is shown in Fig.1a (far view) and Fig.1b (close view). Light is coupled from an optical fiber to the waveguide system and finally received by the BP phototransistor through grating couplers. The SEM image of the grating coupler and optical microscope image of the device before top-gate patterning are shown in Fig. 1c and 1d, respectively.

8" SOI substrate with 400 nm Si layer on 2 μm BOX was employed for the fabrication of waveguides using CMOS compatible processes. Then the wafer went through post-CMOS fabrication to heterogeneously integrate BP phototransistor. A 23-nm BP flake was exfoliated and transferred onto the SiO$_2$ cladding above the output Si grating coupler by a gel stamp. Source and drain electrodes were patterned by electron-beam lithography (EBL), after which 6 nm Ti/100 nm Au were deposited by thermal evaporation. The transistor transport direction is parallel to the TE mode of the waveguide. Following the liftoff process, atomic layer deposition (ALD) was used to grow 20 nm Al$_2$O$_3$ as gate dielectric. Top-gate electrode (6 nm Ti/100 nm Au) was then patterned by EBL and deposited by thermal evaporation.

III. Characterization of Grating Coupler and Material Properties of BP

Fig. 2 presents the transmission spectrum of the grating coupler, where a minimum loss of −10.5 dB/facet is observed at 3.78 μm. The insets display the mode profiles of the grating coupler. While the side view reveals that the light is directed vertically by the grating coupler, the top view shows the confinement of optical mode at the grating coupler/BP interface which promotes light-BP interaction.

The three phonon peaks in Raman spectroscopy (Fig. 3a) represents BP's lattice vibrational modes: armchair (A_g^2), zigzag (B_{2g}) and out-of-plane (A_g^1) vibration [9]. Polarization dependence of the A_g^2 to B_{2g} peak ratio indicates a 50° angle between the transistor transport direction and armchair orientation of BP (Fig. 3b) [10]. Fourier transform infrared (FTIR) measurement was used to obtain the extinction spectrum of 23-nm BP, from which BP's absorption

range is plotted (Fig. 3c). The absorption gradually decreases as wavelength increases, with a steeper roll-off after 3.5 μm. The cut-off wavelength of 4.24 μm corresponds to a bandgap around 0.292 eV. At the wavelength of 3.78 μm, 6% of power is absorbed by the BP flake.

IV. Electrical Properties and Photo-response

The ambipolar behavior of the BP transistor is shown in the transfer curves (Fig. 4). V_0 is defined as the voltage needed to obtain the lowest dark current, which is 2.5 V in our device. The transistor shows a higher hole conductance due to a higher mobility of holes than electrons in BP. I_d-V_d characteristics are shown in Fig. 5a (hole conducting side) and 5b (electron conducting side). The asymmetric I_d-V_d curves for hole conducting side reflects hole barrier modulation by drain bias. Negative drain bias reduces hole barrier near the drain contact and results in higher drain current formed by holes. Contour-plot of the dark drain current I_{dark} versus V_d and V_g is presented in Fig. 6.

To characterize photo-response, a continuous wave MIR laser was used. Photocurrent is defined as $I_{ph} = |I_{light}| - |I_{dark}|$, where I_{light} is the drain current under laser illumination. Photocurrent is strongly dependent on transistor's operation condition, as shown in the I_{ph}-V_g (Fig. 7 and 8) and I_{ph}-V_d (Fig. 9) curves. The V_g and V_d dependence of photo-response is due to the modulation of carrier concentration by gate bias and drain bias as well as hole-barrier height. When V_g starts to increase positively, hole-barrier height and carrier concentration are reduced (insets of Fig. 8). Therefore, I_{ph} rises as $V_g - V_0$ becomes positive initially. However, as V_g continues to increase, higher electron concentration leads to more frequent carrier scattering, giving rise to a lower I_{ph}. A negative drain bias also helps to reduce hole barrier near the drain contact (insets of Fig. 9), enabling more efficient collection of photo-generated holes. Consequently, higher I_{ph} is observed with negative V_d. According to the contour-plot of photocurrent versus V_g and V_d (Fig. 10), maximum photo-response locates at the lightly n-doped region of the transistor, where dark current is relatively low and hole-barrier height is significantly reduced by the positive V_g and negative V_d.

Fig. 11 plots the I_{ph}-V_g and I_{ph}-V_d curves with varying micro-watt level power incident on the phototransistor, showing higher photocurrent when the power is increased. By operating the transistor at the off-state where dark current is minimized, we measured power dependence of photocurrent, from which responsivity $R = \frac{I_{ph}}{Power}$ is calculated (Fig. 12). The responsivity is observed to increase with decreasing power and saturate at around 0.7 A/W when the incident power is below 1 μW. Fig. 13 shows the responsivity benchmarking of waveguide integrated photodetectors for the infrared (IR) wavelengths. Our device is the only demonstration in the MIR with responsivity comparable to the best reported values in other IR regimes.

V. Conclusion

An integrated system featuring BP phototransistor and Si waveguide is demonstrated with high sensitivity MIR detection at a wavelength of 3.78 μm. Through gate and drain bias tuning, the carrier concentration and hole-barrier height can be modulated to suppress dark current and optimize photo-response. The phototransistor achieves a high responsivity of 0.7 A/W for a small drain bias of −1 V, revealing the promising potential of BP for MIR detection even near its cut-off wavelength. Our integrated system paves the way towards on-chip MIR sensing applications such as spectrometer sensor.

References

[1] B. Mizaikoff, *Chem. Soc. Rev.*, 8683, 2013. [2] T. Hu *et al*, *Photonics Res.*, 417, 2017. [3] S. Keuleyan *et al*, *Nat. Photonics*, 489, 2011. [4] R. Soref, *Nat. Photonics*, 495, 2010. [5] Z. Ning *et al*, *Mater. Lett.*, 213, 2016. [6] H. Lee *et al*, *Infrared Phys. Technol.*, 50, 2013. [7] F. Xia *et al*, *Nat. Commun.*, 4458, 2014. [8] F. Xia *et al*, *Adv. Mater.*, 1703748, 2018. [9] R. Fei *et al*, *Appl. Phys. Lett.*, 083120, 2014. [10] X. Ling *et al*, *Nano. Lett.*, 2260, 2016. [11] Y. Bie *et al*, *Nat. Nanotechnol.*, 1124, 2017. [12] T. P. Pearsall *et al*, *EDL.*, 330, 1986. [13] L. Liu *et al*, *J. Nanosci. Nanotechnol.*, 1461, 2010. [14] T. Yin *et al*, *OE.*, 13965, 2007. [15] J. J. Ackert *et al*, *Nat. Photonics*, 393, 2015. [16] R. R. Grote *et al*, *CLEO.*, Stu3G.1, 2014. [17] X. Wang *et al*, *Nat. Photonics*, 888, 2013.

978-1-5386-4219-1/18 $31.00 © 2018 IEEE

2018 Symposium on VLSI Technology Digest of Technical Papers

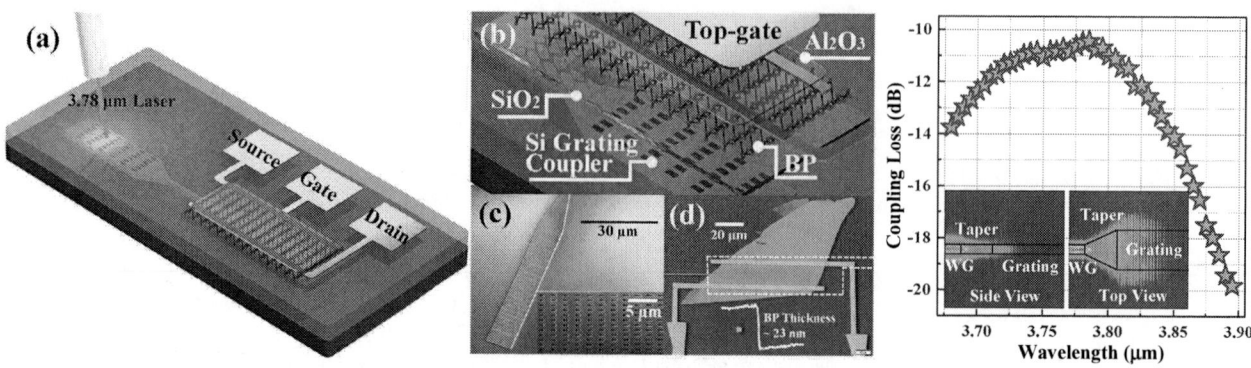

Fig. 1. Device structure. (a) Far view of the waveguide system integrated with BP phototransistor. The light is coupled from an optical fiber, transmitted through a single-mode waveguide, and received by the BP phototransistor through a grating coupler. (b) Close view of the device. (c) SEM image of the grating coupler. Inset: Zoom-in image. (d) Optical microscope image of the device before top gate patterning. The dashed white line outlines the top gate edge. The solid white line indicates the height profile of the 23-nm thick BP flake measured by AFM.

Fig. 2. Transmission spectrum of the grating coupler. A minimum loss of -10.5 dB/facet is observed at 3.78 μm. Insets: E-field distribution in the grating coupler (left: side view; right: top view; WG: waveguide).

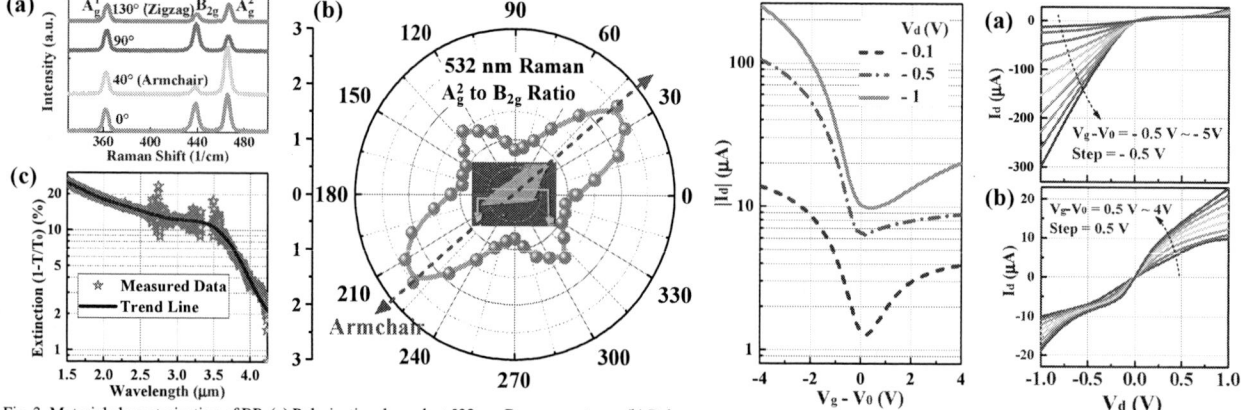

Fig. 3. Material characterization of BP. (a) Polarization dependent 532 nm Raman spectrum. (b) Polar plots of Raman A_g^2 to B_{2g} peak intensity ratio with varying excitation polarization. Zero-degree direction is parallel to propagation direction in waveguide. (c) Extinction spectrum of 23-nm BP from Fourier transform infrared measurement, verifying BP's broadband absorption characteristic. The cut-off wavelength of 4.2 μm corresponds to a bandgap of 0.292 eV.

Fig. 4. Transfer curves of BP phototransistor under different drain biases. Ambipolar behaviour is observed.

Fig. 5. I_d-V_d characteristics on (a) hole conducting side and (b) electron conducting side.

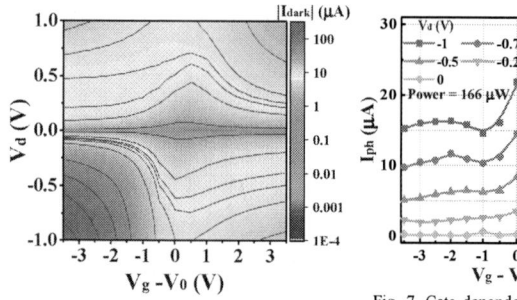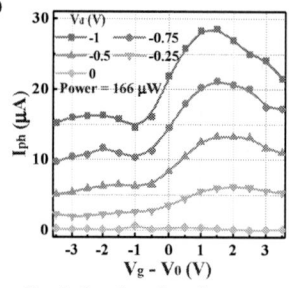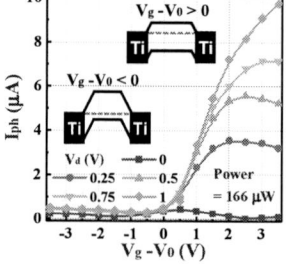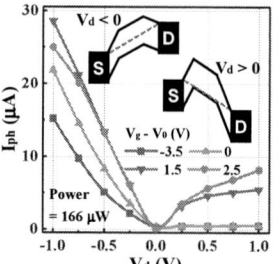

Fig. 6. Contour-plot of dark current versus gate bias and drain bias.

Fig. 7. Gate dependent photocurrent under varying negative drain biases.

Fig. 8. Gate dependent photocurrent under varying positive drain biases. Insets: energy band alignment for hole and electron conducting side.

Fig. 9. Photocurrent as a function of drain bias under different gate voltages. Insets: energy band alignment for positive and negative drain biases.

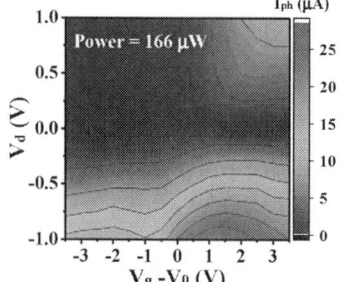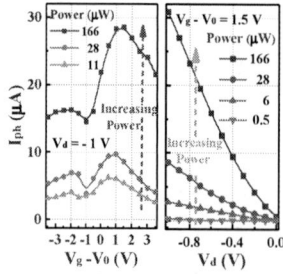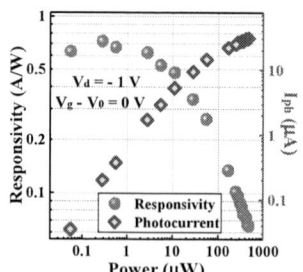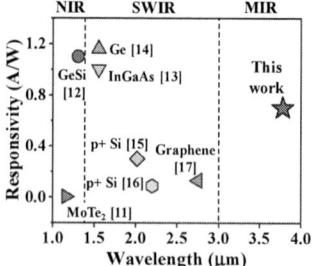

Fig. 10. Optimized photo-response is obtained with negative drain bias at lightly n-doped region where dark current is suppressed and hole-barrier height is significantly reduced.

Fig. 11. Photocurrent as a function of gate bias (left) and drain bias (right) under varying incident power.

Fig. 12. Power dependence of responsivity and photocurrent. Responsivity increases with decreasing power and saturates at 0.7 A/W below 1 μW.

Fig. 13. Benchmarking of responsivities of waveguide integrated photodetectors in infrared region. NIR: near-infrared, SWIR: short-wavelength infrared.

Acknowledgement: This research is supported by A*STAR Science and Engineering Research Council Grant (No. 152-70-00013), NRF Competitive Research Program (NRF-CRP15-2015-01 & NRF-CRP15-2015-02), and by the National Research Foundation, Prime Minister's Office, Singapore under its medium sized center program.

978-1-5386-4219-1/18 $31.00 © 2018 IEEE

Next-generation Fundus Camera with Full Color Image Acquisition in 0-lx Visible Light by 1.12-micron Square Pixel, 4K, 30-fps BSI CMOS Image Sensor with Advanced NIR Multi-spectral Imaging System

Hirofumi Sumi[1),2)], Hironari Takehara[2)], Shunsuke Miyazaki[2)], Daiki Shirahige[2)], Kiyotaka Sasagawa[2)], Takashi Tokuda[2)], Yoshihiro Watanabe[1)], Norimasa Kishi[1)], Jun Ohta[2)] and Masatoshi Ishikawa[1)]

Graduate School of Information Science and Technology, The University of Tokyo[1)], Graduate School of Materials Science, Nara Institute of Science and Technology (NAIST) [2)] <e-mail: hirofumi_sumi@ipc.i.u-tokyo.ac.jp>

Abstract

This paper presents a near-infrared (NIR) multi-spectral imaging system, which can be applied to a CMOS image sensor with fine pixels. Using the multi-spectral technology, NIR1: near 800 nm, NIR2: 870 nm, and NIR3: 940 nm in the NIR wavelength were acquired for a target image. Using this image sensor and imaging system and with the application of interpolation and color correction processing, a color image is reproduced by only multi-NIR signal without visible light (0 lx). We also developed a next-generation fundus camera, which employed this multi-spectral imaging system with a multi-NIR LED illuminator. This multi-NIR LED illumination system, which was also developed, is designed to emit light with high efficiency despite its size of 2.3 mm square in size. We applied this NIR multi-spectral camera module with the multi-NIR LED illuminator to the next-generation fundus camera; the retinal pigment appears progressively more transparent, revealing the underlying choroid.

Introduction

Recently, the CMOS image sensor (CIS) market crossed $11 billion per year. Its Compound Annual Growth Rate has been showing a double-digit growth each year because the digital camera has been widely used globally given that it is attached to almost all smartphones [1]. It is important to consider how best to utilize the small camera installed in the smartphone, which is not only used for taking pictures but also serves more important purposes. Si can perform photoelectric conversion from the ultraviolet to the near-infrared region. Digital still cameras usually have NIR cut filters, which allow the images to be viewed in visible light without acquiring an NIR wavelength [2]. However, a conventional digital camera with a common CIS can acquire a color image with only visible light (wavelength: 400 nm–700 nm). The near-infrared wavelength region is hardly utilized for photoelectric conversion. In these digital cameras, the near-infrared wavelength region is ignored [3]. Recently, some surveillance cameras could capture a black and white image with near-infrared (NIR) light, which uses only mono NIR wavelength [4]. Further, mono NIR wavelength is used in a 3D mapping sensor or in face recognition by KINECT [5]. Lately, in an industrial and a medical camera, multi-band spectral cameras are started to use. Such cameras use all band areas of the wavelength from ultraviolet to near-infrared [6]. However, the pixel size of the CIS should be designed to exceed approximately 5–10 μm square pixel size because of an inherent issue in the realization of spectral imaging technology [7].

On the contrary, a spectral imaging technology has been developed in the NIR wavelength region, which utilizes less than approximately 1.5-micron square pixel size of CIS. This makes it cost effective because of the possibility of designing a small chip size for a high-resolution pixel image sensor. The human eye cannot see the NIR light. When we capture a high

frame rate image under low illumination of visible light, we can use NIR as an auxiliary light source. We propose that a wider wavelength area from 400 nm to approximately 1000 nm is actively used to capture an image with NIR spectroscopy [8]. Especially, this camera module with advanced multi-spectral imaging system, which includes a small multi-LED illuminator, is available in a size that can be mounted on a smartphone. As an application of this small camera, we propose a fundus camera that can capture the health condition of a person anytime anywhere.

NIR multi-spectral imaging system technologies (NANO-tech)

We have developed an advanced NIR multi-spectral technology and its imaging system (NANO-tech); further, a 1.12-μm (micron) square pixel size, 8-megapixel resolution, and 30-fps (frame per second) Back-Side illumination (BSI) CIS, which includes a NANO-tech, has been created entirely new. This advanced NANO-tech film has been arranged in a checkered pattern on the pixels. This NANO-tech film is composed of three types of filters, which include the NIR1, NIR2, and NIR3 films. The NIR1 film can acquire the signal correlated to red, NIR2 can acquire the signal correlated to blue, and NIR3 can acquire the signal correlated to green after signal processing. Fig. 1 shows a cross-sectional view of a pixel area obtained by scanning electron microscopy. Fig. 2 shows the NIR1, NIR2, and NIR3 spectral responses, where the signals are separated by the NIR filters and the spectral imaging system. These NIR1, NIR2, and NIR3 filters are arranged on the 8M pixels in a checkered pattern. In a 0-lx no visible light environment, the color image has been reproduced with only NIR illumination using this image sensor (Fig. 4). Fig. 5 shows a block diagram of the imaging system with NANO-tech. Fig. 3 shows another reproduced image. With a conventional NIR camera, which uses only one wavelength of the NIR spectra, the reflected image is not clear, whereas NANO-tech and its imaging system in the CIS provide a clearer image, thus allowing us to comprehend the image easily. This implies that in an NIR region, it is important for the NIR camera to use an NIR multi-spectral imaging system to be able to reproduce all signals in a captured image. Fig. 6 and Table 1 show a small camera module and its specifications. An NIR camera module was developed and its system included an NIR illuminator. This illuminator can emit light from three kinds of light sources. By using double junction circuits, it enables the higher power LED illuminator despite miniaturization. (Fig. 7). The power consumption is approximately 130 mW (assuming po=50 mA). These systems maintain the size and power consumption that is sufficient to allow mounting on smartphones.

Next Fundus camera with NANO-tech imaging system

We are developing a solution system, which can be used as a camera, and one application of this compact camera is the capturing of health conditions. The fundus of the eye is the only site in the human body, where arteries and capillaries can be directly observed noninvasively [9]. By observing the fundus oculi, it is possible to observe the state of blood vessels and retina / optic papilla and thus diagnose various diseases ranging from glaucoma and retinal detachment to diabetes and arteriosclerosis. By using a next-generation fundus camera with NANO-tech imaging system and NIR wavelength of 800 nm, it is possible to capture the condition of the iris and ciliary body of the anterior uveal. The NIR wavelength of 940 nm, which is used in spectroscopy, can be utilized to capture the image of the choroidal condition of the posterior uvea [10]. To establish this proof of concept, we fabricated the lens and camera system shown in Fig. 8. We also developed three CIS cameras with C mount lenses using the advanced NANO-tech imaging system. These cameras can separately take a photo signal of the multi-band spectral signals NIR1, NIR2, and NIR3 corresponding to visible light R, G, B respectively with near-infrared spectral imaging technology. This filter is a non-organic multilayer, which uses the Fabry-Perot optical principle. Two kinds of films with different refractive indexes are alternately formed. It is possible to transmit specific near-infrared wavelengths by forming the same upper and lower film structures and controlling the film thickness of the high refractive index at the center (Fig. 9). Advanced multi-spectral imaging is obtained, where we control only the center thickness to distinguish the wavelength of a photon. The location of the near-infrared light source was designed to focus the light on a pupil through the fundus lens to achieve Maxwellian illumination. Images corresponding to NIR1, NIR2, and NIR3, as shown in the transmittance spectra in Fig. 9, were captured by using three CIS system. The optical images captured with NIR1, NIR2, and NIR3 are shown in Fig. 10. In addition, the color images of the fundus were obtained by the multi-spectral imaging system, which includes the NANO-tech imaging system, with interpolation and color correction processing of using camera module of NIR1, NIR2, and NIR3. These results which include the result of 1.12-μm square pixel CIS small camera module show that the system can capture the previously mentioned health conditions.

CONCLUSION

Advanced NIR multi-spectral technology has been developed. This is an extreme technology, which can be applied to the smallest pixel cost effectively when compared with conventional technologies [11] [12].

Acknowledgments

We thank Nidek Co. Ltd. for their advice regarding fundus camera systems, and special thanks to Dr. Y. Nagamune (AIST).

Reference

[1] P. Cambou et al., Status of the CMOS Image Sensor Industry 2017 report, Yole.

[2] M. Rosenberger et al., *2016 IEEE imaging Systems and Techniques (IST)*, pp. 7-12.

[3] D. Hertel et al., *2009 IEEE Intelligent Vehicles Symposium*, pp. 273-278.

[4] S. Kawada et al., *SENSORS, 2009 IEEE*, pp 1648-1651.

[5] S. Samoil et al., *2016 IEEE Congress on Evolutionary Computation (CEC)*, pp. 4528-4264.

[6] Y. Fujihara et al., *2016 IEEE SENSORS*, pp 1-3.

[7] M. Perenzoni et al., *2012 IEEE ESSEIRC*, pp 93-96.

[8] H. Sumi et al., IEICE Technical Report OPE2017-114 (2017-12), pp. 125-130.

[9] L. F. F. Ferreira, *Exp. Physiol.* vol. 90, issue 5, pp. 715-726, 2005.

[10] C. Zimmer et al., *Retina Today*, pp. 94-99, Oct. 2014.

[11] J.L.E Honrado, et al., *IEEE GHTC 2017*, pp. 1-7.

[12] M. Martin et al., *IEEE Sensors Letters*, vol. 1, issue 6, No7000404.

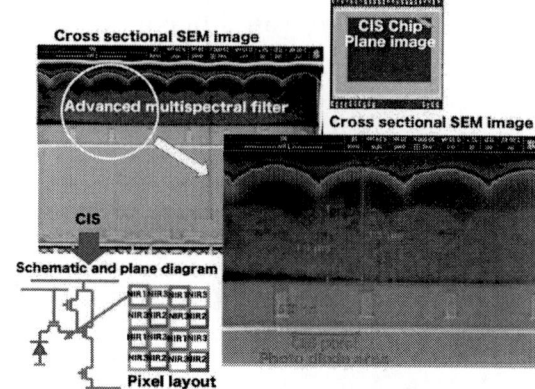

Fig. 1 Developed CIS with the Nano-tech; multi-NIR signal is acquired.

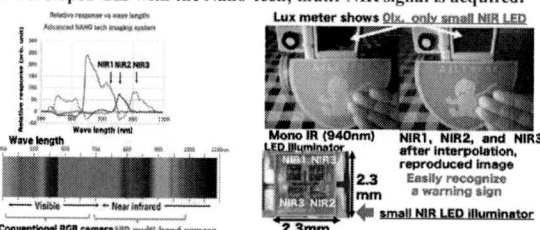

Fig. 2 Nano-tech spectral response   Fig.3 Reproduced image (0 lx)

Fig. 4 No visible light (0 lx); image reproduced only for multi-NIR

Table 1 Specifications
Camera module specification

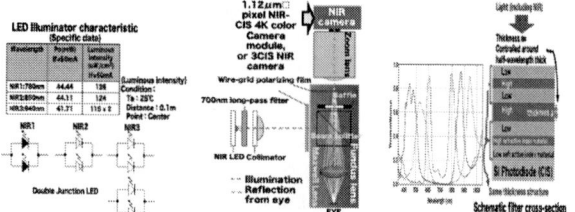

The calculation formula of the color correction matrix is as follows.

$$\begin{bmatrix} R \\ G \\ B \end{bmatrix} = \begin{bmatrix} a_{11} & a_{12} & a_{13} \\ a_{21} & a_{22} & a_{23} \\ a_{31} & a_{32} & a_{33} \end{bmatrix} \times \begin{bmatrix} NIR1 \\ NIR3 \\ NIR2 \end{bmatrix}$$

Fig. 5 Nano-tech imaging system Fig. 6 Multi-NIR spectral camera module

Fig. 7 NIR1, 2, 3   Fig. 8 Developed Fundus-camera   Fig. 9 RGB and
multi-LED Illuminator   NIR1, 2, 3
                                                        multi-band

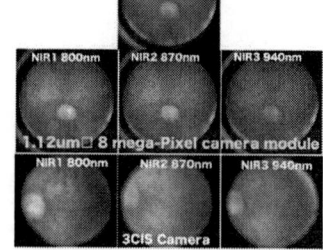

Fig. 10 Fundus observation results with NIR1, NIR2, and NIR3. By using small camera module, we can observe as same as quality image of 3 CIS

InGaAs-on-Insulator MOSFETs Featuring Scaled Logic Devices and Record RF Performance

C. B. Zota,* C. Convertino,* V. Deshpande,* T. Merkle,[†] M. Sousa,* D. Caimi* and L. Czornomaz*

*IBM Research GmbH Zürich Laboratory, Säumerstrasse 4, CH-8803 Rüschlikon, Switzerland

[†]Fraunhofer IAF, Tullastrasse 72, 79108 Freiburg, Germany

E-mail: zot@zurich.ibm.com

Abstract

We demonstrate scaled InGaAs-on-insulator FinFETs and planar MOSFETs on Si substrate for low power logic and RF applications. This Si-CMOS compatible technology implements SiN_x source-drain spacers and doped extensions for reduced overlap capacitances. FinFETs with performance for logic applications matching state-of-the-art are demonstrated. Simultaneously, f_t and f_{max} of 400 and 100 GHz are achieved respectively, the highest reported f_t for a III-V MOSFET on Si. Finally, we explore the use of an extended gate line to reduce gate resistance, offering balanced f_t/f_{max} of 215/300 GHz, the first report of III-V RF devices on Si matching state of the art Si-CMOS.

Introduction

High electron mobility III-V materials such as $In_xGa_{1-x}As$ are considered as replacements for strained Si in nFETs for low power logic applications [1]-[3]. Their superior electron transport properties also make them suitable for RF applications, enabling high-frequency and low-noise functionality [4]. However, a Si CMOS-compatible III-V-on-Si RF-MOSFET technology matching the RF performance of Si-CMOS has not yet been shown. In this work, we demonstrate an InGaAs FET technology enabling scaled FinFETs for low power logic, and RF-MOSFETs for high-frequency applications integrated on the same Si wafer, with performance in both cases matching state-of-the-art.

Device Fabrication

Fig. 1 shows a schematic cross-section of a fabricated InGaAs-on-Insulator RF-MOSFET, and **Fig. 2** shows the layout schematic of RF devices. **Fig. 3** shows STEM images of a fabricated device. The corresponding CMOS-compatible fabrication flow is shown in **Fig. 4** and follows previous work [5] with the addition of three new modules: (1) SiN_x sidewall spacers for reduced parasitic capacitances, (2) doped RSD extensions under the spacers for reduced access resistance, and (3) a gate extension option for balanced RF performance. First, a 20-nm thick InGaAs layer is integrated on a Si wafer using direct wafer bonding [6]. This technique is compatible with large-scale Si substrates and 3D monolithic integration [7]. Subsequently, fins are dry-etched in logic devices, while RF devices remain planar. 3 nm SiN_x spacers (**Fig. 3c**) are formed by ALD and RIE followed by raised source and drain (RSD) epitaxy using a dummy gate. The $Al_2O_3/HfO_2/TiN$ gate stack is subsequently formed by an RMG process. Next, the W gate fill is performed, an interlayer dielectric (ILD0') is deposited, and M1 is patterned. For the RF-MOSFET option, the ILD0' is etched to access the W gate, and the gate extension metal is patterned. Subsequently, ILD1 is deposited, and the source sides of the gate fingers are connected by M2.

Results

Fig. 5 and **6** show output and subthreshold characteristics, respectively, of FinFETs for logic applications, with fin width of $W_{fin} = 25$ nm. At $L_G = 30$ nm, SS in saturation is 83 mV/dec., and at $L_G = 120$ nm devices exhibit $SS_{sat} = 70$ mV/dec. **Fig. 7a** shows minimum SS in the linear region and **Fig. 7b** shows I_{ON}

(at $I_{OFF} = 100$ nA/µm and $V_{DD} = 0.5$ V) versus L_G for planar devices and FinFETs with different W_{fin}. $I_{ON} \approx 250$ µA/µm is demonstrated for $W_{fin} = 25$ nm and $L_G < 30$ nm, matching the state-of-the-art for scaled InGaAs FinFETs on Si [5], which does not utilize sidewall spacers. This shows that the doped RSD extensions introduced in this work effectively mitigate the expected increase of access resistance due to the SiN_x sidewall spacers.

RF characteristics were obtained from S-parameter measurements up to 45 GHz, and small-signal modeling [8]. The pad parasitics were de-embedded using open-short de-embedding at the M2 level [9]. **Fig. 8a** shows a gain plot of a planar RF device with $L_G = 30$ nm at $V_{DS} = 0.9$ V, exhibiting f_t and f_{max} of 400 and 100 GHz, respectively, extrapolated at -20 dB/decade. This is the highest reported f_t for a III-V MOSFET on Si. **Fig. 8b** shows similar for a device with GE, exhibiting a more balanced f_t and f_{max} of 215 and 300 GHz, respectively, in line with Si 14 nm FinFET technology [10][11]. Here, f_{max} is obtained from a small signal model with a good fit to the measured S-parameters. Transconductance is $g_m = 1.5$ mS/µm at $V_{DS} = 0.5$ V. **Fig. 9** shows average values of f_{max} and f_t versus L_G with and without GE at $V_{DS} = 0.7$ V. The increased f_{max} is due to a reduction of R_G, while the decreased f_t is due to an increase of C_{GS}/C_{GD} from parasitic coupling to the GE metal.

Fig. 10 shows R_G versus L_G for devices with and without GE. The apparent increase of R_G at long channels comes from the fact that the modeled R_G contains a contribution from the channel resistance in series. The increase of R_G at very short L_G is in part due to short-channel reduction of g_m and in part by the reduced width of the gate metal. Short gate length also leads to the lower part of the gate being filled only by relatively resistive TiN, rather than both TiN and W as for longer L_G. R_G reaches a minimum of 12 Ω at $L_G \approx 50$ nm. For devices without GE, R_G is increased by ~60 Ω at short L_G. **Fig. 11** shows the gate-source capacitance C_{GS}, versus L_G with and without GE. The y-axis intercept represents the parasitic contribution to C_{GS}, i.e. from the RSD epi to the channel and gate metal and from the gate metal to the source/drain metal and the W plugs. Without GE, this value is $C_{GS,par} = 0.5$ fF/µm, and with GE, it increases to $C_{GS,par} = 0.9$ fF/µm, primarily due to increased coupling between the GE metal and the W plugs. **Fig. 12a** and **b** show f_{max} and f_t, respectively, versus V_{GS} and V_{DS} for a device with GE and $L_G = 40$ nm. **Table 1** shows a benchmark of state-of-the-art Si-CMOS RF as well as III-V-on-Si devices.

Conclusions

We have shown a Si-CMOS compatible InGaAs-On-Insulator MOSFET platform for RF and low power digital applications, as well as a gate extension technology to balance RF performance. Simultaneous f_t/f_{max} of 400/100 GHz as well as 215/300 GHz was demonstrated, together with strong logic performance in FinFETs, with $I_{ON} \approx 250$ µA/µm at $V_{DD} = 0.5$ V and $I_{OFF} = 100$ nA/µm.

Acknowledgement

This work was funded by Horizon 2020 grant agreement no. 688784 (INSIGHT).

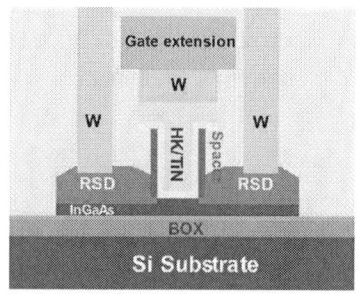

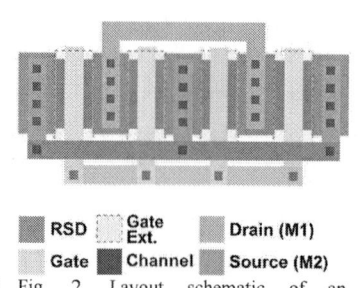

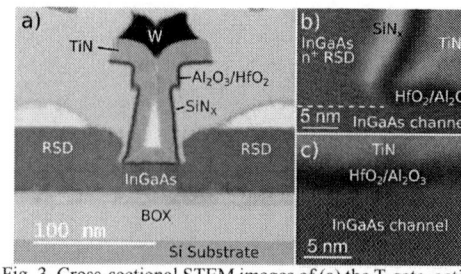

Fig. 1. Cross-sectional schematic of the fabricated InGaAs-on-insulator RF MOSFETs with gate extension. Logic devices include fins in the channel, tighter contact pitch and omission of the T-gate.

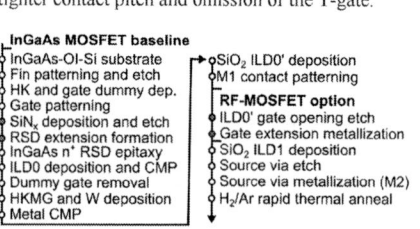

Fig. 2. Layout schematic of an RF-MOSFET with gate extension.

Fig. 3. Cross-sectional STEM images of (a) the T-gate, active region and BOX layer, (b) the RSD epi and the ~4 nm SiN_x spacers and (c) the channel and gate oxide.

InGaAs MOSFET baseline
◊ InGaAs-OI-Si substrate
◊ Fin patterning and etch
◊ HK and gate dummy dep.
◊ Gate patterning
◊ SiN_x deposition and etch
◊ RSD extension formation
◊ InGaAs n⁺ RSD epitaxy
◊ ILD0 deposition and CMP
◊ Dummy gate removal
◊ HKMG and W deposition
◊ Metal CMP

◊ SiO_2 ILD0' deposition
◊ M1 contact patterning
RF-MOSFET option
◊ ILD0' gate opening etch
◊ Gate extension metallization
◊ SiO_2 ILD1 deposition
◊ Source via etch
◊ Source via metallization (M2)
◊ H_2/Ar rapid thermal anneal

Fig. 4. Process flow overview for the fabricated devices up to M2, including an RF-MOSFET option. Newly introduced modules are highlighted in red.

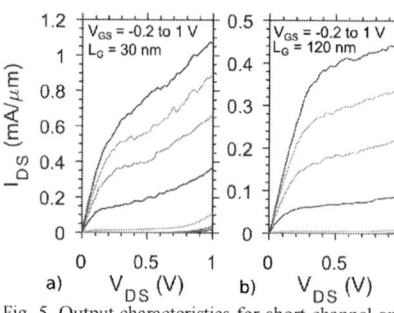

Fig. 5. Output characteristics for short channel and long channel logic devices (FinFETs) with W_{fin} = 25 nm and (a) L_G = 30 nm and (b) L_G = 120 nm.

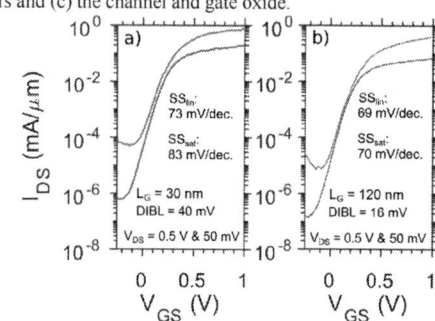

Fig. 6. Subthreshold characteristics for short channel and long channel logic devices with W_{fin} = 25 nm and (a) L_G = 30 nm and (b) L_G = 120 nm.

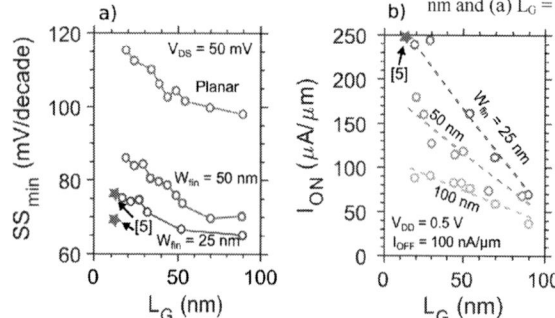

Fig. 7. (a) Minimum linear subthreshold slope and (b) on-current versus L_G for logic devices. Performance matches that of [5], which employs a similar technology but without the SiN_x sidewall spacers and doped RSD extension modules introduced in this work.

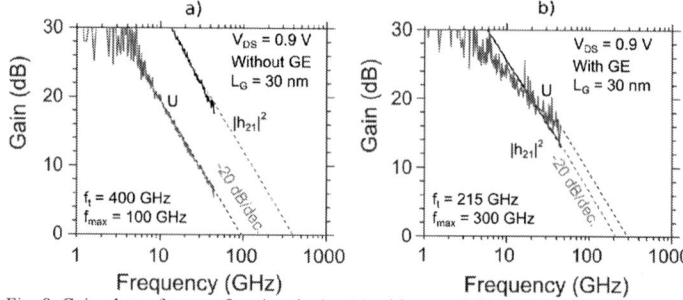

Fig. 8. Gain plots of top performing device (a) without and (b) with gate extension. The former shows f_t = 400 GHz, which is the highest reported value for a III-V MOSFET on Si, while the latter exhibits lower R_G with more balanced f_t/f_{max} matching Si CMOS.

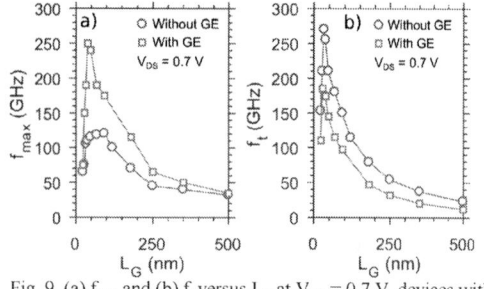

Fig. 9. (a) f_{max} and (b) f_t versus L_G at V_{DS} = 0.7 V, devices with and without gate extensions. The reduction of f_t at short L_G is due to short-channel effects reducing g_m.

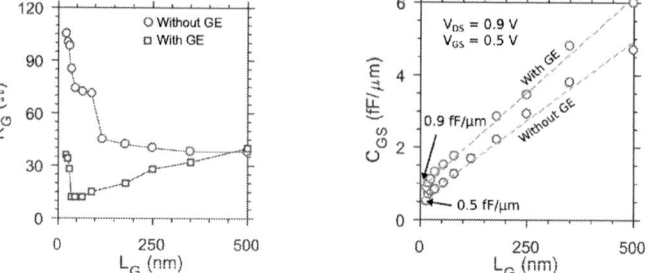

Fig. 10. Gate resistance versus L_G with and without GE.

Fig. 11. C_{GS} versus L_G with and without gate extension. The increased C_{gs} leads to a reduction of f_t.

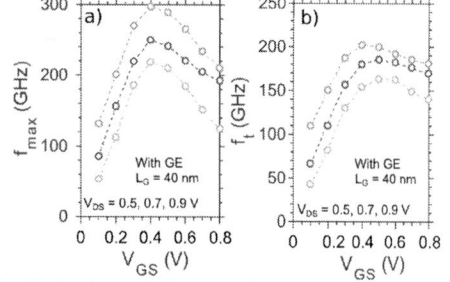

Fig. 12. (a) f_{max} and (b) f_t at different bias conditions for a device with gate extension.

Table 1. Benchmark of state-of-the-art Si and III-V technologies for logic and RF applications.

Technology	Platform	I_{ON} @ V_{DD} = 0.5 V & I_{OFF} = 100 nA/μm	f_t (GHz)	f_{max} (GHz)
This work	InGaAs on Si	250 μA/μm (L_G = 30 nm)	215	300
TSMC FinFET [2]	InGaAs on Si	300 μA/μm (L_G = 120 nm)	-	-
imec GAA [12]	InGaAs on Si	210 μA/μm (L_G = 46 nm)	-	-
Intel 22FFL [11]	Si	-	230	284
GF 14nm FinFET [10]	Si	-	314	180

[1] X. Sun et al., VLSI Tech. Dig., p. T3, 2017; [2] M. L. Huang et al., VLSI Tech. Dig., 2016; [3] C. Zota et al., IEDM Tech Dig. p. 3.2.1, 2016; [4] D. Kim et al., Applied Physics Lett., 101, 2012; [5] H. Hahn et al., IEDM Tech. Dig., 2017; [6] L. Czornomaz et al., IEDM Tech. Dig., p. 23.4.1, 2012; [7] V. Deshpande, IEDM Tech. Dig., p. 8.8.1, 2015; [8] I. Kwon et al., IEEE Trans. Microw. Theory Techn., 50, 2002; [9] M. C. Koolen et al., IEEE BCTM, p. 188, 1991; [10] J. Singh et al., VLSI Tech. Dig., p. T140, 2017; [11] B. Sell et al., IEDM Tech. Dig., p. 685, 2017. [12] X. Zhou et al., VLSI Tech. Dig., 2016.

Gap in pagination due to formatting issues.

Pages 167-168

Neuromorphic Technology Based on Charge Storage Memory Devices

Sung-Tae Lee, Suhwan Lim, Nagyong Choi, Jong-Ho Bae, Chul-Heung Kim, Soochang Lee, Dong Hwan Lee[1], Tackhwi Lee[1], Sungyong Chung[1],
Byung-Gook Park and Jong-Ho Lee

Department of ECE and ISRC, Seoul National University, Seoul 151-742, Korea
[1]R&D Division, SK hynix Inc., Icheon, Gyeongki 467-701, Korea
Phone: +82-2-880-1727; Fax: +82-2-882-4658; E-mail: jhl@snu.ac.kr

Abstract

Four synaptic devices are introduced for spiking neural networks (SNNs) and deep neural networks (DNNs). Unsupervised learning is successfully demonstrated by applying the STDP learning rule reflecting the LTP/LTD characteristics of the fabricated TFT-type NOR flash memory cells. Gated Schottky diode (GSD) and vertical NAND flash cell are proposed as synaptic device for DNNs. Using matched simulation, we obtained higher learning accuracy with GSD and NAND synaptic devices compared to that with a memristor-based synapse. Measured synaptic properties of the vertical NAND cells are reported for the first time.

Introduction

Dense crossbar array of non-volatile memory attracts enormous attention as a promising candidate for highly energy-saving neuromorphic system [1], [2]. Memristors are capable of highly integrated low power operation. However, before widely adopting RRAM crossbar arrays, it is necessary to solve some problems such as high variability and low reliability of the device. To solve these problems, we introduce Si-based synaptic devices such as gated diode, NOR flash cell, gated Schottky diode (GSD), and vertical NAND (VNAND) flash cell. Using TFT-Type NOR flash memory STDP unsupervised learning is successfully demonstrated without additional control circuit [3]. Reconfigurable GSDs have been reported as novel low-power synaptic device with near-linear conductance (G) characteristics [4]. An architecture for implementing on-chip learning DNN using VNAND cells as a high-density synaptic device is proposed, and synaptic characteristics of these cells are reported.

Results and Discussion

Fig. 1 shows the 3-D schematic views of SONOS gated-diode (equivalently TFET) synapse array and a synaptic device [5]. Note that this device is well suited for program and erase for weight updates. The crossbar structure where word-line (WL) and bit-line (BL) cross each other enables NOR type operation, which enables to emulate the weighted sum function of biological synapses. The 3-D schematic views of TFT-type NOR flash synapse array and a synaptic device are shown in Fig. 2. The synaptic characteristics of a TFT-type NOR flash memory array have been reported in our previous work [3]. Fig. 3 shows the schematic illustration of an SNN used in unsupervised pattern learning simulation based on the STDP algorithm. Fig. 4 (a) shows the pattern update process (patterns "8", "4", and "7" are updated in sequence) in a single-neuron based on the characteristic of the SONOS gated-diode synapses in an array. Fig. 4 (b) shows the unsupervised on-chip pattern learning progress with multi-neuron array (784 × 10) using 7,840 TFT-type NOR flash synapses without additional input noise pattern.

To implement DNNs using hardware synaptic devices, adaptive learning rule for hardware-based multi-layer neural networks is devised and compared with that of software-based learning as shown in Table 1. The input signal ($a_i^{(l-1)}$) and the weight (W_{ij}) can be represented by voltage ($V_i^{(l-1)}$) and the conductance (G) difference of a pair of synaptic devices ($G^+_{ij} - G^-_{ij}$), respectively. For backpropagation, error values (δ_i) can be calculated using transposed synapse array, then weight update is performed by single G step as shown in Fig. 5. Note in case of increasing weight, G^+_{ij} is increased in GSD, while G^-_{ij} is decreased in NAND cell because LTP/LTD behavior is opposite. In DNNs, the GSDs can be configured as a NOR-type array. Fig. 6 (b) and (c) show TEM and schematic cross-sections, respectively, cut along B-B' in plane SEM image in Fig. 6 (a). The I_R-V_{BGS} curves are shown in Fig. 6 (d). Fig. 7 (a) shows the I_R-V_O curves and G response of a p-type GSD as 31 identical pulses are applied. Fig. 7 (b) shows

repeated near linear G response in LTP and abrupt G response in LTD. The physical reason for the near linear property in LTP has been reported in [4]. Since the GSDs can be configured as a NOR-type array, they can be represented by an array of resistors whose G is varied by the number of pulses shown in Fig. 8.

NAND cells provide a higher integration density than NOR cells. In Fig. 9, the input values (voltages) are applied to BLs of VNAND cell strings, and forward propagation simply subtracts read current between a pair of bit lines and sum the total currents by using capacitor (not shown), for example. The output currents for all neurons in l^{th} layer are produced sequentially when the read pulse sequentially enters the WLs shown in Fig. 9 (b). When the kth pulse is applied into k^{th} word line, the output current for k^{th} neuron in l^{th} layer is produced. A separate synapse array constructed by transposing the synaptic array used in forward propagation can be used for backpropagation to calculate the error of the previous layer as shown in Fig. 10. By using the sign of the error, we can update the G of synaptic devices. Now we examine the synaptic properties of cells measured in VNAND cell strings composed of >50 cells in a cell string. Fig. 11 (a) shows the decreasing BL current (I_{BL}) in I_{BL}-V_{BL} curves as selected cell is programmed (7 V, 10 μs) 31 times. Fig. 11 (b) shows bidirectional G (=I_{BL}/V_{BL}) response for 32 steps measured at a V_{BL} of 0.2 V. The selected cell is programmed (7 V, 10 μs) 31 times and erased (-7 V, 10 μs) 31 times to represent bidirectional G response. Fig. 12 compares normalized G versus the number of pulses in several devices, using behavior model in [8].

We designed a 3-layer perceptron networks with 319,590 synapses and evaluated classification accuracy for MNIST sets using matched computer simulation. Fig. 13 shows simulated classification accuracy using G response in Fig. 12. As shown in Fig. 13, accuracy of GSD and VNAND cell are 94.68% and 94.5% which are comparable to that (94.69%) obtained by perfect linear device. As shown in Fig. 14, during read process, pass bias disturbance can affect G of cells. A pass bias (V_{pass}) of 3 V changes G by 2.5 nS (quite small), while program bias changes G by 37.5 nS in the first G step of 32 steps. Note a V_{pass} of 3 V is enough because I_{BL} is low and V_{th} of all cells is less than 2V. To check the reliability of VNAND cells, endurance and retention properties are measured. As shown in Fig. 15, we can observe that the G of a cell is almost the same up to 1k cycles. Fig. 16 shows the retention characteristics of G in the last and first steps of 32 steps at 25°C. The change of G in the last step is negligible at <1.5%.

Conclusion

In this paper, we have introduced learning results of four different Si-based synaptic devices. It was confirmed that learning and recognition are possible in single (784 × 1) and multi neuron (784 × 10) arrays based on STDP characteristics of the NOR flash cells. In DNNs, simulated classification accuracy using measured VNAND flash cells and GSDs is comparable to that obtained by perfect linear device. The VNAND cells as synapses showed excellent reliability.

Acknowledgement

This research was supported by the MOTIE (10080583) and KSRC support program and the Brain Korea 21 plus in 2017.

References

[1] D. Kuzum et al., Nanotechnology, vol. 24, 2013. [2] M. Prezioso, et al., Nature, 61-64, 2015. [3] C.-H. Kim et al., Trans. Electron Devices, (to be published). [4] J.-H. Bae et al., Elect. Device Letters, vol. 38, 2017. [5] C.-H. Kim et al., Trans. on Electron Devices, vol. 62, 2015. [6] S. Lim et al., arXiv preprint. arXiv: 1707.06381. [7] P. Pouyan et al., CMOS Variability, pp. 1-6, 2014. [8] D. Querlioz et al., IEEE Transactions on Nanotechnology, vol. 12, pp. 288-295, May 2013.

2018 Symposium on VLSI Technology Digest of Technical Papers

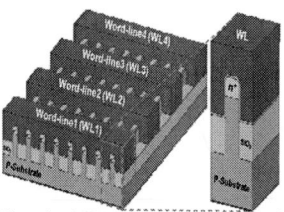

Fig. 1. 3-D schematic views of SONOS gated-diode (equivalently TFET) synapse array (left) and a synaptic device (right).

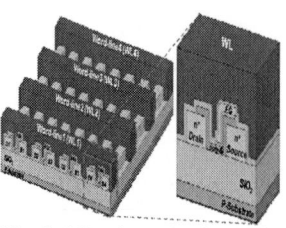

Fig. 2. 3-D schematic views of a TFT-type NOR flash synapse array (left) and a synaptic device (right).

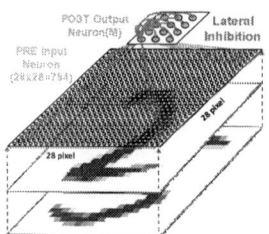

Fig. 3. Schematic illustration of neural network used in STDP unsupervised pattern learning.

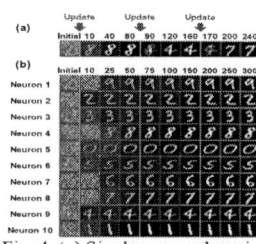

Fig. 4. (a) Single-neuron learning (b) multi-neuron learning based on TFT-type NOR flash.

Table. 1. Learning rule of software-based and hardware-based neural networks.

Target	Software-based	Hardware-based
Weights W_{ij}	W_{ij}	$G_{ij}^+ - G_{ij}^-$
Forward propagation $S_j^{(i)}$	$\sum_i^N W_{ij} a_i^{(i-1)}$	$a_i^{(i-1)} \to V_i^{(i-1)}$ $\sum_i (G_{ij}^+ - G_{ij}^-)V_i^{(i-1)}$
Backward propagation $\delta_i^{(i-1)}$	$\sum_j^M W_{ij}\delta_j^{(i)}\cdot f'(s_i^{(i-1)})$	$\delta_j^{(i)} \to V_j^{(i)}$ $\sum_j^M (G_{ij}^+ - G_{ij}^-)V_j^{(i)}$ $\cdot f'(V_i^{(i-1)})$
Weight updates ΔW_{ij}	$-\eta\cdot\delta_j^{(i)}\cdot f(s_i^{(i-1)})$	$\begin{cases} G_{ij}^+\uparrow \text{ or } G_{ij}^-\downarrow (\Delta W_{ij}>0) \\ G_{ij}^+\downarrow \text{ or } G_{ij}^-\uparrow (\Delta W_{ij}<0) \end{cases}$

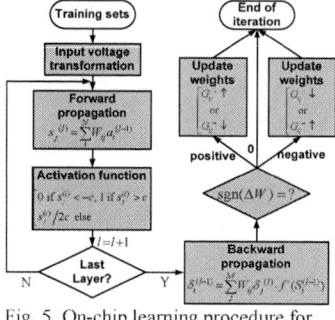

Fig. 5. On-chip learning procedure for hardware-based neural networks.

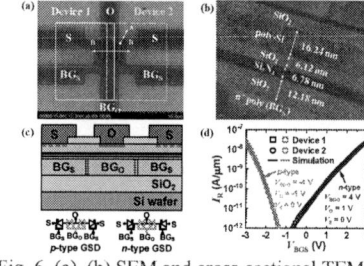

Fig. 6. (a), (b) SEM and cross-sectional TEM views. (c) Schematic view and equivalent circuit diagram for GSDs. (d) I_R-V_{BGS} curves of GSDs. These are reprinted from earlier work [4].

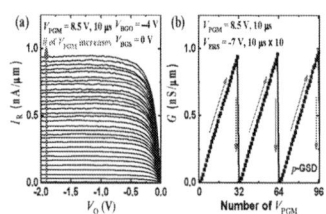

Fig. 7. (a) I_R-V_O (cathode bias) curves of a p-type GSD (b) Conductance response for 3 cycles when selected cell is programmed 31 times and erased at once.

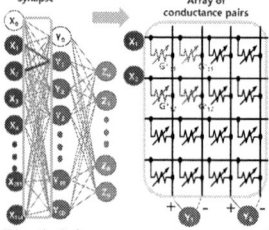

Fig. 8. 3-layer perceptron network in which the synapses in the array can be implemented using GSDs.

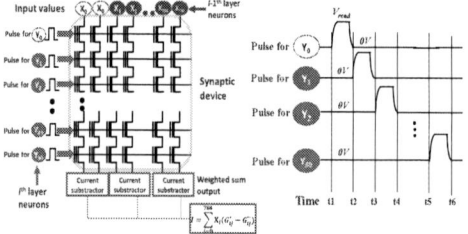

Fig. 9. (a) Schematic of forward propagation of multi-layer neural networks using NAND flash memory. (b) Timing diagram of the pulse applied to word-lines.

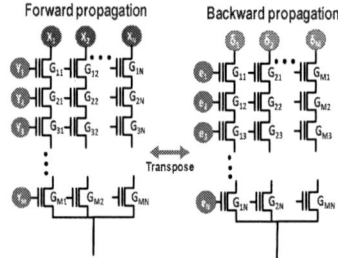

Fig. 10. Method for backward propagation using NAND flash cells as synaptic devices.

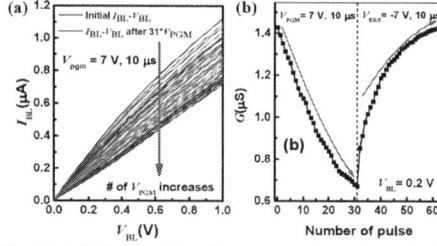

Fig. 11. (a) Measured I_{BL}-V_{BL} curves when selected cell is programmed 31 times. (b) Bidirectional conductance response when selected cell is programmed 31 times and erased 31 times.

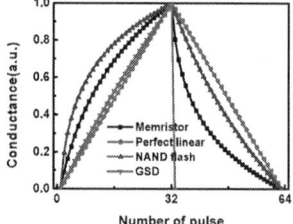

Fig. 12. Conductance responses of memristor [7], perfect linear device, GSD and NAND flash using behavior model in [8].

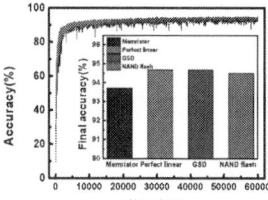

Fig. 13. Simulated classification accuracy.

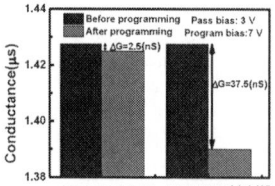

Fig. 14. Pass bias disturbance compared to that of program bias.

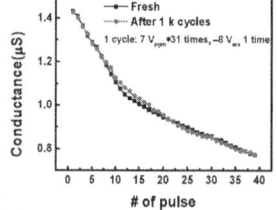

Fig. 15. Conductance response of fresh and 1k cycled cell.

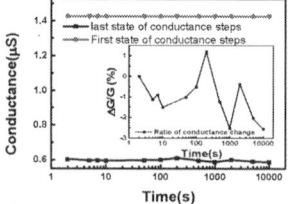

Fig 16. Retention characteristics of conductance states (first and last states). The inset shows $\Delta G/G$.

978-1-5386-4219-1/18 $31.00 © 2018 IEEE

Nonvolatile Circuits-Devices Interaction for Memory, Logic and Artificial Intelligence

Chun-Meng Dou, Wei-Hao Chen, Cheng-Xin Xue, Wei-Yu Lin, Wei-En Lin, Jun-Yi Li, Huan-Ting Lin, Meng-fan Chang

National Tsing Hua University, Hsinchu, Taiwan

E-mail: mfchang@ee.nthu.edu.tw   Phone: +886-3-516-218

Abstract

Emerging nonvolatile memory (eNVM) have aroused extensive attention due to their low power and high speed. Recent advances have further moved eNVM to the forefront as key enablers of nonvolatile logics (nvLogics) for IoT devices and computing-in-memory (CIM) for AI chips. In this paper, we firstly examine the circuit-device-interaction (CDI) issues to implement high-performance memory macro. Then we review examples of emerging eNVM-based nvLogics for nonvolatile processors and CIM macro for AI chips with an emphasis on the challenges required CDI.

Keywords: ReRAM, STT-MRAM, PCM, nvLogics, IoT, computing-in-memory, Artificial Intelligence

Introduction

Memory access has proven a major bottleneck in the pursuit of high-performance and low-power computing based on the von Neumann architecture (Fig. 1) [1]. This issue is of particular importance in systems aimed at IoT and AI applications. IoT edge devices that operate in normally-off/ frequently-on scenarios are impeded by high power consumption and delays associated with the need for the serial transfer of data between logic circuits and NVM to backup/recovery the system for power-off operations (Fig.2). Figure 3 shows a typical deep neural networks (DNN) processor [1]. It is tasked with parallel processing large amounts of data though heavy cross-memory-hierarchy access and generates a massive volume of intermediate data among the various layers in DNN, which greatly increases the power and latency of the system. Recent advances have demonstrated that eNVMs can serve not only as memory macro [2-9], but also in the development of nvLogics and CIM to progress beyond von Neumann architecture [10-17]. In the following, we examine the challenges in the development of these technologies stressing on CDI.

Circuit-Device Interaction (CDI) for Memory Macros

Figure 4 outlines the major eNVMs, including STT-MRAM, PCM, and ReRAM, and their high/low resistive states (HRS/LRS). Although they provide advantages in terms of power and speed, eNVMs suffer from relatively low resistance ratios (R-ratios) (Fig.5), resulting in a small read margin [2-4]. Figure 6 shows (a) the cell current of 1T1R cell and (b) the ratio between the common-mode (I_{COM}) and differential-mode current (I_{DIFF}) as a function of LRS resistance (R_{LRS}) in a typical current-mode sensing scheme. An excessively small R_{LRS} can reduce the ratio I_{DIFF}/I_{COM} required for a good sensing yield, whereas an excessively large R_{LRS} tends to increase the read access time because of reduced cell current (Fig. 6 (c)). This means that cell resistance must be optimized taken the trade-off between the read margin and speed into considerations. The wide distribution of write times (T_W) in memory cells is another critical challenge (Fig.7 (a)). Although using the maximum T_W can assure all cells can switching, it results in overs-set/reset and hence degrades device reliability. The write termination schemes (Fig.7 (b)) are therefore indispensable to improve the macro's reliability and energy efficiency [5,6]. Lastly, there is a considerable penalty in the area overhead of memory macros when the operational voltage of memory cell increases because high voltage transistors are required (Fig. 8).

Circuit-Device Interaction for nvLogics

Figure 9 shows from a conceptual perspective the structure of nonvolatile Processors based on nvLogics, including nvFlipflop, nvSRAM, and nvTCAM devices. The ability of nvLogics to store/restore data using local NVM cells makes it possible to backup/recover critical data of logic circuits in a parallel and distributed manner, which greatly reduces the power consumption and latency associated with power-off operations. Figure 10 presents silicon-verified examples based on ReRAM devices [11,14,15]. The relatively small R-ratio of eNVMs is also a major concern here. A large R-ratio can produce large differential input signals at Q and QB nodes to enhance the restore yield of nvSRAM (Fig. 11).

Circuit-Device Interaction for the CIM in AI Chips

On the basis of memory macros, CIM is developed to perform computation within the memory array. Figure 12(a) presents the block diagram of CIM macros that can work in both of memory and CIM modes. To perform logic operations, two word-lines (WLs) are activated to compare the values in the input memory cells (MCs) and the logic results of AND, OR, and XOR can be read out by sensing the bit-line (BL) current (Fig.12(b)) [16]. A large R-ratio can effectively increase the read margin between different states (Fig.13). To process neural network by CIM, weight information is firstly stored inside the memory array and then multiple WLs are activated. The MC current (I_{MC}) gives the multiply results between the stored weight and WL input and then accumulation results can be read from BL (Fig.12(c)). By adopting CIM macro, the power consumption of DNN processors can be substantially improved by reducing the amount of intermediate data (Fig.14) [1, 17]. Figure 15 shows an example of MAC values variations due to number of activated WLs in a CIM macro with 32 inputs [13]. A large R-ratio can effectively increase the read margin between different MAC values by suppressing HRS leakage (Fig. 16).

Conclusions

To fully exploit the advantages of eNVM based memory macro, nvLogics, and CIM macro, an intensive circuit-device interactive design are required to overcome the challenges imposed by their intrinsic characteristics. While innovative device engineering is expected to further increase the R-ratio of eNVM while lower the cell characteristic variations and operational voltages, novel memory circuits for read and write are also indispensable for eNVM-based applications.

References:

[1] M.-F. Chang, *ISSCC tutorial*, 2018. [2] S.-S. Sheu, et al., *ISSCC*, 2011. [3] W. Otsuka, et al., *ISSCC*, 2011. [4] M.-F. Chang, et al., *ISSCC*, 2012. [5] X.-Y. Xue, et al., 2014. [6] M.-F. Chang, et al., *ISSCC*, 2014. [7] T.-H. Yang et al., *ISSCC*, 2018. [8] W.-S. Khwa et al., *ISSCC*, 2016. [9] W.-S. Khwa et al., *IEDM*, 2014. [10] S. Matsunaga, *VLSI*, 2013. [11] A. Lee, et al., *JSSC*, vol. 52, no. 8, 2017.

[12] Y. Liu, et al., *ISSCC*, 2016. [13] F. Su et al., *VLSI*, 2017. [14] P.-F. Chiu, et al., *VLSI*, 2010 [15] M.-F. Chang, et al., JSSC, vol 51, no.11, 2016 [16] W.-H. Chen, et al., *ISSCC*, 2018. [17] W.-H. Chen, et al., *IEDM*, 2017.

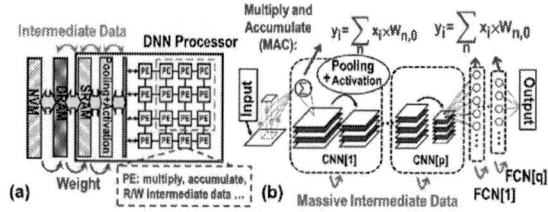

Fig.3 (a) Structure of DNN processor and (b) DNN consists of multiple layers of convolution neural network (CNN) and fully-connected networks (FCN).

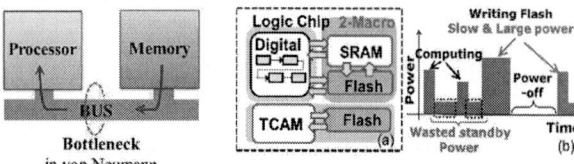

Fig.1 Conceptual view of von Neumann "bottleneck"

Fig.2 (a) SoC based on the two-macro scheme and (b) its power consumption

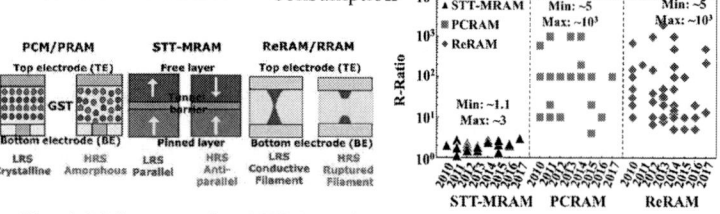

Fig. 4 Major emerging NVMs and their high and low resistive states

Fig. 5 R-ratio of major emerging NVMs (STT-MRAM, PCRAM and ReRAM)

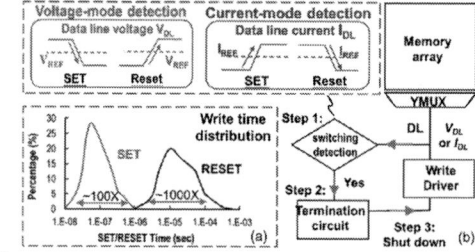

Fig. 7 (a) Write time variation of ReRAM devices, and (b) flow charts of write termination

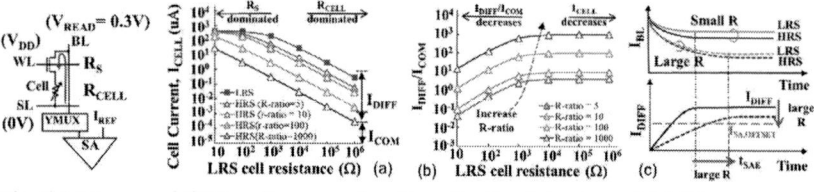

Fig. 6 (a) I_{CELL} and (b) I_{DIFF}/I_{COM} as a function of cell resistance, and (c) BL current develops as a function of time. The inset shows a typical current-mode sensing

Fig.8 Area overhead of 1T+1R cell for different operational voltages

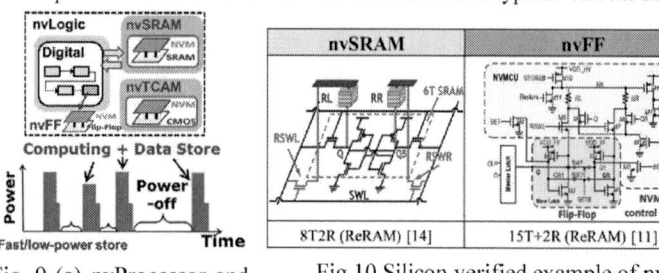

Fig. 9 (a) nvProcessor and (b) its power versus time

Fig.10 Silicon verified example of nvLogic components

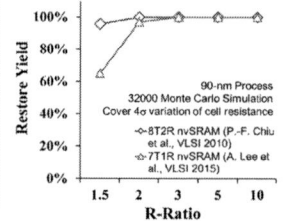

Fig. 11 Restore yield of nvSRAM with different R-ratio

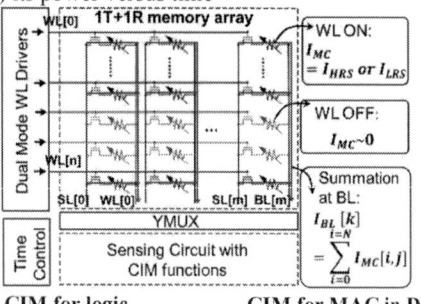

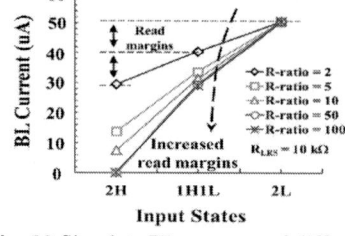

Fig. 13 Simulate BL currents of different input cell states with different R-ratio

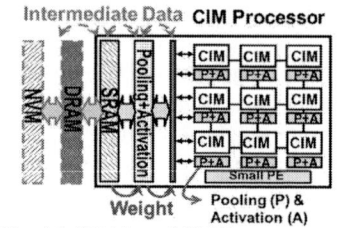

Fig. 14 CIM based DNN processor

CIM for logic

CIM for MAC in DNN

Input MC_1	Input MC_2	Output current (I_{BL})
HRS "1"	HRS "1"	$2I_{HRS}$ (2H)
HRS "1"	LRS "0"	$I_{HRS}+I_{LRS}$ (1H1L)
LRS "0"	HRS "1"	$I_{HRS}+I_{LRS}$ (1H1L)
LRS "0"	LRS "0"	$2I_{LRS}$ (2L)

Input/WL (IN)	Weight/MC (W)	Product (IN×W)	I_{MC}
0	0 (HRS)	0	0
0	+1 (LRS)	0	0
1	+1 (LRS)	+1	I_{LRS}
1	0 (HRS)	0	I_{HRS}

Fig.12 (a) Block diagram of CIM memory Macro, and its applications for (b) logic and (c) CNN/FCN

Fig. 15 (a) voltage-mode sensing and (b) simulated voltage distribution of different MAC values

Fig. 16 Simulated Read margin of MAC values at different R-ratio

978-1-5386-4219-1/18 $31.00 © 2018 IEEE

XNOR-SRAM: In-Memory Computing SRAM Macro for Binary/Ternary Deep Neural Networks

[1]Zhewei Jiang, [2]Shihui Yin, [1]Mingoo Seok, [2]Jae-sun Seo, [1]Columbia University, USA, [2]Arizona State University, USA

Abstract: We present an *in-memory computing* SRAM macro that computes XNOR-and-accumulate in binary/ternary deep neural networks on the bitline without row-by-row data access. It achieves 33X better energy and 300X better energy-delay product than digital ASIC, and also achieves significantly higher accuracy than prior in-SRAM computing macro (e.g., 98.3% vs. 90% for MNIST) by being able to support the mainstream DNN/CNN algorithms.

I. Introduction

Deep Neural Networks (DNNs) and Convolutional Neural Networks (CNNs) have unprecedentedly improved the accuracies in large-scale recognition tasks. However, the arithmetic complexity, memory access, and the associated cost has limited the energy-efficiency and acceleration of DNN hardware. To address this, in recent algorithms, weights and neuron activations are binarized to +1 or -1 [1-2] such that the multiplication between weights and inputs/activations becomes an XNOR operation and the accumulation becomes bitcount of those XNOR results. With such computation reduction, data access can dominate the overall energy of DNN hardware [3]. Conventional data storage (e.g., SRAM) requires row-by-row accesses, and fetching millions of weights in this manner consumes substantial energy and delay.

To avoid this, recent works proposed the in-memory computing scheme, which performs computation on the bitline without reading out each row of memory [4-6], demonstrating large improvements. However, prior in-SRAM computing works are not capable of supporting mainstream DNNs/CNNs. For example, ref. [5] computes multi-input Multiply-and-Accumulate (MAC) but obtains a binary output for the weak-classifier in each column. Due to this limitation, it achieves a low accuracy of ~90% for MNIST dataset.

In this paper, we propose an in-memory mixed-signal SRAM macro (XNOR-SRAM) that not only energy-efficiently computes XNOR-and-Accumulate (XAC) in binary/ternary DNNs, but also supports mainstream DNNs/CNNs with high accuracy. Our XNOR-SRAM performs a 256-input XAC without memory readout, via analog accumulation of bitwise XNOR results on the Read BitLine (RBL) voltage of the SRAM array, and digitizes the RBL voltage (V_{RBL}) using a flash Analog-to-Digital Converter (ADC) embedded in the periphery. XNOR-SRAM supports binary weights (+1, -1) and binary inputs (+1, -1) as well as ternary inputs (+1, 0, -1). The 65 nm prototype achieves 300X better Energy-Delay-Product (EDP) than digital ASIC in computing XAC. DNN/CNN classification using our macro achieves 98.3% accuracy for MNIST (85.7% for CIFAR-10), marking 1.7% error that is 6X smaller than a prior work [5].

II. Architecture and Operation

Fig. 1 presents the proposed XNOR-SRAM architecture, which can map convolutional and fully-connected layers of CNNs and Multi-Layer Perceptrons (MLPs). It consists of a 256-by-64 custom SRAM array, a row decoder, an XNOR-mode WL driver, and a column periphery including a 3.46-b flash ADC. The XNOR-SRAM operates in either of two modes: memory mode and XNOR mode. In memory mode, it performs row-by-row digital read/write as regular memory circuits. In XNOR mode, it performs in-memory XAC computation with all rows asserted simultaneously. Fig. 2 shows the proposed 12T SRAM bitcell. T1 to T6 form a 6T SRAM cell; T7 to T10 form complimentary pull-up/-down (PU/PD) circuits for XNOR mode (and memory mode read); T11 and T12 power-gate the PU/PD circuits when the corresponding column is not enabled.

In XNOR mode, the RWL driver translates each ternary/binary input of an input vector to four RWLs according to Fig. 3. T7 to T10 in each bitcell in a selected column perform XNOR of the RWLs (input/activation) and the weight stored in the bitcell. As shown in Fig. 4, the XNOR output '+1' yields one strong PU by PMOS and

one weak PU by NMOS; XNOR output '-1' results in one strong PD by NMOS and one weak PD by PMOS; XNOR output '0' is implemented by driving even rows of input '0' with strong PU/PD and odd rows of input '0' with weak PU/PD. The output '0' case yields the average strength of the '+1' and '-1' cases, as '0' inputs tend to be randomly distributed in ternary DNNs.

Parallel XNOR-output-controlled PU/PD paths from all bitcells in a column form a voltage divider, where RBL is the output. V_{RBL} therefore becomes a monotonic transfer function of XAC (Fig. 5), and we can obtain the XAC result by digitizing V_{RBL} with the ADC. The embedded ADC plays a key role in speed and DNN accuracy. We chose the flash ADC using strong-arm comparators for speed. We also investigated the required precision based on CIFAR-10 and MNIST datasets, and found that employing 11 levels (3.46 b) with nonlinear quantization using Llyod-Max algorithm based on statistical distribution yields satisfactory accuracy. Fig. 5 shows the optimized references used for 0.6V operation of XNOR-SRAM.

III. Measurements and Comparisons

We prototyped the proposed XNOR-SRAM in 65 nm CMOS (Fig. 6). The power consumption is dependent on the XAC result (Fig. 7), due to the nature of the voltage divider. For random data that corresponds to XAC value near 0 (worst-case), XNOR-SRAM achieves 235.5 pJ and 54.21 ns for 64 operations of 256-input XAC at 1V. Fig. 8 shows the energy and the maximum clock frequency for 0.6-1V, at which similar EDP gains are measured. Note that we perform XAC on a column-by-column basis here, however 64-column parallel operation could improve throughput by ~64X at the penalty of additional ADC circuits and mismatch compensation.

As a comparison, we designed a well-crafted digital accelerator in 65 nm, which computes the same XAC but has to access weights from conventional SRAM. The post-layout simulation based on industrial 6T SRAMs, standard cells, and parasitic-annotated netlists at 1V supply (TT 25°C) shows the digital baseline consumes 7.81 nJ and 514 ns for the same XAC operations, 33X worse in energy and 300X worse in EDP than our macro also operating at 1V (Fig. 9).

The systematic strength imbalance between NMOS and PMOS can skew the transfer function. We addressed this by biasing PMOS N-wells in the bitcell array at marginal area/power penalty (Fig. 10). Fig. 11 shows the measured V_{RBL} variability resulting from process variation and parasitics across different columns and data patterns. The highest variation (20mV standard deviation) occurred at the lowest XAC value of 0, which has the least impact on final accuracy.

Using our macro, we evaluated the accuracy for MNIST and CIFAR-10 datasets. For MNIST, a MLP with three hidden layers, each with 512 neurons, is used. The CNN for CIFAR-10 has six convolutional layers and three fully-connected layers [2]. Starting from the first hidden layer of the MLP/CNN, our macro computes 256-input XACs for MAC/convolution operations. Accumulation of XAC outputs, pooling, and batch normalization are performed in digital simulation with bit precisions of 12, 12, and 10, respectively. The DNNs with our macro (with ideal) achieve 98.3% (98.8%) accuracy for MNIST and 85.7% (90.7%) for CIFAR-10. The power and area breakdowns are presented in Fig. 12. As shown in Fig. 13, our macro improves EDP by 300X over digital hardware performing the same XAC operations while improving classification accuracy significantly over prior in-memory computing hardware [4-6].

[1] M. Rastegari *et al.*, "XNOR-Net: ...," *ECCV*, 2016.
[2] I. Hubara *et al.*, "Binarized Neural Networks," *NIPS*, 2016.
[3] Y. Chen *et al.*, "Eyeriss: ...," *JSSC*, 2017.
[4] Q. Dong *et al.*, "A 0.3V VDDmin 4+2T SRAM...," *VLSI*, 2017.
[5] J. Zhang *et al.*, "In-Memory Computation...," *JSSC*, 2017.
[6] M. Kang *et al.*, "A 19.4 nJ/decision 364K...," *ESSCIRC*, 2017.

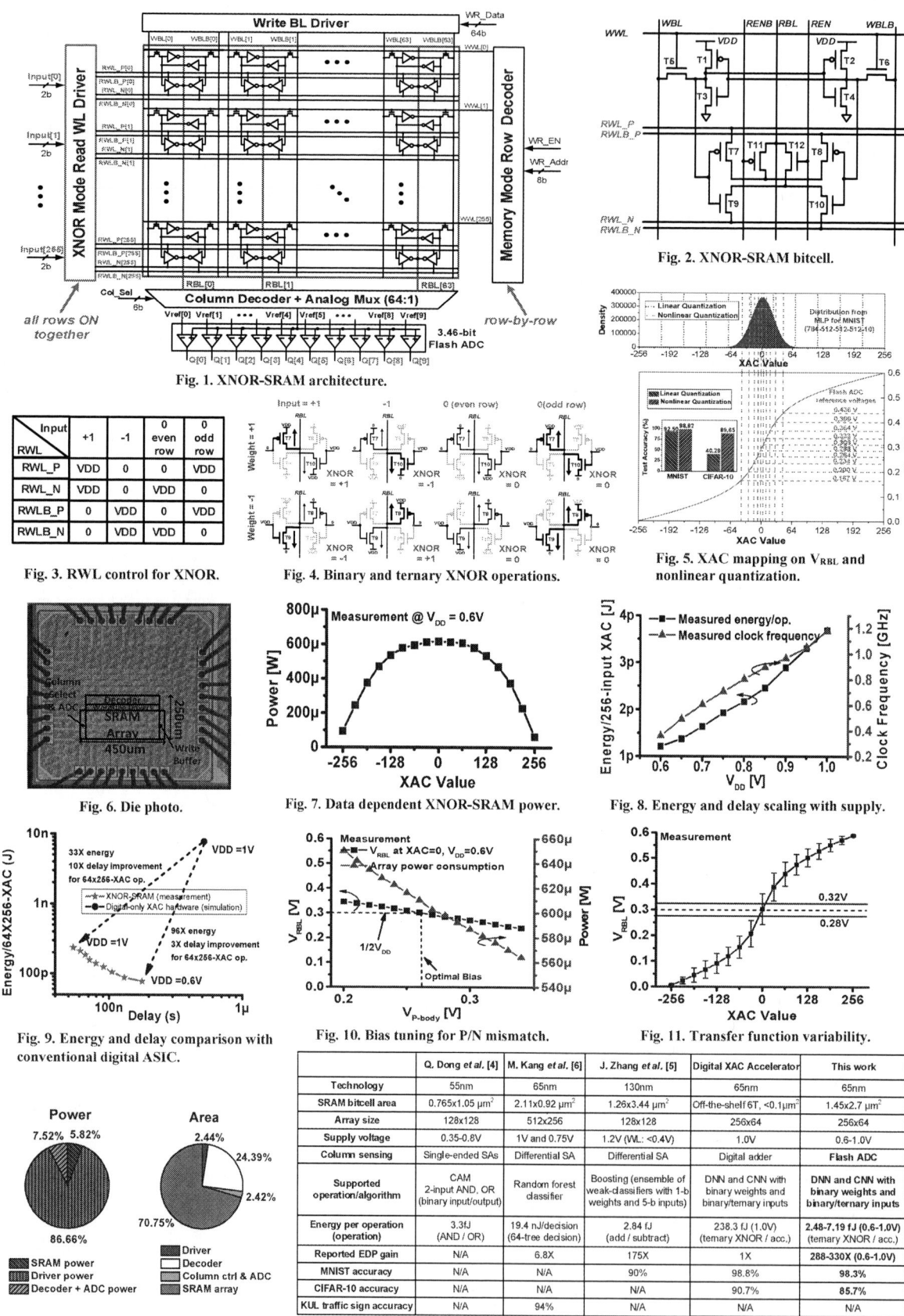

Fig. 1. XNOR-SRAM architecture.

Fig. 2. XNOR-SRAM bitcell.

Fig. 3. RWL control for XNOR.

Fig. 4. Binary and ternary XNOR operations.

Fig. 5. XAC mapping on V_{RBL} and nonlinear quantization.

Fig. 6. Die photo.

Fig. 7. Data dependent XNOR-SRAM power.

Fig. 8. Energy and delay scaling with supply.

Fig. 9. Energy and delay comparison with conventional digital ASIC.

Fig. 10. Bias tuning for P/N mismatch.

Fig. 11. Transfer function variability.

Fig. 12. Area and power breakdown.

Fig. 13. Comparison with recent in-memory computing hardware.

	Q. Dong et al. [4]	M. Kang et al. [6]	J. Zhang et al. [5]	Digital XAC Accelerator	This work
Technology	55nm	65nm	130nm	65nm	65nm
SRAM bitcell area	0.765x1.05 µm²	2.11x0.92 µm²	1.26x3.44 µm²	Off-the-shelf 6T, <0.1µm²	1.45x2.7 µm²
Array size	128x128	512x256	128x128	256x64	256x64
Supply voltage	0.35-0.8V	1V and 0.75V	1.2V (WL: <0.4V)	1.0V	0.6-1.0V
Column sensing	Single-ended SAs	Differential SA	Differential SA	Digital adder	Flash ADC
Supported operation/algorithm	CAM 2-input AND, OR (binary input/output)	Random forest classifier	Boosting (ensemble of weak-classifiers with 1-b weights and 5-b inputs)	DNN and CNN with binary weights and binary/ternary inputs	DNN and CNN with binary weights and binary/ternary inputs
Energy per operation (operation)	3.3fJ (AND / OR)	19.4 nJ/decision (64-tree decision)	2.84 fJ (add / subtract)	238.3 fJ (1.0V) (ternary XNOR / acc.)	2.48-7.19 fJ (0.6-1.0V) (ternary XNOR / acc.)
Reported EDP gain	N/A	6.8X	175X	1X	288-330X (0.6-1.0V)
MNIST accuracy	N/A	N/A	90%	98.8%	98.3%
CIFAR-10 accuracy	N/A	N/A	N/A	90.7%	85.7%
KUL traffic sign accuracy	N/A	94%	N/A	N/A	N/A

A 4M Synapses integrated Analog ReRAM based 66.5 TOPS/W Neural-Network Processor with Cell Current Controlled Writing and Flexible Network Architecture

Reiji Mochida, Kazuyuki Kouno, Yuriko Hayata, Masayoshi Nakayama, Takashi Ono, Hitoshi Suwa, Ryutaro Yasuhara, Koji Katayama, Takumi Mikawa, Yasushi Gohou

Panasonic Semiconductor Solutions Co.,Ltd., 1 Kotari-yakemachi, Nagaokakyo City, Kyoto 617-8520 Japan

E-mail:mochida.reiji@jp.panasonic.com

Abstract

This paper presents low-power neural-network (NN) processor using ReRAM to store weights as analog resistance for future AI computing. We propose ReRAM perceptron circuit for realizing large scale integration, highly accurate cell current controlled writing scheme, and flexible network architecture (FNA) in which any NNs can be configured. Fabricated 180nm test chip shows well-controlled analog cell current with linear 30μA dynamic range and 0.59μA variation of 1 sigma, results in 90.8% MNIST numerical recognition rate. Furthermore, 4M synapses integrated 40nm test chip achieves lower analog cell current and 66.5 TOPS/W power efficiency.

Introduction

The NN carries out enormous calculations of multiply accumulate (MAC) operation between weights and input data, and thus it needs high-performance hardware such as graphics processing unit (GPU), in which it consumes a great amount of power. To overcome this issue, MAC operation circuits equipped with memristor using ReRAM is proposed [1]. However, writing analog cell current accurately to ReRAM is a challenge. In addition, analog based data transfer between perceptrons is difficult, which requires lots of A/D and D/A converters. Thus executing large-scale MAC operations using analog cell current has not been reported yet. This paper presents a newly-developed low-power NN processor using ReRAM to store weights as analog cell current for realizing large scale MAC operations. We call it Resistive Analog Neuro Device (RAND) chip.

ReRAM Perceptron Circuit

The NN is made up of a combination of computation nodes called "perceptron", as shown in Fig. 1. At each perceptron, multiple inputs and weights are multiplied and accumulated and the result of MAC operation is delivered to an activation function to obtain its output. Proposed ReRAM perceptron circuit has a group of word lines (WLs), bit lines (BLs), and source lines (SLs). A memory cell (MC) is composed of one select transistor and one resistive switching element (1T-1R). Because weights in the NN have positive and negative values, two MCs connected to the same WL are used to express one weight. For example, MC connected to BL0 holds positive weight, while MC connected to BL1 holds negative weight. WLs are assigned to corresponding inputs. With these configurations, BL0 transfers results of the MAC operation of positive weights, as analog cell current, to a sense amplifier (SA), and BL1 similarly transfers that of negative weights. SA compares these current and outputs digital value, which is equivalent to a step function output. It allows MAC operation of multiple inputs in a single reading (inference-READ), thus enables MAC operation to be carried out at high speed with lower power consumption. In addition, both input and output are digitized, therefore A/D and D/A converters are unnecessary, which is suitable for large scale MAC operations.

Highly Accurate Cell Current-Controlled Writing

Fig. 2 shows highly accurate cell current-controlled writing circuit, RAND device structure, and evaluation results of analog cell current. Generally, NN operation is improved by writing analog cell current linearly with a wide dynamic range. Analog cell current of ReRAM depends on writing current. The writing current can be set in the following sequence. First, a constant current generated by current supply circuit is sent to weight control circuit, which amplifies the constant current to a desired writing current and then generates the corresponding gate clamp voltage (VCLP). Second, a write driver copies the writing current in current-mirror structure by applying VCLP, and finally it supplies writing current to MC. This analog writing circuit demonstrates the ability of writing analog cell current with linear 30uA dynamic range and 0.59μA variation of 1 sigma.

Flexible Network Architecture

Fig. 3 shows RAND architecture. This architecture can be used as both analog NN processor and digital storage memory by changing cell select control. In NN processor mode, XDRV simultaneously selects multiple WLs, and it includes a pair of latch units LAT1 and LAT2. YMUX selects BLs to be connected to positive and negative input terminals of SA. SA input selector selects BL to be connected as input line to SA.

The inference operation procedure with FNA in the manner of pipeline operation is summarized in Fig. 4. Based on NN information, NN-controller manages the cell addresses, to which weights are assigned respectively, and carries out the following sequence using a single RAND array. At the XDRV, input data A are latched by LAT1 and then copied to LAT2. Next, BLs are selected, and SA obtains data B. Thus, data B are latched by LAT1. The BL selection is changed and SA obtains data C. Data C is latched by LAT1, in addition to Data B. Finally, the data latched by LAT1 are copied to LAT2, which is followed by BL selection and SA obtains output data D, the FNA completes the NN computation. In this manner, the FNA enables a single chip to be applied to various deep NNs.

Measurement Results

Fig. 5 shows results of MNIST handwritten digit datasets recognition. NNs have input layer with 196 nodes and output layer with 10 nodes. Input data are in the form of 14×14-bit images created by compressing the MNIST datasets. The network also includes middle layers with 64 nodes 1, 2, or 3. The processing results gave a maximum accuracy of 87.3% for the NN with 3-middle layer. However, these results are affected by SA's offset and the variation of cell current. To solve this problem, we have developed a circuit for max value search of MAC operation (MSMA). With this MSMA architecture, RAND chip can add current through a fixed resistance to the positive side or to the negative side. MSMA architecture shown in Fig. 5 depicts a case of adding to the negative side. Both SA0 and SA1 output "1" in a normal inference-READ. The difference between positive current

(Ipos) and negative current (Ineg) is larger at SA0 than at SA1, so we use the difference between each current in SA0 (50uA) and SA1 (5uA). In the case of MSMA inference-READ, adding fixed resistance current (Imsma) to negative side eliminates minute difference between Ipos and Ineg. Applying MSMA inference-READ has achieved 90.8% accuracy.

Fig. 6 shows 180nm test chip micrograph, in which we performed NN processing, and 40nm test chip micrograph. Fig. 7 shows results of writing analog cell current demonstrated by both test chips. Since it is possible to scale the filament with fine process technology [5], the cell current of 40nm is lower than that of 180nm, which leads to better power efficiency.

Conclusion

We proposed low-power and high-accuracy NN processor using ReRAM. Table I compares this paper's work with various technologies. RAND chip fabricated by 180nm process consumes power of 15.8mW on a 1024 input inference-READ, achieving power efficiency of 20.7 TOPS/W. In addition, 40nm ReRAM reduces power consumption during an inference-READ to 9.9mW, thus achieving power efficiency of 66.5 TOPS/W.

Acknowledgements

This work is supported by NEDO program. The authors also thank Professor T.Asai of Hokkaido University for valuable discussions.

References

[1] M. Prezioso, et al., Nature, vol. 521, no. 14441, pp.61-64, 2015.
[2] D. Miyashita, et al., ASSCC, pp.25-28, 2016.
[3] B. Moons, et al., ISSCC, pp.246-247, 2017.
[4] K. Ando, et al., VLSI Circuit, pp.24-25, 2017.
[5] Y. Hayakawa, et al., VLSI Technology, pp.14-15, 2015.

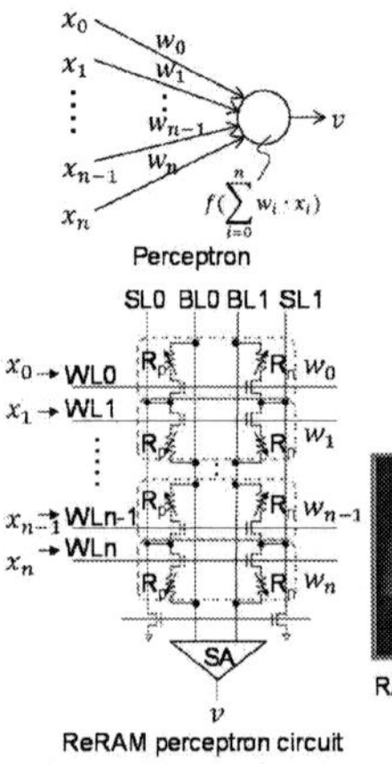

Fig 1. Proposed ReRAM perceptron circuit

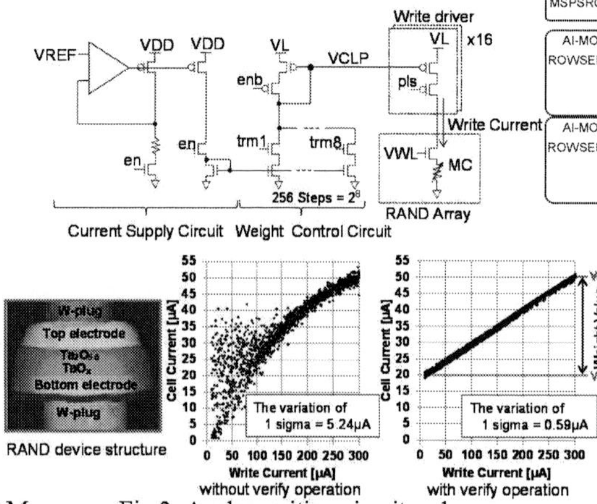

Fig 2. Analog writing circuit and results of analog cell current

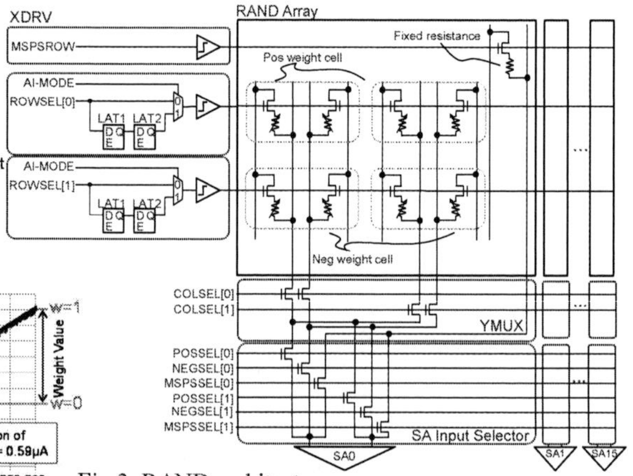

Fig 3. RAND architecture. It can be used as both analog NN processor and digital storage memory

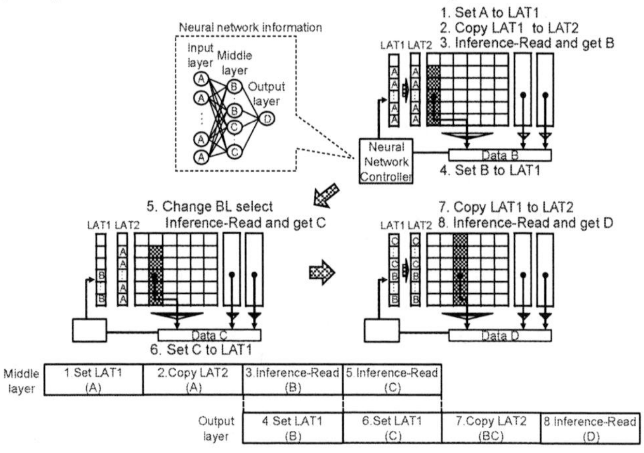

Fig 4. Inference operation procedure with FNA

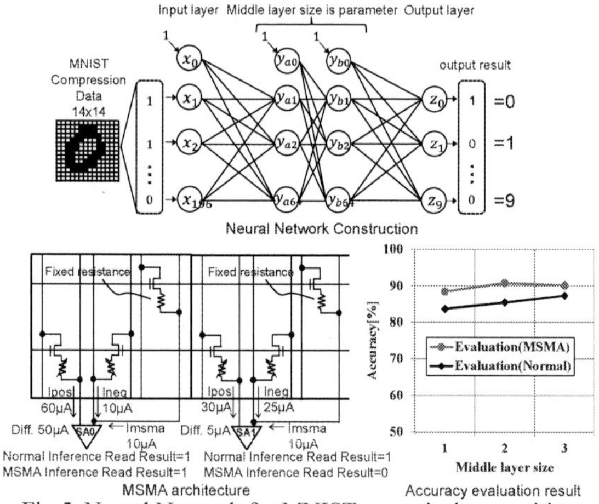

Fig 5. Neural Network for MNIST numerical recognition

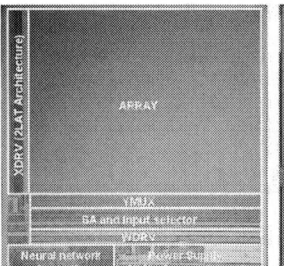
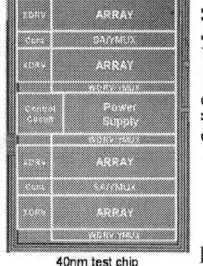

Fig 6. Chip micrograph (scale different)

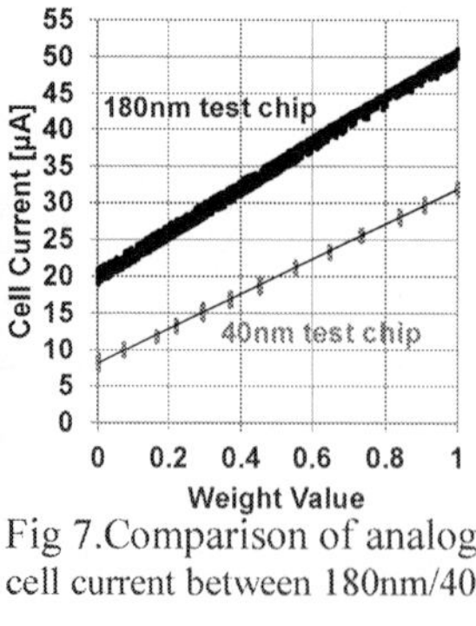

Fig 7.Comparison of analog cell current between 180nm/40nm

TABLE I. Comparison Table

	ASSCC 2016 [2]	ISSCC 2017 [3]	VLSI 2017 [4]	This Work	
Technology	65nm	28nm	65nm	180nm	40nm
Weight Storage	SRAM	SRAM	SRAM	ReRAM	ReRAM
Synapses	32K	128K	0.8M	2M	4M
Area(mm²)	3.61	1.87	3.9	12.6	2.71
Synapses/mm²	0.01M	0.07M	0.21M	0.16M	1.48M
Voltage(V)	N/A	0.65-1.1	0.55-1.0	1.8	1.1
Power(mW)	N/A	7.6	50-600	15.8	9.9
TOPS	N/A	0.076	1.38	0.33	0.66
TOPS/W	48.2	10	6.0	20.7	66.5

A Novel 3D AND-type NVM Architecture Capable of High-density, Low-power In-Memory Sum-of-Product Computation for Artificial Intelligence Application

Hang-Ting Lue, Weichen Chen, Hung-Sheng Chang, Keh-Chung Wang, and Chih-Yuan Lu

Macronix International Co., Ltd., 16 Li-Hsin Road, Hsinchu Science Park, Hsinchu, Taiwan. (e-mail: htlue@mxic.com.tw)

Abstract

An AND-type stackable 3D NVM architecture is proposed to provide an ultra-high density AI computing memory with low power. The advantages are: (1) All memory transistors in the 3D array are connected in parallel, thus enable the sum-of-product operation. (2) The 3D NAND like architecture is possible to stack to > 64 layers, thus provides ultra-high density (>128Gb) AI memory. (3) Many bit lines (>1KB) can operate in parallel for high bandwidth. (4) Uses low-power +/- FN programming/erasing which allows high parallelism, and is bit-alterable thus is ideal for training or transfer learning. (5) Excellent linearity of output current with respect to bitline bias, thus enabling ideal analog computation. (6) Adequate sensing current of the summed product thus permits fast access read for inference device. The proposed memory architecture can achieve TOPS/W>10, which is 10X greater than the conventional von Neumann architecture.

I. Introduction

The conventional "von Neumann architecture" carries out computing in a CPU, which is connected to memory devices (DRAM) though buses. Such design has limited bandwidth and high power consumptions. A memory capable of in-memory computation is highly desired because it can directly carry out the computation in the memory devices to improve the efficiency of computing [1,2].

The most popular computing memory design is to use a NOR-type array, where the transistor is often connected with a resistor memory such as ReRAM or PCM. Such 1T1R structure has some disadvantages. It is not scalable and has limited memory density much below 1Gb. The programming current of ReRAM or PCM is high (>50uA for each cell) thus does not allow highly parallel programming. The resistor devices of ReRAM or PCRAM are often highly non-linear with respect to the applied bias which makes analog computing difficult.

In principle, the 1T NOR Flash without resistor memory is also feasible for AI memory design. However, the conventional NOR Flash has limited scaling capability thus is not cost effective. Also, NOR Flash does not allow bit-alterable operation. Unfortunately, high-density NAND Flash (and 3D NAND) may not be suitable for AI computing because it naturally has slow latency (small read current), and the serially connected NAND string cannot carry out current summation.

II. Structure Explanation of 3D AND-type NOR Flash

Figure 1(a) illustrates the proposed architecture. The structure resembles the SGVC 3D NAND [3], but we introduce vertical N$^+$ buried diffusion lines that connect all memory transistors in parallel. **Figure 1(b)** illustrates the top view layout. Both bitline (connected to drain) and source lines (connected to source) are arranged in parallel direction, thus it should be classified as an "AND-type NVM". It is actually quite similar to the conventional NOR Flash, but AND-type array with parallel BL's and SL's can prevent sneak paths during programming, yet allow +/-FN bit-alterable operation in the 3D array.

Similar to 3D NAND, the memory cells are arranged in a twisted layout so that the density of metal bitline (BL) and source line (SL) are doubled to increase the bandwidth. There are two memory cells inside each trench, which share the same BL and SL by the common connection of a poly plug, but they do not share the same WL. Meanwhile, the memory cell in the nearby adjacent trench shares the same WL, but fortunately they do not share the same BL/SL, thanks to the twisted layout structure that automatically separates the BL/SL for them.

Figure 1 also illustrates an example (type I) of sum-of-product computation. To select a memory cell is to apply bias voltages at the corresponding WL and BL's, while de-selected WL's and BL's=0V. Each memory cell store the data information of weighting factor (W(x,y,z)), which is the conductance (Id/Vd) of the memory cell. The conductance can be programmed/erased to adjust the value. We may parallelly select plural BL's with various BL biases to carry out summed product. The currents are summed in the SL's (selected by SL decoder) for sensing.

The zoom-in structure is illustrated in **Fig. 2(a)**. There are two bits (Bit-1 and Bit-2) inside each trench, which are controlled by the two-side gates separately. The read current flows from the buried diffusion line vertically, and then go into the memory cell horizontally as indicated in the figure. Like 3D NAND, thin-body poly-silicon TFT device is utilized. Lg is the channel length, while Tsi is the thin-body thickness. W is the OP stack poly thickness, which is equal to the effective channel width.

Figure 2(b) and (c) illustrates the processing feasibility. High stacking process of > 64 OP layers like SGVC 3D NAND is doable.

Figure 3 briefly explains the process flow. After a trench etching, the ONO (for charge-trapping device) and thin poly channel are filled-in, followed by the oxide fill-in and top poly plug formation. A hole-type etching is carried out for isolation. Next, a plasma doping method is carried out to dope the edge of the channel into Drain and Source for the N$^+$ buried diffusion line. The metal BL and SL are connected on top of the N$^+$ diffusion. Optimized p-type doping at the thin channel and poly plug are necessary to prevent from punch-through leakage. Sufficient N$^+$ diffusion doping is necessary to reduce the buried diffusion resistance.

Figure 4 illustrates the possible memory design layout schematics for a tile in a chip. Various biases, which stand for the input signals, are applied to the bit lines. Multiple BL's can be operated simultaneously for improving the computing bandwidth. SL's are connected to the sense amplifier via a source line decoder which is designed to allow flexible selection of the address to carry out summed product. A typical input number for a summation unit is 8 (or16) BL's, where the 8 source line currents are summed together for sensing.

III. Electrical Performances and Simulations

Figure 5(a) shows the typical poly silicon TFT device's IdVg characteristics measured in a simplified process. Read current for a single cell can be ~5uA at maximal, with adjustable conductance after programming. **Figure 5(b)** shows that the Id is very linear with respect to Vd, which is important to support the analog computation, where BL's can be applied various bias instead of only digital (binary mode).

Figure 6(a) illustrates the programming simulation for the array. To select the cell, we can apply +FN ISPP voltages on the corresponding WL. The selected BL's=0V to allow programming, while de-selected BL's= +6V for inhibit. **Figure 6(b)** shows the simulated +FN programming, which has excellent program inhibit for de-selected cells.

Similar operations can be applied for erase, but with reverse polarity as shown in **Fig. 7(a)**. Unlike conventional Flash memory, bit-alterable erase is possible because the N$^+$ diffusion can directly pass through the BL bias even when other unselected WL's=0V, without the need of pass gate WL's operation in NAND. **Figure 7(b)** shows that erase is indeed selectable. The bit-alterable operation together with high parallelism of FN P/E offer efficient high-density memory training.

Figure 8 shows the estimated conductance with MLC distribution, which is emulated from the SGVC 3D NAND Vt distribution. Multi-level storage together with multi-level BL input voltages improve the throughputs of computing. **Figure 9** illustrates the range of summed current for an 8-input unit. The summed current range is ~40uA, which is adequate for NOR Flash like sensing to meet the fast random read for inference device. The summed current is almost linear to BL voltages, possible to support analog computation.

IV. Summary:

Figure 10 proposes two types of design method to produce the sum-of-product for 3D AND NVM. Type I (BL input) is suitable for the high-resolution "convolution" operation with analog input, while type II (WL input) is suitable for high-density and high-bandwidth "fully-connected" operation with binary mode. Optimized design may produce TOPS/W ranging from 5~40, which is 10 times greater than conventional von-Neumann architecture.

References: [1] G. Burr, IEDM Tutorial 2, 2017. [2] W. H. Chen, et al, IEDM 2017, session 28-2. [3] H. T. Lue, et al, IEDM session 19-1, 2017.

978-1-5386-4219-1/18 $31.00 © 2018 IEEE

2018 Symposium on VLSI Technology Digest of Technical Papers

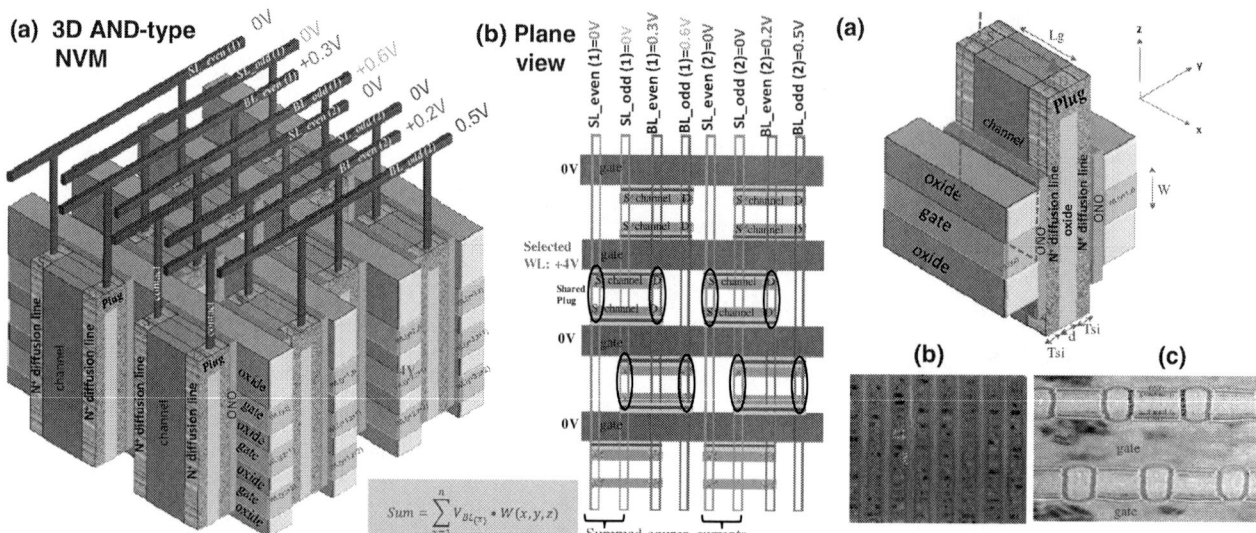

Fig. 1 (a) 3D AND-type NVM architecture. (b) Top-view layout schematics. Current are summed in the source lines.

$$Sum = \sum_{x=1}^{n} V_{BL(x)} * W(x,y,z)$$

Fig. 2 (a) Zoom-in view of the device. (b) Cross-sectional view. (c) Plane-view of the device structure.

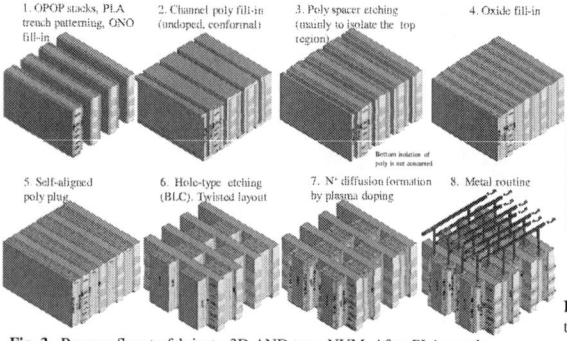

Fig. 3 Process flow to fabricate 3D AND-type NVM. After PLA trench etching, ONO and poly channel fill-in, a BLC hole etching is used to isolate device. Next, a plasma doping method is used to dope the sidewall to form the N+ diffusion line vertically, followed by metal routing.

Fig. 4 Schematic showing how to design the 3D AND-type NVM. BL's are applied various voltages (input), and the current is sensed through the source lines. Address selection for summation is arranged by the source line decoder.

Fig. 5 The experimental data of IdVg and IdVd curves of a typical poly silicon TFT BE-SONOS device fabricated in a simplified 3D process (Lg=80nm, W=40nm, Tsi=6nm). (a) Each single cell can provide ~5uA current at sufficient gate overdrive. (b) The IdVd shows excellent linearity, where Id is almost linear to Vd.

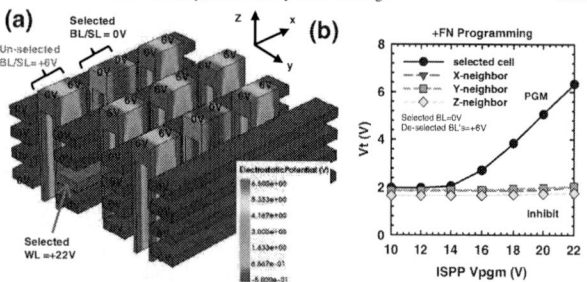

Fig. 6 (a) Programming selection method. Selected WL is applied ISPP +FN voltages. The selected BL's =0V, while de-selected BL's are +6V for program inhibit. The +6V bias can directly pass through the buried diffusion line to inhibit the bottom layers. (b) The selected cell can be programmed, while the X/Y/Z neighbor cells are well inhibited. **Y-neighbor device (back-to-back cell) is free from interference because the channel has shielding effect.**

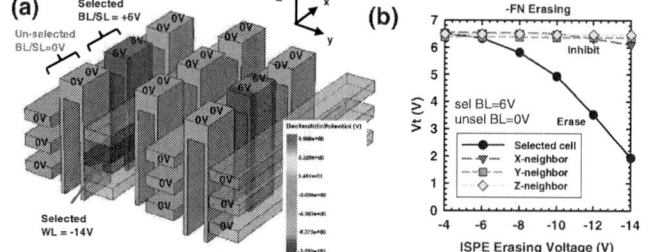

Fig. 7 (a) Erasing selection method. Selected WL is applied ISPE -FN voltages. The selected BL's =6V for erasing, while de-selected BL's are applied 0V for erase inhibit. The +6V bias can directly pass through the buried diffusion line for selected cell to erase. (b) The selected cell can be erased, while the X/Y/Z neighbor cells are well inhibited. This provides the bit-alterable erase operation.

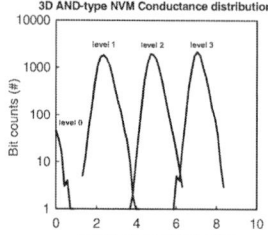

Fig. 8 The conductance (Id/Vd) distribution of 3D AND-type NVM, emulated from the MLC Vt distribution of SGVC 3D NAND chip.

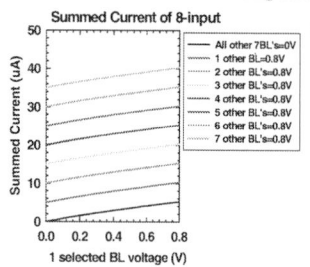

Fig. 9 The summed current of the 8-input unit. The maximal range is around 40uA when all 8 BL's are applied 0.8V. It's linear with respect to the BL voltage.

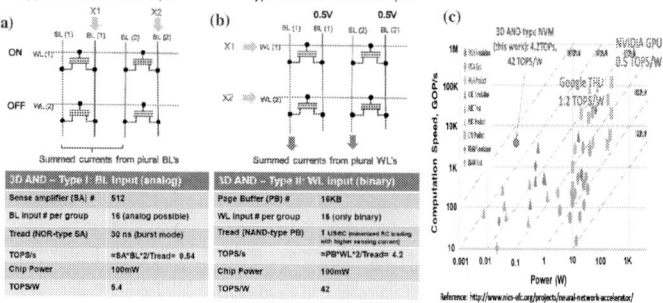

Fig. 10 (a) Type I design: BL as input (analog possible); (b) Type II: WL as input (only binary). (c) Benchmark of TOPS/W. 3D AND may provide TOPS/W ranging from 5-40, greater than conventional von-Neumann architecture.

978-1-5386-4219-1/18 $31.00 © 2018 IEEE   178

Gap in pagination due to formatting issues.

Pages 179-180

Embedded STT-MRAM in 28-nm FDSOI Logic Process for Industrial MCU/IoT Application

Yong Kyu Lee[1, a], *Senior Member, IEEE*, Yoonjong Song[2], JooChan Kim[1], SeChung Oh[2], Byoung-Jae Bae[2], SangHumn Lee[1],
JungHyuk Lee[2], UngHwan Pi[2], Boyoung Seo[1], Hyunsung Jung[2], Kilho Lee[2], HyunChul Shin[2], Hyuntaek Jung[1],
Mark Pyo[1], Artur Antonyan[1], Daesop Lee[1], Sohee Hwang[1], Daehyun Jang[1], Yongsung Ji[1], Seungbae Lee[1], Jungman Lim[1]
Kwan-Hyeob Koh[2], Kihyun Hwang[2], Hyeongsun Hong[2], Kichul Park[1], Gitae Jeong[1], Jong Shik Yoon[1], and ES Jung[1]

[1]Foundry Business, Samsung Electronics Co., Giheung, Korea, Phone: +81-31-209-1993, [a]e-mail: yklee3@samsung.com
[2]R&D Center, Samsung Electronics Co., Hwasung, Korea

Abstract— We demonstrate, for the first time, 28-nm embedded STT-MRAM operating at full industrial temperature range (-40~125 ℃) with >1E+6 endurance and >10 year retention for high speed MCU/IoT application. Robust cell operation is also demonstrated after solder reflow (260℃, 90 second) and during external magnetic disturbance (550-Oe under writing). It is built on 28-nm FDSOI technology in modular format for IP reuse and has great potential to serve wide variety of applications such as IoT, and high performance MCU.

Introduction

New era of IoT (Internet of Things) applications such as smart-car, smart-home and smart-city requires high speed logic with wide-temperature operation (-40~125℃), multi-code programmability and low power wireless connectivity. Therefore high performance logic with large e-NVM capability and, RF capability becomes indispensable.

Since MRAM has logic-process compatibility, and application extendibility as an unified embedded memory from e-Flash [1-3] to storage working memory [4] due to high endurance characteristics with fast write-NV characteristics, there has been researched a lot and also demonstrated some prototype e-MRAM for commercialization.

We previously reported the fabrication of 1T-1MTJ memory based on 28-nm HKMG bulk CMOS process, where the basic functionality of high sensing-margin (TMR 180%) and high temperature (85℃ 10 year) retention for embedded flash application has been verified [1]. In this paper, we firstly demonstrate industrial fully-compatible wide-temperature-range (-40~125℃) operating STT-MRAM with 1E6 cycle endurance and 10 year retention in 28-nm FDSOI CMOS process, which hold robust cell characteristics such as solder reflow data retention (260℃ 90 second) and external magnetic immunity (under 550-Oe write operation at 85℃) for e-NVM application [2, 3]. As a base-line process, 28-nm FDSOI CMOS process is newly adopted from previous 28-nm bulk CMOS process [1] to utilize superior RF performance and better low-power/Analog characteristics compared with those of 28-nm bulk and 14-nm FINFET process.

eMRAM Integration

Since 28-nm FDSOI has a superior RF-performance, low-power, better analog characteristics than 28-nm Bulk and 14-nm FINFET CMOS [Fig. 2], which are advantages for IoT/MCU application, we ported STT perpendicular-MTJ from 28-nm bulk to FDSOI CMOS process. The cell-select transistor and core-periphery logic transistors were replaced with FDSOI transistors instead of bulk transistor, while the MTJ has been improved by process optimization to extend operating range up to higher temperature (125℃) as shown in Fig. 3. It shows stable and symmetric R-V house curve with TMR 195%. MTJ module from bulk CMOS Cu-BEOL layers [1] was easily ported to FDSOI process due to logic compatibility of MTJ.

eMRAM Macro & Characteristics

Fig. 4 shows the block diagram of an 8Mb macro with 64 I/O STT-MRAM, which is a basic-unit for higher density IP, was newly designed based on 28-nm FDSOI process. Repair redundancy and ECC are included for improving yield and reliability. Fig. 5 shows photo image of 8Mb embedded MRAM (eMRAM). Embedded-MRAM can provide the lowest write-power (a ten hundreds) by virtue of faster writing and higher endurance characteristics (several times) compared with previously reported 28-nm e-Flash on bulk-CMOS [5]. Typically, write (read) failure increases at lower (higher) temperature due to MTJ characteristics. Increase of failure is not visible at two worst condition, 40℃ and 125℃ as shown in Fig. 6. We also verified endurance characteristics at wide range temperature (-40℃/25℃/125℃) as shown in Fig. 7. We couldn't see any change of resistance distribution before and after 1E6 cycling MTJ-cells in an 8Mb array. Fig. 8 shows good retention characteristics, which sustain the high pass-rate up to 125℃.

Robust cell operation is also demonstrated in Fig. 10, Fig. 11, Fig.12. Retention characteristics could be increased by simply tuning same MTJ stack (i.e. switching current manipulating) with one or two parameters changing without scarifying high read margin (TMR) as illustrated in Fig.9. The solder-reflow retention and high MTJ-efficiency (delta/Isw) are shown in Fig. 10 and Fig. 11. Even with high retention characteristics which satisfying solder reflow (>200℃ 10 year) [2], we can also achieve high write pass-rate at cold temperature (-40℃) due to high MTJ efficiency. External magnetic field immunity [3] under writing showed up to 550-Oe at 85℃ due to uniform Hc distribution of MTJ (1σ=6.5% in Fig. 12). Even though we have not included data in this paper, our MTJ shows a potential for storage working memory due to high endurance (>1E10) and fast writing (<30ns).

Conclusion

To meet increasingly important new applications, such as high-speed MCU and IoT, we demonstrate, for the first time, embedded STT-MRAM in 28-nm FDSOI process, operating at full industrial temperature range (-40~125℃) with >1E+6 endurance and >10 year retention. Robust cell operation is also demonstrated after solder reflow (260℃, 90 second) and during external magnetic disturbance (550-Oe at 85℃ writing). Since MTJ is built on 28-nm FDSOI technology in modular format, the merged embedded STT MRAM and RF-CMOS process is compatible to existing logic process, enabling reuse of IP.

References

[1] Y.J. Song, et al. IEDM Technical Digest., pp. 27.2.1~27.2.4, 2016
[2] D.Shum, et al. Symposium on VLSI, pp. 208-209, 2017
[3] Chia-Yu, et. al. IEDM Technical Digest., pp. 21.1.1~21.1.4, 2017
[4] S.Kang,C.Park. IEDM Technical Digest., pp. 38.2.1~38.1.4, 2017
[5] Y. K. Lee, et al. Symposium on VLSI Tech., pp. 202-203, 2017

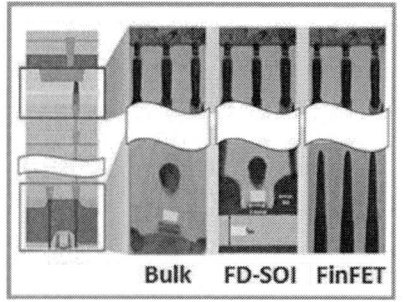

Fig.1. MTJ Compatibility with various logic such as bulk, FDSOI and FINFET because of decoupling with CMOS [1]

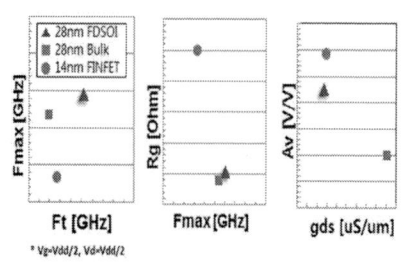

Fig.2 CMOS Performance Comparison, a) Fmax vs. Ft, b) Gate resistance vs. Fmax, c) Gain(Av) vs. Gds

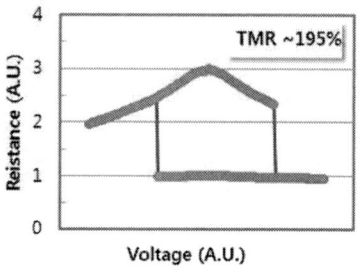

Fig.3. Improved R-V house curve of MTJ cell after full integration (TMR >190%) by process optimization.

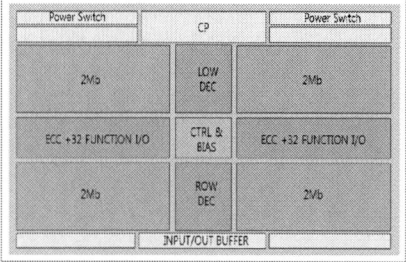

Fig.4. Block diagram of 8Mb STT 1T-1MTJ MRAM macro which has 64 I/O with 2-bit ECC with VDD/Vpower source.

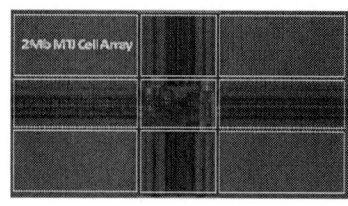

Fig.5. Micrographic View of 8Mb eMRAM macro

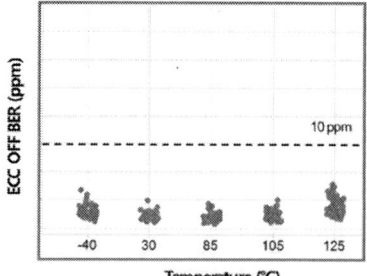

Fig.6. Write/read failure numbers depend on operating temperature at -40℃, 25℃, and 125℃

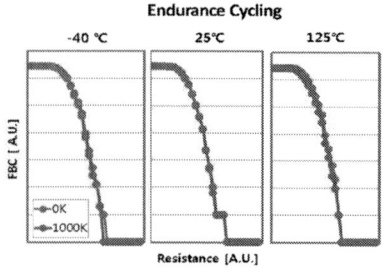

Fig.7. Resistance disribution change after 1E6 cycling at each temperature (-40/25/125℃, ECC Off). Measured in 2048 * 2 (=X/Y) * 64 MTJ-cells of 8Mb.

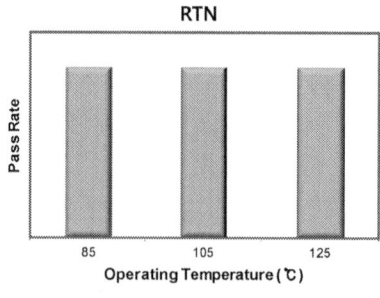

Fig.8. Pass-rate of high temperature retention at 85℃, 105℃ and 125℃ respectively

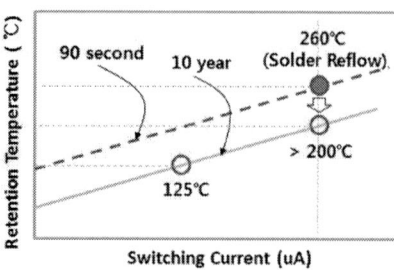

Fig.9. Retention vs. switching current. Solder reflow (260℃ 90second) can be achieved by tuning of MTJ stack (increasing a switching current)

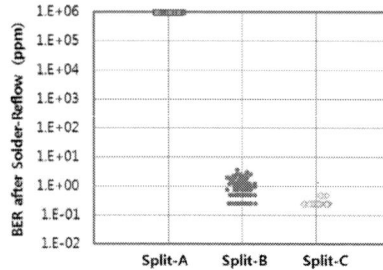

Fig.10. Solder-reflow pass rate after 260℃ 120 second-bake depends on MTJ split condition.

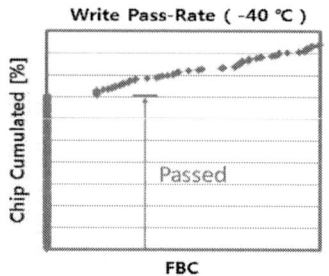

Fig.11. Pass-rate of writing the reflow-stack cells at cold temperature (-40℃), where it is the worst corner (for split-C in Fig. 10, 1-bit ECC On)

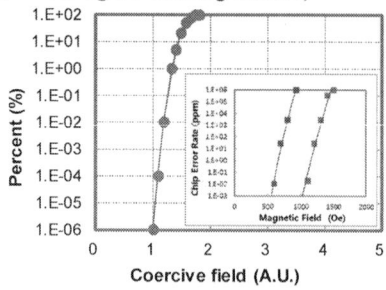

Fig.12. Coercive field distribution of MTJ (1σ=6.5%) and chip-error rate estimated under writing with external field (ECC on). 550-Oe at 85℃ for 0.001ppm CER.

22-nm FD-SOI Embedded MRAM with Full Solder Reflow Compatibility and Enhanced Magnetic Immunity

K. Lee*, K. Yamane, S. Noh, V. B. Naik, H. Yang , S. H. Jang, J. Kwon, B. Behin-Aein, R. Chao, J. H. Lim, S. K., K. W. Gan, D. Zeng, N. Thiyagarajah, L. C. Goh, B. Liu, E. H. Toh, B. Jung, T. L. Wee, T. Ling, T. H. Chan, N. L. Chung, J. W. Ting, S. Lakshmipathi, J. S. Son, J. Hwang, L. Zhang, R. Low, R. Krishnan, T. Kitamura, Y. S. You, C. S. Seet, H. Cong, D. Shum, J. Wong, S. T. Woo, J. Lam, E. Quek, A. See, S. Y. Siah

GLOBALFOUNDRIES Singapore Pte, Ltd., 60 Woodlands Industrial Park D Street 2, Singapore 738406

*Phone: +65-66702589, Email: kangho.lee@globalfoundries.com

Abstract

We demonstrate a fully functional embedded MRAM (eMRAM) macro integrated into a 22-nm FD-SOI CMOS platform. This macro combined with eFlash-flavor MTJ film stacks shows median-die bit error rate (BER) < 1 ppm after 5× solder reflows. It also meets the automotive grade-1 data retention requirement and shows intrinsic stand-by magnetic immunity of 1.4 kOe (BER criteria = 1 ppm) after 1-hr exposure at 25 °C. The results reveal that eMRAM is capable of serving a broad spectrum of eFlash applications at 22 nm or beyond.

Introduction

eMRAM has been considered one of the most promising candidates for a future embedded non-volatile memory (eNVM) technology. However, to establish eMRAM as a mainstream eNVM platform, it is crucial to achieve data retention through package reflow and address automotive eFlash applications beyond 28 nm. Also, the storage element of eMRAM, called magnetic tunnel junction (MTJ), can be disturbed by unintentional external magnetic fields or field tampering, which poses a unique challenge for eMRAM. In this paper, we demonstrate the functionality and reliability of an eMRAM macro integrated onto a 22-nm FD-SOI CMOS platform, highlighting solder reflow compatibility across wafer, automotive grade-1 data retention and outstanding stand-by magnetic immunity.

MTJ Integration and Device Performance

Figure 1 shows the TEM cross section of the eMRAM macro. MTJ was integrated between 1.1x and 2x metal layers, using standard CMOS back-end processes. Figure 2 shows the MTJ resistance distributions of a 15Kb sub-array with $TMR/\sigma(R_{low}) \sim 28$. Coercivity field ($H_c$) measured at the field ramping rate of 1 kOe/sec was > 4 kOe for eFlash-flavor MTJ stacks (Fig. 3). Read/write shmoo data with ECC on show wide operation windows (Fig. 4).

Solder Reflow Compatibility and Data Retention

Figure 5 shows post-reflow BER for five different MTJ stacks. Wafers were exposed to a reflow oven 5 times with the reflow condition that meets the JEDEC standard [1]. Stack E met BER < 10 ppm across wafer with median BER ~ 0.3 ppm. Stack E also improved switching efficiency without compromising reflow compatibility as indicated by lower switching voltage (V_c) compared to stack D (Fig. 6).

To model data retention performance of stack E, we investigated BER after oven-baking at 260-290 °C for 1-24 hours. For the typical CD, 40Mb arrays showed post-bake BER < 1E-6 even after 270 °C 1-hr baking. Figure 7 shows effective energy barrier (E_B) as a function of temperature for different MTJ diameters. E_B was extracted from post-bake BER using the Néel-Brown relaxation time formula [2]. Arrays with 13% and 30% smaller MTJ diameters were also tested. As MTJ diameter scales down, E_B decreased linearly at the rate of ~0.2 k_BT/nm. Still, arrays with 30% smaller MTJ diameter showed E_B > 40 k_BT, marginally meeting the E_B criteria for solder reflow compatibility (post-bake BER = 8.2E-6 after 1-hr baking at 260 °C). This enables a scalable eMRAM solution for eFlash applications beyond 22 nm.

The MTJ stack was optimized to balance post-bake BER at 260 °C for "high" and "low" states. At lower temperatures, data retention BER is dominated by high-state failures due to temperature dependence of magnetostatic coupling fields. Figure 8 shows that the estimated high-state E_B at 150 °C is 73.3 k_BT. This is sufficient to meet the automotive grade-1 data retention requirement (57.1 k_BT for BER criteria of 0.1 ppm after 20 years).

Magnetic Immunity

Stand-by intrinsic magnetic immunity (IMI) is characterized by measuring failure rates as a function of external field for a target exposure time. Perpendicular DC fields of 0.5-1.6 kOe were applied to a 40Mb macro using an electromagnet integrated into a standard wafer sort tester. For BER criteria of 1 ppm after 1-hr exposure, stand-by IMI was 1.39 and 0.6 kOe at 25 and 125°C respectively (Fig. 9). Figure 10 shows E_B at 125 °C as a function of external field. The E_B trend deviates from the uniform-switching model below 900 Oe and is fitted by the domain-wall propagation model [3]. For 1-min exposure to 0.75 kOe at 125 °C, the estimated BER is ~1E-6.

Active-mode IMI is limited by write margin. V_c is modulated linearly by external field (Fig. 11). Package-level magnetic shielding can be employed to lower effective field for the macro and prevent excessive V_c increase. Finite-element simulations (ANSYS Maxwell 3D) for a typical wire-bonding package show that magnetic shielding with 300 μm-thick shield can provide ~65% shielding efficiency at 500 Oe with shielding material and geometry properly optimized (Fig. 12). With this magnetic shielding, active-mode IMI of 500 Oe corresponds to 3% V_c increase. The active-mode IMI requirement needs to be incorporated into bitcell and macro designs.

Conclusion

eMRAM is a viable solution for an eNVM platform at advanced CMOS nodes. 40Mb array data from 22-nm FD-SOI eMRAM macros showed that solder reflow compatibility and magnetic immunity are not fundamental challenges for commercialization of eMRAM. The MTJ stack optimized for data retention through package reflow guarantees automotive grade-1 data retention, enabling eMRAM for high-performance automotive MCU markets.

References

[1] J-STD-020E (2014) for reflow condition from https://www.jedec.org/

[2] W. F. Brown, Phys. Rev., 130, 1677 (1963)

[3] L. Thomas et al., IEDM Tech. Dig., pp 26.4.1-4 (2015)

2018 Symposium on VLSI Technology Digest of Technical Papers

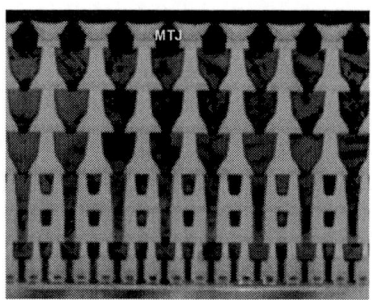

Figure 1. TEM cross section of 22-nm eMRAM macro

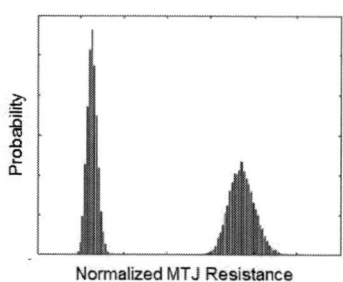

Figure 2. MTJ resistance distributions. $TMR/\sigma(R) \sim 28$.

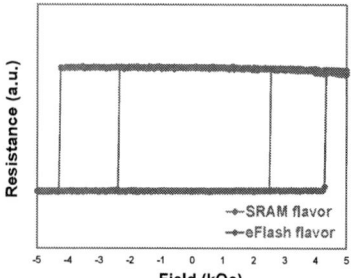

Figure 3. RH loops of SRAM-flavor and eFlash-flavor stacks

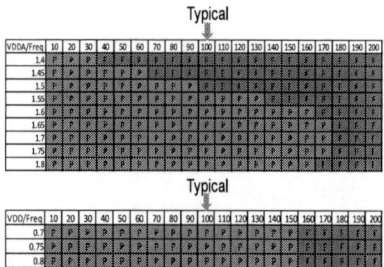

Figure 4. Read/write shmoo data with ECC on showing wide operation windows. System frequency in MHz.

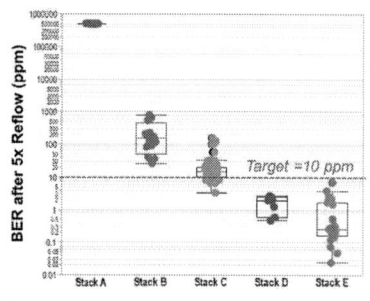

Figure 5. Post-reflow BER for different MTJ stacks. Stack D and E met BER < 10 ppm after 5x reflow across wafer.

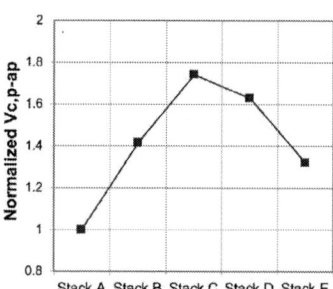

Figure 6. Normalized Vc for different MTJ stacks.

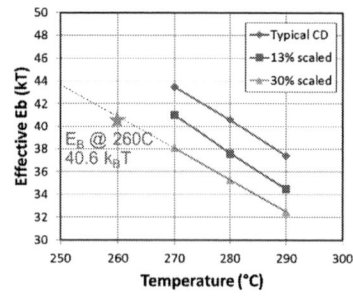

Figure 7. Effective energy barrier (E_B) as a function of temperature for different MTJ diameters. E_B was calculated from post-bake array BER.

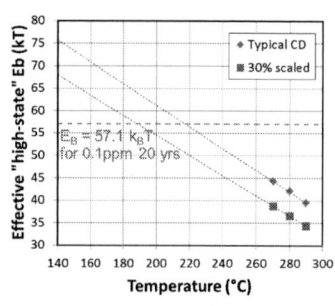

Figure 8. "High-state" energy barrier as a function of temperature. The array with 30% smaller MTJ diameter passed the automotive data retention requirement.

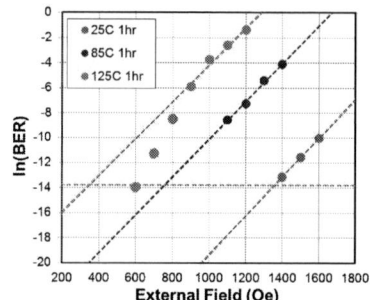

Figure 9. BER as a function of external field for 1-hr exposure. The macro shows < 1 ppm after 1-hr exposure to 600 Oe at 125 °C.

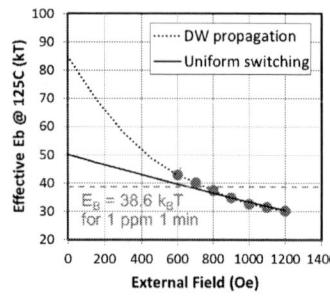

Figure 10. E_B at 125 °C as a function of external field with model fitting. For 1-min exposure to 750 Oe, estimated BER ~1 ppm.

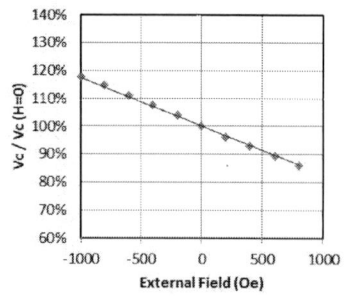

Figure 11. V_c modulated linearly by external field. The sensitivity is ~18%/kOe.

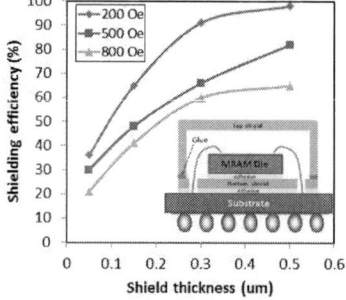

Figure 12. Magnetic shielding efficiency as a function of shield thickness for typical wire-bonding package geometry.

Acknowledgement - Authors acknowledge Everspin for valuable technical discussions.

978-1-5386-4219-1/18 $31.00 © 2018 IEEE   184

Low RA Magnetic Tunnel Junction Arrays in Conjunction with Low Switching Current and High Breakdown Voltage for STT-MRAM at 10 nm and Beyond

C. Park[1,a], H. Lee[1], C. Ching[2], J. Ahn[2], R. Wang[2], M. Pakala[2], and S. H. Kang[1]

[1] Corporate Research and Development, Qualcomm Technologies, Inc., San Diego, CA, USA. E-mail: [a]chandop@qti.qualcomm.com
[2] Silicon Systems Group, Applied Materials, Inc., Sunnyvale, California 94085, USA

Abstracts

The scaling of STT-MRAM for deeply scaled nodes (e.g. sub-10 nm CMOS) requires low resistance-area-product (RA) magnetic tunnel junctions (MTJs) to contain switching voltage (V_c) and to assure high endurance. In contrast to various reports, we demonstrate systematic engineering of low-RA MTJs without trading off key device attributes and remarkably, with higher barrier reliability. The MTJs integrate an ultra-thin synthetic antiferromagnetic layer (tSAF) with a Co/Pt pseudo-alloy pinned layer. By reducing RA from 10 to 5 $\Omega\mu m^2$, significantly reduced V_c and reliable switching at 5 ns have been achieved. Furthermore, the breakdown voltage (V_{BD}) has been improved. The results suggest that the tunability of MTJ is extended to sub-10 nm CMOS for high-performance and high-reliability MRAM.

Introduction

STT-MRAM is a new class of memory and an enabler of advanced system architectures for pervasive applications [1]. Product prototypes have already been demonstrated, proving that MRAM can be a fast, high-endurance, and low-power embedded memory solution [1,2]. To extend its advantages at more advanced nodes, small MTJ sizes (< ~40 nm) [1] and low V_c are necessary. These allow low switching current (I_c) and sufficient TDDB margin for practically unlimited endurance, respectively. Achieving these requires low RA. While it is straightforward to tune RA itself, it was known very challenging to preserve critical device attributes such as I_c, tunneling magnetoresistance ratio (TMR), coercivity (H_c), thermal stability factor (Δ), and V_{BD} as RA is lowered. This work demonstrates that such parameters can be kept comparable or even improved with a combination of tSAF that integrates a pseudo-alloy pinned layer and a co-designed free layer. In particular, to our best knowledge, this is the first report that higher V_{BD} can be achieved for low-RA MTJ arrays (40 nm diameter) in conjunction with low I_c.

RA Scaling: Benefits and Challenges

Scaling down the diameter of MTJ is of clear benefit in reducing I_c. However, the increased resistance of smaller MTJ limits the supply currents (I_{on}) for write operations (Fig.1). To alleviate this problem, RA of this work was reduced from 10 to 5 $\Omega\mu m^2$ for 40 nm MTJs (R_p from 8 to 4 kΩ). Accordingly, I_{on} increased by 40% for an improved write margin (i.e. larger I_{on}-I_c). Moreover, for high-speed applications (<10 ns), V_c increases sharply as pulse widths are reduced. Hence, lowering RA is a necessary, yet effective means of reducing V_c, which facilitates improved write margins and TDDB reliability. In addition, low RA can help increase read speed because the discharge rate for the sensing is inversely proportional to the MTJ resistance (Fig.2). Low RA MTJs are commonly fabricated by reducing the thickness of the MgO barrier. In Fig.3, I_c, H_c, and Δ, measured from an array of 40 nm MTJs, did not change significantly with the RA reduction. The V_c reduction at the lower RA (Fig. 3 (d)) is naturally expected since V_c is linearly proportional to RA. However, reducing RA simply via thinning the MgO barrier causes significant reductions in TMR and V_{BD} (Fig.3 (e) and (f)), which negatively impacts the design and reliability margins. The impact of V_{BD} reduction is illustrated in Fig. 4. Note that

TMR and V_{BD} are primarily determined by the microstructural quality and interfaces of the MgO barrier [3,4], which must improve as follows.

tSAF with a Pseudo-alloy Pinned Layer

A tSAF integrated with a Co/Pt pseudo-alloy pinned layer has been designed to form smooth underlayers with sharp interfaces on which a robust MgO barrier can be deposited. An alloy-like pinned layer (2~4 nm) was achieved by forming Co and Pt monolayers alternatively beyond the conventional superlattice process conditions. This promotes a smoother MgO barrier interface as well as a much thinner MTJ stack (Fig.5). The periodicity of the Co/Pt interfaces is clearly visible in the standard Co/Pt multilayer structure, but absent in the case of the pseudo-alloy structure. The total thickness of the pMTJ was reduced by ~30 % from 14 to 10 nm. TMR is also improved in the case of tSAF with the pseudo-alloy pinned layer (Fig.6). The average V_{BD} increased significantly from 1.26 to 1.40 V (Fig.7), which should accordingly increase the reliability margin between V_c and V_{BD} (Fig.4). This clearly suggests that RA of scaled MTJ can be lowered without previously perceived tradeoffs through the tuning of the SAF thickness and processing methods.

Free Layer Co-optimization

Device and circuit models have shown that, for high-performance applications at sub-10 nm, I_c needs to be reduced to <~50 μA at 10 ns. Since I_c is primarily determined by the properties of the free layer (FL), three types of FLs with different magnetic moments were investigated in this work. FL1 was a typical CoFeB layer, while FL2 and FL3 were formed by modulating the composition and thickness of FL1. A noticeable I_c reduction was observed for FL3 (Fig.8). Even though H_c was reduced to 1300 from 2000 Oe, FL3 could maintain relatively high Δ of 56 for 40 nm MTJs (Fig.9). As the pulse width decreases, the write-error-rate (WER) slopes decrease (Fig. 10) for both FL1 and FL3. However, this trend is more pronounced for FL1. The WER slope of FL3 at 5 ns is more than twice steeper than that of FL1, indicating a much smaller temporal variation of FL3. Based on the results, I_c of 36 μA at 10 ns is projected for 30 nm MTJ, which should meet the requirement for high-performance MRAM application in sub-10 nm CMOS. The overall benefits of reduced RA are summarized in Table 1.

Conclusions

In contrast to prior reports and prevalent perceptions, low RA MTJs can be engineered without tradeoffs through a systematic optimization of SAF thickness and processing methods. Furthermore, V_{BD} can even be improved significantly (from 1.26 to 1.40 V at RA=5 $\Omega\mu m^2$). With a co-optimized FL, low I_c can also be achieved at short pulses (65 μA at 10 ns for 40 nm MTJ arrays) with relatively high Δ. The reported results suggest that the tunability of MTJ can be extended to deeply scaled nodes for high-performance and high-reliability MRAM.

References

[1] S. H. Kang and C. Park, IEDM 38.21-38.24 (2017).
[2] Y.J. Song et al., IEDM 27.2.2-27.2.4 (2016).
[3] C. Park et al., IEDM 26.2.1-26.24 (2015).
[4] M. Gottwald et al., Appl. Phys. Lett. 106, 032413 (2015).

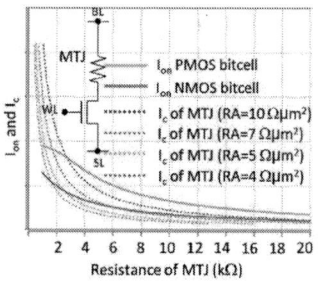

Fig. 1. I_{on} and I_c as a function of MTJ resistance. RA determines the MTJ diameter at a given resistance and hence I_c. Write margin (I_{on}-I_c) increases with reducing RA.

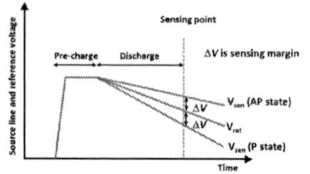

Fig. 2. An illustration of a sensing window as a function of sensing time. A steeper discharge slope can reduce the sensing time.

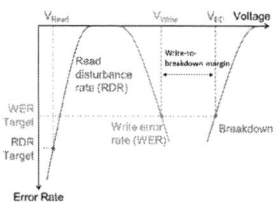

Fig. 4. An illustration of respective design margins for read, write, and breakdown voltages. High V_{BD} and steep slopes of WER and RDR (read disturbance rate) curves are required.

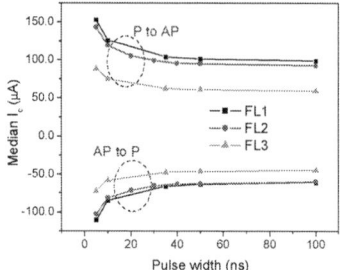

Fig. 8. I_c vs. pulse width of MTJs with different FLs. The MTJ size is 40 nm.

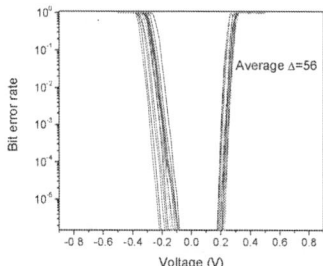

Fig. 9. RDR plots for FL3 at room temperature. Δ was extracted using $\ln(RDR)=\ln(t_p/\tau_0)-\Delta(1-V/V_c)$. The MTJ size is 40 nm.

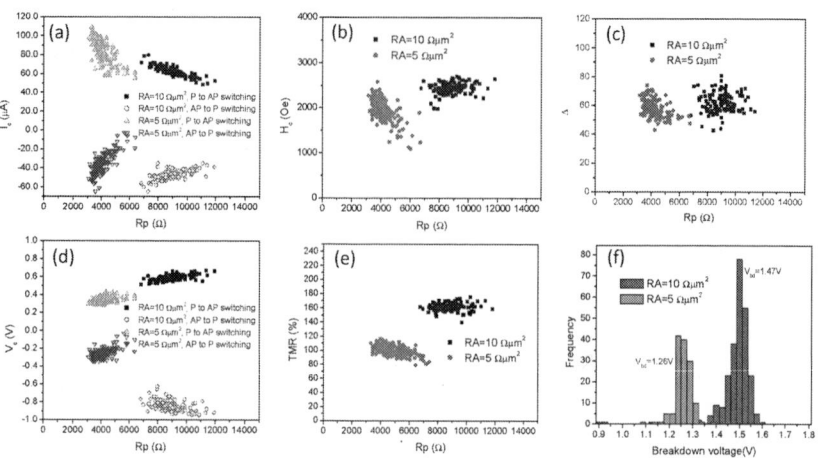

Fig. 3. (a) I_c (switching current), (b) H_c (coercivity), (c) Δ (thermal stability factor), (d) V_c (switching voltage) and (e) TMR as a function of R_p (low resistance state). (f) V_{BD} (breakdown voltage) distributions with two different RAs. Note that Δ was extracted from the V_c vs. pulse width plot. The MTJ size is 40 nm.

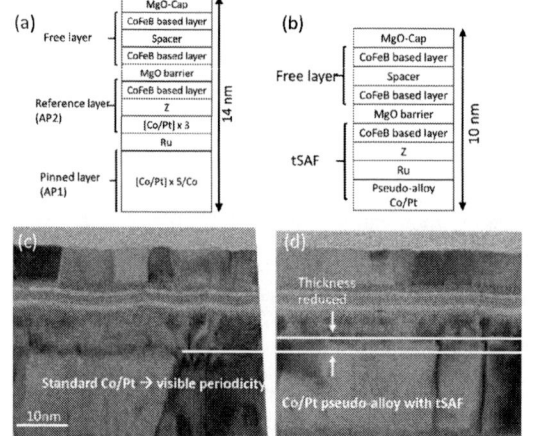

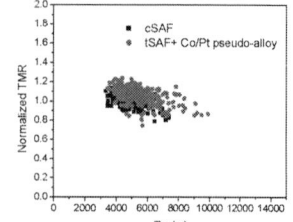

Fig. 6. TMR vs. Rp of MTJs with cSAF and tSAF with a Co/Pt pseudo-alloy (RA=5 $\Omega\mu m^2$).

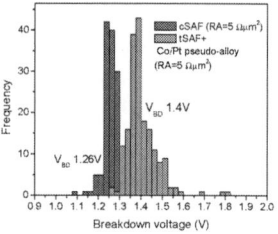

Fig. 7. Breakdown voltage distributions of MTJs with cSAF and tSAF with a Co/Pt pseudo-alloy (measured with 10 µs pulses).

Fig. 5. MTJs with (a) conventional SAF(cSAF) and (b) tSAF with a Co/Pt pseudo-alloy pinned layer. Cross-section TEM images of MTJs with (c) cSAF and (d) tSAF. Z is a texture breaking insertion layer in (a) and (b).

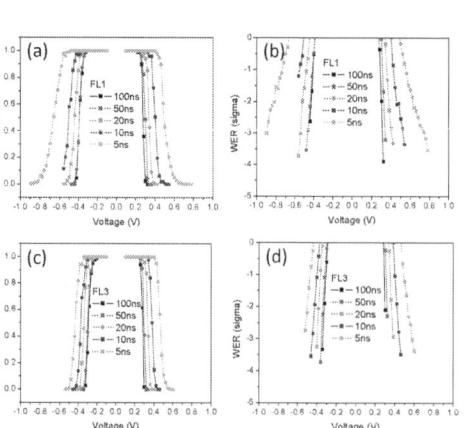

Fig. 10. (a) and (c) WER plots, (b) and (d) quantile plots of WERs as a function of write voltage at pulse widths of 5-100 ns for FL1 and FL3.

Table 1. Estimated gains from lowering RA from 10 to 5 $\Omega\mu m^2$ for 40 nm MTJs.

Parameter	Performance Impact	Estimated gains
I_{on}	Write margin	40% higher I_{on}; 150% higher write margin
R_p	Read speed	$\geq$20% higher read speed
V_c	Reliability (TDDB, endurance)	50% lower V_c; 33% higher write-to-breakdown margin

Rare-Failure Oriented STT-MRAM Technology Optimization

Nuo Xu[1,*], Fan Chen[1], Dmytro Apalkov[1], Weiyi Qi[1], Jing Wang[1], Zhengping Jiang[1], Woosung Choi[1], Dae Sin Kim[2]

[1]Samsung Semiconductor Inc., San Jose, CA 95134, USA Tel: +1-408-544-5273, *E-mail: nuo.xu@samsung.com
[2]Semiconductor R&D center, Samsung Electronics, Korea

Abstract

A rare-failure oriented optimization methodology for state-of-the-art STT-MRAM technology has been proposed. Physics-based device models and novel rare event sampling algorithms are used for massively parallel *Monte Carlo* simulations to identify the critical process variability sources and to evaluate the Write Error Rate (WER) at the resolution of 1E-9. New rare-failure figure-of-merits (FoMs) and design guidelines are suggested for optimizing the operation conditions of STT-MRAMs so that the energy-delay product can be minimized at satisfactory WER level.

Introduction

Spin-transfer torque (STT) magnetic random access memory (MRAM) has now become an early-stage commercialized technology for stand-alone and embedded memory applications, due to its significant advantages of low *Write* voltage (V_W) and short switching delay over other emerging non-volatile memory candidates [1, 2]. However, technology challenges still remain, among which the relatively high failure rates caused by the intrinsic (*i.e.* thermal noise) and the extrinsic (*i.e.* process variability) sources hinder the device and large-array yield improvement for state-of-the-art STT-MRAMs [3]. Although existing experimental and theoretical works have shown the impacts of different variability sources on STT-MRAM's Write Error Rate (WER) and other figure-of-merits (FoMs), a design and optimization flow targeted at the very rare failure rate has not yet been proposed. In this work, a comprehensive framework is developed to perform massively parallel *Monte Carlo* (MC) simulations to evaluate the WER, followed by the statistical parameter extractions on rare-failure FoMs and the energy-delay product (EDP) optimizations with operation conditions. New insights are revealed to better design the STT-MRAM technology subject to a low WER. The impact of MTJ cell area scaling on rare-failure FoMs is studied as well as the interplay with process variability.

STT-MRAM Technology Modeling

The physics-based macro-spin model developed in [4] is used in this work. Fig.1 shows the STT-MRAM device structure, consisting of the free layer (FL)/MgO/polarization enhancement layer (PEL) MTJ and the synthetic anti-ferromagnet (SAF) stack as pinned layers. Interfacial perpendicular magnetization anisotropy (PMA) with the easy-axis along z represents the modern technology and is studied herein. The offset fields are assumed to be well compensated to have symmetrical *Write* (P→AP *vs.* AP→P) conditions. Fig.2 plots the simulated and experimental [3] WER *vs.* V_W of the nominal MTJ devices under increasing pulse widths from 10ns to 100ns, and for different cell areas. In practice, the edge damages occurred at the FL are considered as the major defective source in small area MTJs [5, 6] and are modeled to match experimental results (Fig.3 and 4(a)). Geometry variations on MTJ's widths (W) and lengths (L) can well explain the measured parallel-state resistance (R_p) fluctuations while interfacial PMA energy (K_s) variations are included to better capture the thermal stability factor (Δ) fluctuations, as shown in Fig.3(b). Fig.4(b) summarizes the calibrated MTJ device, edge damage and variability parameters, together with projected values for improved technology.

Rare Failure Oriented STT-MRAM Design

The rare-failure oriented STT-MRAM technology evaluation and design framework is proposed in Fig.5. The overall flow takes the experimentally determined 3σ process variability corners and high-σ (>6) failure rate results as inputs and evaluates the standard FoMs (RA product, Δ, *etc.*) as well as the rare-failure FoMs (WER, *etc.*). GPU-based computing grids are used to massively parallelize MC simulations to capture the thermal noise induced intrinsic failures. Further, the stochastic importance sampling [7] method is customized for simulating the LLG-equation based magnetic trajectories to reduce the sampling variance so that the failure rates can be calculated down to 1E-9 with good confidence. Process variations can be included at the same level as the thermal noise in the MC sampling, in which way the averaged WER values under a certain operation condition can be evaluated. Another approach is to create random process samples using the design-of-experiment (DoE) method and evaluate the WER values at each individual sample, which enables the process variation induced WER distribution profiling as well as the statistical parameter extraction procedure. Finally, the *Pareto*-front based optimization [8] is performed on the EDP of STT-MRAM operations while maintaining a satisfactory WER level.

Fig.6 shows the simulated WER *vs.* V_W and *Write* pulse width (t_{pulse}) for MTJ devices w/ and w/o process variability sources, indicating that material parameter variations in FL dominate the WER degradation. Note that although the WER at low voltages is reduced slightly under process variation (as shown in Fig.6(a) inset), the worsening rare-failure FoMs $S_{V,WER}$ and $S_{t,WER}$ (*i.e.* the required dV_w and dt_{pulse} to lower WER by 1 decade, as defined at Fig.6 top) exacerbate the WER degradation under process variation, demanding higher V_w and longer t_{pulse}. To further investigate the causes of $S_{V,WER}$ and $S_{t,WER}$ degradations, 10^3 MC samples are generated and the intrinsic WER is evaluated on each case. The WER quantile plots in Fig.7(a) suggests that applying higher V_w may not be effective to suppress the failures of tail bits, explaining these swing degradations at rare failure regimes. Also, the required V_W to achieve a WER of 1E-8 ($V_{W,1E-8}$) becomes more fluctuated as t_{pulse} decreases, as shown in Fig.7(b). Fig.8 plots the standard and rare-failure FoMs' distributions and their co-variance using statistical parameter extractions. $S_{V,WER}$ and $S_{t,WER}$ are well correlated and their strongly non-*Gaussian* distributions contribute as the major sources to WER degradations at rare failure regions. To calculate the EDP accurately, the hold time (t_{hold}, *a.k.a. Write* interval) cannot be ignored as it affects the WER remarkably at rare failure regions through the initial magnetizations, as shown in Fig.9. Fig.10 performs the *Pareto*-front optimizations to minimize the EDP while maintaining the WER below 1E-8. It is found that the EDP increases drastically when the WER is approaching ~5E-9. As determined by the overlap between *Pareto* set and (WER-)satisfactory set, the optimal t_{pulse}/t_{hold} should be ~40/15ns and a $V_{W,1E-8}$ below 0.55V can be achieved for a 30nm-diameter MTJ. Fig.11 predicts the impact of MTJ area scaling on $V_{W,1E-8}$ for minimum delay and optimal EDP operations. The rapid increasing of $V_{W,1E-8}$ overhead under process variation for smaller area MTJs results from the severe $S_{V,WER}$ degradations as suggested in Fig.12.

Conclusion

An STT-MRAM technology optimization flow has been developed focusing on rare failure rates evaluation. Material parameter fluctuations in the FL are regarded as the dominant process variability sources in degrading the rare-failure FoMs, such as $V_{W,1E-8}$ and $S_{V,WER}$. Proposed optimization methodologies minimize the EDP subject to low WER values, enabling the performance and large-array yield co-optimizations for STT-MRAM technology.

References: [1] Y. J. Song *et al.*, *IEDM Tech. Dig.*, pp. 663-666, 2016. [2] S.-W. Chung *et al.*, *IEDM Tech. Dig.*, pp. 659-662, 2016. [3] J.J. Nowak *et al.*, *IEEE Magnetic Letters*, 3102604, 2016. [4] N. Xu *et al.*, *IEDM Tech. Dig.*, pp. 735-738, 2015. [5] J.-H. Kim *et al.*, *Symp. VLSI Tech.*, pp.60-61, 2014. [6] J.H. Jeong *et al.*, *Symp. VLSI Tech.*, pp.158-159, 2015. [7] O. Mazonka *et al.*, *Nuclear Physics A*, pp.335-354, 1998. [8] K. Miettinen, *Nonlinear Multiobjective Optimization*, 2012.

978-1-5386-4219-1/18 $31.00 © 2018 IEEE

2018 Symposium on VLSI Technology Digest of Technical Papers

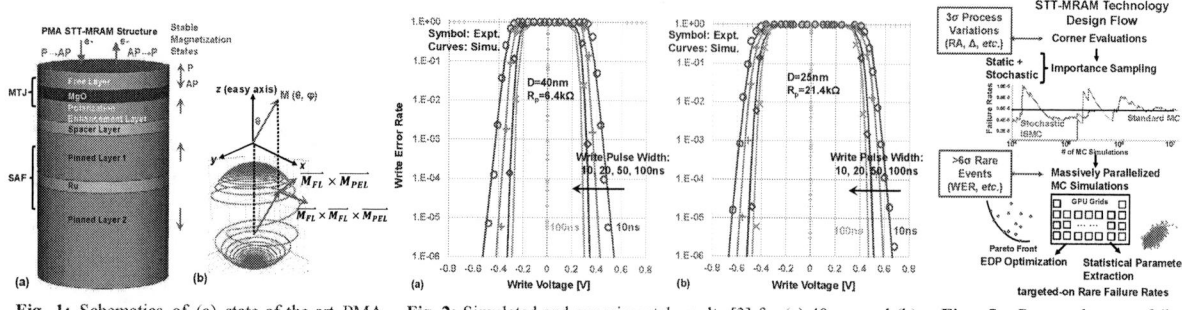

Fig. 1: Schematics of (a) state-of-the-art PMA STT-MRAM device and (b) switching trajectories of free layer (FL)'s magnetization under STT.

Fig. 2: Simulated and experimental results [3] for (a) 40nm- and (b) 25nm-diameter MTJ's Write Error Rate (WER) vs. Write pulse voltage, under different pulse widths.

Fig. 5: Proposed rare failure oriented STT-MRAM technology design and optimization flow.

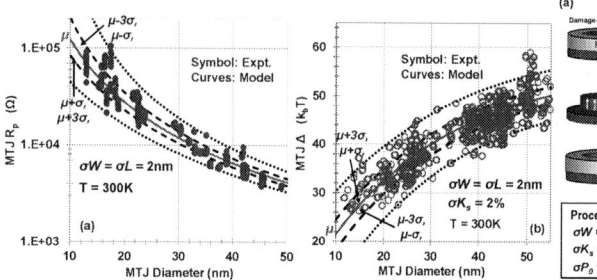
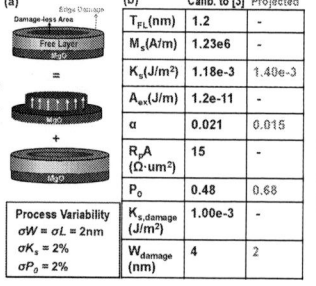
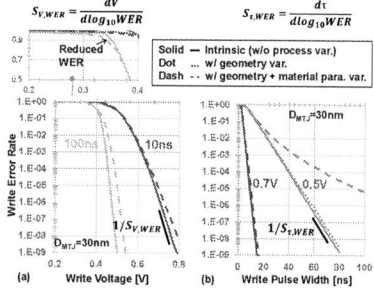

Fig. 3: Modeled and experimental (~650 samples) results [3] for (a) parallel state resistance (R_p) vs. the MTJ diameter, in which geometry variations are sufficient to capture the range of R_p variations; (b) thermal stability factor (Δ) vs. the MTJ diameter, in which both geometry and interfacial uniaxial anisotropy energy (K_s) variations are included for good model predictability.

Fig. 4: (a) Illustration of edge damage deffects in MTJ structures and (b) calibrated and projected key parameters for evaluating STT-MRAM technology.

Fig. 6: Simulated WER vs. (a) Write voltage and (b) Write pulse widths for a 30nm-diameter MTJ, w/ and w/o process variability. The effective voltage swing ($S_{V,WER}$) and pulse width swing ($S_{\tau,WER}$) to reduce WER are defined at the top.

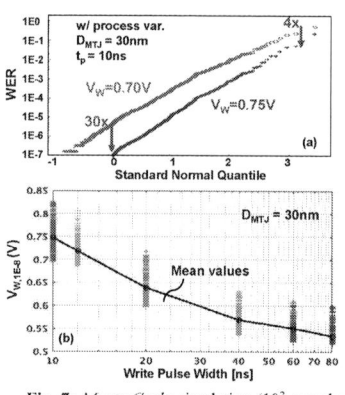
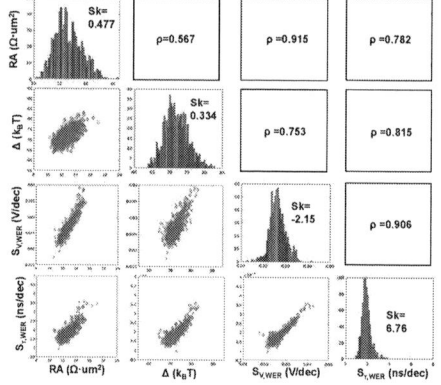
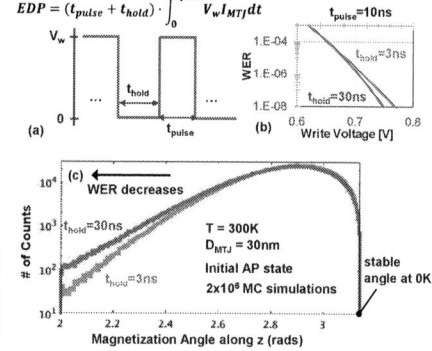

Fig. 7: Monte Carlo simulation (10^3 samples) results of (a) WER quantile plots and (b) Write voltage at WER=1E-8 ($V_{W,1E-8}$) vs. Write pulse width in a 30nm-diameter MTJ under process variation.

Fig. 8: Key MTJ figure-of-merits (FoMs) to capture process variability through statistical parameter extraction procedure. Non-Gaussian distributions are clearly seen in rare failure related FoMs ($S_{V,WER}$, $S_{\tau,WER}$).

Fig. 9: (a) Formula to calculate MTJ's Write energy-delay product (EDP); (b) impact of hold time (t_{hold}) on WER; and (c) FL's initial magnetization distributions (m_z) under different t_{hold}.

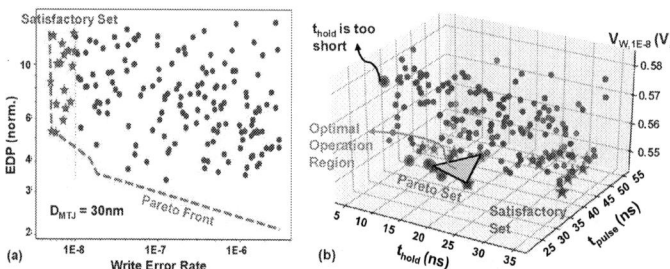
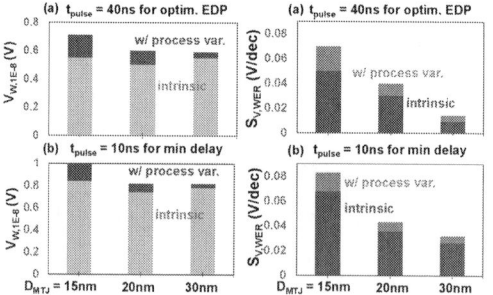

Fig. 10: (a) EDP vs. WER for a 30nm-diameter MTJ. Pareto front optimization has been applied to identify the designs achieve the minimum EDP subject to WER<1E-8. (b) Results suggest the optimal hold time and Write pulse width should be ~15ns and 40ns, respectively.

Fig. 11: Impact of MTJ area scaling on $V_{W,1E-8}$, with optimal EDP and minimum delay designs.

Fig. 12: Impact of MTJ cell area scaling on $S_{V,WER}$, with optimal EDP and minimum delay designs.

978-1-5386-4219-1/18 $31.00 © 2018 IEEE   188

Gap in pagination due to formatting issues.

Pages 189-190

Significant Performance Enhancement of UTB GeOI pMOSFETs by Advanced Channel Formation Technologies

W. H. Chang, T. Irisawa, H. Ishii, H. Hattori, N. Uchida and T. Maeda

National Institute of Advanced Industrial Science and Technology (AIST),
1-1-1 Umezono, Tsukuba, Ibaraki 305-8568, Japan   E-mail: wh-chang@aist.go.jp

Abstract

Advanced channel formation technologies, such as precise control of GeOI body thickness (T_{body}), surface roughness and interfacial quality, utilizing Si-passivation/Ge-channel/SiGe hetero-epitaxy and Ge digital etching (DE) techniques were implemented for UTB GeOI structure. Si passivation for Ge/BOX interface has been verified to suppress Coulomb scattering owing to better interfacial quality. Insertion of SiGe etching stop (ES) layer and dozens DE (DDE) were found to be quite effective to reduce T_{body} fluctuation as well as surface roughness, resulting in the significant improvement of mobility. As a result, we have demonstrated record high hole mobility of ~200 cm^2/Vs in UTB GeOI pMOSFETs without the strain technology, which outperforms Si universal mobility by 2 times even under T_{body} of 9 nm.

Introduction

Although Ge possesses 4 times higher hole mobility than that of Si, the hole mobility demonstrated in GeOI pMOSFETs with body thickness (T_{body}) less than 10 nm was still much lower than its Si counterpart. [1, 2] Poor Ge/BOX interface and Ge T_{body} fluctuation limited the device performance of GeOI pMOSFETs. Therefore, more sophisticated fabrication methods have been strongly desired. Recently, we have developed advanced HEtero-Layer-Lift-Off (HELLO) technology to obtain UTB GeOI structures down to a few nm thick, [3] demonstrating unique electron mobility enhancement owing to the well-defined GeOI structures. [4] Besides, Si passivation for Ge/BOX interface was highly effective in mitigating unpleasant hole mobility degradation in UTB GeOI regime. [5] In this work, in addition to advanced HELLO technology utilizing SiGe hetero-epitaxial growth method, we employ dozens DE (DDE) process to realize extremely flat UTB GeOI structure with record high hole mobility. Mechanism on hole mobility enhancement has been also discussed through low temperature electrical measurement and TCAD simulation.

Advanced GeOI channel formation processes and material characterization of UTB GeOI: Ge buffer (150 nm)/SiGe etch stop (6.5 nm)/Ge channel/Si passivation (0.5 nm)/Al$_2$O$_3$ (5 nm)/SiO$_2$ (10 nm)/Si hetero-structure was prepared by advanced HELLO technology. The fabricated Ge/SiGe/Ge/Si/BOX/Si structure has been examined by high resolution scanned EDX analysis as given in Fig. 1, showing the existence of SiGe etch stop (ES) and Si passivation layers on top of BOX. Fig. 2 shows schematic pictures of ultrathin GeOI channels formed by different processes. The UTB GeOI channel was trimmed from 50 nm to 9-15 nm through DDE (Fig. 3) after removing SiGe ES in Process C, while in Process B T_{body} was determined without DDE. The GeOI surface morphology measured by AFM before and after 60 cycles DDE is shown in Fig. 4. RMS surface roughness clearly reduces from 1 nm to 0.4 nm after DDE, indicating the effectiveness of DDE for smoothing Ge surface. Fig. 5 shows TEM images of 9-nm-thick GeOI structures fabricated by (a) Process A, [6, 7] (b) B and (c) C, respectively, and their statistical thickness distribution is shown in Fig. 6. Smoother front surface and resultant much smaller T_{body} fluctuation were achieved thanks to both SiGe ES and DDE. The detailed process for fabricating GeOI pMOSFETs can be found elsewhere. [5]

Mechanism of hole mobility enhancement by advanced HELLO: We have already demonstrated that HELLO is beneficial to achieve high hole mobility in UTB GeOI pMOSFETs, [5] and here we discuss its physical origins. Fig. 7 shows the temperature dependence of effective mobility in 9-nm-thick UTB GeOI pMOSFETs fabricated by (a) Process A and (b) B. The mobility of devices fabricated by Process B increases with decreasing temperature in entire range of N_s, suggesting suppression of thickness fluctuation scattering, which should limit the mobility in the devices fabricated by Process A. Smaller interfacial state density (D_{it}) owing to Si passivation of BOX interface was also confirmed by the I_D-V_G characteristics of back gate mode operation as shown in Fig. 8, where subthreshold slope (SS) of back gate mode operation was drastically improved by advanced HELLO from 247 to 109 mV/dec. This reduced D_{it} should suppress Coulomb scattering. Therefore, the suppression of both thickness fluctuation and Coulomb scattering owing to advanced HELLO is considered to contribute to the mobility enhancement. To further study the effect of advanced HELLO on hole mobility, hole distributions in GeOI channel with and without Si passivation were simulated by TCAD, with T_{body} of 9 nm and a fixed N_s of 10^{13} cm^{-2}, as shown in Fig. 9. It is seen for GeOI channel with Si passivation that holes exist not only at front interface but also at back interface owing to Ge/Si band offset (inset of Fig. 8). This modulated carrier distribution decreases the electric field in the channel and can also enhance overall mobility, which is similar to the case of volume inversion.

Effect of dozens digital etching (DDE) on hole mobility: Comparison of effective hole mobility of 9-nm-thick UTB GeOI pMOSFETs with different fabrication processes is shown in Fig. 10. The devices fabricated by Process C with extremely smooth surface lead to the highest hole mobility especially at high N_s region, indicating the effectiveness of DDE for hole mobility enhancement. About 2 times mobility enhancement can be achieved at high N_s region for the GeOI devices fabricated by Process C as compare with Si universal mobility. Fig. 11 shows the T_{body} dependence of effective mobility of GeOI pMOSFETs fabricated by Process C. Although mobility degradation was observed as T_{body} scaling down, record high hole mobility above 200 cm^2/Vs maintained in 9-nm-thick GeOI devices, which can be attributed to both SiGe ES and DDE. Fig. 12 summarized the T_{body} dependence of mobility at N_s of 2×10^{12} cm^{-2} of present devices fabricated by Process A, B and C. The mobility of reported GeOI [1, 2] and SOI devices [8] are also plotted for comparison. With extra care of ultrathin channel formation, device performance of GeOI devices with Process C is significantly improved and distinctly outperforms its SOI counterpart even below 10 nm.

Conclusion

High quality UTB GeOI pMOSFETs have been fabricated by advanced channel formation technologies, resulting in better T_{body} fluctuation, surface roughness and back interfacial quality. The highest hole mobility in UTB GeOI devices with T_{body} scaling down below 10 nm was obtained, which also outperforms its Si counterpart. Si passivation contributes to suppression of Coulomb scattering, while SiGe ES and DDE reduce thickness fluctuation scattering and surface roughness scattering, respectively.

Acknowledgement

The authors would like to thank the support from Sumitomo Chemical Co., Ltd. and Hitachi Kokusai Electric Inc. for Ge layer transfer technology and AIST-NPF for device fabrication.

References

[1] C. H. Lee et al., *IEEE Int SOI Conf.* 1 (2011). [2] X. Yu et al., *IEDM* 20 (2015). [3] T. Maeda et al., *APL* **109**, 262104 (2016). [4] W. H. Chang et al., *Symp. VLSI Technol.* 192 (2017). [5] W. H. Chang et al., *APEX* **9**, 091302 (2016). [6] T. Maeda et al., *ME* **109**, 133 (2013). [7] E. Mieda et al., *JJAP* **54**, 036505 (2015). [8] K. Uchida et al., *IEDM* 47 (2002).

2018 Symposium on VLSI Technology Digest of Technical Papers

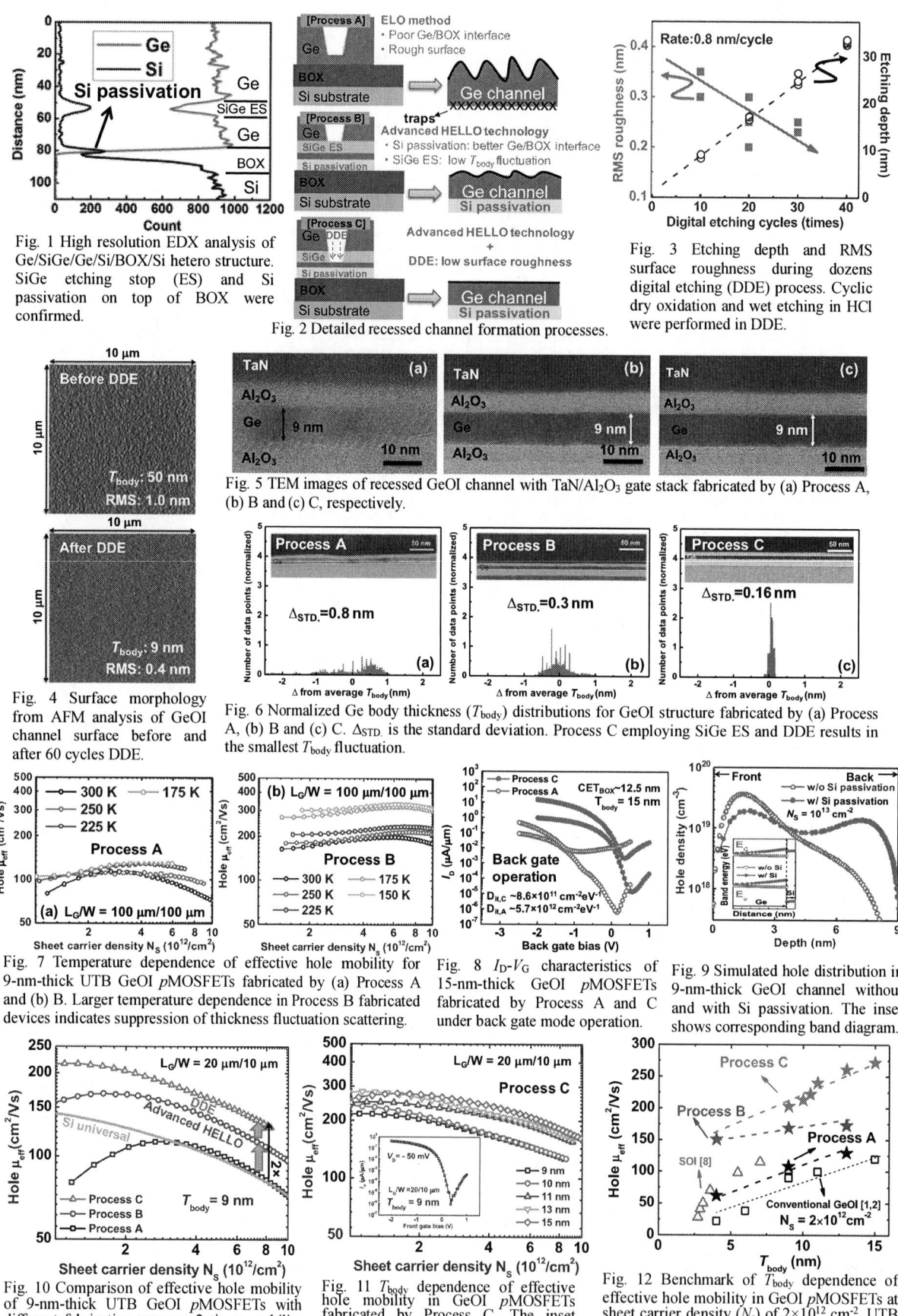

Fig. 1 High resolution EDX analysis of Ge/SiGe/Ge/Si/BOX/Si hetero structure. SiGe etching stop (ES) and Si passivation on top of BOX were confirmed.

Fig. 2 Detailed recessed channel formation processes.

Fig. 3 Etching depth and RMS surface roughness during dozens digital etching (DDE) process. Cyclic dry oxidation and wet etching in HCl were performed in DDE.

Fig. 4 Surface morphology from AFM analysis of GeOI channel surface before and after 60 cycles DDE.

Fig. 5 TEM images of recessed GeOI channel with TaN/Al$_2$O$_3$ gate stack fabricated by (a) Process A, (b) B and (c) C, respectively.

Fig. 6 Normalized Ge body thickness (T_{body}) distributions for GeOI structure fabricated by (a) Process A, (b) B and (c) C. $\Delta_{STD.}$ is the standard deviation. Process C employing SiGe ES and DDE results in the smallest T_{body} fluctuation.

Fig. 7 Temperature dependence of effective hole mobility for 9-nm-thick UTB GeOI pMOSFETs fabricated by (a) Process A and (b) B. Larger temperature dependence in Process B fabricated devices indicates suppression of thickness fluctuation scattering.

Fig. 8 I_D-V_G characteristics of 15-nm-thick GeOI pMOSFETs fabricated by Process A and C under back gate mode operation.

Fig. 9 Simulated hole distribution in 9-nm-thick GeOI channel without and with Si passivation. The inset shows corresponding band diagram.

Fig. 10 Comparison of effective mobility of 9-nm-thick UTB GeOI pMOSFETs with different fabrication process. 2 times mobility enhancement from Si universal mobility was obtained in devices with Process C.

Fig. 11 T_{body} dependence of effective hole mobility in GeOI pMOSFETs fabricated by Process C. The inset shows transfer curve at V_D of -50 mV of 9-nm-thick device. A high I_{on}-I_{off} ratio of 10^6 was obtained.

Fig. 12 Benchmark of T_{body} dependence of effective hole mobility in GeOI pMOSFETs at sheet carrier density (N_s) of 2×10^{12} cm^{-2}. UTB SOI devices are also plotted for comparison. Record high mobility above 200 cm^2/Vs was obtained in Process C fabricated devices.

978-1-5386-4219-1/18 $31.00 © 2018 IEEE   192

First demonstration of vertically-stacked Gate-All-Around highly-strained Germanium nanowire p-FETs

E. Capogreco, L. Witters, H. Arimura, F.Sebaai, C. Porret, A. Hikavyy, R. Loo, A. P. Milenin, G. Eneman, P. Favia, H. Bender, K. Wostyn, E. Dentoni Litta, A. Schulze, C. Vrancken, A. Opdebeeck, J. Mitard, R. Langer, F. Holsteyns, N. Waldron, K. Barla, V. De Heyn, D. Mocuta, N. Collaert

Imec, Kapeldreef 75 Leuven Belgium   E-mail: Elena.Capogreco@imec.be

Abstract

This paper reports on strained p-type Ge Gate-All-Around (GAA) devices on 300mm SiGe Strain-Relaxed-Buffers (SRB) with improved performance as compared to our previous work. The Q factor is increased to 25, I_{on}=500µA/µm at I_{off}=100nA/µm is achieved, approaching the best published results on Ge finFETs. Good NBTI reliability is also maintained. By using the process flow developed for the single nanowire (NW), vertically stacked strained Ge NWs featuring 8nm channel diameter are demonstrated for the first time. A systematic analysis of the strain evolution is conducted on both single and double Ge NWs, demonstrating for the first time 1.7GPa uniaxial-stress along the Ge wire, which originates from the lattice mismatch between the Ge S/D and the $Si_{0.3}Ge_{0.7}$ SRB.

Introduction

Thanks to its higher intrinsic hole mobility, which can be further enhanced by applying compressive stress, Ge represents a promising candidate to replace Si as channel material for pMOSFETs. In order to overcome challenges like reduced electrostatic control and high OFF-state leakage, which are caused by the larger permittivity and the lower band gap of Ge, the GAA device architecture is gaining a lot of interest for future technology nodes [1,2]. In this work the previously reported 14/16nm-node strained p-type Ge GAA process flow [1] is improved to achieve better electrical performance. The same process is adapted to demonstrate vertically stacked Ge NWs. Furthermore, physical analysis is performed with the aim of studying the strain evolution along the Ge wire.

Single Ge nanowire

The single Ge NWs are fabricated starting from the deposition of the SiGe/Ge/SiGe stack, with the concentration of the SiGe sacrificial layer varying between 55% and 70%, on 300mm (100) $Si_{0.3}Ge_{0.7}$ SRB wafers provided by Siltronic. Following improvements have been implemented over the steps of the process flow described in [1]: 1. the spacer and the recess etch are optimized, resulting in a shorter distance between embedded S/D epi and gate. 2. The selective etch of the SiGe sacrificial layer, which uses a chemistry from Entegris, Inc. [3], is improved, resulting in better controlled and wider NW, while the time delay between selective etch and the gate deposition is minimized to avoid wire oxidation. 3. Ge S/D epi is used instead of SiGe S/D, leading to a reduced contact resistance [4], as well as increased uniaxial stress in the Ge wire. The TEM image of the final Ge NW featuring 13nm channel diameter, is shown in Fig. 1. These process optimizations lead to a reduction of the external resistance (R_{ext}) from ~500Ω·µm to ~190Ω·µm (Fig. 2.a) with a consistent performance improvement (Fig. 2.b) as compared to our previous work [1]. Although the I_{on}/I_{off} performance of the devices fluctuates due to possible wire dimension variations that require further process tuning, the I_{on} at I_{off}=100nA/µm is increased from 300µA/µm [1] to ~500µA/µm (Fig. 3); this value is comparable to the best data

reported in literature for high-Ge content SiGe and Ge finFETs [6,7]. Extrinsic G_{mSAT}/SS_{SAT} benchmarking shows a significant G_{mSAT} improvement, leading to a Q factor increase from 15 to 25 (Fig. 4). Excellent NBTI reliability is also demonstrated (Fig. 5), thanks to the use of the Si passivated gate stack: Ge GAA devices have a V_{ov} similar to the one of Ge finFETs with same gate stack and metal electrode [8], indicating a conformal Si cap on our Ge GAA devices, as also shown by TEM (Fig. 1).

Double Ge nanowire

The process flow developed for the fabrication of the single Ge NW is adjusted to fabricate stacked NWs, starting from the repetition of the SiGe/Ge/SiGe layers [1]. Double Ge NWs featuring ~8nm channel diameter, are successfully demonstrated for the first time (Fig. 6). The electrical performance is reported in Fig. 7: we believe that non-optimized junction formation by extension implantation is the root cause for sub-threshold slope (SS_{SAT}) roll-off below L_G=40nm (Fig. 7.b) [10,11]. We believe further optimization of the junctions will show the potential for achieving nanowire electrostatic control down to much smaller L_G values.

Strain investigation on Ge nanowire

Previously we have shown that after STI the Ge remains fully strained along the fin with respect to the underlying $Si_{0.3}Ge_{0.7}$ SRB [1]. In this work, a systematic study of the strain evolution within the Ge throughout the subsequent processing steps is conducted by means of nanobeam diffraction (NBD) on a 110nm contact poly pitch (CPP) structure (Fig. 8).

After the spacer recess etch we found the Ge to have identical lattice parameters in x- and y-direction (~0.8% larger with respect to the SRB), indicating that the Ge is fully relaxed after recess etch (Fig. 8.a). Such strain loss was reported earlier [12-14] and is solely due to elastic relaxation and therefore not associated with the formation of extended defects. After growing embedded Ge S/D regions, the mismatch between the Ge channel and the underlying SRB in the x-direction reduces to ~0 again (Fig.8.b). This verifies that the Ge S/D grown on the $Si_{0.3}Ge_{0.7}$ SRB acts as an efficient stressor thereby recovering the stress along the channel [12-14]. We could furthermore validate that the strain is maintained in the middle of the Ge channel even after wire release and gate dielectric deposition (Fig. 8.c). The strain is also monitored on the double Ge NW (Fig. 9) by means of Geometric Phase Analysis (GPA). These additional results verify that the strain induced by the Ge S/D stressors is maintained after wire release for single and double Ge NW devices.

Conclusions

We demonstrated single and stacked strained Ge GAA pFETs on a 14/16nm node platform with excellent performance at reduced power (V_{DD}=-0.5V) matching state of the art Ge finFETs. The Ge GAA device is a promising candidate as high performance pFET transistor for future nodes, provided that the junctions are further optimized to allow good electrostatic control down to the targeted L_G values.

978-1-5386-4219-1/18 $31.00 © 2018 IEEE

2018 Symposium on VLSI Technology Digest of Technical Papers

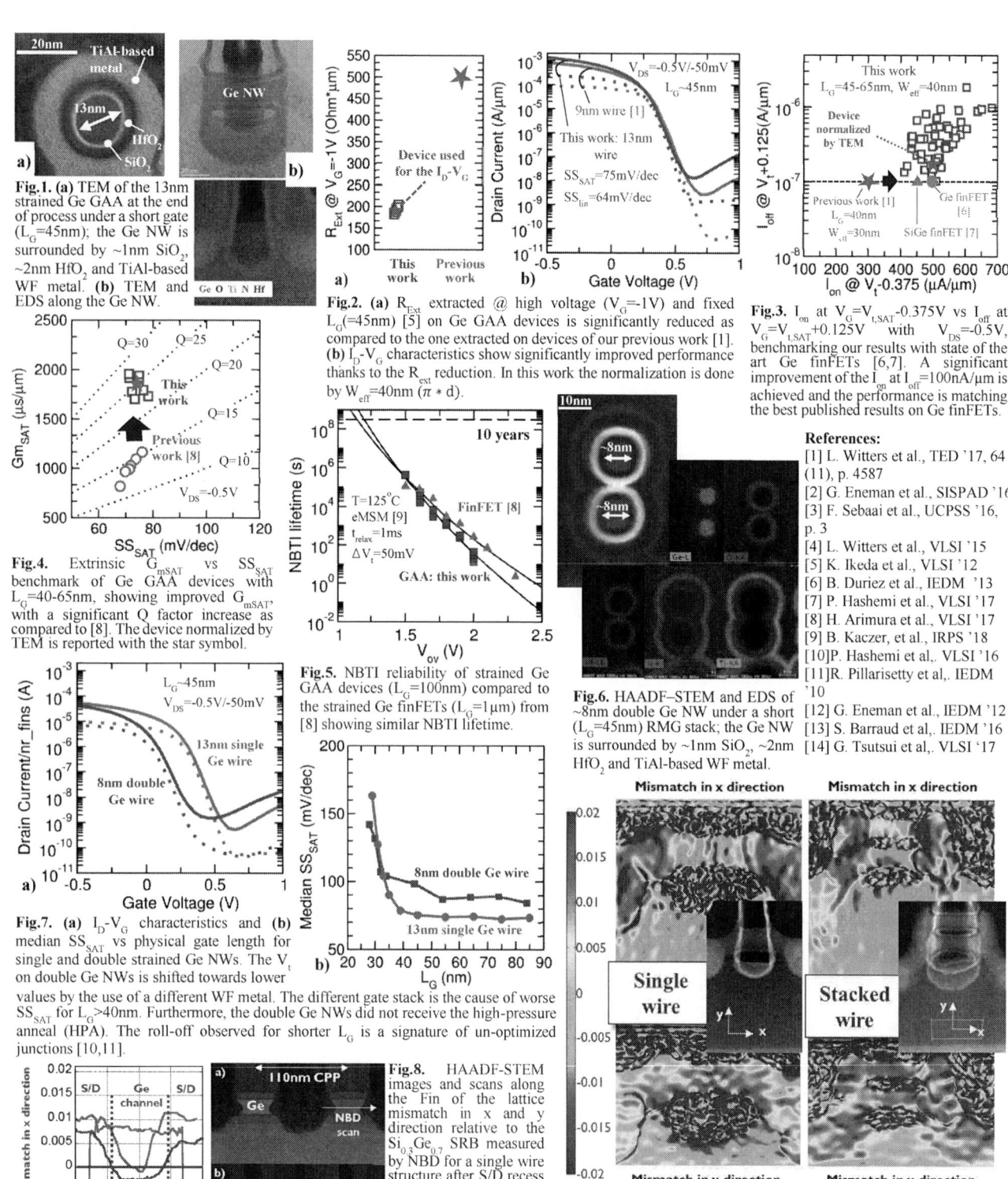

Fig.1. (a) TEM of the 13nm strained Ge GAA at the end of process under a short gate (L_G=45nm); the Ge NW is surrounded by ~1nm SiO_2, ~2nm HfO_2 and TiAl-based WF metal. **(b)** TEM and EDS along the Ge NW.

Fig.4. Extrinsic G_{mSAT} vs SS_{SAT} benchmark of Ge GAA devices with L_G=40-65nm, showing improved G_{mSAT}, with a significant Q factor increase as compared to [8]. The device normalized by TEM is reported with the star symbol.

Fig.7. (a) I_D-V_G characteristics and **(b)** median SS_{SAT} vs physical gate length for single and double strained Ge NWs. The V_t on double Ge NWs is shifted towards lower values by the use of a different WF metal. The different gate stack is the cause of worse SS_{SAT} for L_G>40nm. Furthermore, the double Ge NWs did not receive the high-pressure anneal (HPA). The roll-off observed for shorter L_G is a signature of un-optimized junctions [10,11].

Fig.2. (a) R_{Ext} extracted @ high voltage (V_G=-1V) and fixed L_G(=45nm) [5] on Ge GAA devices is significantly reduced as compared to the one extracted on devices of our previous work [1]. **(b)** I_D-V_G characteristics show significantly improved performance thanks to the R_{ext} reduction. In this work the normalization is done by W_{eff}=40nm ($\pi * d$).

Fig.5. NBTI reliability of strained Ge GAA devices (L_G=100nm) compared to the strained Ge finFETs (L_G=1μm) from [8] showing similar NBTI lifetime.

Fig.6. HAADF–STEM and EDS of ~8nm double Ge NW under a short (L_G=45nm) RMG stack; the Ge NW is surrounded by ~1nm SiO_2, ~2nm HfO_2 and TiAl-based WF metal.

Fig.3. I_{on} at V_G=$V_{t,SAT}$-0.375V vs I_{off} at V_G=$V_{t,SAT}$+0.125V with V_{DS}=-0.5V, benchmarking our results with state of the art Ge finFETs [6,7]. A significant improvement of the I_{on} at I_{off}=100nA/μm is achieved and the performance is matching the best published results on Ge finFETs.

References:
[1] L. Witters et al., TED '17, 64 (11), p. 4587
[2] G. Eneman et al., SISPAD '16
[3] F. Sebaai et al., UCPSS '16, p. 3
[4] L. Witters et al., VLSI '15
[5] K. Ikeda et al., VLSI '12
[6] B. Duriez et al., IEDM '13
[7] P. Hashemi et al., VLSI '17
[8] H. Arimura et al., VLSI '17
[9] B. Kaczer, et al., IRPS '18
[10] P. Hashemi et al,. VLSI '16
[11] R. Pillarisetty et al,. IEDM '10
[12] G. Eneman et al., IEDM '12
[13] S. Barraud et al,. IEDM '16
[14] G. Tsutsui et al,. VLSI '17

Fig.8. HAADF-STEM images and scans along the Fin of the lattice mismatch in x and y direction relative to the $Si_{0.3}Ge_{0.7}$ SRB measured by NBD for a single wire structure after S/D recess etch (a), Ge S/D epitaxy and dummy gate removal (b), and after wire release and gate dielectric deposition (c) for L_G~30nm and 110nm CPP.

Fig.9. HAADF-STEM images and corresponding map of the mismatch in x and y direction relative to the $Si_{0.3}Ge_{0.7}$ SRB measured by GPA along the Ge nanowire for a single wire and a double stacked wire after gate dielectric deposition. L_G~30nm and CPP is 110nm. Embedded epitaxial Ge source/drain was used. Inside the Ge wire, the mismatch in x direction is close to zero, while the mismatch in y direction approaches 1.5%, indicating the presence of ~1.7GPa uniaxial channel stress in the wires.

Acknowledgement:
The authors thank the imec core CMOS program members, the European commission, local authorities and the imec pilot line for their support. All epitaxial depositions in this work were done on ASM intrepid®. GPA of Strain in HRSTEM images are elaborated with ImageEval software, UniBremen, Electron Microscopy Group Prof A. Rosenauer.

978-1-5386-4219-1/18 $31.00 © 2018 IEEE

Hole mobility enhancement in extremely-thin-body strained GOI and SGOI pMOSFETs by improved Ge condensation method

K.-W. Jo, W.-K. Kim, M. Takenaka, and S. Takagi

Department of Electrical Engineering and Information Systems, The University of Tokyo, 7-3-1 Hongo, Bunkyo-ku, Tokyo 113-8656, Japan, Tel: +81-3-5841-6733, Fax: +81-3-5841-8564, E-mail: jkw@mosfet.t.u-tokyo.ac.jp

Abstract We demonstrate high performance extremely-thin-body (ETB) Ge-on-insulator (GOI) and SiGe-on-insulator (SGOI) pMOSFETs with the body thickness ranging from 10 to 2 nm by applying the improved Ge condensation process with slow cooling to initial substrates with thinner SiGe layers. When we employ Si/40-nm-thin $Si_{0.75}Ge_{0.25}$/SOI structures as starting substrates for Ge condensation, the high compressive strain of ~1.75% is maintained in GOI, leading to the operation of 10-nm-thick GOI pMOSFETs with hole mobility (μ_h) of 467 cm²/Vs. Furthermore, by thinning the fabricated GOI and SGOI films, we demonstrate the operation of ETB GOI and SGOI pMOSFETs with the body thickness down to 2 nm without losing high compressive strain. Comparing with the reported results, the record-high μ_h is obtained in GOI pMOSFETs in the GOI thickness ranging from 10 to 2 nm.

Introduction GOI and SGOI MOSFETs have stirred much attention as p-channel devices, because of the high hole mobility (μ_h). These structures can provide low leakage current and suppression of short channel effects. Here, strain engineering plays a key factor to enhance the performance of ETB pMOSFET. However, compressive strain (ε_c) can be easily relaxed in high Ge fractions because of various crystal defects [1]. As a result, it is difficult to achieve high ε_c in high Ge fractions, where the Ge-like-band structure dominates μ_h. In order to suppress this strain relaxation, we have proposed a new Ge condensation process including slow cooling [2]. By employing this process, μ_h of 301 and 138 cm²/Vs was obtained for 15- and 4.5-nm- thick GOI pMOSFETs, respectively, with ε_c of ~1.5 % [2]. Also, we have reported that a thinner SiGe film in a staring substrate can mitigate the strain relaxation, leading to higher μ_h [3].

In this study, we further improve the fabrication process under the modified Ge condensation using starting substrates with thinner SiGe films to achieve much higher μ_h in GOI pMOSFETs. Also, the impacts of the improved condensation on the characteristics of GOI and high-Ge-fraction SGOI pMOSFETs are systematically examined. In addition, further thinning of GOI and SGOI films is conducted down to 1 nm and the impact of ETB thickness on the electrical characteristics is quantitatively studied. As a result, we successfully realize 10-nm-thick GOI pMOSFETs with ε_c of ~1.75% and μ_h of 467 cm²/Vs, which is the record-high value at this GOI thickness (t_{GOI}). Also, compressively- strained 2-nm-thick ETB GOI and SGOI pMOSFETs are shown to operate with higher μ_h than the reported ones, while strain is relaxed at 1-nm thickness.

Highly-strained GOI fabrication Fig. 1 schematically shows the Ge condensation process flow and the recipe. The Ge condensation is performed by stepwise high temperature oxidation and intermixing annealing under N_2 ambient. Furthermore, the continuous thermal process and 4-hour slow cooling after condensation were combined to suppress the strain relaxation [2]. In this study, epitaxially-grown Si(10nm)/$Si_{0.75}Ge_{0.25}$(40 and 60nm)/SOI(10nm) stacks were prepared as the starting substrates. The TEM images are shown in Fig. 2. Fig. 3 shows the ε_c behaviors during condensation as function of the Ge fraction. Here, 40- and 60-nm-thick SiGe with 4-hour slow cooling, and 40-nm- thick SiGe with quick cooling are compared. It is confirmed that strain relaxation is effectively mitigated for slow cooling. In contrast, the conventional quick cooling causes almost full

relaxation [1, 2]. It is found that 40-nm-thick SiGe provides higher ε_c than 60-nm-thick SiGe. Fig 4 quantitatively shows ε_c at 2500 points over an area of 100 μm². The starting substrates including 40-nm-thick SiGe, combined with slow cooling, can yield higher and more uniform ε_c. Fig. 5 shows the change of surface roughness during the Ge condensation and the AFM images of the three GOI. The cross-hatched patterns are also suppressed by 40-nm-thick SiGe with slow cooling. These results indicate that higher ε_c in the 40-nm-thick SiGe with slow cooling can be realized by the lower dislocation/defect density, attributable to reduction in the total amount of strain energy by thinning SiGe.

Characteristics of strained GOI and SGOI pMOSFETs

In order to evaluate the electrical characteristics of the fabricated (S)GOIs, back-gate pMOSFETs with Ni S/D and Al back-gate metals were fabricated, as shown in Fig. 6. The important process improvement is the top surface passivation of GOI and SGOI by 10 nm-thick Al_2O_3 with post plasma oxidation [4], which is effective in suppressing μ_h degradation observed in the previous results [3]. Fig. 7 shows the transfer curves of 10-nm-thick GOI pMOSFETs formed by (a) 60- and (b) 40-nm-thick SiGe with slow cooling, in order to study the effect of the initial SiGe layer thickness. Also, the transfer curves for 40-nm- thick SiGe with quick cooling are compared as Fig. 7(c). GOI MOSFETs formed by 40-nm-thick SiGe with slow cooling exhibit the highest drain current and on/off ratio. Fig. 8 shows extracted μ_h as function of N_s. The 40-nm SiGe with slow cooling provides the μ_h enhancement of 1.4 against 60-nm SiGe and slow cooling. The importance of the surface passivation is also evident. Fig. 9 shows peak μ_h and ε_c as a function of the Ge fraction. These results indicate that much higher μ_h for the 40-nm SiGe with slow cooling is attributed to higher ε_c and resulting better crystal quality in high Ge fractions. As a result, we have achieved 10-nm-thick GOI pMOSFETs with ε_c of ~1.75% and μ_h of 467 cm²/Vs.

Next, in order to realize ETB (S)GOI pMOSFETs, 100 % GOI and SGOI with Ge fractions of 49 to 96 % were thinned by using ECR plasma oxidation/etching and TMAH etching respectively, with atomic-scale thickness controllability. The TEM images (Fig. 10) clearly show 2.0-nm-thick GOI and 2.1-nm-thick SGOI (Ge 70 %). Fig. 11 shows measured ε_c as a function of GOI and SGOI thickness. It is found that there is no strain relaxation down to 3 nm, while strain starts to get relaxed around 2 nm and is fully relaxed at 1 nm. Fig. 12 shows the μ_h-N_s characteristics of the (S)GOI MOSFETs with the Ge fractions from 100 to 49 % as a parameter of $t_{(S)GOI}$ down to 2 nm. It is confirmed that higher μ_h in higher Ge fractions is maintained down to $t_{(S)GOI}$ of 2 nm. Fig. 13 shows the benchmark of μ_h in GOI pMOSFETs as a function of the GOI thickness [2-9]. It is found that the present devices provide the highest μ_h at t_{GOI} of 10 nm and 2 nm, where the enhancement factor of 1.7 and 5.8, respectively, is obtained, thanks to higher ε_c. Fig. 14 also shows the benchmark of μ_h in SGOI pMOSFETs. It is found that pure GOI can provide higher μ_h in studied $t_{(S)GOI}$ than SGOI.

Conclusion We have demonstrated ETB (S)GOI MOSFETs with the body thickness from 10 to 2 nm with high μ_h due to high ε_c by the combination of the improved Ge condensation using slow cooling with the thinner initial SiGe layers in the initial substrates before condensation.

Acknowledgments This work was supported by JST-CREST Grant Number JPMJCR 1332, Japan and the Grant in-Aid for Scientific Research

through the MEXT (17H06148)
References [1] S. Nakaharai et al., APL **83** (2003) 3516 [2] W.-K. Kim et al, VLSI Symp. (2017) T124 [3] K.-W Jo et al, SSDM (2017) 217 [4] R. Zhang et al., TED **59** (2014) 335 [5] X. Yu et al., IEDM (2015) 20 [6] X. Yu et al., Micro Electron Eng. **147** (2015) 196 [7] C. H. Lee et al., SOI Conf., (2011) 1 [8] F. Rozé et al., ECS Trans., **75(8)** (2016) 67 [9] W.-H. Chang et al, APEX **9** (2016) 0913

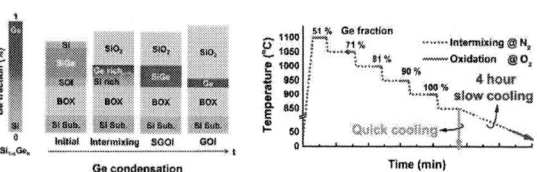

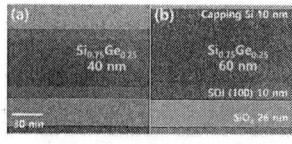

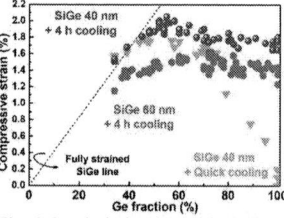

Fig. 1 The schema of the Ge condensation with the process flow chart, which composed oxidation and intermixing with Ge %

Fig. 2 TEM image of starting substrates with (a) SiGe 60nm-thick, and (b) 40 nm-thick

Fig. 3 Strain behavior with Ge %

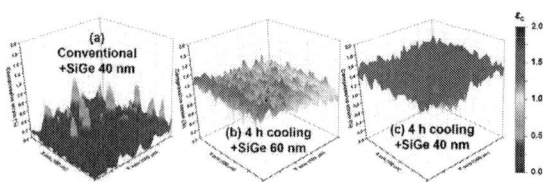

Fig. 4 Monitoring on ε_c on resulting GOI films

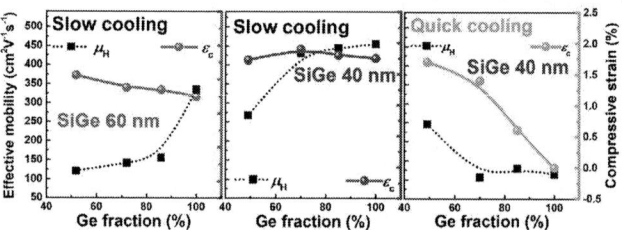

Fig. 5 The roughness of SGOI layer and AFM image during the process

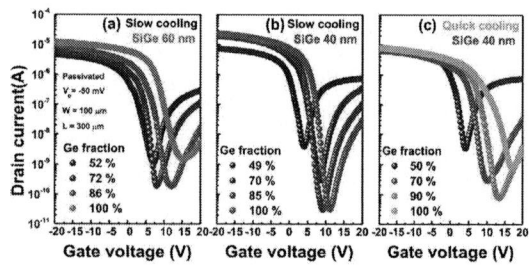

Fig. 6 The device structure, and fabrication process

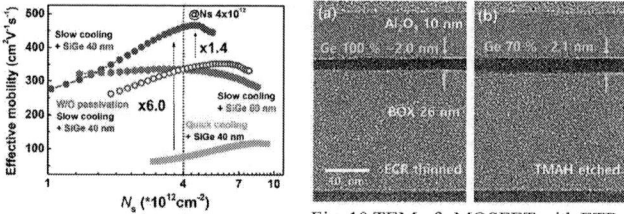

Fig. 9 The relationship between μ_h and ε_c as function of Ge fraction

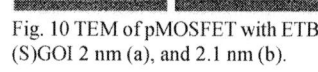

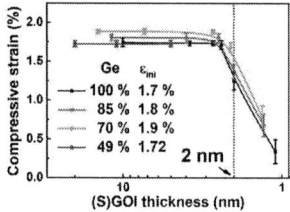

Fig. 7 Id-Vg of (S)GOI pMOS FETs dervied from the starting substrate with 40 or 60 nm-thick

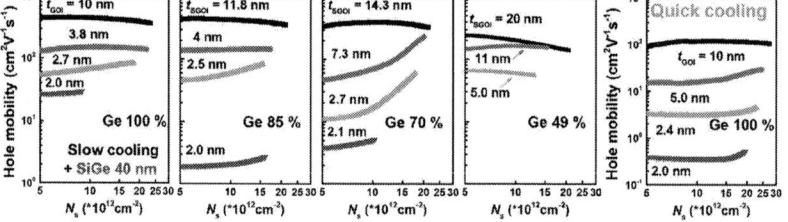

Fig. 8 μ_h of GOI MOSFETs as function of N_s with different method.

Fig. 10 TEM of pMOSFET with ETB (S)GOI 2 nm (a), and 2.1 nm (b).

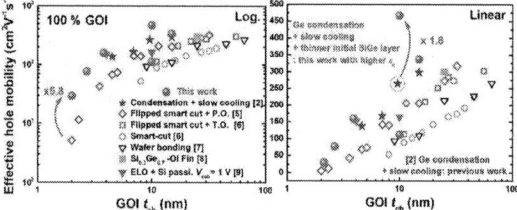

Fig. 11 t_{SGOI} dependence of compressive strain after thinning process.

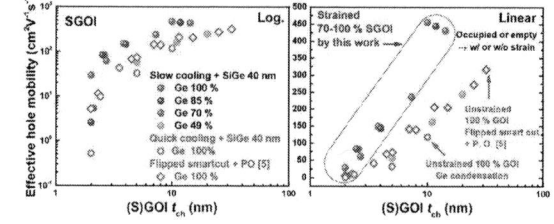

Fig 12 t_{SGOI} dependence of effective mobility, as function of N_s, to (S)GOI MOSFET

Fig. 13 Benchmarking of μ_h of GOI MOSFETs as function t_{GOI}.

Fig. 14 Benchmarking of μ_h of (S)GOI MOSFETs as function of $t_{(S)GOI}$.

2018 Symposium on VLSI Technology Digest of Technical Papers

GeSn p-FinFETs with Sub-10 nm Fin Width Realized on a 200 mm GeSnOI Substrate: Lowest SS of 63 mV/decade, Highest $G_{m,int}$ of 900 µS/µm, and High-Field μ_{eff} of 275 cm²/V·s

Dian Lei,[1] Kaizhen Han,[1] Kwang Hong Lee,[2] Yi-Chiau Huang,[3] Wei Wang,[1] Sachin Yadav,[1] Annie Kumar,[1] Ying Wu,[1] Huiquan Heliu,[1] Shengqiang Xu,[1] Yuye Kang,[1] Yang Li,[1] Eugene Y.-J. Kong,[1] Chuan Seng Tan,[2] and Xiao Gong.[1,*]

[1] Department of Electrical and Computer Engineering, National University of Singapore (NUS), Singapore.
[2] School of Electrical and Electronic Engineering, Nanyang Technological University (NTU), Singapore.
[3] Applied Materials Inc. Sunnyvale, California, United States.
*Phone: +65 6516-7871, Fax: 65 6516-1689, E-mail: elegong@nus.edu.sg

ABSTRACT

We report the first GeSn p-FinFETs with sub-10 nm fin width (W_{Fin}) enabled by the formation of the first 200 mm GeSn-on-insulator (GeSnOI) substrate and a self-limiting digital etch for accurate control of the fin dimension, achieving a fin with a top width of 5 nm. Owing to the excellent gate control using extremely scaled GeSn fin and the good GeSn fin quality maintained using a device fabrication process with low thermal budget, an SS of 63 mV/decade was achieved at channel length (L_{CH}) of 50 nm, which is a record low for Ge-based p-FETs. Furthermore, record high $G_{m,int}$ of 900 µS/µm (V_{DS} of -0.5 V) and $G_{m,int}/S_{sat}$ of 10.5 for GeSn p-FETs were achieved. A high high-field hole mobility μ_{eff} of 275 cm²/V·s (at inversion carrier density N_{inv} of 8×10^{12} cm⁻²) was also obtained.

I. INTRODUCTION

GeSn p-FET has higher hole mobility than Ge p-FET and is a promising candidate for future high performance applications [1-7]. Recently, GeSn p-FinFETs [8] and nanowire p-FETs [9] have been experimentally demonstrated. To explore the potential for GeSn to be used in future technology nodes, evaluation of the transistor performance at a fin dimension of less than 10 nm is very important, but has not been done before. At the same time, the thermal stability of GeSn fins or nanowires can easily degrade due to greater exposed surface area [10]. In order to maintain good material quality in the GeSn channel, a fabrication process with low thermal budget is thus preferred.

In this work, for the first time, we demonstrate GeSn p-FinFETs with fin width (W_{Fin}) less than 10 nm on a 200 mm GeSnOI substrate. By employing a digital etch (DE) process with a dimension control accuracy of ~1 nm, a GeSn fin with top width as small as 5 nm was fabricated. The dark-field TEM image in Fig. 1(a) shows 5 GeSn fins surrounded by gate dielectric and metal gate while the HRTEM image in Fig. 1(b) shows the width of the fin is ~5 nm at the top and ~9 nm at the middle. Using an interlayer-free gate stack module and a low thermal budget process with temperature not exceeding 250 °C post fin formation, a GeSn p-FinFET with channel length L_{CH} of 50 nm achieves the smallest SS of 63 mV/decade, highest $G_{m,int}$ of 900 µS/µm (V_{DS} of -0.5 V) and highest $G_{m,int}/S_{sat}$ of 10.5 for all reported GeSn p-FETs. In addition, the effect of W_{Fin} on the effective hole mobility and the scaling characteristics of GeSn p-FETs are investigated.

II. SUB-10 NM GESN FIN FORMATION USING DIGITAL ETCH

Fig. 2 (a)-(c) show a simple illustration of the steps in the DE process [10]. The initial GeSn W_{Fin} is 15 nm, defined by electron beam lithography (EBL) and a Cl-based dry etch. The DE process involves two key steps. First, a low-power oxygen plasma is applied to form Ge and Sn oxides in a self-limiting process. Second, the oxides are removed using diluted HCl solution (HCl:H₂O = 1:10) for 1 minute. Each cycle reduces W_{Fin} by ~2 nm and H_{Fin} by ~1 nm. After 5 cycles of DE, the 15 nm GeSn fin was trimmed down to ~5 nm at the top part of the fin. The reduction of the fin dimension after DE can be clearly observed from the tilt-view SEM images in Fig. 2 (d) and (e).

III. FABRICATION OF GESN P-FINFET

A. 200 mm GeSnOI Substrate Formation

To realize the first 200 mm GeSnOI substrate, a donor wafer with a ~100 nm of CVD-grown fully-strained (around -0.7%) $Ge_{0.95}Sn_{0.05}$ film on Si substrate was used. The same methods reported in [8] and [11] were employed for the GeSnOI formation. Fig. 3(a) shows a photograph of a 200 mm GeSnOI substrate. In Fig. 3(b), HRXRD (004) rocking scans of the GeSnOI substrate measured at different locations across the 200 mm wafer from center to edge show excellent uniformity in GeSn thickness and Sn composition. The existence of a Ge peak is due to the presence of a 200-nm-thick Ge cap layer on the surface during the XRD measurement. The Ge cap was then removed using a F-based dry etch and the GeSn layer was further thinned down to ~35 nm to complete the substrate formation. The TEM images in Fig. 4 show the complete GeSnOI layer stack with excellent crystalline quality in the GeSn layer and a sharp GeSn/SiO₂ interface.

B. Fabrication of GeSn p-FinFETs

The key process steps for fabricating the $Ge_{0.95}Sn_{0.05}$ FinFETs are listed in Fig. 5. Small pieces were cut from the 200 mm GeSnOI substrate for device fabrication. The channel region was defined and capped with HSQ using EBL followed by Boron implantation in the source/drain (S/D) regions. The HSQ was stripped after dopant activation at 400 °C for 60 s. GeSn fins were

patterned with EBL and dry-etched using a Cl-based plasma, followed by HSQ cap removal. 5 cycles of DE were then carried out to trim the fin width to sub-10 nm. After surface cleaning, the sample was immediately loaded into an ALD system where 10 cycles of TMA cleaning were performed, followed by 4-nm-thick HfO₂ deposition. Mo/W gate metal was then deposited by sputtering, and the gate was patterned using ICP dry etching. Finally, metal S/D contacts were formed. The tilt-view SEM image in Fig. 6 shows a completed GeSn p-FinFET device with one fin.

IV. ELECTRICAL CHARACTERIZATION AND BENCHMARKING

A. Electrical Characterization at Room Temperature

The fabricated GeSn FinFETs have L_{CH} from 50 to 200 nm, W_{Fin} from ~9 to 20 nm, and fin height H_{Fin} of 30 nm. I_{ON} and G_m are normalized by the total perimeter of the fin. Fig. 7 shows the transfer and output characteristics of two devices with L_{CH} of 50 nm and 60 nm and W_{Fin} of ~9 nm. I_{ON}/I_{OFF} ratio of 10⁴ and SS of 75 mV/decade were obtained. A high I_{ON} of 420 µA/µm is achieved at |V_{GS} - V_{TH}| of 1 V and V_{DS} of -1 V for the device with L_{CH} of 50 nm. The device also shows a high transconductance $G_{m,ext}$ of 490 µS/µm at V_{DS} of -0.5 V (Fig. 8). A low gate leakage current of 3×10^{-3} A/cm² is achieved with the interlayer-free gate stack (Fig. 9). Fig. 10 shows the smallest SS value of 63 mV/decade achieved by a GeSn FinFET with L_{CH} of 50 nm. Fig. 11 compares the I_D-V_G curves of two GeSn p-FinFETs with different W_{Fin}. The device with smaller W_{Fin} shows better control of short channel effects (SCEs). SS vs. L_{CH} is plotted in Fig. 12. Data from non-trimmed devices are also shown for comparison [8]. Much smaller SS values with less severe SS degradation are observed for devices with smaller W_{Fin}, reflecting the improved gate control. Fig. 13 shows the scaling metrics of $G_{m,int}$. The highest $G_{m,int}$ achieved is 900 µS/µm (V_{DS} of -0.5 V).

B. Low-Temperature Study

The measured I_D-V_G curves of a device with L_{CH} of 100 nm (Fig. 14) show clear improvements in SS and I_{ON} with decreases in temperature T. Fig. 15 shows the measured C-V curves of long-channel devices (L_{CH} = 4 µm) at frequencies of 20 kHz to 1 MHz for temperatures of 300 K, 150 K, 77 K, and 5 K. The negligible frequency dispersion indicates excellent gate stack quality. The I_D-V_G curves of these long-channel devices were also measured (Fig. 16). Effective hole mobility (μ_{eff}) at different temperatures was extracted using the split C-V method. In Fig. 17, the μ_{eff} extracted at 300 K decreases with decreasing W_{Fin}. The μ_{eff} extracted at different temperatures is shown in Fig. 18. Obvious μ_{eff} increase is observed at lower T from the mid-field to high-field due to the suppression of phonon scattering. A high μ_{eff} of 275 cm²/V·s is extracted at N_{inv} of 8×10^{12} cm⁻² (300 K).

C. Benchmarking

Fig. 19 benchmarks μ_{eff} at N_{inv} of 8×10^{12} cm⁻² as a function of Sn composition for GeSn p-FETs with CVD-grown GeSn channel. The GeSn p-FinFETs in this work have the highest μ_{eff} of 275 cm²/V·s at N_{inv} of 8×10^{12} cm⁻². The benchmark plots in Figs. 20 and 21 also show that this work achieves the smallest SS of 63 mV/decade and best $G_{m,int}/S_{sat}$ (V_{DS} of -0.5 V) among all GeSn p-FETs in the literature. The SS is also the lowest for all Ge-based p-FETs.

V. CONCLUSION

GeSn p-FinFETs with extremely-scaled sub-10 nm W_{Fin} were fabricated on 200 mm GeSnOI substrate with high GeSn film quality and good uniformity. Excellent electrical characteristics were achieved at L_{CH} of 50 nm in terms of SS, $G_{m,int}$, $G_{m,int}/S_{sat}$, and mobility, showing the promise of GeSn p-FETs for future high performance applications.

Acknowledgement. We acknowledge fund support from MOE Tier 1 (R-263-000-C58-133) and NUS Hi-FES Program (R-263-501-006-731).

References

[1] G. Han et al., IEDM2011, p. 402. [2] S. Gupta et al., IEDM2011, p. 398. [3] P. Guo et al., JAP 114, 044510 (2013). [4] X. Gong et al., EDL 34, 339 (2013). [5] D. Lei et al., JAP 119, 024502 (2016). [6] X. Gong et al., VLSI2012, p. 99. [7] X. Gong et al., VLSI2013, p. T34. [8] D. Lei et al., VLSI2017, p. T198. [9] Y.-S. Huang et al., IEDM2017, p. 832. [10] W. Wang et al., Scientific Report 7, 1835 (2017). [11] D. Lei et al., APL 111, 252103 (2017). [12] D. Lei et al., APL 109, 022106 (2016). [13] S. Gupta et al., EDL 34, 831 (2013). [14] Y.-S. Huang et al., IEDM2016, p.822. [15] L. Wang et al., SSE 83, 66 (2013). [16] M. Liu et al., VLSI2014, p. 100. [17] C. Zhan et al., VLSI-TSA2013, p. 8.

2018 Symposium on VLSI Technology Digest of Technical Papers

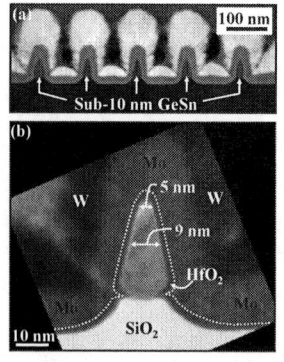

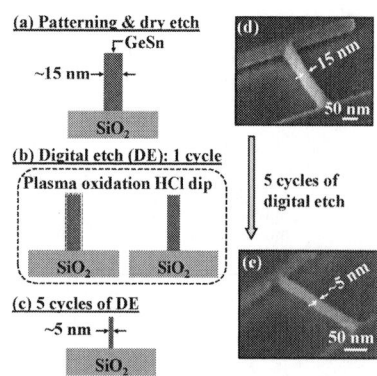

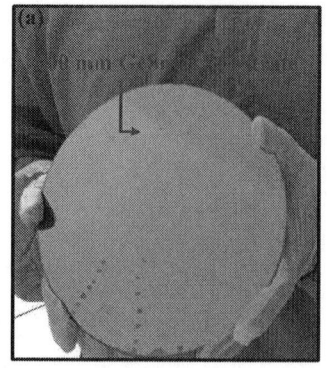

Fig. 1. (a) Dark field TEM image of GeSn p-FinFETs with sub-10 nm W_{Fin}. (b) HRTEM image of a GeSn fin surrounded by high-k/MG.

Fig. 2. (a)-(c) Schematic illustration of the formation of extremely-scaled GeSn fins using digital etch. After 5 cycles of digital etch, the width of GeSn fin top reduces from (d) 15 nm to (e) 5 nm.

Fig. 3. (a) The photograph of the 200 mm GeSnOI substrate formed using direct wafer bonding technique for the fabrication of GeSn p-FinFETs. (b) (004) HRXRD scans at different locations across the 200 mm GeSnOI substrate from the center to edge indicate good uniformity in thickness and Sn composition of the GeSn film.

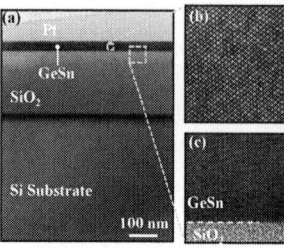

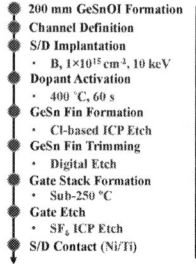

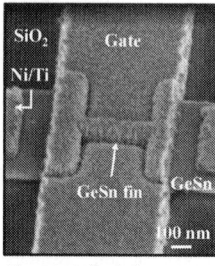

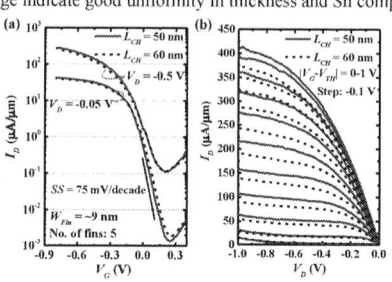

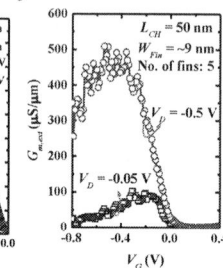

Fig. 4. (a) XTEM image of the formed 200 mm GeSnOI substrate. HRTEM images show (b) good GeSn crystalline quality and (c) sharp GeSn/SiO2 interface.

Fig. 5. Key process flow for the fabrication of GeSn p-FinFETs.

Fig. 6. Tilted-view SEM image of a completed GeSn p-FinFET.

Fig. 7. Transfer and output characteristics of two devices with L_{CH}=50 nm and 60 nm at V_{DS} of -0.05 V and -0.5 V. Highest I_{ON} of 420 µA/µm is obtained in the 50 nm device at V_G-V_{TH}= -1 V.

Fig. 8. $G_{m,ext}$ vs V_G for the 50 nm L_{CH} device shows high $G_{m,ext}$ of 490 µS/µm at V_{DS} of -0.5 V.

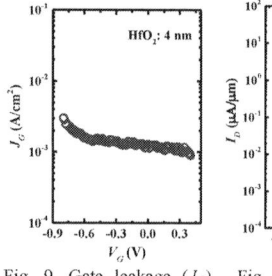

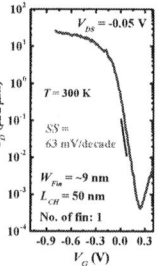

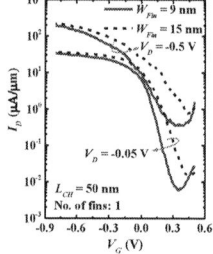

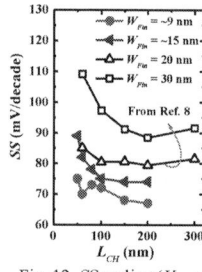

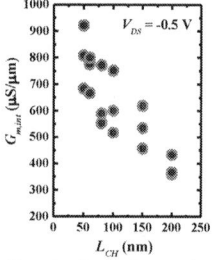

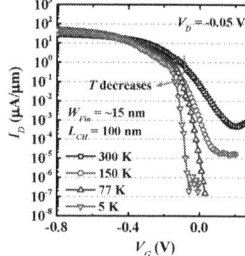

Fig. 9. Gate leakage (J_G) characteristics vs V_G. Low J_G of 3×10^{-3} A/cm² was maintained with the sub-1 nm EOT.

Fig. 10. I_D-V_G curve of a GeSn p-FinFET achieves the smallest SS value of 63 mV/decade.

Fig. 11. I_D-V_G of GeSn p-FinFETs with W_{Fin} of 9 nm and 15 nm. The device with a smaller W_{Fin} shows better control of SCEs.

Fig. 12. SS scaling (V_{DS}= -0.05 V) metrics. GeSn fins with smaller W_{Fin} achieve lower SS values.

Fig. 13. $G_{m,int}$ values scale well with L_{CH}. The highest $G_{m,int}$ of 900 µS/µm was achieved at V_{DS} of -0.5 V and L_{CH} of 50 nm.

Fig. 14. Low temperature I_D-V_G characteristics of GeSn p-FinFETs (W_{Fin} = ~15 nm). Improved SS and I_{ON} can be observed at low T.

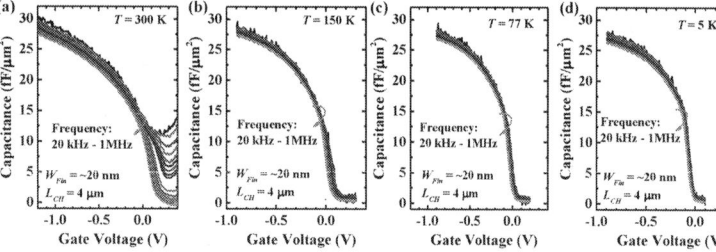

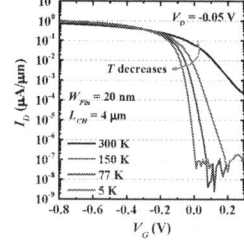

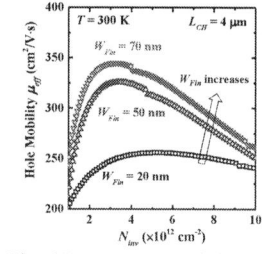

Fig. 15. Inversion C-V measured with the frequency between 20 kHz and 1 MHz at the temperature of (a) 300 K, (b) 150 K, (c) 77 K, (d) 5 K. The negligible frequency dispersion at all frequencies and temperatures indicates the good gate stack quality and low D_{it} across the bandgap.

Fig. 16. Low temperature I_D-V_G curves of long channel GeSn p-FinFETs which are used for mobility extraction.

Fig. 17. μ_{eff} vs N_{inv} of GeSn p-FinFETs. μ_{eff} increases with larger W_{Fin}.

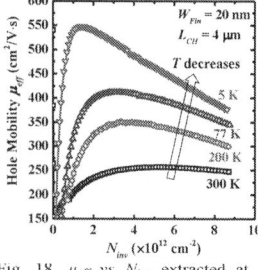

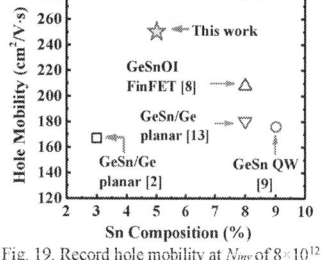

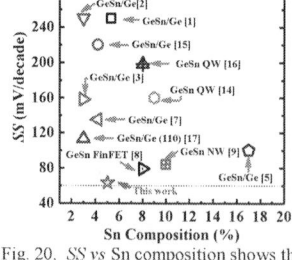

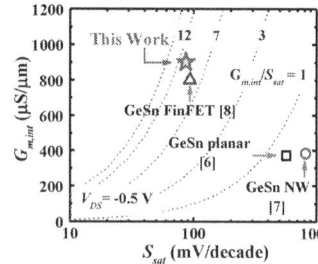

Fig. 18. μ_{eff} vs N_{inv} extracted at various temperatures.

Fig. 19. Record hole mobility at N_{inv} of 8×10^{12} cm⁻² is achieved in this work.

Fig. 20. SS vs Sn composition shows the smallest SS achieved in this work.

Fig. 21. GeSn FinFETs realized in this work have the best $G_{m,int}/S_{sat}$ for any reported GeSn p-FETs.

978-1-5386-4219-1/18 $31.00 © 2018 IEEE

Gap in pagination due to formatting issues.

Pages 199-200

2018 Symposium on VLSI Technology Digest of Technical Papers

Space Program Scheme for 3-D NAND Flash Memory Specialized for the TLC Design

Ho-Jung Kang, Nagyong Choi, Dong Hwan Lee[1], Tackhwi Lee[1], Sungyong Chung[1], Jong-Ho Bae, Byung-Gook Park and Jong-Ho Lee

Department of ECE and ISRC, Seoul National University, Seoul 151-742, Korea

[1]R&D Division, SK hynix Inc., Icheon, Gyeongki 467-701, Korea

Phone: +82-2-880-1727; Fax: +82-2-882-4658; E-mail: jhl@snu.ac.kr

Abstract

A new space program (PGM) scheme is proposed to achieve reliable triple-level-cell (TLC) 3-D NAND flash memory. Considering the lateral diffusion issue of stored electrons in the nitride storage layer, the proposed scheme stores electrons in the nitride layer of the space region between adjacent cells to suppress the lateral movement of trapped electrons in the programmed target cells. The effect of the space PGM can be sustained until 10^4 s at 90 °C and up to 1k read cycles at 25 °C. The programmed space region of the nitride layer improves the retention characteristics of the cells in the PGM state by 40% and remarkably reduces the V_{th} redistribution.

Introduction

Because of high density and low bit cost, 3-D NAND flash memory using a word-line (WL) stacked structure and charge trap layer has become mainstream [1]-[3]. Although 3-D NAND flash memory with SiN as a charge trap layer has many merits, the lateral diffusion issue resulting in a data retention problem is still a problem. Therefore, many studies have been done to solve the lateral diffusion caused by the continuous charge trap layer [4]-[8]. In this work, we propose a new space PGM scheme for controlled lateral migration of the stored electrons in the nitride storage layer of 3-D TLC NAND flash memory.

Measurement Result and discussion

Fig. 1 shows the TEM cross-section image and simulation structures of 3-D NAND flash memory used in this work [4]. Fig. 2 explains schematically the three electron loss mechanisms of the programmed target cells during the retention time. In 3-D NAND flash memory, the lateral charge distribution resulting from the continuous charge trap layer between the cells is an important issue in terms of reliability. Moreover, because a cell must be able to store multiple bits for a TLC implementation in NAND flash memory, the threshold voltage (V_{th}) verify margin is reduced. Therefore, the reduced V_{th} verify margin has a high possibility of a read fail occurring at each of the programmed verify cell states due to the V_{th} redistribution after some retention time shown in Fig. 3 (a). To alleviate the problem, we propose a space PGM scheme (Fig. 4) that can suppress the V_{th} redistribution in the PGM state shown in Fig. 3 (b). Fig. 4 (a) shows the signal timing diagram of the proposed scheme. The proposed scheme consists of all gate PGM, odd (or even) gate erase (ERS), and even (or odd) gate ERS in sequence, and the time after each pulse applied to the control gate is depicted as t1, t2, and t3. Fig. 4 (b) shows schematic cross-sectional views showing the trapped electron distribution in the nitride layer at t1, t2, and t3. t1 is the time after the PGM pulse is applied to all gates, so that electrons are trapped in the nitride layer except the DSL and SSL. t2 is the time after an ERS operation is performed to the odd gates; thus, the electrons trapped in the nitride under WL1, WL3, and WL5 escape into the channel. t3 is the time after an ERS operation is performed to the even gates. When the cells are sequentially erased by dividing the odd gates and the even gates in the cell strings, the fringing field from the ERS pulse applied to the space region between adjacent cells is dispersed, and the electrons programmed in the space region (called the space PGM) remain. The space PGM effectively suppresses the V_{th} redistribution in the PGM state. To verify the space PGM, we analyzed the measured BL current (I_{BL})-read voltage (V_{read}) curves of WL3, WL4, and electrically tied these two WLs shown in Fig. 5. Note that both WL cells have a V_{th} of 0 V. The V_{th} of the tied WLs is higher than that for each cell of WL3 and WL4, because the space region between WL3 and WL4 is more difficult to be inverted by the V_{read}. As a result, the space region between WL3 and WL4 in the tied WLs becomes a part of the channel controlled by the V_{read}. Thus the channel in the space region determines the V_{th} of the tied WLs.

Therefore, the space PGM shown in Fig. 5 (b) makes the V_{th} of the tied WLs larger than that of the tied WLs with the space ERS shown in Fig. 5 (a). The simulated I_{BL}-V_{read} curves in Fig. 6 show the same result as the measured data. To check the endurance of the space PGM, we measured the V_{th} shifts (ΔV_{th}) of the tied WLs over the retention time at different temperatures and the ΔV_{th} with the number of read cycles shown in Fig. 7. The effect of the space PGM can be sustained until 10^4 s at 90 °C and up to 1k read cycles at 25 °C. To see the effect of the proposed scheme, the retention characteristics of the target WL were measured shown in Fig. 8. Fig. 8 (a) and (b) show the ΔV_{th}s over the retention time measured from the target WL cell using two different schemes at 25, 90, and 200 °C in the PGM and ERS states, respectively. The proposed scheme has a smaller ΔV_{th} in the PGM state compared to that of the conventional scheme but a larger ΔV_{th} in the ERS state because the electrons trapped by the space PGM diffuse into the target WL. The V_{th} redistribution of cells in the PGM state is effectively suppressed; however, in the ERS state, it is a bit degraded. Because the TLC design requires a finely divided V_{th} in the PGM state, the retention characteristics of the programmed cells is more important than that of the erased cells in 3-D NAND flash memory. To check the effectiveness of the retention properties in the PGM state improved by suppressing lateral diffusion, ΔV_{th}s over the retention time are measured with different pass biases for adjacent WLs (V_{adj}) at 90°C shown in Fig. 9. The electrons trapped in the nitride layer by the space PGM can interfere with the inversion in the Si region, thereby increasing the V_{th} of the target WL cell, which may interfere with reading the V_{th} reduction due to a charge loss in the nitride of the target cell. However, as shown in Fig. 9, the ΔV_{th}s over the retention time in the PGM state are quite similar under different V_{adj}s regardless of the proposed and conventional schemes. The I_{BL}-V_{BL} curves of the target cell are measured as a parameter of V_{adj} in the PGM and ERS states shown in Fig. 10. As the V_{adj} decreases, I_{BL} decreases due to the channel resistance increase of the pass cells. The I_{BL} decrease of the erased target cell is more significant at a V_{adj} of 6 V when the proposed scheme is applied (see Fig. 10 (c)), because the programmed space regions in a cell string have a relatively high channel resistance. These results show that the space PGM does not affect the channel of the target WL cell in the PGM state. Fig. 11 compares the V_{th} distribution of the programmed cells after a retention time of 10^4 s at 25°C when the proposed and conventional schemes are applied. The histogram and Gaussian profiles verify that the V_{th} redistribution is remarkably improved by the proposed scheme shown in Fig. 11 (a) and (b).

Conclusion

We have proposed a new space program scheme, which suppresses the lateral migration of the stored electrons in the programmed cells, for reliable TLC 3-D NAND flash memory. Most electrons trapped in the nitride of the space region could be sustained until 10^4 s at 90 °C and up to 1k read cycles at 25 °C. With the proposed scheme, the retention characteristics of the programmed cells after 10^4 s were improved by ~40% and the V_{th} redistribution also was improved. Our approach is very useful for the TLC design that requires a finely split V_{th} in the PGM state.

Acknowledgement

This work was partially supported by the Brain Korea 21 plus in 2017 and SK hynix Inc. in 2017.

References

[1] S.-H. Lee, *IEDM*, pp. 1-8, 2016. [2] C. Kim *et al.*, *ISSCC*, pp. 202-203, 2017. [3] R. Yamashita *et al.*, *ISSCC*, pp 196-197, 2017. [4] H.-J. Kang *et al.*, *VLSI*, pp. 182-183, 2015. [5] B. Choi *et al.*, *VLSI*, pp. 78-79, 2016. [6] Y.-H. Liu *et al.*, *EDL*, pp. 48-51, 2017. [7] J. Wu *et al.*, *IEDM*, pp. 95-98, 2017. [8] K. Mizoguchi *et al.*, *IEDM*, pp. 465-468, 2017.

978-1-5386-4219-1/18 $31.00 © 2018 IEEE

2018 Symposium on VLSI Technology Digest of Technical Papers

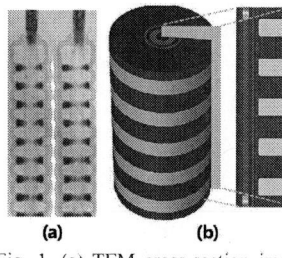

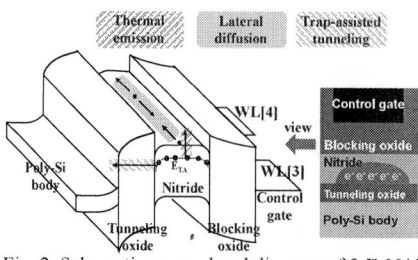

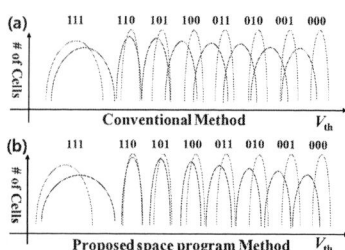

Fig. 1. (a) TEM cross-section image and (b) simulation structure of a 3-D NAND flash memory cell array.

Fig. 2. Schematic energy band diagram of 3-D NAND flash memory cells in the program (PGM) state, explaining the trapped charge loss mechanisms during the retention time.

Fig. 3. Schematic of the V_{th} redistributions of triple-level-cell (TLC) NAND flash memory cells using the (a) conventional and (b) proposed methods after the retention time.

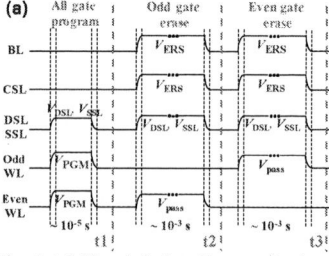

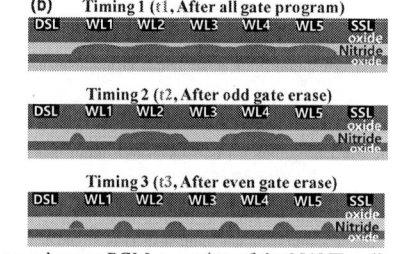

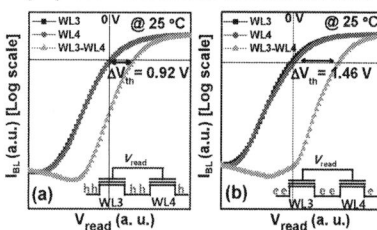

Fig. 4. (a) Signal timing diagram for the proposed space PGM operation of the NAND cells. Proposed scheme consists of all gate PGM, odd (or even) gate erase (ERS) and even (or odd) gate ERS in sequence. (b) Schematic cross-sectional views showing trapped electron distribution in the nitride layer at t1, t2, and t3 of (a) when the proposed scheme is applied.

Fig. 5. I_{BL}-V_{read} curves of WL3, WL4, and tied two WLs with the (a) space ERS and (b) space PGM. The V_{th}s of WL3 and WL4 are 0 V and the V_{th}s of the tied two WLs are different with charge polarity in the space region.

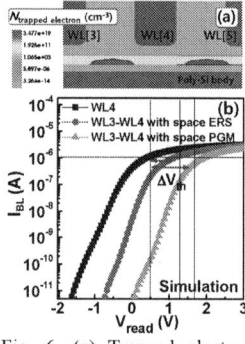

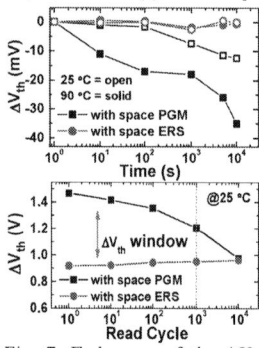

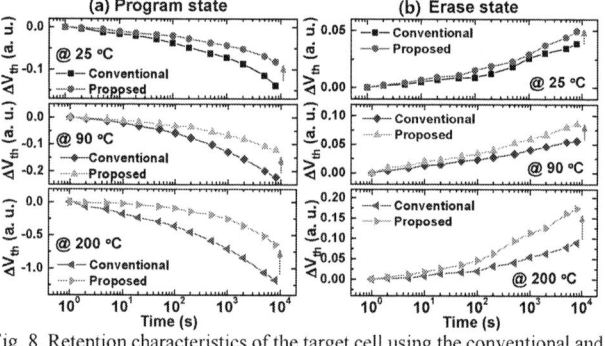

Fig. 6. (a) Trapped electron contours after the space PGM from simulation. (b) Simulated I_{BL}-V_{read} curves of the WL4 (squares), tied two WLs with the space ERS (circles) and PGM (triangles).

Fig. 7. Endurance of the ΔV_{th} with the different space charges depicted in Fig. 5. The electrons trapped by the space PGM are effective up to 10^4 s at 90 °C and 1k read cycles at 25 °C.

Fig. 8. Retention characteristics of the target cell using the conventional and proposed schemes at different temperatures (25, 90 and 200 °C) in the (a) space PGM and (b) ERS states. The retention properties of the programmed cells after 10^4 s with the proposed space PGM method are improved by 40% compared to that of the conventional method.

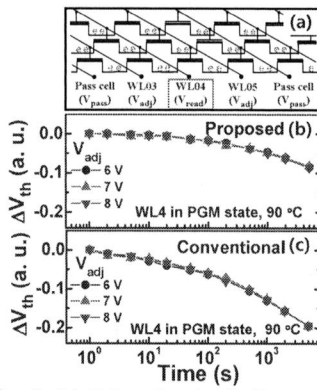

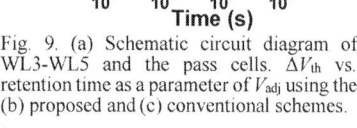

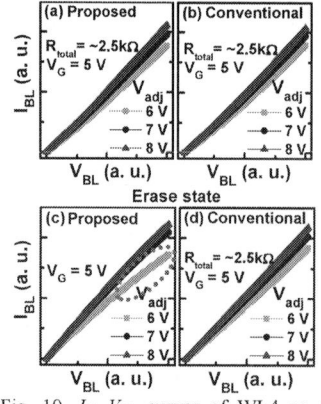

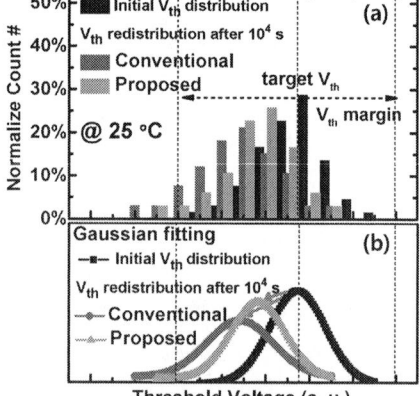

Fig. 9. (a) Schematic circuit diagram of WL3-WL5 and the pass cells. ΔV_{th} vs. retention time as a parameter of V_{adj} using the (b) proposed and (c) conventional schemes.

Fig. 10. I_{BL}-V_{BL} curves of WL4 as a parameter of V_{adj} in the PGM and ERS states. (a), (b) and (d) show nearly the same tendency with different V_{adj}s.

Fig. 11. V_{th} redistribution of the programmed cells using the proposed (green) and conventional (red) schemes after 10^4 s at 25 °C. (b) Gaussian profiles extracted from (a). The V_{th} redistribution is improved by the proposed space PGM scheme.

978-1-5386-4219-1/18 $31.00 © 2018 IEEE

202

First demonstration of monocrystalline silicon macaroni channel for 3-D NAND memory devices

R. Delhougne, A. Arreghini, E. Rosseel, A. Hikavyy, E. Vecchio, L. Zhang, M. Pak, L. Nyns,
T. Raymaekers, N. Jossart, L. Breuil, S. S. V-Palayam, C.-L. Tan, G. Van den bosch, A. Furnémont

imec, Kapeldreef 75, B3001 Leuven, Belgium. Email : romain.delhougne@imec.be

Abstract: We are demonstrating for the first time epi-based monocrystalline silicon macaroni channel 3-D NAND devices. The highly controllable channel replacement process sequence leads to > 95% yield, with excellent uniformity and reproducibility, proving its potential for manufacturability. The electron mobility of the channel is improved by a factor 30 compared to the polycrystalline macaroni Si channel, together with a reduction of the off state leakage. Furthermore, this channel replacement fabrication process does not affect memory performance and reliability. The performance benefits of this channel replacement technique make it a potential candidate for fabricating future 3-D NAND devices.

Introduction: 3-D NAND flash memory is currently the mainstream technology for large density storage applications. The bit density increase is mostly driven by a combination of increasing the number of bit/cell and aggressive z-stacking [1,2]. In this race towards higher densities, the use of a macaroni channel is necessary to guarantee a good gate control over the channel and limit the off state leakage. The poor conduction of the polysilicon channel is a potential limiting factor for further z-stacking. The polysilicon channel is subject to mobility degradation due to highly-defective grain boundary regions [3]. To enable further vertical stacking, it is required to increase the drive current, i.e. to improve the channel conduction, either by boosting the mobility of the silicon channel or by using alternative high mobility channel materials. In this paper, we are demonstrating the first time epi-based monocrystalline silicon macaroni channel 3-D NAND, displaying an overall improved electrical performance compared both to polysilicon macaroni channel [4] and to monocrystalline silicon full channel [5].

Process flow description: In this study we have used a 3-fold poly gate test vehicle, with conventional Si_3N_4 trapping layer, SiO_2 tunnel barrier and macaroni channel [4]. We have engineered the process flow to form an epitaxially grown monocrystalline silicon (EPI) macaroni channel, in place of the state of the art polysilicon macaroni channel (poly macaroni). This innovative channel replacement process scheme is presented in Fig. 1. After the formation of the memory hole by dry etching (Fig. 1a), the ONO stack is deposited by LPCVD, and subsequently followed by the deposition of an amorphous silicon (a-Si) protection layer in a vertical furnace (Fig. 1b). The ONO and protection layer are anisotropically etched to open the bottom of the memory hole (Fig. 1c), while the ONO remains protected at the sidewalls. A 10 nm a-Si sacrificial layer is then deposited as dummy channel (Fig. 1d). The a-Si bottom opening etch needs to be sufficiently long to ensure 50 nm minimum recess in the substrate. This 50 nm recess (Fig. 1e) will help to strongly anchor the central oxide pillar used as template during EPI growth. The remaining hole is filled with PEALD oxide (Fig. 1f), followed by a controlled recess to uncover the top of the sacrificial dummy channel

(Fig. 1g). The channel replacement processing was performed in an Intrepid XP™ RPCVD epi cluster from ASM. After a low-temperature removal of the native oxide, the sacrificial a-Si was etched with Cl_2 at 575 °C for ~10 min (Fig. 1h) [6], down to the silicon substrate, forming typical [111] facets (Fig. 5). The Cl_2 etch has been carefully tuned to remove all a-Si but at the same time avoid any undercut of the oxide pillar (Fig. 3). The tilt of the central oxide pillar has been characterized by top view CDSEM for various CD and has been found to be less than +/- 2 nm in X and Y directions (Fig. 6). The void thickness left after Cl_2 recess of the dummy channel has been measured to be 10-15 nm across the wafer, independently of the memory hole CD (Fig. 7). The subsequent selective epitaxial Si regrowth was done at 810 °C for ~15 min using a conventional $SiCl_2H_4/HCl$ process (Fig. 1i). As shown in Fig. 4, the tunnel oxide remains pristine after the replacement channel process. Fig. 2 further highlights the high quality bottom interface, and the symmetrical replacement EPI macaroni channel. To test electrically the devices, the top Drain contact is then formed by using doped polysilicon (Fig. 1j), followed by staircase patterning, passivation, metallization and finally forming gas anneal.

Electrical results: Single cell devices with CD of 80 nm have been used for in-depth electrical testing. A yield of > 95% has been obtained on all wafers. The typical I_D-V_G characteristics show clear improvements for the EPI macaroni replacement channel compared to the poly macaroni (Fig. 8). The ON current values of the EPI macaroni channel are 2 orders of magnitude higher than the poly macaroni, and on par with the EPI full channel (Fig. 9). STS follows the same trend, with a 0.3 V/dec decrease for the EPI macaroni with respect to the poly macaroni (Fig. 10). The Program/Erase characteristics for the EPI macaroni are comparable to the poly macaroni (Fig. 11). The EPI macaroni channel is further improving the trade-off between the electron mobility and the off state leakage (Fig. 12). The electron mobility is 30x higher for the EPI macaroni channel compared to the poly macaroni, reaching the same level as the EPI full channel. As expected, the off-state leakage of the EPI macaroni is at an intermediate value between the poly macaroni and EPI full channel. The endurance of the EPI macaroni channel remains similar to the poly macaroni reference (Fig. 13). The data retention at both 85 °C and 150 °C is improved for the EPI macaroni channel, confirming that the tunnel oxide is not damaged by the replacement channel process scheme (Fig. 14).

Conclusions: We have successfully demonstrated the first monocrystalline silicon macaroni in 3-D NAND devices. The replacement channel process scheme was shown to be highly reproducible and controllable, with potential for manufacturability. A large improvement in device performance has been obtained, which opens up opportunities for further 3-D NAND scaling.

2018 Symposium on VLSI Technology Digest of Technical Papers

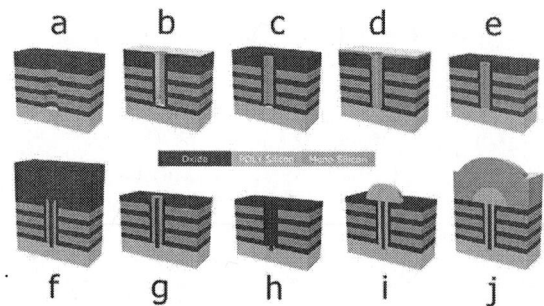

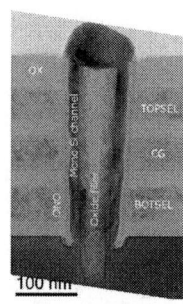

Fig 1: Replacement channel process scheme: (a) memory hole patterning; (b) ONO and a-Si deposition; (c) ONO and a-Si bottom opening; (d) dummy a-Si channel deposition; (e) a-Si bottom opening; (f) memory hole oxide fill deposition; (g) oxide fill recess; (h) dummy a-Si channel removal (Cl_2 etch in epi tool); (i) silicon channel epi growth; (j) Polysilicon drain formation

Fig. 2: TEM cross section after the silicon channel epi growth showing the high quality bottom interface, the pristine tunnel oxide and the symmetrical replacement epi silicon channel

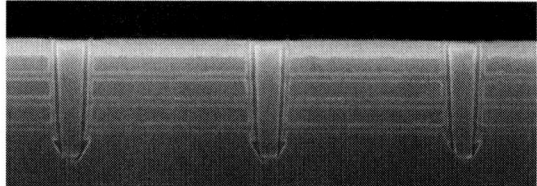

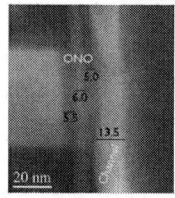

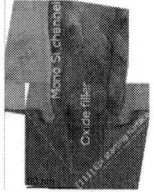

Fig. 3: SEM cross section after dummy a-Si channel removal (Cl_2 etch in epi tool). The dummy channel and substrate recess is uniform both across adjacent memory holes, and across the wafer

Fig. 4: The tunnel oxide remains intact after the Cl_2 dummy channel recess and the silicon channel epi regrowth

Fig. 5: [111] facets act as starting surfaces for EPI Si macaroni channel regrowth

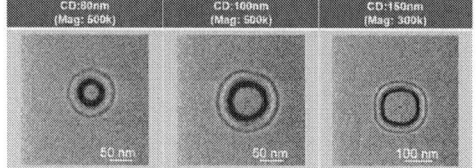

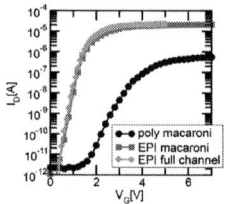

Fig. 6: Top view SEM after the Cl_2 in situ recess of the dummy Si channel. The minimal tilt of the central pillar leads to a uniform void thickness (future EPI channel thickness)

Fig. 7: Void size uniformity along the radius of the wafer for three different memory hole CD

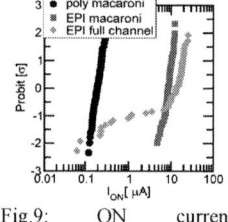

Fig.8: Typical I_D-V_G characteristics for EPI macaroni and EPI full channel, compared to poly macaroni channel

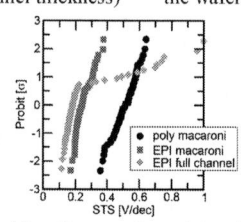

Fig.9: ON current distributions showing the large improvement for the EPI macaroni channel

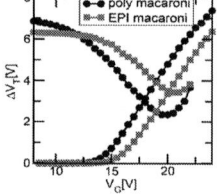

Fig.10: Improved subthreshold swing (STS) for EPI macaroni compared to poly macaroni channel

Fig.11: ISPP/ISPE characteristics are comparable for EPI macaroni and poly macaroni channel

References

[1] A. Goda et al., Invited paper, IEDM 2012, pp.13-16

[2] J.Lee et al., Invited paper, IEDM 2016, pp. 284-287

[3] R. Degraeve et al., IEDM 2015, pp. 121–124

[4] A.Arreghini et al., IMW 2017, pp. 115-118

[5] A. Subirats et al., IEDM 2017, pp. 517-520

[6] A. Hikavyy et al., Semi. Sci. Technol. 32 (2017) 114006

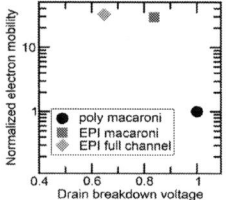

Fig. 12: The mobility vs. normalized drain breakdown voltage tradeoff is largely improved for the EPI macaroni channel compared to the poly macaroni channel

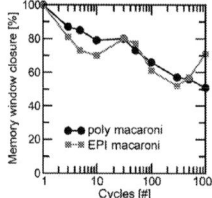

Fig. 13: The window closure with cycling of the EPI macaroni channel is on par with the poly macaroni reference

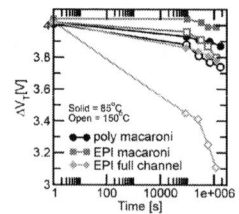

Fig. 14: The data retention of the EPI macaroni channel is improved compared to poly macaroni channel and EPI full channel

Acknowledgments: This work is supported by imec's Industrial Affiliation Program on Advanced Flash Memory devices. Authors acknowledge imec's pilot line, the various engineering, support, process and characterization teams in imec and the support from Hitachi, ASM and Synopsys.

978-1-5386-4219-1/18 $31.00 © 2018 IEEE

High Endurance Self-Heating OTS-PCM Pillar Cell for 3D Stackable Memory

C.W. Yeh[1], W.C. Chien[1], R.L. Bruce[2], H.Y. Cheng[1], I.T. Kuo[1], C.H. Yang[1], A. Ray[2], H. Miyazoe[2],
W. Kim[2], F. Carta[2], E.K. Lai[1], M. BrightSky[2], and H.L. Lung[1]

IBM/Macronix Phase Change Memory Joint Project

[1]Macronix International Co., Ltd., [1]Emerging Central Lab, 16 Li-Hsin Rd., Science Park, Hsinchu, Taiwan
[2]IBM T.J. Watson Research Center, 1101 Kitchawan Road, Yorktown Heights, NY, 10598, USA
TEL: +1-914-945-2031, Email: cwyeh@mxic.com.tw; cwyeh@us.ibm.com

Abstract

For the first time published, high endurance OTS (ovonic threshold switch, here, TeAsGeSiSe-based) is integrated with PCM (here, doped Ge2Sb2Te5) to form a 3D stackable pillar type device. With the help of an etch buffer layer and a damage-free pillar RIE process, we achieved 100% array yield without OTS/PCM composition modification. Anneal tests show this one-selector/one-resistor (1S1R) pillar device is BEOL-compatible.

We report excellent electrical performance by 1S1R OTS-PCM device; selector provides the fast turn on/off speed which enables 10ns fast RESET speed, program endurance is 10^9 cycles, and read endurance is higher than 10^{11} cycles.

Introduction

SCM (storage class memory) is being pursued aiming to boost system performance while DRAM is facing scaling challenges. Large density, bit addressability and high endurance are the main requirements and also the biggest challenges for SCM. PCM is an ideal candidate that meets these requirements. Self-heating PCM [1] provides benefits of good program power efficiency and simple process without an additional heater. To realize large density memory capacity, transistor-free structure is proposed to achieve $4F^2$ foot print and 3D stackability [2, 3].

In this work, TeAsGeSiSe-based OTS [4] is incorporated with GeSbTe (GST) based PCM that fulfills the strict requirements for SCM. We overcame the challenge of integrating PCM and OTS into a self-heating pillar structure with a damage-free RIE process. With its simple structure and stackability, it is cost effective to use this 1S1R unit in a 3D cross-point memory. The TeAsGeSiSe-based OTS provides adequate matching electrical characteristics for the operating current and voltage of PCM. Furthermore, this 1S1R structure showed very good program cycle endurance, read cycle endurance and thermal stability.

Device Fabrication and Process Improvement

Fig. 1 shows device structure of testable 1S1R pillar device built on top of the contact layer of standard CMOS logic process. The layers of 1S1R including OTS/buffer/PCM/buffer/top electrode are deposited with a PVD tool and patterned into pillars. The TEM shows the critical dimension is approximately 50nm at the top electrode and 80nm at PCM.

Carefully controlling the etch processes is critical to integrating the 1S1R OTS-PCM devices into a self-heating pillar. Both chalcogenide materials, OTS and PCM, are extremely sensitive to RIE chemistry and can easily be modified. Fig. 2 (a) shows PCM has an obvious composition change after etch. With buffer layer, damage-free RIE is demonstrated in Fig. 2 (b), showing the etch chemistry can be switched between electrode etch and PCM etch to prevent the etch damage issue.

RESET/SET Operation of 1S1R Cell

The equivalent circuit and I-V characteristic of the 1S1R device is shown in Fig 3. The OTS is in series with PCM which can be RESET (high resistance) or SET (low resistance). The I-V behavior shows that the threshold voltage required to turn on the device when in the RESET state (VtR) is higher than when in the SET state (VtS). Utilizing this difference in the turn-on behaviors, a read voltage (Vread) can be chosen between VtS and VtR and results in >100 times difference in reading current between a RESET bit and a SET bit.

Fig. 4 shows the 1S1R cell can be programmed to RESET state by adjusting the input pulse current. OTS is able to provide more than sufficient current for the RESET operation. Fig 5 shows transient current of 1S1R device with a 10ns wide input pulse. The selector is able to be turned on/off with such a short input pulse which demonstrates the potential of a fast read/write operation. In addition, OTS ON-current can reach 850uA which means a high current density of 8.9MA/cm^2. Fig. 6 further proves the RESET speed can be as fast as 10ns. Fig. 7 shows SET operation by 1us box pulse. The lowest VtS can be achieved at 90uA current. Distribution of delta Vth (VtR - VtS) shows more than 0.5V read window (Fig. 8). This 1S1R self-heating cell is also promising for scaling. In Fig. 9, the RESET currents of devices with different sizes are predicted by COMSOL simulations. Smaller devices show better heating efficiency and have substantially lower RESET current (980K temperature is required for PCM RESET). Fig. 10 shows that the Vread window enlarges with device scaling (due to higher PCM RESET resistance).

Endurance and Thermal Stability

Optimized RESET and SET pulses enabled program endurance up to 10^9 cycles (Fig. 11). Program failure happens after 10^9 cycles due to stuck to SET as shown in the post failure I-V curve (Fig. 12). The read endurance is further checked on a cell in the SET state (Fig. 13). With 3.3V read, OTS is turned on and allows 30 uA current flowing through the device. A read pulse at lower voltage 1V is then applied to the device to verify the OTS is properly in the off state. The device remains functional after 10^{11} read on/off cycles. Table I summarizes the results of previous publications and this work; Our TeAsGeSiSe-based OTS + doped-Ge2Sb2Te5 PCM shows the better endurances. Thermal stability is checked by 400°C/30 mins anneal (Fig. 14). The device shows nearly identical I-V characteristics before and after anneals, thus demonstrates compatibility with further BEOL processing.

Summary

We publish for the first time a pillar 1S1R OTS-PCM cell representing a promising scalable solution for a stackable cross point memory. Superb endurance (10^9 programming, $>10^{11}$ read), good thermal stability (400°C/30min) and a simple structure for stacking are demonstrated.

References

[1] T.D. Happ, et al., *VLSI, p 120-121, 2006*
[2] Yi-Chou Chen, et al., *IEDM Technical Digest, p 905-908*, 2003
[3] DerChang Kau,et al., *IEDM, p 617-620*, 2009
[4] H. Y. Cheng, et al., *IEDM, p 28-31*, 2017
[5] G. Navarro, et al., *VLSI, p T94-T95*, 2017

2018 Symposium on VLSI Technology Digest of Technical Papers

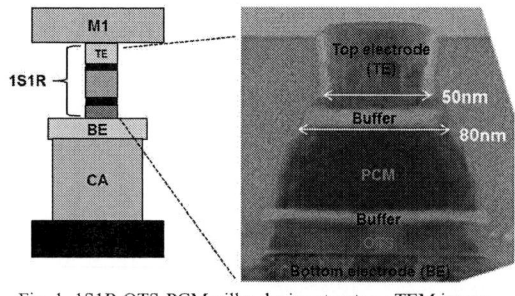

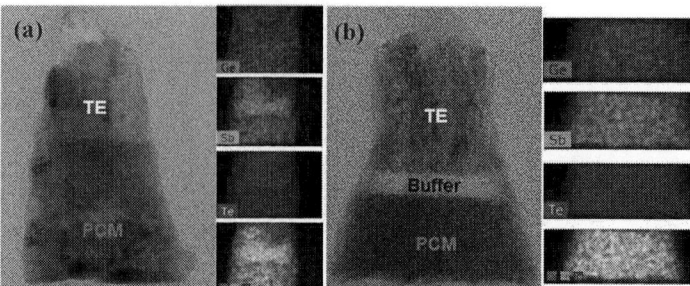

Fig. 1 1S1R OTS-PCM pillar device structure. TEM image shows a close look of 1S1R device after pillar RIE, encapsulation and CMP. PCM and OTS along with buffer layers are sandwiched by top electrode and bottom electrode.

Fig. 2 TEM and EDX mapping images of different PCM etch schemes. (a) Etch without buffer layer results in composition modification especially on top of PCM which is caused by RIE chemistry for TE. (b) Etch with buffer layer results in uniform composition across the entire PCM region.

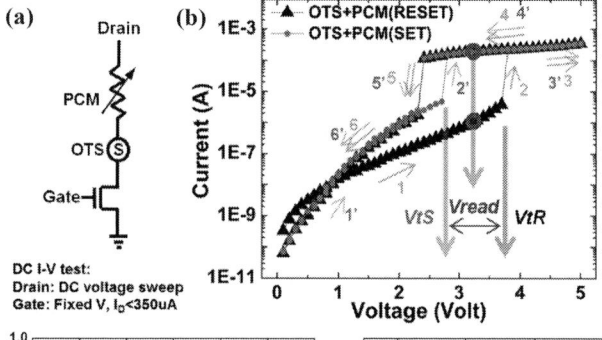

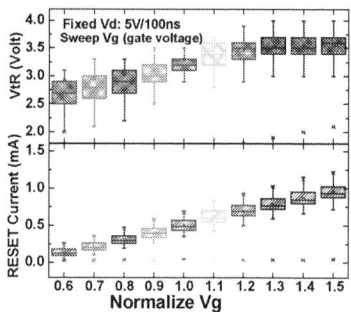

Fig. 3 (a) Equivalent circuit of 1S1R device . The PCM is in series with OTS. A transistor is used as a local current compliance. The gate is turned on with a fixed voltage allowing < 350uA current. (b) I-V characteristic for 1S1R device in RESET state and SET state. (Forming is not shown. ~3.5V) The threshold voltage of RESET state (VtR) is higher than SET state (VtS). More than 100X resistance difference can be read out by read voltage (Vread) which is between VtS and VtR.

Fig. 4 VtR increases as RESET current increases. VtR saturated at 3.5V.

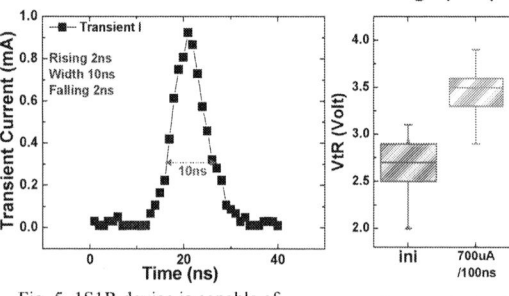

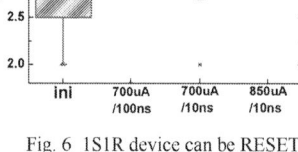

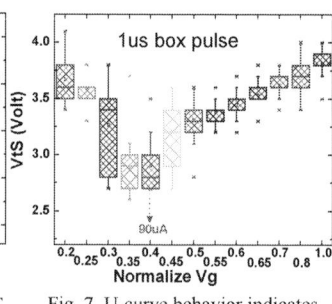

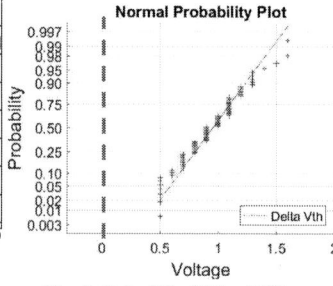

Fig. 5 1S1R device is capable of responding to an AC pulse as short as 10 ns and turns on with a current up to 850 uA (8.9MA/cm^2).

Fig. 6 1S1R device can be RESET well by using 10ns pulse.

Fig. 7 U-curve behavior indicates PCM can be SET with a current of approximately 90uA.

Fig. 8 Delta Vth (VtR – VtS) distribution shows more than 0.5V read window.

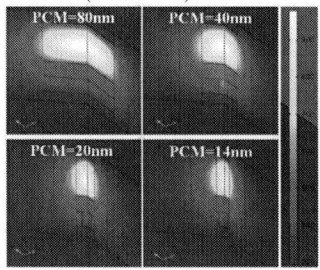

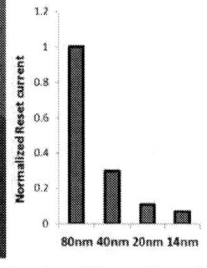

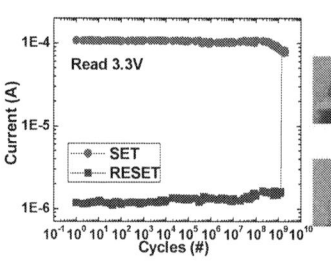

Fig. 9 RESET currents of different device sizes (80nm, 40nm, 20nm and 14nm) are simulated by COMSOL. To achieve the PCM melting temperature of 980K, the scaled device shows better heating efficiency and requires lower RESET current.

Fig. 10 Delta Vth (VtR – VtS) simulation of 80nm and 14nm devices. Read window increases as the device scales down.

Fig. 11 Program endurance of 1S1R device can achieve 1E9 cycles until PCM is stuck in the SET state. TEM images shows 1S1R cells after programing to RESET state and SET state.

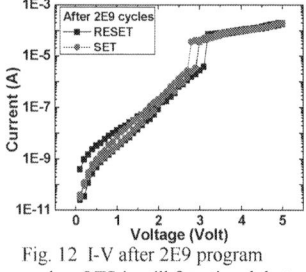

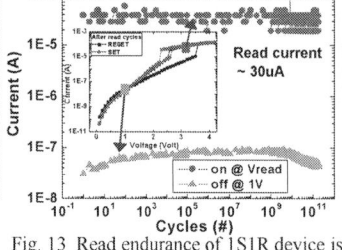

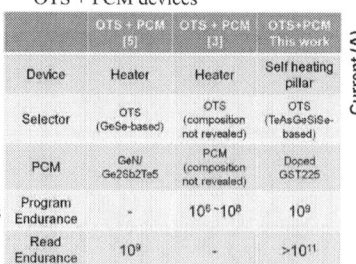

Fig. 12 I-V after 2E9 program cycles. OTS is still functional, but the PCM is stuck in the SET state.

Fig. 13 Read endurance of 1S1R device is > 1E11 cycles. Inset shows that the I-V after the endurance test is still good.

Table I Endurance comparison of OTS + PCM devices

	OTS + PCM [5]	OTS + PCM [3]	OTS+PCM This work
Device	Heater	Heater	Self heating pillar
Selector	OTS (GeSe-based)	OTS (composition not revealed)	OTS (TeAsGeSiSe-based)
PCM	GeN/ Ge2Sb2Te5	PCM (composition not revealed)	Doped GST225
Program Endurance	-	10^6~10^8	10^9
Read Endurance	10^9	-	>10^{11}

Fig. 14 I-V characteristic before and after 400°C 30 minutes anneal.

978-1-5386-4219-1/18 $31.00 © 2018 IEEE

Te-based binary OTS selectors with excellent selectivity ($>10^5$), endurance ($>10^8$) and thermal stability ($>450°C$)

Jongmyung Yoo, Yunmo Koo, Solomon Amsalu Chekol,
Jaehyuk Park, Jeonghwan Song and Hyunsang Hwang
Dept. of Mat. Sci. and Eng., Pohang University of Science and Technology (POSTECH), Pohang, Korea
Phone: +82-54-279-5123, Fax: +82-54-279-5122, e-mail: hwanghs@postech.ac.kr

Abstract: We have investigated various Te-based binary materials for Ovonic Threshold Switch (OTS) selector application. We found that both Te composition and difference in atomic radius of elements composing the telluride film are the key control parameters to maximize the OTS characteristics such as low leakage current (<5 nA for device area of 30 nm^2), good switching endurance (10^8), and thermal stability ($450°C$).

Introduction: To implement X-point memory array devices, excellent selector devices with low leakage current, fast switching speed, scalability and thermal stability are required. Ovonic Threshold Switching (OTS) devices have the potential for meeting these requirements [1-3]. However, the thermal stability of the devices remains a concern because they are based on low crystallization temperature (T_c) of telluride materials whose amorphous phase induces OTS behavior [3]. To overcome this problem, devices based on telluride doped with various elements [4] and those based on selenide materials are recently reported [5,6] but they suffer from large threshold voltage (V_{Th}) and complex material composition which hinder their feasibility as a selector device in X-point array.

In this study, investigation of various binary OTS material systems to understand the origin of difference in their device properties is conducted to maximize selector characteristics of the OTS devices.

Device fabrication: To investigate diverse binary telluride systems with varying Te composition, we adopted a co-sputtering method for fabrication of the film and prepared sputter-targets made of various elements, which is illustrated in Fig. 2. All of the telluride films were deposited at room temperature on top of patterned W bottom electrodes, whose cell size varies from 150 nm^2 to 30 nm^2. The film thickness was also controlled to monitor its effect. W top electrode was deposited after the deposition of telluride films.

Composition, device area, and thickness effects on various types of OTS devices: Te composition effect on the fabricated devices are shown in Fig. 3. In B-Te and Al-Te based devices, decrease in Te ratio results in increased V_{Th} and decreased I_{OFF}. This tendency is shown because the reduction in Te amount results in diminished number of Te-induced traps through which the field-induced carriers are hopping [7]. In contrast, C-Te based device shows decrease in I_{OFF} and increase in V_{Th} with increasing amount of Te. This happens because the increase of C amount in the film causes formation of more C-C sp^2 hybridized conductive bonds which can act as defect sites just like Te-induced traps [8]. The device area and thickness effects on each of the three fabricated devices are demonstrated in Fig. 4. All of the fabricated devices show tendency of decreasing I_{OFF} and increasing V_{Th} as the device area is scaled down or the film thickness increases, which are consistent with previous reports [5,9]. By selecting proper composition and thickness of films, we found that all of the fabricated binary devices showed good selectivity ($>10^5$) and low I_{OFF} (<5 nA) at cell size of (30 nm)^2.

Analysis on conduction mechanism: To understand the conduction mechanism of the fabricated devices, we adopted Poole-Frenkel based analytical model with deep traps inhibiting leakage current at voltage condition lower than V_{Th}, following the equation described in Fig. 5 (a). By assuming that $\tau_0 = 10^{-15}$ s and N_T is proportional to $(\Delta z)^{-3}$, the selected analytical model explains the OFF-state conduction mechanism of the devices well with parameters shown in Fig. 5 (b). According to the analysis, the low I_{OFF} of the devices is

due to their large E_a values higher than 0.45 eV. Moreover, as it is expected, the device with lower I_{OFF} exhibits higher E_a.

Reliability test results: Experimental data showing the stability of the OTS devices under electrical or thermal stress are shown in Fig. 6 and 7. Fast operating speed and high endurance were confirmed (Fig. 6). The devices based on C-Te and B-Te can repeatedly switch more than 10^8 cycles in maintaining low I_{OFF} while the Al-Te based device shows I_{OFF} degradation after 10^7 cycles. Similarly, high thermal stability up to $450°C$ is shown for C-Te and B-Te based devices while the Al-Te based device cannot endure a temperature of $400°C$ (Fig. 7). The strong stability of the C-Te and B-Te based devices can be due to two points. One is their strong bond strengths that enable the films to delay the telluride crystallization [10]. The other point is the existence of small atoms (B and C) which make a big atomic radius difference in the system. The big atomic difference makes the tellurides difficult to form a crystalline structure [11] and therefore helps stabilization of the film.

Discussion: Performances of binary OTS devices studied in this work and those of Si-Te reported in our previous work [2] are illustrated in Fig. 8. First of all, we found that all of the binary devices with Te composition close to 70 at% showed low leakage current with E_a higher than 0.45 eV. This indicates that adjusting the amount of Te incorporated in the telluride film, which is related to the formation of lone-pairs and traps that affect the OFF-state conduction in OTS devices [7], is more important than selecting elements bonding with the Te in terms of lowering I_{OFF} of the devices. In addition, tellurides with high bond strength and small size of atoms showed good endurance properties (10^8). In particular, tellurides including small atoms showed good OTS device properties even at a temperature higher than $400°C$, which was where even the Si-Te system with strong bond strength started to show I_{OFF} degradation [2]. This indicates adopting telluride systems including atoms with a small radius, which make the whole system hard to form a crystalline structure, is more effective than selecting a system with strong bond strength in order to obtain devices highly resistant to either electrical or thermal stress. The proportional relation between atomic radius difference and thermal stability of various types of Te-based binary OTS devices visualized in Fig. 9 also supports our argument on the importance of choosing telluride system with big atomic size difference to obtain high thermal stability of OTS devices.

Conclusion: We demonstrate that a simple binary telluride OTS devices with high thermal stability and extremely low leakage can be obtained by adopting telluride system with small atomic size of elements and appropriate Te composition. Our findings will give an insight to design of simple binary OTS device with excellent performance without a need of complex material systems.

References: [1] Y. Koo et al., VLSI, p. 86-87, 2016. [2] Y. Koo et al., Electron Device Letters, vol. 38, no. 5, 2017. [3] A. Velea et al., Scientific Reports 7, 8103; 10.1038/s41598-017-08251-z, 2017 [4] H. Y. Cheng et al., IEDM, p. 28-31, 2017 [5] B. Govoreanu et al., VLSI, p. 92-93, 2017 [6] G. Navarro et al., VLSI, p. 94-95, 2017 [7] D. Ielmini et al., Materials Today, vol. 14, no. 12, 2011. [8] N. Avasarala et al., European Solid-State Device Research Conference, p. 168-171, 2017. [9] M. Lee et al., Nature Communications 4, 2629; 10.1038/ncomms3629, 2013. [10] M.H.R. Lankhorst, Journal of Non-Crystalline Solids 297, p. 210-219, 2002. [11] Y. S. Yun et al., Metals and Materials International, vol. 20, no. 1, p. 105-111, 2014

2018 Symposium on VLSI Technology Digest of Technical Papers

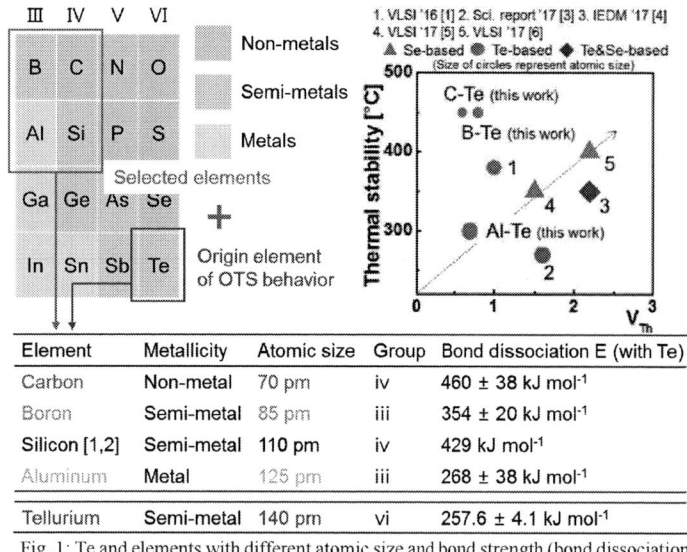

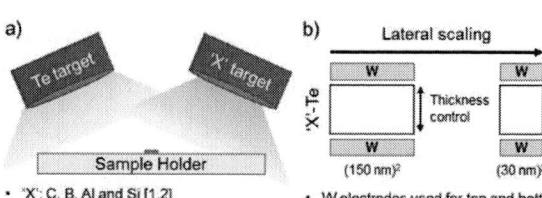

Fig. 2: Co-sputtering method is adopted for telluride film fabrication on various cell sizes of W wafer to investigate binary OTS materials.

Element	Metallicity	Atomic size	Group	Bond dissociation E (with Te)
Carbon	Non-metal	70 pm	iv	460 ± 38 kJ mol⁻¹
Boron	Semi-metal	85 pm	iii	354 ± 20 kJ mol⁻¹
Silicon [1,2]	Semi-metal	110 pm	iv	429 kJ mol⁻¹
Aluminum	Metal	125 pm	iii	268 ± 38 kJ mol⁻¹
Tellurium	Semi-metal	140 pm	vi	257.6 ± 4.1 kJ mol⁻¹

Fig. 1: Te and elements with different atomic size and bond strength (bond dissociation energy) are selected for investigation of various binary OTS materials. The Inset graph shows trade-off between thermal stability and V_{Th} of Te and Se-based OTS devices.

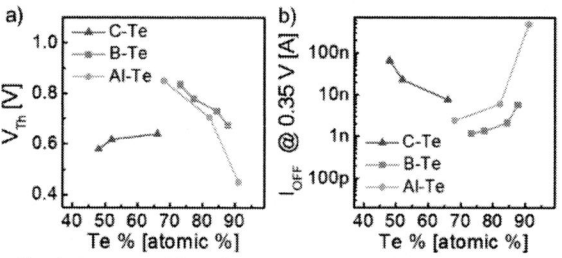

Fig. 3: Te composition effects on V_{Th} and leakage current of OTS devices fabricated in this study are illustrated. All the systems with Te amount close to 70 atomic % results in low leakage current <5 nA.

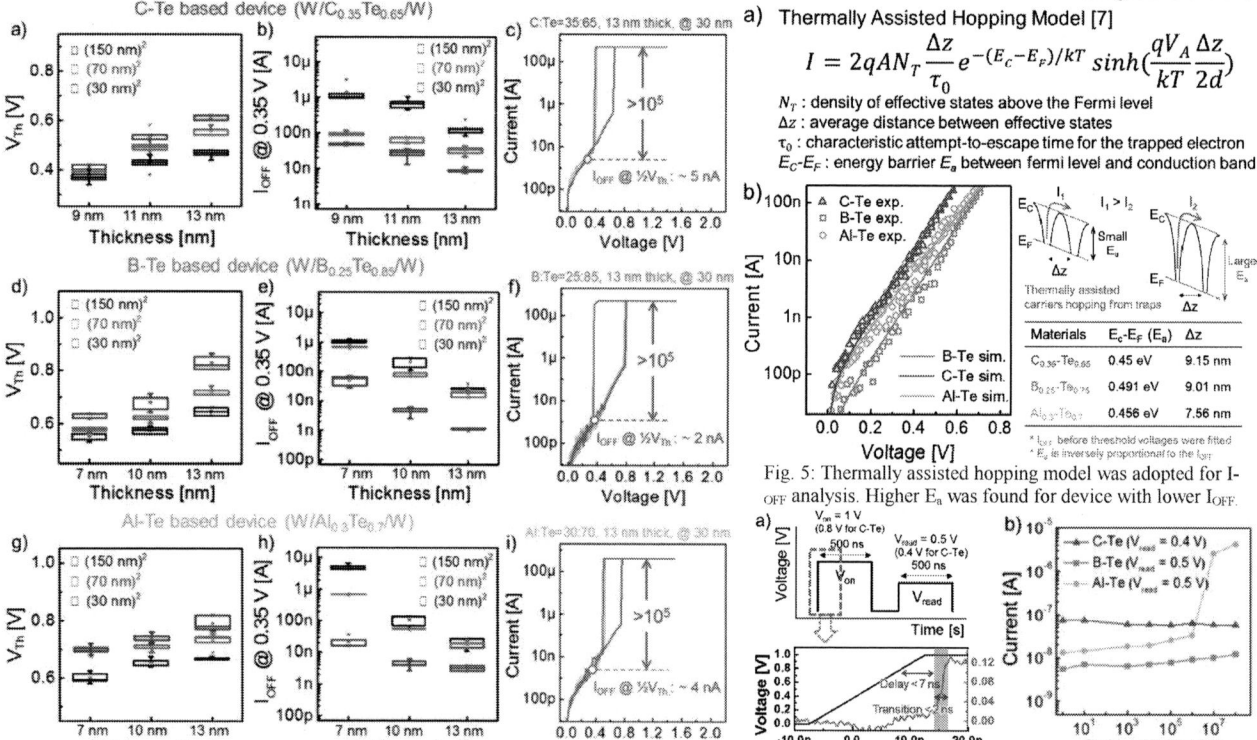

a) Thermally Assisted Hopping Model [7]

$$I = 2qAN_T \frac{\Delta z}{\tau_0} e^{-(E_c - E_F)/kT} sinh(\frac{qV_A}{kT}\frac{\Delta z}{2d})$$

N_T : density of effective states above the Fermi level
Δz : average distance between effective states
τ_0 : characteristic attempt-to-escape time for the trapped electron
$E_C - E_F$: energy barrier E_a between fermi level and conduction band

Materials	$E_c - E_F$ (E_a)	Δz
$C_{0.35}$-$Te_{0.65}$	0.45 eV	9.15 nm
$B_{0.25}$-$Te_{0.75}$	0.491 eV	9.01 nm
$Al_{0.3}$-$Te_{0.7}$	0.456 eV	7.56 nm

* I_{OFF} before threshold voltages were fitted
* E_a is inversely proportional to the I_{OFF}

Fig. 5: Thermally assisted hopping model was adopted for I$_{OFF}$ analysis. Higher E_a was found for device with lower I_{OFF}.

Fig. 4: Device area and thickness dependency of fabricated devices are shown. Devices with thicker film and smaller area show larger V_{Th} and lower I_{OFF}. All of the fabricated devices exhibit extremely low leakage current of <5 nA with high selectivity of >10⁵.

Fig. 6: a) Pulse scheme used for AC endurance test. Fast operating speed is confirmed b) Endurance test results shown for the OTS devices fabricated in this study

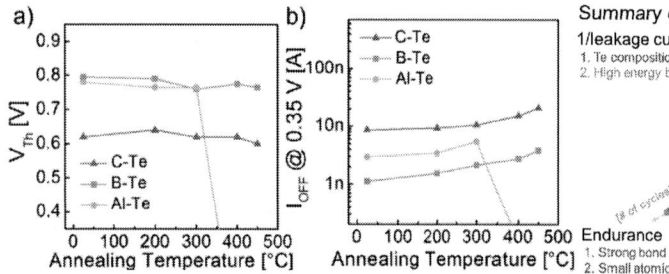

Fig. 7: V_{Th} and I_{OFF} of the OTS devices after thermal treatments at different temperatures are shown. C-Te and B-Te systems show excellent thermal stability of 450°C with minor I_{OFF} degradation.

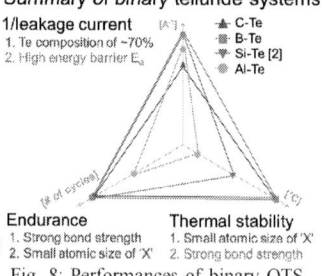

Summary of binary telluride systems
1/leakage current
1. Te composition of ~70%
2. High energy barrier E_a

Endurance
1. Strong bond strength
2. Small atomic size of 'X'

Thermal stability
1. Small atomic size of 'X'
2. Strong bond strength

Fig. 8: Performances of binary OTS devices studied in our group are summarized as a radar plot.

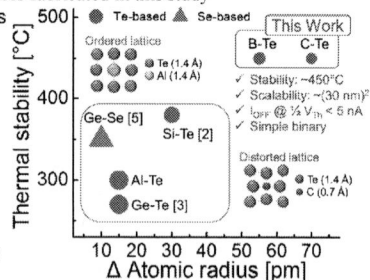

Fig. 9: Relation between thermal stability and atomic radius difference is shown based on this study and reports on binary systems.

978-1-5386-4219-1/18 $31.00 © 2018 IEEE

Half-threshold bias I_{off} reduction down to nA range of thermally and electrically stable high-performance integrated OTS selector, obtained by Se enrichment and N-doping of thin GeSe layers

Naga Sruti Avasarala[#1], G. L. Donadio, T. Witters, K. Opsomer, B. Govoreanu, A. Fantini, S. Clima,
H. Oh, S. Kundu, W. Devulder, M. H. van der Veen, J. Van Houdt[1], M. Heyns[1], L. Goux, G. S. Kar
imec, Leuven, Belgium, [1] also with KU Leuven, Leuven, Belgium [#] Contact: naga.sruti.avasarala@imec.be

Abstract. We report on the reduction of leakage current at half threshold bias ($I_{off1/2}$) down to the 1nA range achieved using Se-enriched or N-doped GeSe. Integrated 50nm OTS devices demonstrated excellent thermal stability up to 600°C, as well as electrical stability (V_{th}, $I_{off1/2}$) when operated at a high on current density of $23MA/cm^2$ for 10^8 cycles.

Introduction. Selector devices are essential for the implementation of dense memory arrays to avoid sneak path issues and disturb of unselected (half-biased) resistance states [1]. In the latter case, they require low off-state (half-bias) current ($I_{off1/2}$), while when selected, they must sustain a high current drive (I_{on}) required to operate the memory devices. These requirements translate into a Non-Linearity ($NL_{1/2}$) factor that will directly determine the array size *(Fig.1)*. Ovonic threshold switching (OTS) [2] materials have gained interest for selector applications because they can sustain high I_{on} [3]. A key attention point of OTS devices is their thermal stability during processing and operation [4].

*Previously we had reported on functional integrated Ge_xSe_{1-x} selector devices [5]. In this work, we report both on reduced $I_{off1/2}$ and drastic improvement of thermal stability, as obtained by using **Se-enriched** composition and by **doping with N**. We also show the impact of composition and operating conditions on electrical performance of 50nm integrated cells.*

Thermal Stability. Ge_xSe_{1-x} films are deposited by sputtering at room temperature while N is incorporated from N_2 in the gas flow. The thermal stability of these amorphous films was measured by in-situ XRD and XRF *(Fig.2a)*. Following the trend shown in [5], 50nm-thick Ge_x1Se_{1-x1} (ref. GeSe composition) layers crystallize at 350°C (T_c). T_c increases further above 400°C, the BEOL integration limit, by doping with N [6]. T_c also increases above 400°C through film thickness (t) reduction below t=20nm. Increasing Se content allows further increase in T_c **above 600°C** *(Fig.2b,c)*. Finally, XRF characterization revealed no composition change after anneal upto 400°C for 90 minutes *(Fig.2c, inset)*.

Electrical Performance. TiN BE was patterned as a pillar which defines the device size down to 50nm. The top stack, TiN/Ge_xSe_{1-x} (t=10nm, 20nm)/TiN *(Fig.3)* is patterned and passivated. Process optimization such as surface treatments improved further the OTS cell module quality and roughness as compared to [5], leading to improved symmetry in bipolar operation and short cycling stabilization phase.

In [7], ab-initio simulations reported that N-doping of Ge-Se layers results in increased mobility gap. This trend is also predicted for increasing Se content, as simulated *(Fig.4a)*. Consistently, *Fig. 4b,c* show that pristine leakage of integrated cells decrease with increasing Se content (also included is Ge-richer (Ge_x0Se_{1-x0}) [5]) or N doping of Ge_x1Se_{1-x1}. Bipolar switching characterization was systematically performed using a triangular pulse with 100ns rise and fall times, using a matched impedance test setup. From this study, threshold voltages (V_{th}) follow the inverse trend of leakage, increasing

with Se content and N doping *(Figs 5,6)*. $Ge_x1Se_{1-x1}N$ displays stable OTS switching at T=25°C, 55°C and 85°C with marginal decrease in V_{th} with increasing temperature.

$I_{off1/2}$-V_{th} trade-off determines the optimum composition and thickness suited for the selected memory element and array size. 20nm $Ge_x1Se_{1-x1}N$ leads to an *$I_{off1/2}$ ~2nA*, resulting in **$NL_{1/2}$ of 10^5** (I_{on}=450µA), an order of magnitude increase compared to the value for Ge_x0Se_{1-x0} reported in [5] and one of the highest values reported so far. This $I_{off1/2}$ makes it possible to realize dense (>64kbit) arrays with Iread ~ 10uA *(Figs.7,8)*. V_{th} can be reduced to ~2V by reducing thickness while confining $I_{off1/2}$ <10nA.

1T1S Current Control. 1Transistor/1Selector (**1T1S**) structures where the transistor precisely controls current drive of the selector (I_{on}) during threshold switching were used to evaluate dependence of I_{off}@0.5V on the maximum operating current I_{on}. Ge_x2Se_{1-x2} shows *I_{on}-independent leakage* while Ge_x0Se_{1-x0} exhibits *leakage proportional to I_{on} (Fig. 9)*. This implies that the Ge_x2Se_{1-x2} material exhibits *electrical stability*, being more robust against operating current stress and thus allows us to further improve $NL_{1/2}$ if using I_{on} > 450µA. Good within wafer uniformity of V_{th} and I_{off}@1V is demonstrated for this Ge_x2Se_{1-x2} composition *(Fig. 10)*.

Cycling. 50nm devices with large on chip resistance (**1S1R**) limiting I_{on} to ~450uA were cycled, while *monitoring the drift in V_{th} and resistance*. $Ge_x1Se_{1-x1}N$ can be cycled upto 10^8 cycles with a 100ns square pulse of 4.5V with majority of devices retaining their resistance state (negligable I_{off} drift) with a $NL_{1/2}$ of 10^5 *(Fig.11)* after cycling and mean V_{th} drift of 62mV/decade (Initial decrease in V_{th} of 1V occurs during the first 5 cycles after which it stabilises). Ge_x2Se_{1-x2} is cycled with a 100ns square pulse of 3.5V with devices retaining their initial I_{off} up to 10^6 cycles, followed by a increase in I_{off} which remains in the nA range (I_{off} drift = 0.1nA/decade) and continued cycling till 10^8 cycles with drift in V_{th} of 37mV/decade *(Fig.12)*.

Conclusions. We demonstrated half-threshold bias leakage well below 10nA and thermal stability of at least 600°C using Se-enriched or N-doped GeSe OTS bipolar selectors with device sizes scaled to 50nm displaying 10^5 non-linearity ($NL_{1/2}$), $23MA/cm^2$ current drive and 10^8 cycle endurance with low V_{th} drift of 37mV/ decade during cycling. Increase in V_{th} generally associated with lower $I_{off1/2}$ can be reduced by reducing thickness. Se-enriched Ge-Se devices were shown to have leakage independent of I_{on} making them ideal for memories requiring high write currents.

REFERENCES

[1] G. W. Burr, J. Vac. Sci. Technol. B, vol. 32, no. 4, 2014; [2] S. R. Ovshinsky, Phys. Rev. Lett., 21(2), 1450-52, 1968; [3] H. W. Ahn et al, ECS Solid St Lett, 2(9) N31-33, 2013; [4] H. Y. Cheng et al, IEDM 2017; [5] B. Govoreanu et al, VLSI 2017; [6] N.S. Avasarala et.al, ESSDERC 2017; [7] S. Clima et.al, IEDM Tech. Digest, 2017.

2018 Symposium on VLSI Technology Digest of Technical Papers

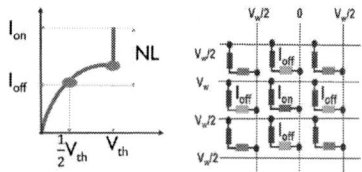
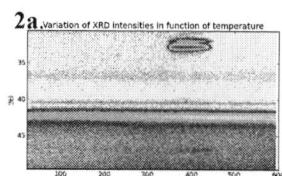
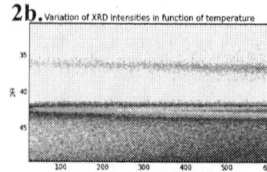
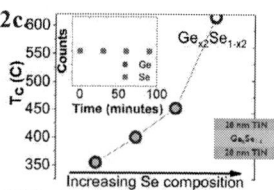

Fig.1. Selector restricts the array leakage while providing current drive to program the selected memory cell. Cell scalability is limited by Ion while array size is limited by $I_{off1/2}$ @ half-bias

Fig.2. Thermal stability measured for TiN capped Ge-Se samples by in-situ XRD; **(a)** $Ge_{x1}Se_{1-x1}$ films shows first instability at 350°C; **(b)** No visible crystallization in XRD until 600°C for the highest Se content tested $Ge_{x2}Se_{1-x2}$; **(c)** Increasing Tc with Se content; (Inset-top): XRF performed after upto 90 minuted at 400°C for capped $Ge_{x2}Se_{1-x2}$ evidencing composition stability for $Ge_{x2}Se_{1-x2}$; (Inset-bottom) Schematic of sample

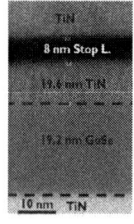
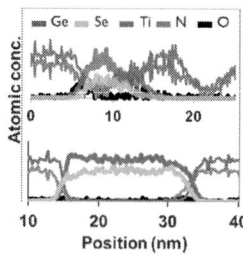
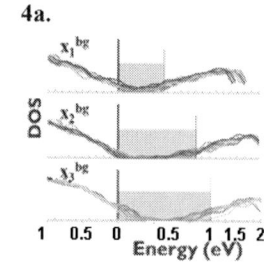
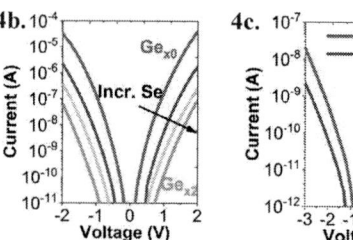

Fig.3a HRTEM displaying smooth chalcogenide layer and sharp top and bottom TiN interfaces

Fig. 3b EDS line scans before (top) and after (bottom) surface treatment showing reduced oxygen at the bottom interface and increased composition homogenity

Fig.4a. Ab-initio simulations predict increasing mobility gap for increasing Se content (Ge: x_1^{bg} > x_2^{bg}> x_3^{bg})

Fig 4b,c. Pristine leakage of 1um-cells decreases with **(b)** Increasing Se content ranging from Ge rich composition ($Ge_{x0}Se_{1-x0}$) to Se-rich ($Ge_{x2}Se_{1-x2}$) **(c)** N doping of $Ge_{x1}Se_{1-x1}$, consistently with increasing mobility gap

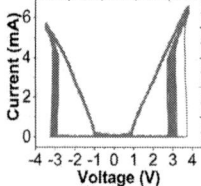
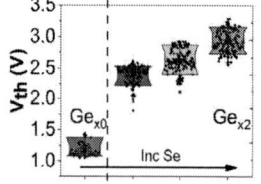
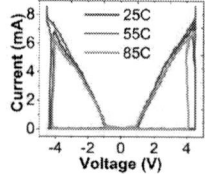
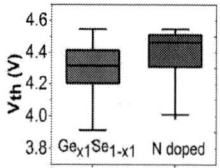
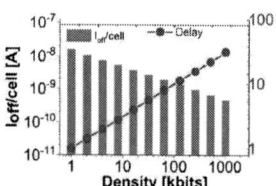

Fig.5a. Switching curves of 10nm $Ge_{x2}Se_{1-x2}$ for 50 bipolar cycles

Fig. 5b. Box plots indicate increase in V_{th} with increasing Se content. Composition ranges from $Ge_{x0}Se_{1-x0}$ to Se richer $Ge_{x2}Se_{1-x2}$ for 10nm films

Fig. 6a. Median switching curves of 20nm $Ge_{x1}Se_{1-x1}N$ at different T (~4.5V at 25°C)

Fig. 6b. There is an increase in V_{th} with N doping

Fig.7: Estimated $I_{off1/2}$ requirement per cell for array size up to 1Mbit. Delay is calculated by cell per bit line relative to value for 1kbit

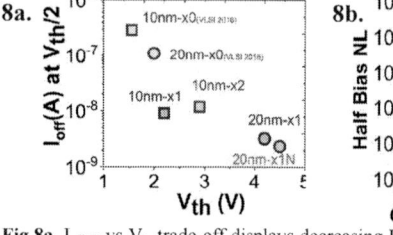
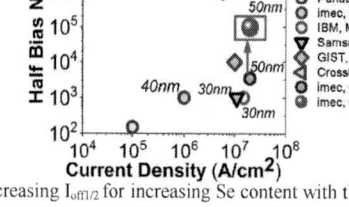
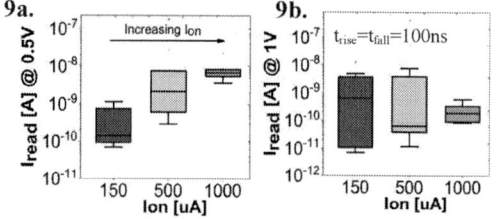

Fig.8a. $I_{off1/2}$ vs V_{th} trade-off displays decreasing $I_{off1/2}$ for increasing Se content with the lowest value for 20nm N doped $Ge_{x1}Se_{1-x1}N$. The increase in V_{th} can be offset by reducing thickness while restricting $I_{off1/2}$ <10nA; **8b.** Benchmarking of selector performance clearly indicating the improvement in $NL_{1/2}$ achieved in this work using 20nm $Ge_{x1}Se_{1-x1}N$ and Ion=450μA

Fig.9. Effect of current drive (I_{on}) on leakage of 1T1S devices **(a)** $Ge_{x0}Se_{1-x0}$- Leakage increases with I_{on} **(b)** $Ge_{x2}Se_{1-x2}$- Leakage is independent of I_{on} allowing it to increase further $NL_{1/2}$ using higher I_{on}

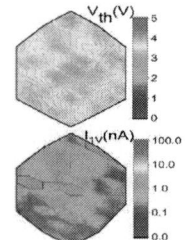
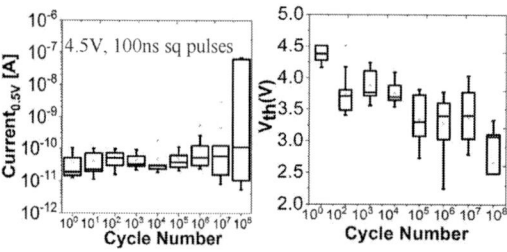
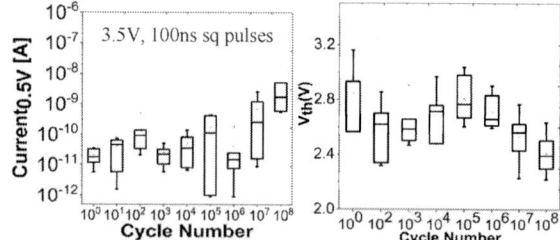

Fig. 10: Distribution of V_{th}, and leakage for 10nm $Ge_{x2}Se_{1-x2}$

Fig.11 Endurance test for 20nm thick 50nm size $Ge_{x1}Se_{1-x1}N$ devices. Most devices retain their original resistance state after 10^8 cycle with a drift in V_{th} of 0.5V after 10^8 cycles. (An initial drop in V_{th} of 1V occurs during the first 5 cycles).

Fig.12: Endurance test for 10nm thick 50nm size $Ge_{x2}Se_{1-x2}$ devices. Devices retain their original resistance state for 10^6 cycles and exhibit a slight decrease in I_{off} upto the nA level while V_{th} drifts by 0.3V after 10^8 cycles

978-1-5386-4219-1/18 $31.00 © 2018 IEEE

Gap in pagination due to formatting issues.

Pages 211-212

Highly Manufacturable Low Power and High Performance 11LPP Platform Technology for Mobile and GPU Applications

H.-J. Kim, B.H. Choi, Y.H. Lee, J.H. Ahn, Y.S. Bang, Y.D. Lim, J.H. Do, J.H. Jung, T.J. Song,

Y. Yasuda-Masuoka, K.C. Park, S.D. Kwon, and J.S. Yoon

Foundry Division, Samsung Electronics, Yongin City, Korea, email: hyunjo_kim@samsung.com

Abstract

11nm bulk FinFET process employing 3rd generation 14nm FEOL and 10nm BEOL process has been successfully demonstrated with updated design rules for optimal design kit support with 6.75T library. Compared to 14nm 1st generation FinFET, device performance has been improved by 25% in ring oscillator AC frequency at same Iddq or 42% power reduction is achieved. Adopting already mature 14nm and 10nm process technology, we can setup and demonstrate fast yield ramp.

Introduction

FinFET-based logic process technologies have been successfully in production thanks to superior scalability, low power and high performance benefits at 14nm [1]. 2nd generation 14nm FinFET process (14LPP) has improvement in device performance by ~10% compared to 1st generation (14LPE). In 11LPP, with 3rd generation Fin process combined with S/D engineering, DC performance gain becomes ~36% vs. 1st generation. We provide most competitive device performance and the most aggressive gate pitch (78nm) at >10nm technology nodes for foundry business.

Process Architecture

11LPP is defined as the combination of 3rd generation 14nm FinFET and advanced BEOL process for Metal-2 (M2) layer pitch reduction to emphasize DTCO to enable smaller track without process risks. This combination is especially useful for standard cell library scaling since M2 layer is used for signal pick-up. 48nm pitched M2 layer made 6.75T standard cell library enabled.

Table 1 summarizes the key features of 11LPP CMOS FinFET technology in comparison with 14nm technology. Key process knobs used for 11LPP transistor performance are Fin width scaling and highly-doped S/D epitaxy. The 3rd generation 14nm Fin becomes more vertical and narrow compared to 2nd generation. (Fig.1) Narrow Fin provides better short channel behavior to support shorter gate length. Fig.2 shows core device DIBL becomes less than 30mV in 11LPP. An upgraded multi-eWF gate stack is also adapted to enable ultra-low-power (ULP) HVT transistors. 11LPP FEOL design rules are compatible to those of 14LPP. However, some minor changes are needed for BEOL 1x layers, where 10nm metallization process is used to support 48nm M2 pitch [2]. Since 11LPP uses volume-production verified process from both 14nm and 10nm, no yield risk is expected and already verified <2% gap to the current 14nm mass production.

Device performance

Ring oscillator AC performance improvement is clearly observed in Fig.3. Compared to 2nd generation 14nm FinFET process, 11LPP provides 14% higher frequency at same Iddq.

Another feature is that SLVT device is positioned in faster side compared to 14nm. We can reach higher frequency than 2nd generation by 20% using 11LPP SLVT. Fig. 4~5 show DC nFET and pFET Idoff vs. Ieff correlation. Compared to 14LPP, 17% (nFET) and 21% (pFET) DC performance gain is observed. HVT is an optional device which can provide ~10pA/um Idoff at 0.7V Vdd, which is at comparable level to the previously published ULP devices [3]. SRAM cell characteristics are also improved. Fig.6 shows 25% higher Iread than that of 14nm. This gain results in better Vmin performance in SRAM circuit blocks. (Fig.7) An improved Vth mismatch also contributes to Vmin gain. (Fig.8)

Thanks to those transistor improvements, product level gain is confirmed in Vmin characteristics. Fig.9 shows that ~70mV Vmin gain is achieved from a CPU block of a mobile AP (application processor) product. In-wafer device variation measured from CPU Iddq distribution is depicted in Fig.10, which shows better uniformity in chip level. Even with significant performance boosting, variation is controlled tighter than 2nd generation due to better DIBL and within-wafer process control.

6.75T Standard Cell Library

To provide the optimum power, performance, and area (PPA) to designers, 11LPP offers the ultra-high-density (UHD) standard cell of 6.75T with 25% height scaling versus 14LPP 9T. To implement UHD standard cell, design rules are optimized for Via and Metal. Moreover, M2 pitch is shrunk from 64nm to 48nm, which is essential to support the enough routing resource with smaller pin solution. As a result, UHD shows more than 15% block-scaling versus 14LPP 9T standard cell. Table 2 shows a benchmark evaluation result for 6.75T library. The combination of speed-oriented 9T and area-and-power-oriented 6.75T library would provide a better chance to maximize PPA in 11LPP technology.

Summary

3rd generation 14nm FinFET process adopted new performance knobs including scaled Fin width and gate length, and higher doping in S/D epitaxy. In addition to those knobs, 11LPP adopted 48nm pitch at M2 layer which is used in 10nm process. 11nm UHD standard cell library features 0.85X logic area and 35% power reduction. It provides superior performance and power consumption advantage for next generation high-end mobile computing, network, consumer, and GPU applications.

References

[1] E.-Y. Jeong et al., VLSI Tech. Dig., pp. 142-143, 2017.
[2] H.-J. Cho et al., VLSI Tech. Dig., pp. 1-2, 2016.
[3] C.-H. Jan et al., VLSI Tech. Dig., pp. 12-13, 2015.

	14LPE	14LPP	11LPP
Fin, SD process	1st gen	2nd gen	3rd gen
Gate pitch	78nm	78nm	78nm
Mx pitch	64nm	64nm	64/48nm
Device perf.(AC)	100	108	123

Table.1 Key features of 11nm technology. 11LPP uses reduced Metal-2 pitch for scaling and 3rd generation 14nm Fin process for device performance boost.

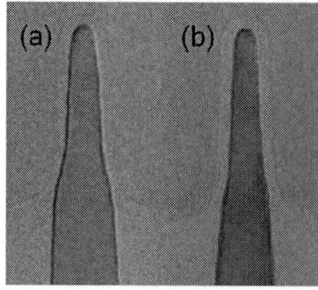

Fig.1 (a) 2nd generation 14nm Fin, (b) 3rd generation 14nm Fin

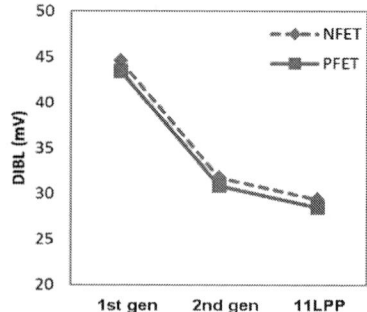

Fig.2 Core device DIBL.

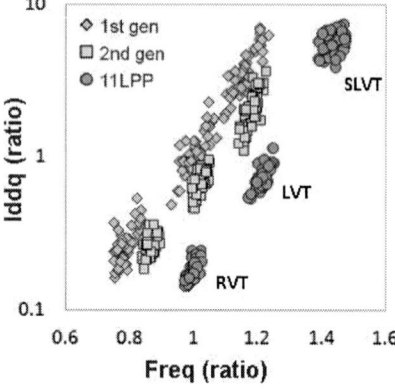

Fig.3 AC performance improvement in 11LPP (INV RO, F/O=3)

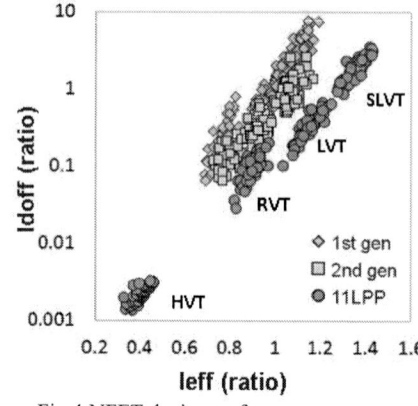

Fig.4 NFET device performance.

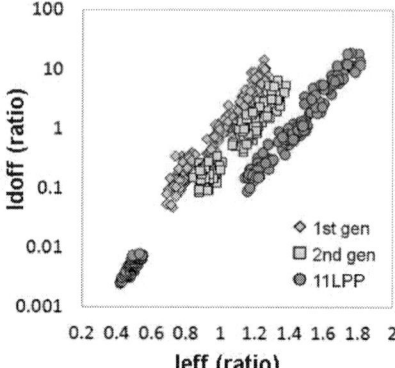

Fig.5 PFET device performance

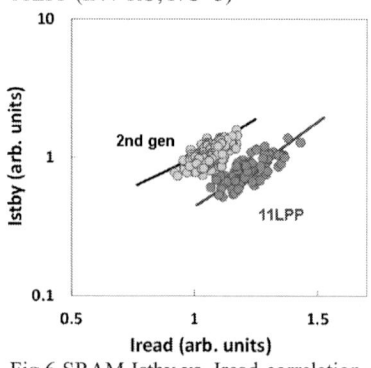

Fig.6 SRAM Istby vs. Iread correlation, showing performance boost in 11LPP.

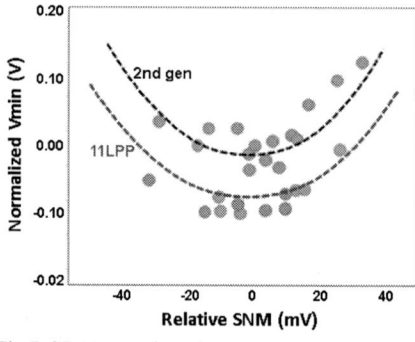

Fig.7 SRAM Vmin gain obtained from SRAM memory block characteristics.

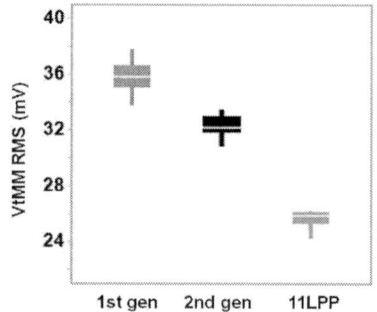

Fig.8 SRAM Vth mismatch data from bit-cell type test keys.

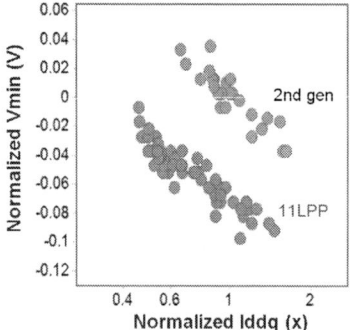

Fig.9 Chip Vmin and Iddq characteristics measured from a mobile processor CPU.

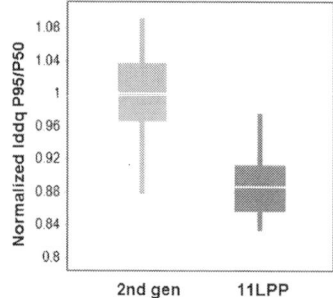

Fig.10 Device variation comparison between 14nm and 11nm device. 95percentile divided by 50percentile ratio is used as a metric.

Item	14LPP		11LPP
CPP [nm] (X)	78	78	84
Std. cell track (Y)	9T	9T	6.75T
Unit cell area (X*Y)	1.00	1.00	0.81
Power	1.00	0.86	0.65

Table 2. Standard cell comparison. UHD (6.75T) standard cell provides power and area saving option in 11LPP.

A 12nm FinFET Technology Featuring 2nd Generation FinFET for Low Power and High Performance Applications

H.C. Lo, D. Choi, Y. Hu, Y. Shen, Y. Qi, J. Peng, D. Zhou, M. Mohan, C. Yong, H. Zhan, H. Wei, X. He, D. Kang, A. Sirman, Y. Wang, H. Zang, S.Y. Mun, A. Vinslava, W.H. Chen, C. Gaire, J. Liu, X. Dou, Y. Shi, P. Zhao, B. Zhu, A. Jha, X. Zhang, X. Wan, E. Lavigne, C. Kyono, M. Togo, J. Versaggi, H. Yu, O. Hu, J.G. Lee, S. B. Samavedam, D.K. Sohn

Advanced Technology Development Div, GLOBALFOUNDRIES, Malta, USA, E-mail: hsien-ching.lo@globalfoundries.com

Abstract

We present a state-of-art 12LP FinFET technology with PPA (Performance, Power, and Area) improvement over 14LPP. 12LP enables >10% area reduction including a 7.5T library and 16% power reduction at fixed frequency or a 15% performance improvement at given leakage over 14LPP with comparable reliability and yield. In addition, SRAMs benefit from a 30% leakage reduction at the same Iread. 12LP extends the 14nm technology with compelling performance and area scaling.

Keywords: FinFET, 12LP, 7.5T library, PPA

Introduction

16/14nm FinFET technologies are expected to be long-lived manufacturing nodes being at the sweet spot for performance, scaling and complexity relative to 28nm and 10nm/7nm. Multiple market segments will benefit from enhanced performance, area scaling, new features and cost reduction on these offerings. Starting with the 14LPP technology platform [1-4], the 12LP technology demonstrates a significant DC and AC performance improvement over 14LPP with device architecture and process enhancements. A new 7.5T library with special constructs and pushed rules was created to enable further logic scaling. The 12LP processes were optimized to ensure equivalent reliability and product yield to 14LPP.

12LP Features

The reference for 12LP device architecture enhancements was 14LPP CMOS baseline described in previous publications [1-4]. The transistors with performance elements described in sections below achieve 15% faster ring oscillator AC performance (Fig.1) and 6%/22% higher NFET/PFET DC performance compared to 14LPP (Fig. 2). Fig. 3 shows that 12LP 7.5T standard cells consume 16% less total power compared to 14LPP 9T cells at equivalent speed. In addition, 12LP logic library achieves 12% area reduction over 14LPP (Table 1). 12LP enables new constructs and pushed rules, resulting in an area shrink, but maintains compatibility with 14LPP.

Performance Enhancement Elements

A. Fin optimization: Fin profile optimization enables the thinner and taller fins for higher drive current and improved short channel control. In addition, fin surface roughness reduction resulted in carrier mobility increase by 6% and 9% for NFET and PFET respectively (Fig. 4).

B. PFET source-drain cavity optimization: The benefit of optimized source-drain eSiGe cavity on PFET is illustrated by the TCAD simulations. Fig. 5a shows the device performance of the control cavity and the optimized cavity as a function of the cavity depth. For the control group, the device performance

first increases and then decreases from combined effects of larger eSiGe and higher leakage as cavity depth increases. The optimized cavity achieves a 4% higher device performance at comparable leakage (Fig. 5b).

C. PEFT source-drain SiGe epitaxy optimization: Similar to planar technology, in-situ boron-doped embedded SiGe (eSiGe) epitaxy suffers from pattern-loading effects in FinFET technology. 40-fin and SDB (single poly diffusion break) devices experience relatively lower performance as SiGe grows slower in these dense macros. In this work, by carefully optimizing the process conditions of SiGe epitaxy, the loading effects were significantly improved. 40-fin and SDB devices show 4% and 5% improvement respectively while 2-fin devices have similar performance (Fig 6).

D. NFET source-drain Si-P epitaxy optimization: Phosphorus doping concentration is key for source-drain (S/D) resistance reduction. The active dopant density was increased by optimizing the Si-P epitaxial process, which leads to S/D resistance reduction. This process enabled 6% NFET Ron reduction at same device overlap (Fig. 7).

E. Contact Resistance Reduction: Contact resistance optimization is critical overall external resistance reduction for high drive current. The trench contact profile was optimized to have an enlarged bottom contact size for a bigger contact area while maintains the contact to gate isolation. The doping profile under the trench contact was optimized to reduce the contact barrier height. Through combination of these two processes, contact resistance reduction of 12% for NFET and 35% for PFETs was achieved (Fig. 8).

SRAM, Reliability and Product Yield

12LP SRAM FETs shows comparable Vt mismatch (Vtmm) to 14LPP and 30% Istby reduction at same Iread (Fig. 9). The reliability for thin oxide and thick oxide (I/O) devices on 12LP are comparable or better than 14LPP (Table2). 12LP based product yields are comparable to 14LPP as shown in Fig. 10.

Conclusion

A highly manufacturable and reliable 12LP FinFET technology is presented. 15% AC performance improvement and 30% SRAM leakage improvement over 14LPP came from optimizing device architecture in the fin, junctions, and contact modules. A new 7.5T library with special constructs and aggressive rules enables 12% logic area scaling and 16% reduction in total power compared to 14LPP.

References

[1] J. Singh, et al, VLSI Tech. Symp. Dig., pp140-141, 2017
[2] Y. Qi, et al, ECS Trans., Vol.75, issue 8, pp256-272, 2016
[3] H. Lo, et al, Solid State Sci. Technol., Vol. 6 issue 8, pp137-141, 2017
[4] J. Peng, et al, Semicond. Sci. Technol., Vol. 32 No.9, pp1-6, 2017

2018 Symposium on VLSI Technology Digest of Technical Papers

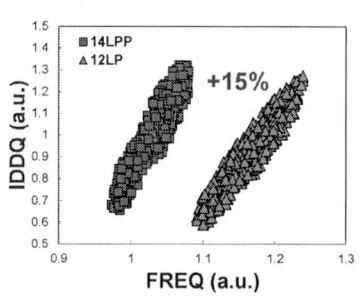

Fig. 1 12LP ring oscillator performance achieves 15% improvement over 14LPP technology.

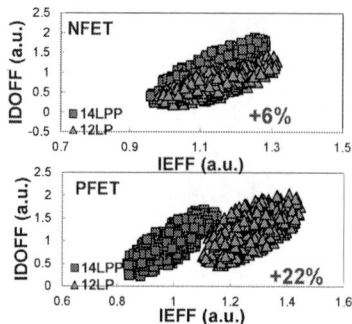

Fig. 2 NFET and PFET DC performance improve 6% and 22% with all elements.

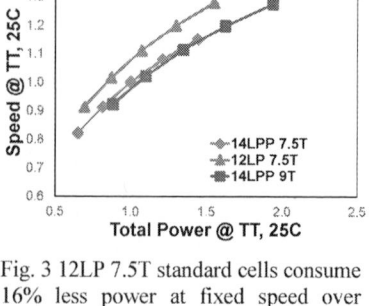

Fig. 3 12LP 7.5T standard cells consume 16% less power at fixed speed over 14LPP 9T.

TABLE 1
14LPP 9T v.s. 12LP 7.5T STD CELL

Parameter	14LPP	12LP
CPP (nm)	84	84
M2 pitch (nm)	64	64
Track	9T	7.5T
	(576nm)	(480nm)
Logical area (%)	1	0.88

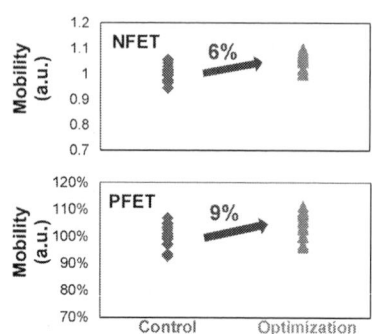

Fig. 4 NFET and PFET mobility improves by 6% and 9%, respectively with fin surface roughness reduction.

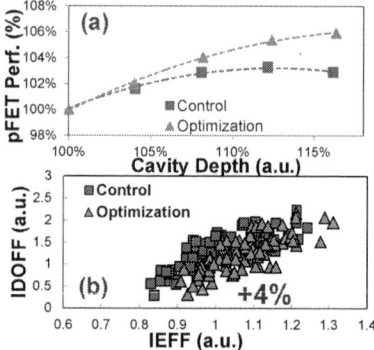

Fig. 5 The optimized cavity minimizes the source/drain leakage and improves PFET DC performance. (a) TCAD (b) Silicon data.

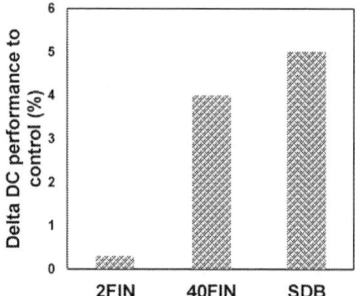

Fig. 6 2-fin, 40-fin and SDB PFET devices performance with optimized eSiGe

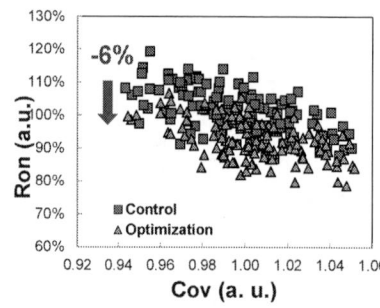

Fig. 7 NFET Ron reduction by 6% with eSD process optimization.

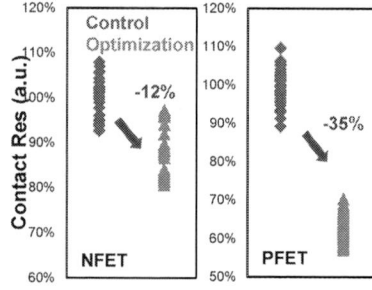

Fig. 8 Contact resistance improves 12% and 35% after optimization.

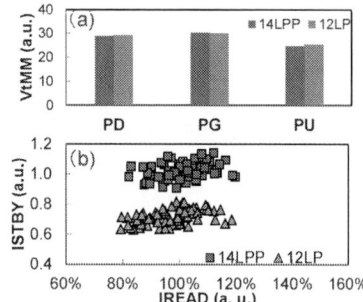

Fig. 9 SRAM Vtmm (a) and Istby vs. Iread (b) between 12LP and 14LPP.

TABLE 2
RELIABILITY FOR 12LP AND 14LPP

Parameter	14LPP	12LP
NFET TDDB (a.u.)	1	1.02
PFET TDDB (a.u.)	1	1.01
NFET BTI (a.u.)	1	1.00
PFET BTI (a.u.)	1	0.98
I/O NFET HCI (a.u.)	1	0.95
I/O PFET HCI (a.u.)	1	0.94

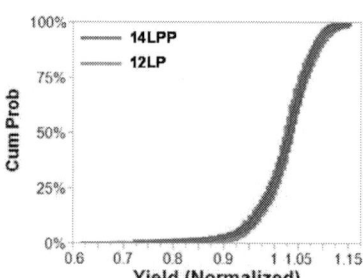

Fig. 10 Comparable product yield between 12LP and 14LPP

978-1-5386-4219-1/18 $31.00 © 2018 IEEE

8LPP Logic Platform Technology for Cost-Effective High Volume Manufacturing

Hwasung Rhee, Ilryong Kim, Jaehun Jeong, Nakjin Son, Heebum Hong, Sungil Cho, Yongmin Park, Dongwoo Kim, Yunki Choi, Jeonghoon Ahn, Sung Gun Kang, Kyunghwan Yeo, Jungtae Kim, Euncheol Lee, Jong Mil Youn, Jong Shik Yoon

Foundry Division, Samsung Electronics, Yongin City, Korea, email: hsrhee@samsung.com

Abstract

8LPP logic platform technology supports mobile and high-performance and lower power application especially for mobile, artificial intelligence (AI), and cryptocurrency devices. 8LPP is employing the evolutionary generation of bulk FinFET FEOL and 44nm EUV-less multi-patterning BEOL process, resulting in 7% power reduction and ~15% area scaling compared with the previous 10LPP. The cost-effective high volume manufacturing is achieved with the minimum additional critical layers and the comparable process steps over the current high volume 10nm production.

Introduction

FinFET-based logic process technologies have been successfully used for logic platform architecture thanks to superior scalability, low power and high performance benefits [1]. Recently, the further scaling below 10nm meets a lot of challenges for multi-patterning complexity before EUV solution. In order to support cost-effective high-volume production, 8LPP is introduced to maximize the merit of well-established 10LPP production line. 44nm pitch 1x BEOL usage showed 15% area reduction by uHD cell [2]. When combining this extended multi-patterning BEOL and the further scaled FEOL process, 8LPP logic platform strongly satisfy the increasing market needs for the cost-effective, high performance, and lower power consumption. 8LPP also can fill the technology gap between the next cutting-edge technology node with high-cost production and the current 10nm line. We provide most competitive device performance and its ultra-lower operation voltage characteristics down to 0.35V, required for newly growing IOT and cyptocurrency application.

Process Architecture

8LPP key dimension feature frame is using the extended 44nm pitch 1x BEOL, resulting in 15% cell area reduction (table 1). It also has 10LPP HD cell as well, to use the previous 10LPP IP again with minor adjustment or re-characterization for fabless cost-benefit. Local layout effect (LLE) aware diffusion break usage is provided as well. The evolutionary FEOL process is used to improve device performance and to manage variation from the further scaled FinFET dimension (Table 2). The power reduction is the key benefit of 8LPP, mainly coming from DC resistance and capacitance reduction. The gate length is kept scaling by ~5% to get capacitance reduction and Fin shape is adjusted more vertically to control sub-threshold leakage, 4th generation RMG stack is used for Vt centering and increasing drive-current. Contact resistance is decreased by thither N and PFET optimization, separately.

Device performance

Fig.1 and Fig.2 show NFET and PFET Ieff-Idoff performance correlations, resulting in 8% and 4% DC gain, respectively. NFET is lifted up by source-drain and contact engineering. Higher doped eSD and hybrid contact reduces Rch and Rext at the same time. PFET is boosted by optimizing eSiGe and contact process. Fig. 3 shows contact resistance reduction in 8LPP, 15 and 9% less Rc for N and PFET, respectively. The hole mobility enhancement is quite limited on narrow and vertical Fin even with higher SiGe:B. We utilized 4th HK/MG stack to increase drive-current as well. Fig. 4 shows 8LPP ring oscillator (RO) AC performance improvement compared with 10LPP under the exactly same testing layout. RO gain 8% is more than DCR gain (~6%), which comes from capacitance reduction. Ceff is suppressed by 4%. Scaled Lg strongly contributes to Ceff reduction. Then, Lg scaling might cause larger variation. NFET Vt variation is slightly increased according to Lg scaling. However, the increased NFET Vt variation is much lower compared with PFET. Fig. 6 shows that PFET Vt variation is much higher degraded with Lg scaling. The uniformity knob dramatically reduced PFET Vt variation even lower than 10LPP. Since PFET is the main contributor for leakage and variation, the variation control is the key factor for the further scaling. The process knobs were designed to optimize source/drain etch and eSiGe on local and global variation.

Lower Voltage Operation

To provide the lower power, operation voltage reduction can be one of solution. Therefore, the lower Vop characteristic is very important. Fig. 7 and Fig. 8 show NFET and PFET Ieff with the different Vdd . The model describes 0.3V and Si data is correlated within model target. In addition to the lower voltage operation, single Fin usage can effectively reduce power. Since single Fin has larger current fluctuation, its variation should be carefully managed. Fig. 9 and Fig. 10 shows NFET and PFET Idsat variation depending on the number of Fin, respectively. Fig. 11 shows 10% RO gain with lower operation voltage 0.4V, indicating the strong benefit on 8LPP. Vmin and Ids correlation plot is shown in Fig. 12, demonstrating the enhanced Vmin 10% gain (power gain) over 10LPP. 8LPP process gain on Vmin is compared with the same 10LPP design, resulting in the same level of product D0 yield. 8LPP new product will show much better than this gain with design optimization.

Summary

8LPP logic platform technology is introduced to provide mobile and high-performance and lower power devices. 8LPP is employing the evolutionary generation of bulk FinFET FEOL and 44nm 1x BEOL process, demonstrating in 7% power reduction and ~15% area scaling. 8LPP is the promising logic platform technology to fill the gap between 10nm and 7nm, with the strong cost-effective merit of high volume production

References

[1] H.-J. Cho et. al., VLSI Tech. Dig., pp. 1-2, 2016.
[2] W. C. Jeong et. al., VLSI Tech. Dig., T11-3, 2017.

2018 Symposium on VLSI Technology Digest of Technical Papers

FINFET Logic Technology	8nm (8LPP)	10nm (10LPP)	14nm (14LPP)
Minimum Gate pitch	64nm	64nm	78nm
Minimum Contact Pitch	64nm	64nm	78nm
Minimum BEOL Pitch	44nm	48nm	64nm
Area Reduction	X0.59	X0.65	Reference
Standard Cell	uHD (-15%), HD	HD	HD
Diffusion Break	LLE Aware DB	SDB/DDB	SDB/DDB

Table 1. 8LPP key platform dimension feature comparison with the previous low power performance logic technology.

Process Architecture	8nm (LPP)	10nm (LPP)	14nm (LPP)
FIN	5th Gen VF	4th Gen	2nd Gen
Gate /Vt	Scaled Lg 4th HK/MG	3rd HK/MG	2nd HK/MG
S/D Engineering	5th Gen	4th Gen	2nd Gen
Contact	3rd Gen Hybrid N/P	2nd Gen Hybrid N/P	1st Gen

Table 2. 8LPP process architecture to maximize the device performance and power reduction. The scaled gate length is the key driver to achieve lower capacitance.

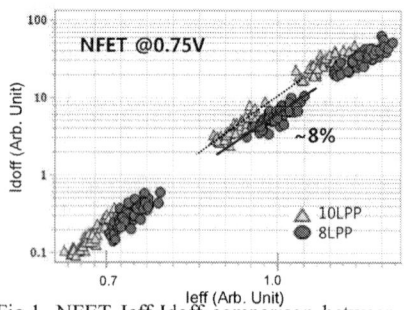

Fig.1. NFET Ieff-Idoff comparison between 8LPP and 10LPP, improving 5% from 10LPP.

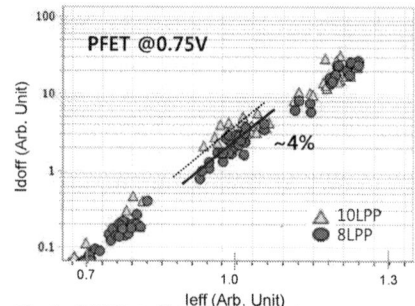

Fig.2. PFET Ieff-Idoff comparison between 8LPP and 10LPP, improving 5% from

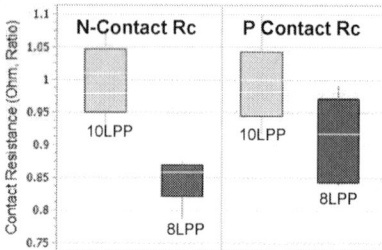

Fig.3. Contact resistance reduction for 8LPP compared with 10LPP, showing 15% and 9% reduction N and PFET, respectively.

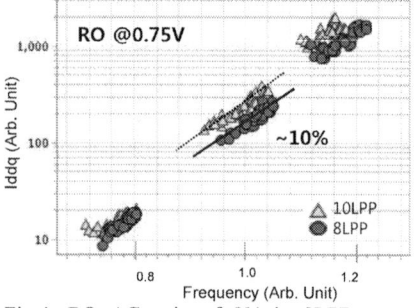

Fig.4. RO AC gain of 8% in 8LPP over 10LPP, reducing DCR and Ceff.

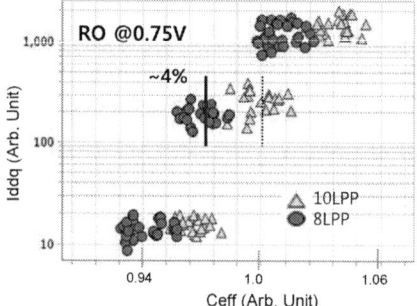

Fig.5. Ceff reduction by scaled gate length (gate cap) and Fin optimization (parasitic cap).

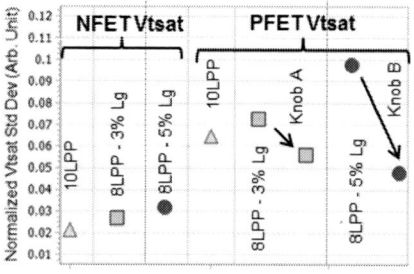

Fig.6. Normalized Vt sigma comparison from 10LPP to 8LPP. PFET Vt variation is significantly reduced for 8LPP by S/D tuning.

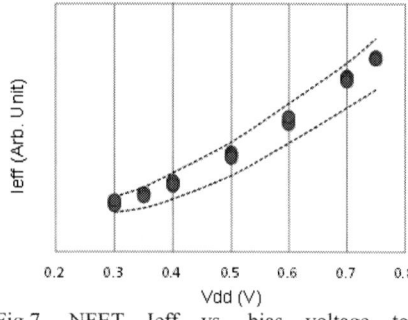

Fig.7. NFET Ieff vs. bias voltage to correspond with model, supporting lower Vdd.

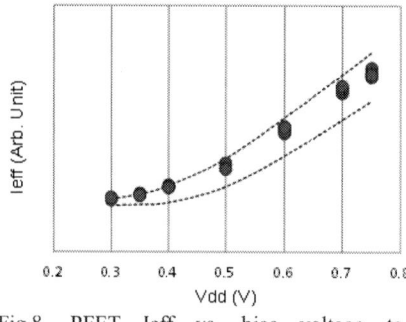

Fig.8. PFET Ieff vs. bias voltage to correspond with model, supporting lower Vdd

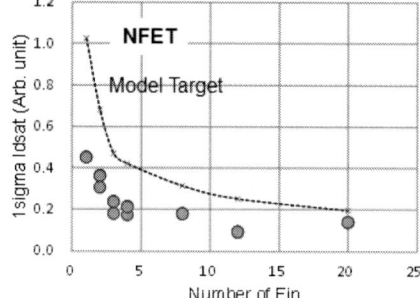

Fig.9. NFET Idsat variation for discrete fin numbers including single fin.

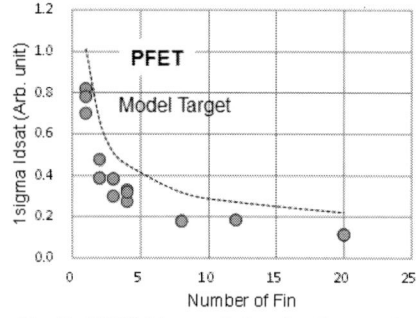

Fig.10. PFET Idsat variation for discrete fin numbers including single fin.

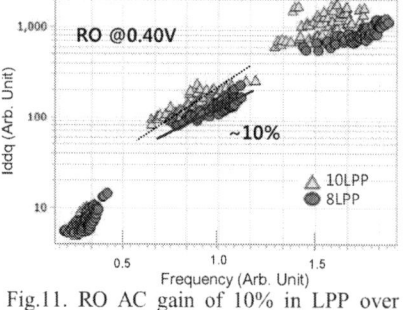

Fig.11. RO AC gain of 10% in LPP over 10LPP, indicating much better benefit for lower Vdd operation.

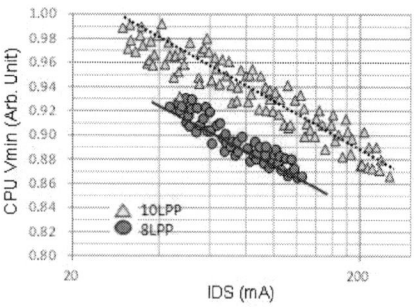

Fig.12. Product CPU Vmin vs. IDS correlation, resulting in lower Vmin gain.

978-1-5386-4219-1/18 $31.00 © 2018 IEEE

High Performance Mobile SoC Productization with Second-Generation 10-nm FinFET Technology and Extension to 8-nm Scaling

Jun Yuan, Ken Rim, Ying Chen, Ming Cai, Youseok Suh, Jihong Choi, Jie Deng, Jerry Bao, Zhimin Song, Lixin Ge, Hao Wang, Xiao-Yong Wang, Vicki Lin, Chihwei Kuo, Sam Yang, Ashwin Rabindranath, Shrihari Siva, Prasad Bhadri, Sungwon Kim*, Kwon Lee*, Soon Cho*, Sunggun Kang*, Saechoon Oh*, S. D. Kwon*, Xiangdong Chen, Paul Penzes, Parag Agashe, William Miller, and P. R. Chidambaram

Qualcomm Technologies Incorporated, 5775 Morehouse Drive, San Diego, CA 92121, USA; *Samsung Electronics, Korea
Email: juny@qti.qualcomm.com

Abstract

We report on Snapdragon™ SDM845 mobile SoC in mass production with a second-generation 10-nm finFET technology. SDM845 exhibits 30–40% CPU/GPU performance gain over SDM835 (first-generation 10-nm finFET process) together with ~10% battery life increase driven by new design features and technology improvements in both transistor performance and uniformity, enabling high performance and low power solution for both mobile and computing/AI applications. Extending the technology scaling further, ~15% logic circuit area scaling over 10 nm has been realized in an 8-nm node with gate and BEOL pitch scaling enabled by quadruple patterning (LE^4). Yield equivalence to 10 nm has been demonstrated in 8-nm IP chips.

Introduction

Mobile SoC has become the main driver for CMOS technology scaling starting from 28nm [1]. High-performance mobile SoC with finFET technology has become a powerful platform for immersive virtual reality, machine learning, and computing applications. The SDM835 mobile SoC with first generation 10-nm finFET technology (10LPE) has been reported, integrating octa-core CPU, AI-ready GPU, and first-generation gigabit LTE modem, with 16% faster speed and 30% lower power than its 14-nm predecessor [2], [3].

To further enhance user experience with neural network-based deep learning, computing, and stronger security on most advanced mobile devices, Snapdragon™ SDM845 was designed with a second-generation gigabit LTE modem (1.2Gbps), new octa-core CPU core (Kryo385), GPU A630 and DSP Hexagon 685, and was fabricated with a second-generation 10-nm finFET technology (10LPP).

10LPP: Second-Generation 10-nm Technology

10LPP technology has the same physical design rules as the first 10-nm finFET technology (10LPE), offering 8 V_{th} flavors by RMG process with low leakage variants. Compared to 10LPE, 10LPP adopted more advanced process elements to boost device dc and ac performance through parasitic resistance reduction with source/drain engineering and parasitic capacitance reduction with fin shape and gate profile optimization. 10LPP process also tightens within-wafer device uniformity through RMG module and source/drain optimization. In addition, advanced STI module process has been adopted to reduce STI-related defect.

Figs. 1(a) and 1(b) show NMOS/PMOS I_{on}–I_{off} improvement by ~5% and 12% respectively compared to 10LPE through eS/D and eSiGe process optimization for reduced parasitic resistance and enhanced channel stress. PMOS contact resistance has been reduced by 30% as shown in the Fig. 1(b) insert. Fig. 2 shows the ~10% RO ac performance improvement in 10LPP in which 3% RO C_{eff} reduction was achieved with gate profile optimization together with improved fin sidewall profile, while maintaining the same short channel control (DIBL). Fig. 3 shows ~20% improvement in NMOS and PMOS V_{th} variation in 10LPP vs. 10LPE mainly from the RMG module and eSiGe process optimization.

10LPP Snapdragon™ SDM845 Productization

Shown in Fig. 4, the flagship Snapdragon™ SDM845 IP includes the second-generation gigabit LTE modem (x20, 1.2Gbps), new octa-core CPU (Kryo385), GPU A630, DSP Hexagon 685 and ISP280 for mobile and computing/AI applications. Fig. 5 shows the high-performance CPU V_{min} @Freq has been reduced by 50 mV at the same leakage current for lower power operation by using 10LPP vs. 10LPE process. Fig. 6 shows SDM845 CPU performance @power has been improved by 25% than previous 10LPE SDM835 due to new CPU design feature plus process performance boost. As the third-generation mobile AI platform, SDM845 SOC design (shown in Fig.7) enables heterogeneous computing by deploying workload efficiently from CPU to GPU/DSP including intensive computing and signal detect/track which accelerates AI/machine learning process. Fig. 8 shows 30% GPU performance/power gain achieved in SDM845 A630 vs. SDM835 A540.

Chip leakage variation has been studied in Fig.9 with CPU cores. $Iddq$ 99-to-50 percentile ratio has been improved by 30% with 10LPP process. Chip sleep mode leakage current is illustrated in Fig. 10 where >50% reduction has been achieved in SDM845 over SDM835 due to improved design feature to extend battery life. Fig. 11 shows 10% longer chip battery life (also 10% better battery life distribution) together with 40% higher CPU performance achieved in SDM845, enabled by new SOC design features together with advanced 10LPP finFET process.

The 10LPE process parametric- and defect-limited yield was analyzed. Fig 12(a) shows that BJT V_{be} variation (p99/50) has been dramatically reduced by 3×, which helps suppress analog-related parametric yield loss by 10× as shown in the Fig. 12(a) insert. 20× defect reduction was achieved with STI module optimization by reducing gate-to-S/D bridge defect (Fig. 12(b)).

Extension to 8LPP Technology

In 8LPP technology which is an extension node of the 10-nm node, gate pitch scaling plus dense metal pitch scaling enabled up to 15% logic circuit area reduction compared to 10 nm. Gate pitch of the dense standard cell library scaled from 68nm to 64nm, and BEOL Mx pitch scaled from 48nm to 44nm with LE^4 (LELELE-LE) patterning [4]. Fig 13 compares the 8LPP ultra-high density (uHD) standard cell with 10-nm high density cell (HD). The technology definition also features aggressive metal tip-tip space ground rule for improved place-and-route efficiency. 8LPP transistor process builds on all 10LPP performance boost elements. By optimizing the BEOL process, comparable defect density has been achieved in IP blocks built with 8LPP uHD cell compared to 10-nm HD cells integrated in the same test chip (Fig. 14). This confirms manufacturing readiness of 8LPP technology for next generation SOC products.

Conclusion

Snapdragon™ mobile SOC SDM845 achieved 30–40% CPU/GPU performance gain together with 10% battery life increase compared to previous SDM835 by using new SOC design features together with the second-generation 10-nm finFET technology. Uniformity improvement in 10LPP helped reduce parametric yield loss by 10×. 8LPP technology extension with ~15% logic circuit area scaling has been verified for future SOC product design.

References
[1] P.R. Chidambaram et al., IEDM Tech. Dig., pp. 27.3.1–4, 2010.
[2] S. Yang et al., VLSI Symp. Tech. Dig., pp. 90–91, 2014.
[3] H.-J. Cho et al., VLSI Symp. Tech. Dig., pp.12–13, 2016.
[4] W.C. Jeong et al., VLSI Symp. Tech Dig., pp. 144–145, 2017.

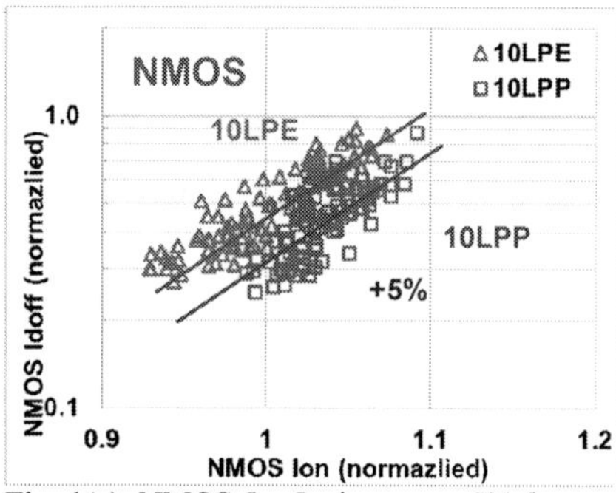

2018 Symposium on VLSI Technology Digest of Technical Papers

Fig. 1(a). NMOS I_{on}–I_{doff} improves 5% from 10LPE to 10LPP.

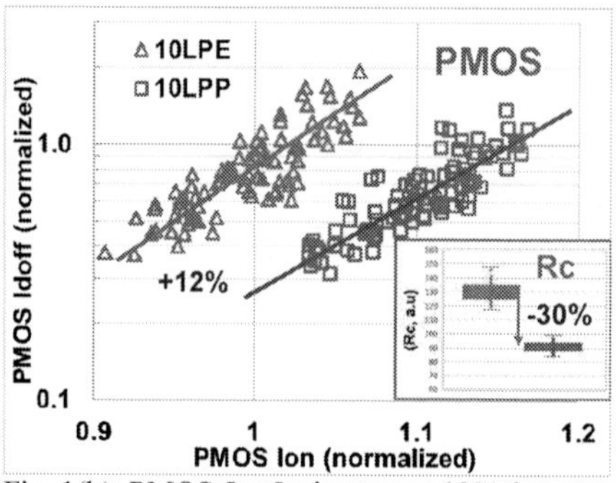

Fig. 1(b). PMOS I_{on}–I_{off} improves 12% from 10LPE to 10LPP with $R_{contact}$ reduced by 30%.

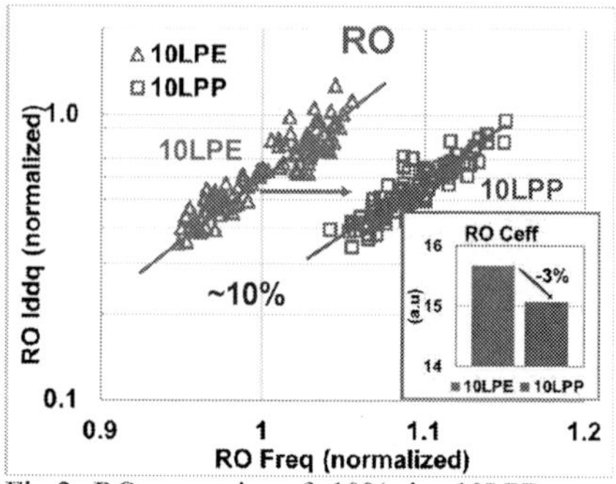

Fig.2 RO ac gain of 10% in 10LPP vs. 10LPE. RO C_{eff} reduces 3% with gate and fin profile optimization shown in the insert

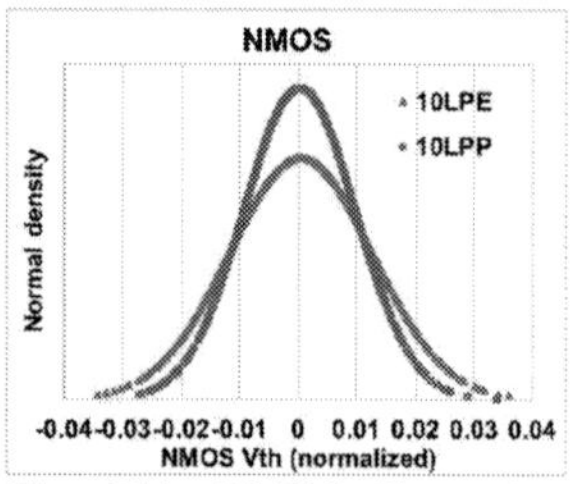

Fig. 3(a). NMOS V_{th} σ improves by ~20% from 10LPE with RMG optimization

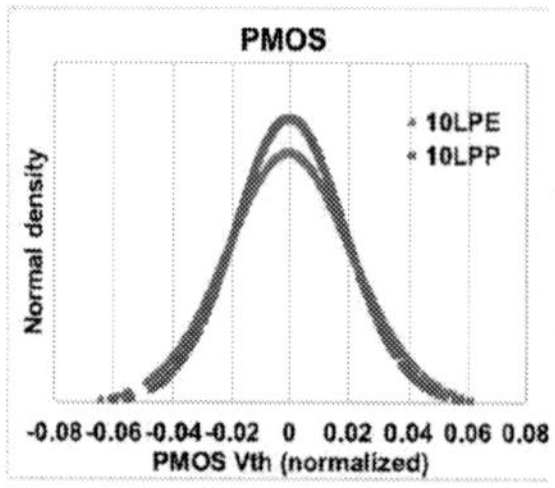

Fig. 3(b). PMOS V_{th} σ improves by ~20% from 10LPE with optimized S/D.

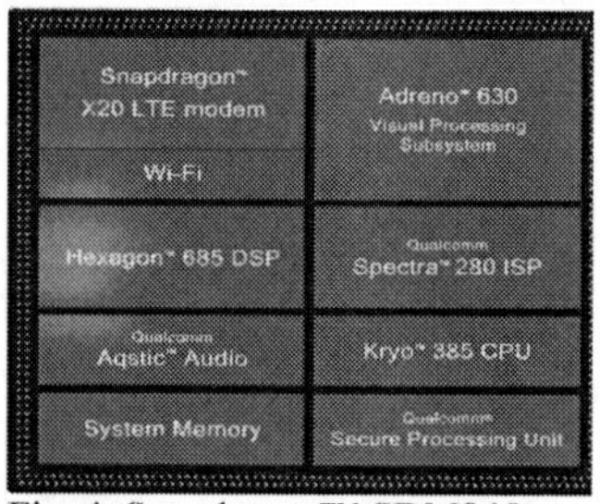

Fig. 4. Snapdragon™ SDM845 IP with Modem/CPU/GPU/ISP.

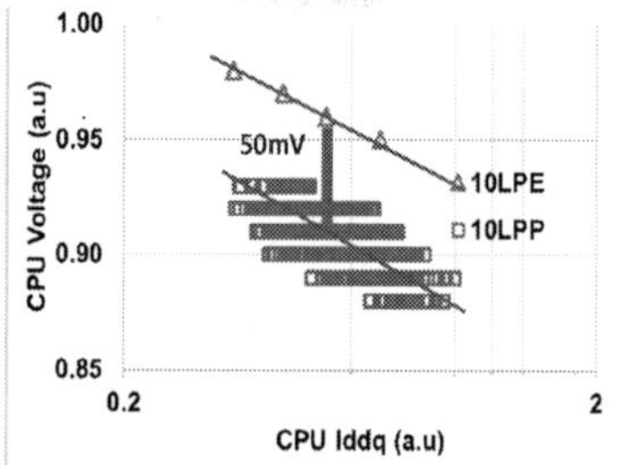

Fig. 5. SDM845 CPU voltage reduces 50 mV in 10LPP vs. 10LPE.

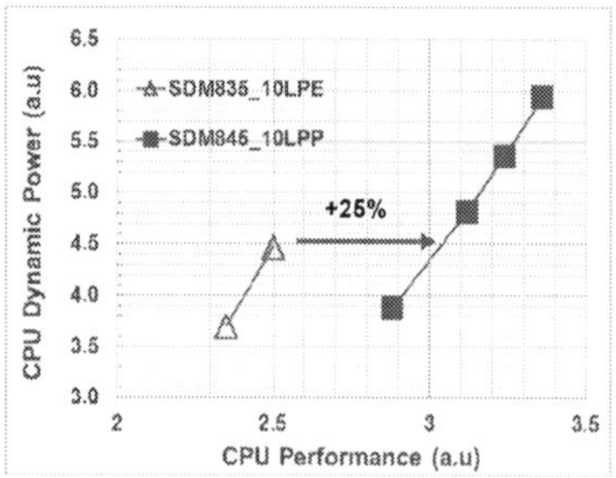

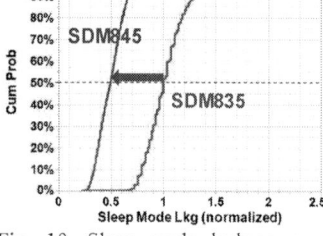

Fig.6 SDM845 CPU performance @power improves 25% vs. SDM835.

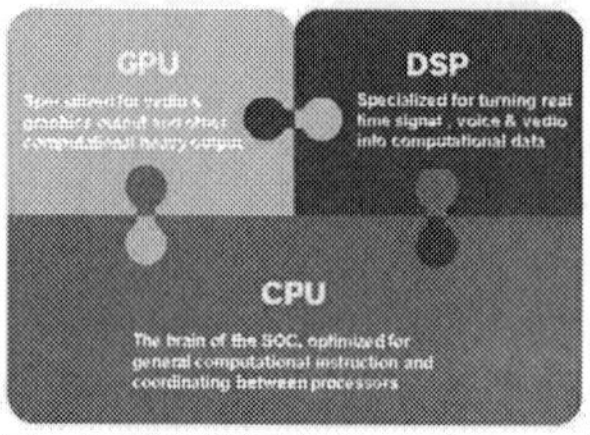

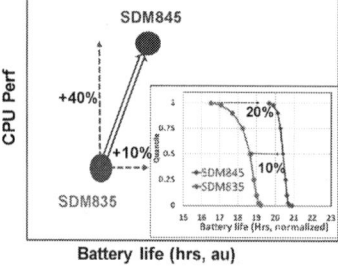

Fig. 7. SDM845 SOC design deploys workload efficiently from CPU to GPU/DSP for computing and AI applications.

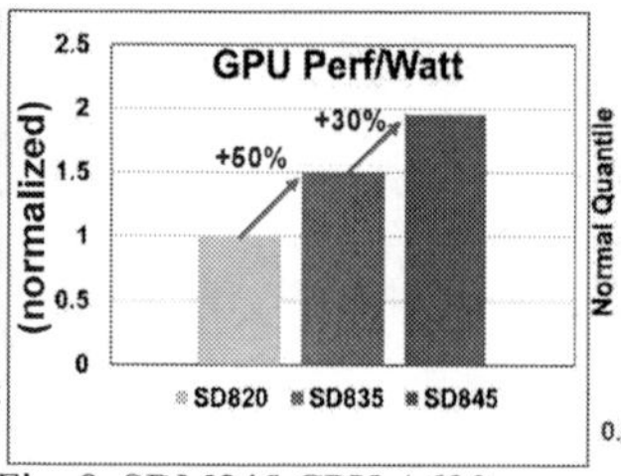

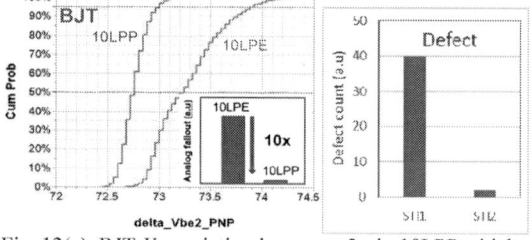

Fig. 8. SDM845 GPU A630 performance improves 30% vs.10LPE SDM835 A540

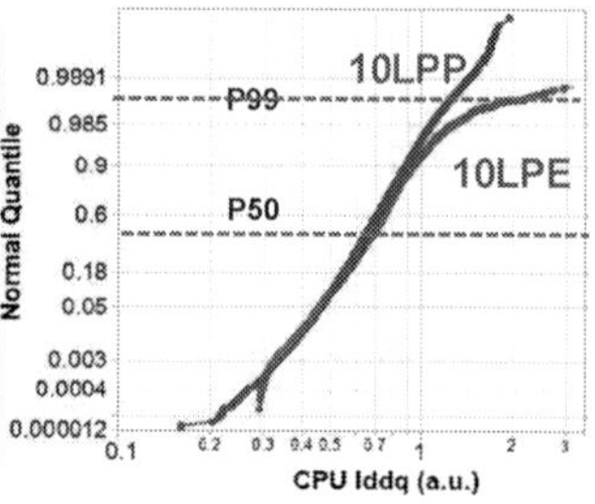

Fig. 9. $Iddq$ variation P99/p50 reduces 30% with 10LPP process vs. 10LPE.

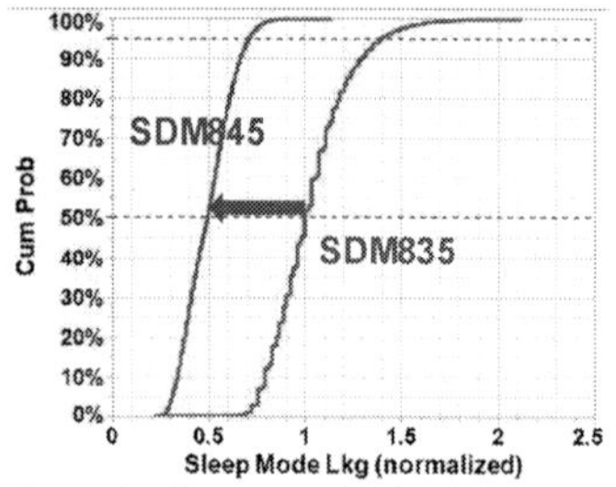

Fig. 10. Sleep mode leakage reduced by >50% in SDM845 vs. SDM835 due to new design feature.

Fig. 11. CPU perf +40% and battery life +10% & better uniformity in SDM845.

Fig. 12(a). BJT V_{be} variation improves 3× in 10LPP which reduced analog yield fallout by 10×.

Fig. 12(b). 20× defect reduction with new STI module in 10LPP.

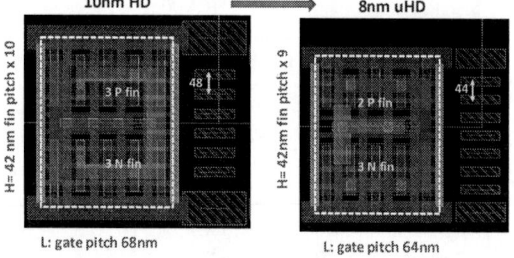

Fig. 13. Standard cell scaling from 10-nm HD to 8-nm uHD cell, gate pitch scaled from 68 to 64 nm, Mx pitch from 48 to 44 nm with LELELELE process.

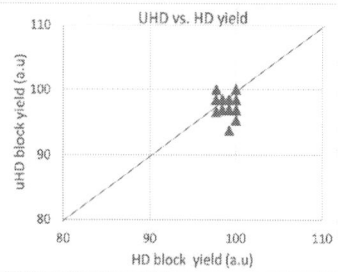

Fig. 14. 8LPP uHD cell achieved similar yield as 10-nm HD cell.

978-1-5386-4219-1/18 $31.00 © 2018 IEEE   220

Hybrid 14nm FinFET - Silicon Photonics Technology for Low-Power Tb/s/mm² Optical I/O

M. Rakowski*, Y. Ban, P. De Heyn, N. Pantano, B. Snyder, S. Balakrishnan, S. Van Huylenbroeck, L. Bogaerts, C. Demeurisse, F. Inoue, K. J. Rebibis, P. Nolmans, X. Sun, P. Bex, A. Srinivasan, J. De Coster, S. Lardenois, A. Miller, P. Absil, P. Verheyen, D. Velenis, M. Pantouvaki, J. Van Campenhout

imec, Kapeldreef 75, 3001 Leuven – Belgium; *now with GlobalFoundries – USA

jvcampen@imec.be

Abstract

We demonstrate a microbump flip-chip integrated 14nm-FinFET CMOS-Silicon Photonics (SiPh) technology platform enabling ultra-low power Optical I/O transceivers with 1.6Tb/s/mm² bandwidth density. The transmitter combines a differential FinFET driver with a Si ring modulator, enabling 40Gb/s NRZ optical modulation at 154fJ/bit dynamic power consumption in a 0.015mm² footprint. The receiver combines a FinFET trans-impedance amplifier (TIA) with a Ge photodiode, enabling 40Gb/s NRZ photodetection with -10.3dBm sensitivity at 75fJ/bit power consumption, in a 0.01mm² footprint. High-quality data transmission and reception is demonstrated in a loop-back experiment at 1330nm wavelength over standard single mode fiber (SMF) with 2dB link margin. Finally, a 4x40Gb/s, 0.1mm² wavelength-division multiplexing (WDM) transmitter with integrated thermal control is demonstrated, enabling Optical I/O scaling substantially beyond 100Gb/s per fiber.

Keywords: optical interconnect, silicon photonics, FinFET

Introduction

Exponentially growing demand for I/O bandwidth in datacenter switches and high-performance computing nodes drives the need for tight co-integration of optical interconnects with advanced CMOS logic, enabling high-density, power efficient and low-latency optical I/O for a wide range of interconnect distances (1m-500m+) [1]. Silicon photonics is a prime technology platform to realize the desired scaling in optical I/O cost and performance, by leveraging established CMOS manufacturing [2] and advanced 3-D integration methods. Here, we present a hybrid FinFET CMOS-SiPh technology realizing optical I/O transceiver density beyond 1Tb/s/mm², combined with record low dynamic power consumption of 230fJ/bit at 40Gb/s single-lane non-return-to-zero (NRZ) data rates. This is achieved through high-density, low-parasitic microbump flip-chip integration of compact, high-speed monolithic SiPh modulators and photodetectors, with co-designed highly power-efficient, high-speed circuits in bulk FinFET CMOS.

Technology Description

A. Silicon Photonics Chip

The SiPh chips (20mm²) are manufactured on 300mm SOI wafers in an R&D CMOS fab [2]. They contain a dense 8-channel transmitter (TX) array and an 8-channel receiver (RX) array, both operating at 1330nm wavelength (Fig. 1). Each TX unit cell (0.015mm²) contains a 7.5µm radius, 45fF, 30GHz Si ring modulator (RM) with integrated heater, whereas each RX unit cell (0.01mm²) contains a 15x2µm², 30fF, 50GHz Ge waveguide photodetector (PD) with 0.9A/W responsivity. Fiber grating couplers are implemented on the north side of the chip for interfacing with a 12-channel, 250-µm pitch SMF V-groove array. Wirebond and RF probe pads are included for delivery of dc and high-speed signals respectively.

B. 14nm FinFET CMOS Chip

The 14nm FinFET chips (1.5mm²) are manufactured in GlobalFoundries 14LPP technology, and mirror the design of the SiPh chip, having dense arrays of ultra-compact invertor-based modulator drivers and TIAs. The drivers are designed to produce a differential output swing of $1.6V_{DD}$, maximizing the optical modulation amplitude (OMA) of the depletion-type ring modulator. The circuit architecture (270µm²) is shown in Fig. 2. The design targets 40Gb/s NRZ data rate at the nominal 0.8V supply voltage, for a total capacitive load of 100fF. The TIA circuit architecture (<50µm²) is shown in Fig. 2, and targets low-power operation with high sensitivity at 40Gb/s for a 120fF total input load, by balancing the transimpedance gain (1.6kΩ) and bandwidth (26GHz). A buffer amplifier is added to drive the electrical signal across a 50Ω on-chip transmission line into a high-speed RF probe for off-chip analysis on a high-speed sampling oscilloscope.

C. Flip-Chip Assembly and Packaging

The FinFET dies are flip-chipped onto known-good SiPh dies through microbumps with 50µm pitch and 30µm pad size, minimizing the interface parasitic capacitance to below 30fF. Next, the CMOS-SiPh assemblies are glued on test boards, wirebonded, and finally the SMF V-groove array is actively aligned and glued to the assembly (Fig. 1), achieving 4.5dB fiber-to-chip coupling losses.

Results

Fig. 3 shows the measurement setup used to test the transceiver. Fig. 4 illustrates single-lane TX performance: at nominal 0.8V supply and optimized RM heater tuning, 40Gb/s NRZ modulation is achieved with 4.2dB extinction ratio (ER) and 3.9dB insertion loss (IL), which is equivalent with a transmitter penalty $TP=OMA/(2P_{IN})$ of 9dB. The energy consumption (E_{bit}) is 154fJ/bit. Increasing the supply voltage to 0.9V enables modulation rates up to 52Gb/s. The RX performance, as measured with an external 44Gb/s LiNbO₃ modulator (ER>15dB), is summarized in Fig. 5. At 0.8V supply, and, wide open electrical eye diagrams with signal-to-noise (SNR) ratio above 9 are obtained at 40Gb/s data rate for average photocurrents as low as 42µA, equivalent with a waveguide-referred OMA sensitivity of -10.3dBm. The TX and RX are also tested in a loop-back configuration. In this experiment, 13dBm laser power is coupled into the SiPh chip, resulting -4.2dBm fiber-coupled OMA at the TX output. The modulated light beam is subsequently attenuated off-chip by 2dB and coupled back into the RX. The resulting 32Gb/s electrical eye diagram recorded at the RX output (Fig. 6) is wide open with SNR>6. Finally, 4x40Gb/s WDM transmission is demonstrated from a cascaded SiPh RM array in Fig. 7. By appropriately tuning the integrated heater currents, the

978-1-5386-4219-1/18 $31.00 © 2018 IEEE 221

operation wavelengths of the individual SiPh ring modulators can be tuned (11mW/nm) to the 2nm wavelength grid of a WDM laser source. Very uniform power efficiencies and OMA have been obtained for all tested channels. Finally, Table 1 summarizes the measured performance and compares with previous reports of CMOS-SiPh transceiver demonstrations. The presented results clearly stand out in terms of power efficiency, bandwidth density or both. Optimizations in SiPh device designs and CMOS circuit layouts are possible in future demonstrations, and are expected to further improve modulation rates and link margin.

Acknowledgements

This work has been supported by imec's industry-affiliation R&D program on Optical I/O.

References

[1] I.A. Young, et al., "Optical I/O Technology for Tera-Scale Computing," JSSC, pp.235,248, Jan. 2010.

[2] M. Pantouvaki et al., "Active Components for 50 Gb/s NRZ-OOK Optical Interconnects in a Silicon Photonics Platform," in Journal of Lightwave Technology, vol. 35, no. 4, pp. 631-638, Feb.15, 2017.

[3] C Sun, MT Wade, Y Lee, JS Orcutt, L Alloatti, MS Georgas, AS Waterman, "Single-chip microprocessor that communicates directly using light," Nature 528 (7583), 534, 2015

[4] E. Timurdogan et al., "An ultra-low power 3D integrated intra-chip silicon electronic-photonic link," OFC, 2015, PDP Th5B.8

[5] M. Rakowski et al., "22.5 A 4×20Gb/s WDM ring-based hybrid CMOS silicon photonics transceiver," ISSCC, 2015, pp. 1-3.

[6] F.Y. Liu, et al., "10-Gbps, 5.3-mW optical transmitter and receiver circuits in 40-nm CMOS," JSSC, pp. 2049–2067, Sep. 2012.

[7] J. Buckwalter, et al., "A Monolithic 25-Gb/s Transceiver with Photonic Ring Modulators and Ge Detectors in a 130-nm CMOS SOI Process," JSSC, pp.1309–1322, Jun 2012.

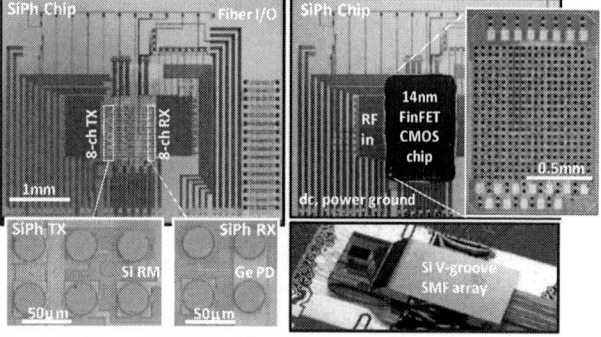

Fig. 1. SiPh and FinFET chip micrographs, including SiPh TX and RX unit cell details, and images of the fiber-packaged assembly.

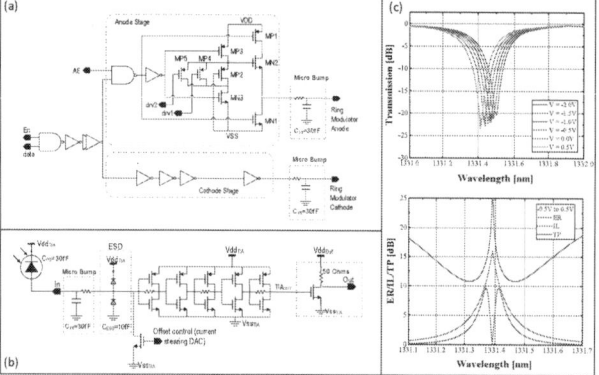

Fig. 2. (a) TX driver, and (b) RX TIA schematic. (c) SiPh RM: static optical transmission vs V_{bias} (top), and ER, IL, TP spectra for 1Vpp swing (bottom).

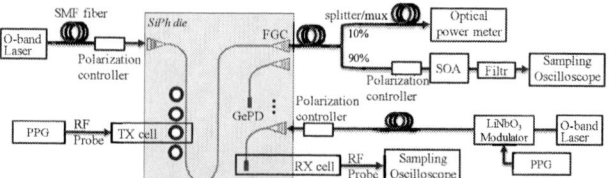

Fig. 3. Measurement setup schematic

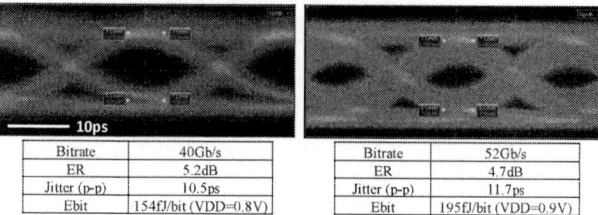

Bitrate	40Gb/s		Bitrate	52Gb/s
ER	5.2dB		ER	4.7dB
Jitter (p-p)	10.5ps		Jitter (p-p)	11.7ps
Ebit	154fJ/bit (VDD=0.8V)		Ebit	195fJ/bit (VDD=0.9V)

Fig. 4. TX eye diagrams: left 40Gb/s (PRBS31), right 52Gb/s (PRBS7).

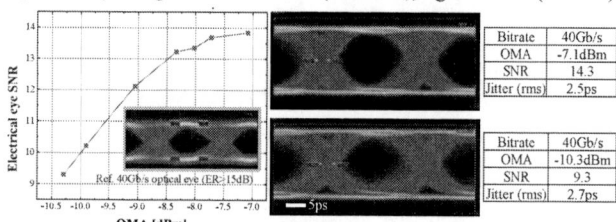

Bitrate	40Gb/s
OMA	-7.1dBm
SNR	14.3
Jitter (rms)	2.5ps

Bitrate	40Gb/s
OMA	-10.3dBm
SNR	9.3
Jitter (rms)	2.7ps

Fig. 5. RX at 40Gb/s (PRBS31). Left: Electrical eye diagram SNR vs. OMA Right: eye diagrams at OMA = -7.1dBm and OMA = -10.3dBm.

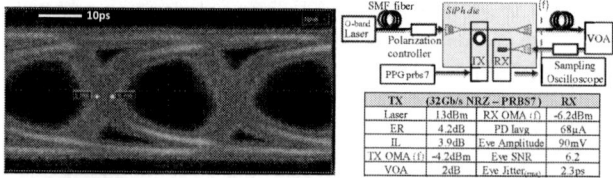

TX	(32Gb/s NRZ – PRBS7)	RX	
Laser	1.3dBm	RX OMA (f)	-6.2dBm
ER	4.2dB	PD Iavg	68μA
IL	3.9dB	Eye Amplitude	90mV
TX OMA (f)	-4.2dBm	Eye SNR	6.2
VOA	2dB	Eye Jitter(rms)	2.3ps

Fig. 6. TX-to-RX loop-back performance at 32Gb/s.

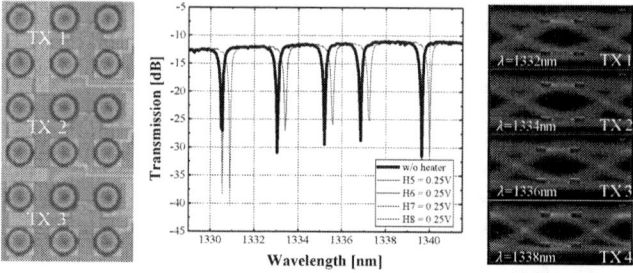

Fig. 7. WDM TX. Left: chip micrograph. Middle: static TX optical transmission. Right: 4λ x 40Gb/s eye diagrams (PRBS31).

TABLE I. Performance overview and benchmarking

	This work	[3]	[4]	[5]	[6]	[7]
Year of Publication	2018	2015	2015	2015	2012	2012
Integration Method	50um pitch Cu Flip Chip	Monolithic	Wafer-to-Wafer Through-oxide Via	150um pitch C4 Flip Chip	25um pitch microsolder Flip Chip	Monolithic
Technology (CMOS)	14nm LPP bulk FinFET CMOS	45nm CMOS SOI	65nm CMOS bulk	40nm LP CMOS	40nm CMOS	130nm CMOS SOI
Technology (Si Photonics)	28nm SOI	"zero change"	65nm SOI	130nm SOI	130nm SOI	
I/O Bandwidth Density	1.6Tb/s/mm²	400Gb/s/mm²	500Gb/s/mm²	90Gb/s/mm²	1.6Tb/s/mm²	12.5Gb/s/mm²
Wavelength	1330nm	1183nm	1529nm	1550nm	1550nm	1550nm
Supply Voltage	0.8V	1V	1.2V	1.3V	1V,1.3V,2V	1.2V-1.2V
Single-Lane Data Rate	40Gb/s NRZ	5Gb/s NRZ	5Gb/s	20Gb/s NRZ	10Gb/s NRZ	25Gb/s NRZ
TX (Ring Modulator)						
Driver Power	154fJ/bit	30fJ/bit	10fJ/bit	1.3pJ/bit	135fJ/bit	8.3pJ/bit
ER	4.2dB	6.5dB	5dB	>7dB	7dB	6.9dB
IL	3.9dB	4dB	1dB	n/a	n/a	n/a
#WDM channels	4	1-11	1	4	8	n/a
Ring Tuning Power	11mW/nm	2.5mW	n/a	6.2mW/nm	5.3mA/nm	n/a
Bandwidth per fiber	160Gb/s	5-55Gb/s	5Gb/s	80Gb/s	80Gb/s	25Gb/s
RX (Ge WPD)						
TIA power	75fJ/bit	~400fJ/bit	240fJ/bit	580fJ/bit	120fJ/bit	1.92pJ/bit
Rx sensitivity	-10.3dBm OMA	-5dBm OMA	-16dBm OMA	-7.2dBm Pavg	-15dBm Pavg	-6dBm Pavg

AUTHOR INDEX

Absil, P. 221
Agashe, Parag 219
Ahn, Hyun Jun 73
Ahn, J. 185
Ahn, J.H. 213
Ahn, Jaesoo 117
Ahn, Jeonghoon 217
Akinaga, Hiroyuki 63
Akita, Ippei 41
Alian, Alireza 133, 159
Amisse, A. 125
Ando, T. 115
Andrieu, F. 153
Ang, Kah-Wee 159, 161
Antonyan, Artur 181
Apalkov, Dmytro 187
Arimura, H. 83, 193
Arreghini, A. 203
Asheghi, Mehdi 17
Asra, R. 81
Avasarala, Naga Sruti 209
Ayele, Getenet Tesega 97
Azarkish, H. 15
Aziz, A. 129
Bae, Byoung-Jae 181
Bae, Jong-Ho 169, 201
Baek, Gwanghyeon 151
Baert, R. 143
Bai, X. 127
Balakrishnan, S. 221
Balasubramaniam, P. 81
Ban, Y. 221
Ban, Yongchan 137
Bang, Y.S. 213
Bangar, Mangesh 117
Banghart, E. 81
Banno, N. 127
Bao, Jerry 219
Bao, R. 115
Bao, Ruqiang 87
Bao, T. Huynh 141
Bardon, M. Garcia 143
Barla, K. 193
Barrau, J. 15
Barraud, S. 125
Batude, P. 75
Bäuerle, C. 125
Bazizi, E. 81
Beckrich-Ros, H. 15
Bedell, Stephen 87
Behin-Aein, B. 183
Bender, H. 83, 193
Bermak, Amine 39
Berthelon, R. 153
Bertrand, B. 125

Besnard, G. 69
Bex, P. 221
Bhadri, Prasad 219
Boemmels, J. 141
Boeuf, Frederic 97
Bogaerts, L. 221
Bohuslavskyi, H. 125
Boisseau, P. 35
Boisseau, S. 35
Bon, Romain 97
Bouche, G. 141
Boudier, D. 85
Bourdet, L. 125
Boussaid, Farid 39
Breuil, L. 203
Brightsky, M. 205
Brown, James 95
Bruce, R. L. 205
Brunet, L. 75
Bu, Huiming 87
Bufler, F.M. 69
Burnett, D. 81
Cai, Ming 219
Caimi, D. 165
Campbell, Jason P. 89
Cao, L. 61
Capogreco, E. 83, 193
Carta, F. 205
Cassé, M. 75
Chan, B. T. 69
Chan, T. H. 183
Chandrashekar, A. 61
Chang, Chien-Min 21
Chang, Ching-Chih 37
Chang, Chun-Yen 139
Chang, E. Y. 47
Chang, Hung-Sheng 177
Chang, Meng-Fan 171
Chang, Tun-Jen 139
Chang, W. H. 191
Chang, Y. L. 157
Chang, Yen-Chung 37
Chang, Yuhua 161
Chang, Yu-Ming 37
Chao, R. 183
Chatterjee, Korok 49, 53
Chaudhuri, R. 61
Chekol, Solomon Amsalu 207
Chen, Bing 105
Chen, Fan 187
Chen, H. W. 29
Chen, Hao 117
Chen, Hsin 37
Chen, J. 29
Chen, Jiezhi 95

AUTHOR INDEX

Chen, Kuen-Yi	119
Chen, Nan	161
Chen, S.-H.	85
Chen, Shih-Hsin	37
Chen, T. P.	29
Chen, W.H.	215
Chen, Wan-Hsin	139
Chen, Weichen	177
Chen, Wei-Hao	171
Chen, Xiangdong	219
Chen, Yen-Peng	37
Chen, Ye-Ting	37
Chen, Ying	219
Chen, Yu-Chieh	37
Chen, Yu-Hsin	37
Cheng, Chun-Hu	139
Cheng, H. Y.	205
Cheng, Osbert	29
Cheng, Ran	105
Cheung, Kin P.	89
Chidambaram, P. R.	219
Chien, W. C.	205
Ching, C.	185
Ching, Chi	117
Cho, Byung Jin	73
Cho, H.-J.	59
Cho, Soon	219
Cho, Sungil	217
Choe, M.	121
Choi, B.H.	213
Choi, D.	215
Choi, Jihong	219
Choi, K.	121
Choi, Nagyong	169, 201
Choi, Woosung	187
Choi, Yunki	217
Chuang, Ji-Feng	37
Chung, Chia-Che	113
Chung, N. L.	183
Chung, Steve S.	29
Chung, Sungyong	169, 201
Chung, Wonil	89
Cibrario, G.	153
Clima, S.	209
Cloarec, Jean-Pierre	97
Cohen, B.	81
Colinet, E.	35
Collaert, N.	69, 83, 193
Collaert, Nadine	133, 159
Collin, L.-M.	15
Colonna, J.-P.	15
Cong, H.	183
Convertino, C.	165
Corna, A.	125
Coudrain, P.	15

Cretu, B.	85
Crippa, A.	125
Crowford, Jacob	95
Czomomaz, L.	165
Damjanovic, D.	61
Datta, S.	129
Datta, Suman	131
De Coster, J.	221
De Franceschi, S.	125
De Heyn, P.	221
De Heyn, V.	69, 193
Deboer, Scott	3
Deguchi, Yoshiaki	109
Dekkers, H.	83
Delapierre, G.	35
Delhougne, R.	203
Demeurisse, C.	221
Demuynck, S.	141, 147
Deng, Jie	219
Deshpande, V.	69, 165
Devriendt, K.	69
Devulder, W.	209
Do, H. B.	47
Do, J.H.	213
Do, JH	59
Donadio, G. L.	209
Dong, Bowei	161
Dong, Wenfeng	105
Dou, Chun-Meng	171
Dou, X.	215
Drouin, Dominique	97
Duarte, Juan Pablo	49
Durfee, Curtis	87
Ecoffey, Serge	97
El Kazzi, Salim	133
Eneman, G.	83, 193
Ernst, T.	35
Fan, Chia-Chi	139
Fan-Chiang, C. C.	47
Fantini, A.	209
Favia, P.	193
Fenouillet-Béranger, C.	75
Franco, J.	69
Fréchette, L.G.	15
Fujii, Satoru	63
Fujii, Shosuke	107
Fujimoto, Akira	41
Fukai, T.	59
Fukuyama, Shouhei	109
Fukuzawa, Hideaki	65
Furnemont, A.	203
Gaillard, F.	75
Gaire, C.	215
Gan, K. W.	183
Gao, Bin	31, 103

AUTHOR INDEX

Gao, Rui 95
Garros, X. 75
Gaudin, G. 69
Ge, Lixin 219
Geiss, E. 81
George, T. 157
Gerbelot-Barillon, R. 35
Ghibaudo, G. 75
Goh, L. C. 183
Gohou, Yasushi 175
Gokmen, T. 25
Gong, Xiao 77, 197
Goodson, Kenneth E. 17
Goux, L. 209
Govoreanu, B. 209
Gribelyuk, M. 61
Guillemaud, R. 35
Guo, D. 115
Guo, Dechao 87
Guo, Xin 161
Guo, Zheng 151
Gupta, Dinesh 87
Gupta, S. K. 129
Ha, DW 59
Ha, M. T. H. 47
Hada, H. 127
Haensch, W. 25
Han, Kaizhen 77, 197
Haq, Jesmin 65
Harada, M. 71
Hashimoto, S. 93
Hassan, M. K. 81
Hassan, Sajjad 117
Hatcher, R. 51
Hattori, H. 191
Hayata, Yuriko 175
He, Liuhuiquan 77
He, Renren 65
He, X. 215
Heliu, Huiquan 197
Hellings, G. 69, 85
Hentz, S. 35
Heylen, N. 69
Heyns, M. 209
Hikavyy, A. 69, 83, 193, 203
Himeda, Y. 93
Hoentschel, J. 153
Holsteyns, F. 193
Homma, Kazunari 63
Hong, Heebum 217
Hong, Hyeongsun 181
Hong, P. Y. 157
Horiguchi, N. 83, 85, 141
Hou, F.-J. 45
Howarth, James 117

Hsieh, Chih-Cheng 37
Hsieh, E. R. 29
Hsieh, Ping-Hsuan 37
Hsieh, Sung-En 37
Hu, C. 47
Hu, Chenming 49, 53
Hu, O. 81, 215
Hu, Y. 215
Huang, G.-W 45
Huang, Li 159, 161
Huang, P. 47
Huang, S. A. 29
Huang, Xin 161
Huang, Yen-Hua 119
Huang, Yi-Chiau 197
Hutin, L. 125
Huynh, S. H. 47
Hwang, Hyunsang 207
Hwang, J. 183
Hwang, Kihyun 181
Hwang, Sohee 181
Iguchi, N. 127
Ikeda, H. 93
Im, Sung Gap 73
Inokawa, H. 93
Inoue, F. 69, 221
Inoue, Y. 71
Irisawa, T. 191
Ishii, H. 191
Ishikawa, Masatoshi 163
Isono, S. 71
Ito, Satoru 63
Ivanov, Tsvetan 133
Iwai, H. 47
Iwata-Harms, Jodi 65
Jagannathan, H. 115
Jallon, P. 35
Jamieson, G. 69
Jan, Guenole 65
Jan, Sun-Rong 113
Jang, D. 83, 143
Jang, Daehyun 181
Jang, M.S 59
Jang, S. H. 183
Jehl, X. 125
Jeong, Gitae 181
Jeong, Jaehun 217
Jeong, T. 121
Jeong, WC 59
Jerry, M. 129
Jerry, Matthew 131
Jha, A. 215
Ji, Yongsung 181
Ji, Zhigang 95
Jiang, Zhengping 187

AUTHOR INDEX

Jiang, Zhewei 173
Jiang, Zizhen 107
Jin, Y. D. 47
Jo, K.-W. 195
Jossart, N. 203
Jung, B. 183
Jung, Es 181
Jung, Hyunsung 181
Jung, Hyuntaek 181
Jung, J.H. 213
Jung, Jonghoon 149
Jung, Ki Wook 17
Jung, S.-M 59
Kamakura, Y. 93
Kang, D. 215
Kang, H.K. 59
Kang, Ho-Jung 201
Kang, S. H. 185
Kang, Sung Gun 217
Kang, Sunggun 219
Kang, Yuye 77, 197
Kao, K.-H. 45
Kao, Ruei-Wen 119
Kar, G. S. 209
Karl, Eric 151
Kassim, J. 61
Katayama, Koji 63, 175
Kawai, Ken 63
Kelly, James 87
Kim, BS 59
Kim, Chris H. 19
Kim, Chul-Heung 169
Kim, Dae Sin 187
Kim, Daeyeon 151
Kim, Dongwoo 217
Kim, H.-J. 213
Kim, Hoonki 149
Kim, HT 59
Kim, Ilryong 217
Kim, J. 121
Kim, Joochan 181
Kim, Jungtae 217
Kim, N. 85
Kim, S. 25
Kim, Sungwon 219
Kim, W. 121, 205
Kim, W.-K. 195
Kishi, Norimasa 163
Kitamura, T. 183
Kittl, J. A. 51
Kleemeier, Walter 87
Knorr, Andreas 87
Koenig, A. 35
Koh, Kwan-Hyeob 181
Kojima, Akihiro 41

Kolar, Pramod 151
Konar, Aniruddha 87
Kong, Eugene Y.-J. 197
Kontos, Alex 117
Koo, Yunmo 207
Koswatta, S. 25
Kouno, Kazuyuki 175
Koyanagi, T. 71
Krishnan, B. 61
Krishnan, R. 183
Krivokapic, Zoran 49
Kumada, T. 93
Kumar, Annie 197
Kundu, S. 209
Kuo, Chihwei 219
Kuo, I. T. 205
Kuo, J. L. 29
Kuo, M. H. 157
Kuo, Po-Yi 21
Kurui, Yoshihiko 41
Kwon, Daewoong 49, 53
Kwon, DJ 59
Kwon, J. 183
Kwon, O. 81
Kwon, S. D. 213, 219
Kye, Jongwook 149
Kyono, C. 215
Laguna, G. 15
Lai, E. K. 205
Lai, Wei-Chih 37
Lakshmipathi, S. 183
Lam, J. 183
Lam, Vinh 65
Langer, R. 193
Lardenois, S. 221
Lavigne, E. 215
Lazar, H. 81
Le, Son 65
Lee, C. H. 115
Lee, Chengkuo 161
Lee, Daesop 181
Lee, Dong Hwan 169, 201
Lee, Dongjin 107
Lee, EC 59
Lee, Euncheol 217
Lee, H. 185
Lee, HJ 59
Lee, J.G. 81, 215
Lee, Jong-Ho 169, 201
Lee, Junghyuk 181
Lee, K. 183
Lee, Kilho 181
Lee, KW 59
Lee, Kwang Hong 197
Lee, Kwon 219

AUTHOR INDEX

Lee, Rinus T.P. 61
Lee, Sanghumn 181
Lee, Seungbae 181
Lee, Seungyoung 149
Lee, Shiuh-Wuu 105
Lee, Soochang 169
Lee, Sung-Tae 169
Lee, Tackhwi 169, 201
Lee, TJ 59
Lee, Y.H. 213
Lee, Y.-J. 45
Lee, Yong Kyu 181
Lee, Yuan-Jen 65
Lee, YW 59
Lei, Dian 77, 197
Leobandung, E. 25
Levin, Theodore 87
Li, Haitong 107
Li, Juntao 87
Li, Jun-Yi 171
Li, P. W. 157
Li, W. 69
Li, Y. 25
Li, Yang 77, 197
Li, Yida 159
Liang, Shurong 117
Liao, P. H. 157
Liao, Yan 31
Liao, Yu-Hung 49, 53
Liao, Yu-Te 37
Lim, J. H. 183
Lim, Jinyoung 149
Lim, Jungman 181
Lim, Suhwan 169
Lim, Y.D. 213
Lin, Dennis 133, 159
Lin, H. C. 157
Lin, H. H. 113
Lin, Huan-Ting 171
Lin, Vicki 219
Lin, Wei-En 171
Lin, Wei-Yu 171
Lin, Y. C. 47
Lin, Y. K. 47
Lin, Yan-Xiao 119
Lin, Yen-Kai 49
Linder, B. P. 115
Ling, T. 183
Linten, D. 85
Litta, E. Dentoni 193
Liu, B. 183
Liu, C. W. 113
Liu, Chien 139
Liu, H. 61
Liu, Huanlong 65

Liu, J. 61, 215
Liu, Po-Tsun 21
Liu, Tanya 17
Liu, Tzu-Hao 37
Liu, Wei 105
Lo, H.C. 81, 215
Loo, R. 83, 193
Low, R. 183
Lu, Chih-Yuan 177
Luc, Q. H. 47
Lue, Hang-Ting 177
Lung, H. L. 205
Machillot, J. 85
Maeda, S. 59
Maeda, T. 191
Mahalingam, A. S. 61
Mahmoodi, Mohammad Reza 99
Mailley, P. 35
Mallik, A. 141
Masudy-Panah, Saeid 77
Matsukawa, T. 93
Matsuki, T. 93
Mattii, L. 143
Maurand, R. 125
Merkle, T. 165
Mertens, H. 83, 85, 141, 143
Mesaki, K. 93
Meunier, T. 125
Michailos, J. 15
Mikawa, Takumi 175
Milenin, A. 83
Milenin, A. P. 193
Milenin, Alexey 133
Miller, A. 221
Miller, William 219
Mitard, J. 83, 193
Miyamura, M. 127
Miyano, Satoru 7
Miyazaki, Shunsuke 163
Miyazoe, H. 205
Mo, R. 25
Mochida, Reiji 175
Mochizuki, Shogo 87
Mocuta, A. 83, 85, 141, 143, 147
Mocuta, D. 69, 83, 141, 193
Mocuta, Dan 133, 159
Mody, J. 61
Mohan, M. 215
Monfray, Stephane 97
Morgan, J. 153
Morioka, A. 127
Mun, S.Y. 215
Naik, V. B. 183
Naitoh, Yasuhisa 63
Nakayama, Masayoshi 175

AUTHOR INDEX

Nalam, Satyanand 151
Nam, KJ 59
Narayanan, V. 115
Nebashi, R. 127
Nguyen, B.-Y. 69
Nguyen, T. A. 47
Ni, K. 129
Ni, Kai 131
Nili, Hussein 99
Niquet, Y.-M. 125
Nishimura, T. 27
Noh, S. 183
Nolmans, P. 221
Numata, H. 127
Nyns, L. 203
Oba, S. 93
Obradovic, B. 51
Oh, H. 209
Oh, Saechoon 219
Oh, Sechung 181
Ohta, Jun 163
Okamoto, K. 127
Oniki, Y. 147
Ono, Takashi 175
Opdebeeck, A. 193
Opsomer, K. 209
Pae, S. 121
Pak, Kwanyong 73
Pak, M. 203
Pakala, M. 185
Pakala, Mahendra 117
Pantano, N. 221
Pantisano, L. 81
Pantouvaki, M. 221
Park, Byung-Gook 169, 201
Park, C. 185
Park, C.H. 59
Park, DW 59
Park, J. 121
Park, Jaehyuk 207
Park, JS 59
Park, K.C. 213
Park, Kichul 181
Park, Yongmin 217
Parvais, B. 69, 83, 85, 143
Patel, Sahil 65
Pauliac-Vaujour, E. 35
Pena, V. 85
Peng, J. 215
Peng, L. 69
Peng, Xiaochen 103
Penzes, Paul 219
Perumel, Ramesh 37
Petrov, N. 61
Petykiewicz, Jan 137

Pi, Unghwan 181
Pinna, Nicolò 133
Pirro, L. 153
Plantier, C. 35
Polizzi, J.P. 35
Porret, C. 193
Prieto, R. 15
Pyo, Mark 181
Qi, Weiyi 187
Qi, Y. 215
Qian, He 31, 103
Qin, Shengjun 107
Quek, E. 183
Rabindranath, Ashwin 219
Radisic, D. 147
Radu, I. 69
Ragnarsson, L.-A. 85
Rakowski, M. 221
Rakshit, T. 51
Ramanathan, E. 61
Rassoul, N. 69
Ray, A. 205
Ray, S. 61
Raymaekers, T. 203
Rebibis, K. J. 221
Ren, Z. 25
Rhee, Hwasung 217
Richard, O. 83
Rim, Ken 219
Rim, WJ 59
Ritzenthaler, R. 69, 83
Rodder, M. S. 51
Rosseel, E. 69, 203
Ryckaert, J. 141, 147
Sachid, Angada B. 53
Sadana, Devendra 87
Sagong, H. C. 121
Saito, Tomohiro 41
Sakamoto, T. 127
Salahuddin, Sayeef 49, 53
Samavedam, S. B. 81, 215
Sanquer, M. 125
Saoutieff, E. 35
Sasagawa, Kiyotaka 163
Savelli, G. 15
Schmidt, Daniel 87
Schram, T. 85
Schuddinck, P. 141, 143
Schulze, A. 193
Schwarzenbach, W. 69
Sebaai, F. 83, 193
See, A. 183
Seet, C. S. 183
Seo, Boyoung 181
Seo, Jae-Sun 173

AUTHOR INDEX

Seok, Mingoo 173
Serrano-Guisan, Santiago 65
Shah, Jinal 151
Shen, Dongna....................65
Shen, T. 61
Shen, Y. 215
Sherazi, Y. 141, 143
Shi, Qi 95
Shi, Y. 215
Shibata, Hideki....................41
Shieh, J.-M. 45
Shim, H. 121
Shima, Hisashi....................63
Shima, K. 93
Shin, Hyunchul....................181
Shin, S. 121
Shirahige, Daiki....................163
Shirazi, M. 15
Shrestha, Pragya R. 89
Shrivastava, Ayush 151
Shukla, N. 129
Shum, D. 183
Si, Mengwei 89
Siah, S. Y. 183
Simicic, M. 85
Simoen, E. 85
Sirman, A. 215
Siva, Shrihari 219
Smith, J. 141
Smith, Jeffrey A. 131
Snyder, B. 221
Sohn, D.K.....................81, 215
Solomon, P. 25
Son, J. S. 183
Son, Nakjin 217
Song, Jeonghwan....................207
Song, T.J. 213
Song, Taejoong 149
Song, Yoonjong 181
Song, Zhimin 219
Souifi, Abdelkader 97
Sousa, M. 165
Southwick, R. G. 115
Sporer, R. 61
Srinivasan, A. 221
Strinati, E. Calvanese 35
Strukov, Dmitri. B....................99
Struss, Q. 15
Su, C.-J. 45
Subirats, A. 85
Sugibayashi, T. 127
Suh, Youseok 219
Sumi, Hirofumi....................163
Sun, S. 85
Sun, X. 25, 221

Sun, Z. 61
Sundar, Vignesh 65
Sung, P.-J. 45
Suwa, Hitoshi 175
Suzuki, Y. 93
Tada, M. 127
Takagi, S. 195
Takase, M. 71
Takehara, Hironari 163
Takenaka, M. 195
Takeuchi, Ken 109
Takezawa, H. 93
Tan, Ava J. 49, 53
Tan, C.-L. 203
Tan, Chuan Seng 197
Tang, Kea-Tiong 37
Tang, T. J. 61
Tang, Y.-T 45
Teng, Zhongjian Jeffrey 65
Teugels, L. 69
Thean, Aaron V-Y 159
Thiyagarajah, N. 183
Thomas, Luc 65
Thomas, O. 35
Tien, C. W. 157
Ting, J. W. 183
Togo, M. 81, 215
Toh, E. H. 183
Tokuda, Takashi....................163
Tokuhara, T. 71
Tomekawa, Y. 71
Tomita, M. 93
Tomizawa, Hideyuki 41
Tong, Ru-Ying 65
Toriumi, A. 27
Triantopoulos, K. 15, 75
Tsai, H. 25
Tseng, Hsin-Wei 117
Tsiara, A. 75
Tsuda, K. 93
Tsui, Cy 39
Tsuji, Y. 127
Tsutsui, Gen 87
Tu, Chun-Yuan 139
Uchida, N. 191
Urdampilleta, M. 125
Van Campenhout, J. 221
Van Den Bosch, G. 203
Van Der Veen, M. H. 209
Van Houdt, J. 209
Van Huylenbroeck, S. 221
Van Marcke, P. 83
Vandooren, A. 69
Vanherle, W. 69
Vecchio, E. 69, 203

AUTHOR INDEX

Velenis, D. 221
Veloso, A. 141
Verbinnen, G. 69
Verheyen, P. 221
Verhulst, Anne 133
Verkest, D. 143
Versaggi, J. 81, 215
Vilarrubí, M. 15
Vincent, B. 141
Vinet, M. 75, 125, 153
Vinslava, A. 215
Vivet, P. 15
Vladimirova, K. 15
V-Palayam, S. S. 203
Vrancken, C. 193
Waldron, N. 69, 193
Walke, A. 69
Wan, W. K. 113
Wan, Weier 31
Wan, X. 215
Wang, C.-J. 45
Wang, H. C. 47
Wang, Hao 219
Wang, Hong 161
Wang, Jing 187
Wang, Keh-Chung 177
Wang, Miaomiao 87
Wang, Po-Kang 65
Wang, R. 185
Wang, Rongjun 117
Wang, Wei 77, 197
Wang, Xiaodong 117
Wang, Xiaofei 151
Wang, Xiao-Yong 219
Wang, Y. 215
Wang, Y.-H. 45
Wang, Y.-S. 45
Wang, Yu-Jen 65
Watanabe, Koji 87
Watanabe, T. 93
Watanabe, Yoshihiro 163
Weckx, P. 141, 147
Wee, T. L. 183
Wei, H. 215
Wei, Zhiqiang 63
Wen, Shi-Jie 19
Weng, Yi-Chin 37
Whig, Renu 117
Wiatr, M. 153
Widiez, J. 15
Wiedemer, Jami 151
Witters, L. 69, 83, 193
Witters, T. 209
Wong, H.-S. Philip 31, 107
Wong, J. 183

Wong, Richard 19
Wong, Simon 107
Woo, S. T. 183
Wostyn, K. 193
Wu, Chuang-Yi 37
Wu, Heng 87
Wu, Huaqiang 31, 103
Wu, Jixuan 95
Wu, T.-L. 45
Wu, W.-F. 45
Wu, Wei 103
Wu, Ying 77, 197
Wu, Yung-Hsien 119
Wu, Z. 69
Xu, M. 93
Xu, Nuo 187
Xu, Shengqiang 77, 197
Xu, Shun 105
Xue, Cheng-Xin 171
Xue, Lin 117
Yadav, Ajay K. 49, 53
Yadav, Sachin 197
Yajima, T. 27
Yakimets, D. 143
Yamaga, Yusuke 109
Yamaguchi, S. 81
Yamane, K. 183
Yamato, R. 93
Yan, Jhih-Yang 113
Yang, C. H. 205
Yang, Chun Ju 87
Yang, H. 183
Yang, H. S. 81
Yang, J. 61
Yang, M.-T. 113
Yang, Sam 219
Yang, Yi 65
Yao, Peng 103
Yasuda-Masuoka, Y. 213
Yasuhara, Ryutaro 175
Ye, Peide D. 89
Yeap, K.B. 61
Yeh, C. C. 25
Yeh, C. W. 205
Yeh, M.-S. 45
Yeh, Shih-Rung 37
Yeh, W.-K. 45
Yeo, Kyunghwan 217
Yeo, Yee-Chia 77
Yin, Shihui 173
Yoneda, Shinichi 63
Yong, C. 215
Yoo, Jongmyung 207
Yoon, Alexander 73
Yoon, J.S. 213

AUTHOR INDEX

Yoon, Jong Shik 181, 217
Yoon, Seong Jun 73
Yoshida, N. 85
You, Y. S. 183
Youn, Jong Mil 217
Yu, H. 81, 215
Yu, Qian 39
Yu, Shimeng 103
Yuan, Jun 219
Zainuddin, A. N. 61
Zang, H. 215
Zeng, D. 183
Zeng, Jia 137
Zhan, H. 215
Zhan, T. 93
Zhang, Chi 17
Zhang, H. 93
Zhang, Jianfu 95
Zhang, K. Y. 47
Zhang, L. 183, 203
Zhang, Weidong 95
Zhang, Wenqiang 31
Zhang, X. 81, 215
Zhang, Xiang 103
Zhang, Yi 105
Zhang, Zheng 77
Zhao, P. 215
Zhao, Yi 105
Zheng, T. 69
Zhong, Tom 65
Zhong, Xiaopeng 39
Zhou, Bo 95
Zhou, Chen 19
Zhou, D. 215
Zhou, H. 115
Zhou, Hong 53
Zhou, Huimei 87
Zhu, B. 215
Zhu, Jian 65
Zhu, Xuelian 137
Zota, C. B. 165

IEEE
445 Hoes Lane
Piscataway, NJ 08854-4141

ISBN 978-1-5386-4219-1